VOLUME FIVE HUNDRED AND SEVENTY THREE

METHODS IN ENZYMOLOGY

Enzymes of Epigenetics, Part A

METHODS IN ENZYMOLOGY

Editors-in-Chief

ANNA MARIE PYLE
Departments of Molecular, Cellular and Developmental Biology and Department of Chemistry
Investigator, Howard Hughes Medical Institute
Yale University

DAVID W. CHRISTIANSON
Roy and Diana Vagelos Laboratories
Department of Chemistry
University of Pennsylvania
Philadelphia, PA

Founding Editors

SIDNEY P. COLOWICK and NATHAN O. KAPLAN

VOLUME FIVE HUNDRED AND SEVENTY THREE

METHODS IN ENZYMOLOGY

Enzymes of Epigenetics, Part A

Edited by

RONEN MARMORSTEIN
*Abramson Family Cancer Research Institute,
Perelman School of Medicine at the University
of Pennsylvania,
Philadelphia, PA, United States*

ELSEVIER

AMSTERDAM • BOSTON • HEIDELBERG • LONDON
NEW YORK • OXFORD • PARIS • SAN DIEGO
SAN FRANCISCO • SINGAPORE • SYDNEY • TOKYO
Academic Press is an imprint of Elsevier

Academic Press is an imprint of Elsevier
50 Hampshire Street, 5th Floor, Cambridge, MA 02139, USA
525 B Street, Suite 1800, San Diego, CA 92101–4495, USA
The Boulevard, Langford Lane, Kidlington, Oxford OX5 1GB, UK
125 London Wall, London, EC2Y 5AS, UK

First edition 2016

ISBN: 978-0-12-805365-2
ISSN: 0076-6879

For information on all Academic Press publications
visit our website at https://www.elsevier.com

Publisher: Zoe Kruze
Acquisition Editor: Zoe Kruze
Editorial Project Manager: Sarah Lay
Production Project Manager: Magesh Kumar Mahalingam
Cover Designer: Greg Harris

Typeset by SPi Global, India

CONTENTS

Contributors *xiii*
Preface *xvii*

Part I
Chromatin Structure and Histones

1. In Vitro Chromatin Assembly: Strategies and Quality Control **3**

U. Muthurajan, F. Mattiroli, S. Bergeron, K. Zhou, Y. Gu, S. Chakravarthy, P. Dyer, T. Irving, and K. Luger

1. Introduction 4
2. Chromatin Reconstitution Strategies 7
3. Quality Control Strategies for In Vitro Reconstituted Chromatin 16
4. Summary and Conclusion 36

References 38

2. Preparation, Crystallization, and Structure Determination of Chromatin Enzyme/Nucleosome Complexes **43**

R.K. McGinty, R.D. Makde, and S. Tan

1. Introduction 44
2. Prepare Nucleosome Core Particles 44
3. Prepare Chromatin Protein 47
4. Purify Chromatin Complex 47
5. Concentrate Chromatin Complex 54
6. Crystallize Chromatin Complex 56
7. Postcrystallization Soaks to Improve Diffraction 61

References 63

3. Preparation of Recombinant Centromeric Nucleosomes and Formation of Complexes with Nonhistone Centromere Proteins **67**

N. Sekulic and B.E. Black

1. Introduction 68
2. Preparation of Histones for CENP-A Nucleosome Assembly 69
3. Preparation of DNA for CENP-A Nucleosome Assembly 79
4. CENP-A Nucleosome Assembly 84
5. Complex of CENP-A Nucleosome with CENP-C 90

Acknowledgments 95
References 95

4. Functional Characterization of Histone Chaperones Using SNAP-Tag-Based Imaging to Assess De Novo Histone Deposition **97**

C. Clément, I. Vassias, D. Ray-Gallet, and G. Almouzni

1. Introduction 98
2. The SNAP-Tag System to Visualize in vivo New Histone Incorporation into Chromatin: A De Novo Histone Deposition Assay in Cells 100
3. De Novo Histone Deposition Assay Coupled with Histone Chaperone Depletion and Labeling of Replicating Cells 106
4. De Novo Histone Deposition Assay Coupled with Histone Chaperone Depletion in Response to UV-DNA Damage 108
5. Imaging and Quantifying SNAP Labeling 111
6. Interpreting the De Novo Histone Deposition Assay Coupled with Histone Chaperone Depletion 113
7. Conclusion and Perspectives 113

Acknowledgments 115
References 115

5. Analysis of Nucleosome Sliding by ATP-Dependent Chromatin Remodeling Enzymes **119**

C.Y. Zhou and G.J. Narlikar

1. Introduction 119
2. Reconstituting Nucleosomes for FRET-Based Method 121
3. ATP-Dependent Chromatin Remodeling Assays 128
4. Data Analysis 134
5. Adaptation to Other Types of Remodeling Reactions 134

Acknowledgments 134
References 135

Part II
Posttranslational Histone Modification Enzymes and Complexes

6. In Vitro Activity Assays for MYST Histone Acetyltransferases and Adaptation for High-Throughput Inhibitor Screening **139**

C.E. McCullough and R. Marmorstein

1. Introduction 140

2. Preparation of Recombinant MYST Proteins 142
3. Kinetic Analysis 144
4. Thermofluor Assay to Investigate Cofactor and Inhibitor Binding 147
5. Inhibitor Screening 149
6. Prospects and Conclusions 157
Acknowledgments 158
References 158

7. Preparation and Biochemical Analysis of Classical Histone Deacetylases 161

A. Villagra, E. Sahakian, and E. Seto

1. Introduction 162
2. Generation of HDAC Stable Cell Lines Using Lentivirus 164
3. Conventional HDAC Activity Assay 167
4. Assessment of Selective Inhibition of HDACs Using HDAC Knockout Cells 169
5. HDAC Chromatin Immunoprecipitation Assay 174
6. Identification of Specific HDAC Substrates 179
Acknowledgments 180
References 180

8. Recombinant Preparation, Biochemical Analysis, and Structure Determination of Sirtuin Family Histone/Protein Deacylases 183

B. Suenkel and C. Steegborn

1. Sirtuins—Evolutionary Conserved Epigenetic Regulators 184
2. Recombinant Sirtuin Preparation—Constructs and Expression Systems 188
3. Purification of a Mammalian Sirtuin Recombinantly Expressed in *E. coli* 189
4. Biochemical Characterizations Using Purified, Recombinant Sirtuins 198
Acknowledgments 205
References 205

9. Preparation, Biochemical Analysis, and Structure Determination of SET Domain Histone Methyltransferases 209

E. Bergamin and J.F. Couture

1. Introduction 210
2. Experimental Procedures 218
3. Conclusions 231
Acknowledgments 231
References 232

10. Expression, Purification, and Biochemical Analysis of the LSD1/KDM1A Histone Demethylase **241**

B. Laurent and Y. Shi

1. Introduction 242
2. Expression and Purification of LSD1 Proteins 245
3. In Vitro Enzymatic Assays 251
4. Analysis of LSD1 Demethylase Activity 253

Acknowledgments 257
References 258

11. LSD1 Histone Demethylase Assays and Inhibition **261**

D. Hayward and P.A. Cole

1. Introduction and Background 262
2. LSD1 Assays 264
3. Inhibitors of LSD1 268
4. Applications of LSD1 Inhibitors 272
5. Summary and Future Directions 274

Acknowledgments 275
References 275

12. Purification, Biochemical Analysis, and Structure Determination of JmjC Lysine Demethylases **279**

S. Krishnan and R.C. Trievel

1. Introduction 280
2. Purification 282
3. KDM Assay Methodology 287
4. Structure Determination of JmjC KDMs 293
5. Summary and Future Perspectives 295

Acknowledgment 296
References 296

13. Preparation and Analysis of Native Chromatin-Modifying Complexes **303**

Y. Doyon and J. Côté

1. Introduction 304
2. Methods 305
3. Conclusion 317

Acknowledgments 317
References 318

Part III
Histone Modification Binders (Readers)

14. Preparation, Biochemical Analysis, and Structure Determination of the Bromodomain, an Acetyl-Lysine Binding Domain **321**

C. Ren, L. Zeng, and M.-M. Zhou

1. Introduction 322
2. General Equipment 324
3. General Materials 324
4. Preparation of BrDs 326
5. Biochemical Analysis of BrDs 330
6. Structural Determination of BrDs 334
7. Conclusions and Perspectives 338

Acknowledgments 340
References 340

15. Preparation, Biochemical Analysis, and Structure Determination of Methyllysine Readers **345**

C.A. Musselman and T.G. Kutateladze

1. Introduction 345
2. Preparation of Methyllysine Readers 347
3. Screening for Methyllysine Recognition 355
4. Measuring Affinity for Methylated Substrate 356
5. Analysis of Binding Using NMR Spectroscopy 358
6. Structure Determination by X-ray Crystallography 360

Acknowledgments 361
References 361

Part IV
DNA Modifications and Nucleic Acid Regulators

16. Quantification of Oxidized 5-Methylcytosine Bases and TET Enzyme Activity **365**

M.Y. Liu, J.E. DeNizio, and R.M. Kohli

1. Introduction 366
2. Analysis of Cytosine Modifications in Cellular DNA 366
3. Analysis of Cytosine Modifications In Vitro 373

Acknowledgments 383
References 383

17. Characterization of How DNA Modifications Affect DNA Binding by C2H2 Zinc Finger Proteins **387**

A. Patel, H. Hashimoto, X. Zhang, and X. Cheng

1. Introduction 387
2. Soluble Expression and Purification of ZnF Proteins 391
3. Fluorescence Polarization Assay for Analysis of DNA Binding 393
4. Crystallization of ZnF Proteins in Complex with DNA 396

Acknowledgments 397
References 398

18. Crystallographic Studies of Telomerase **403**

H. Hoffman and E. Skordalakes

1. Introduction 404
2. Crystal Structure Determination of the Catalytic Subunit of *Triobolium castaneum* Telomerase (tcTERT) 407
3. Crystal Structure Determination of a Partial Telomerase Elongation Complex 414
4. Conclusions 417

Acknowledgments 417
References 417

19. Detection and Analysis of Long Noncoding RNAs **421**

F. Lai, E. Blumenthal, and R. Shiekhattar

1. Introduction 422
2. Traditional Methods for lncRNA Detection 423
3. Genome-Wide Detection of lncRNAs Through Direct RNA Sequencing Approaches 423
4. Indirect Sequencing Methods Coupled with Transcription 428
5. Protein-Based Approaches for Genomic Characterization of lncRNAs 435
6. Future Directions and Conclusion 441

References 442

20. Identifying Centromeric RNAs Involved in Histone Dynamics In Vivo **445**

D. Quénet, D. Sturgill, and Y. Dalal

1. Introduction 446

2. Materials 446
3. Methods 450
Acknowledgments 464
References 464

Author Index *467*
Subject Index *497*

CONTRIBUTORS

G. Almouzni
Institut Curie, PSL Research University, CNRS, UMR3664, Equipe Labellisée Ligue contre le Cancer; Sorbonne Universités, UPMC Univ Paris 06, CNRS, UMR3664, Paris, France

E. Bergamin
Ottawa Institute of Systems Biology, University of Ottawa, Ottawa, ON, Canada

S. Bergeron
University of Colorado at Boulder, Boulder, CO, United States

B.E. Black
Perelman School of Medicine, University of Pennsylvania, Philadelphia, PA, United States

E. Blumenthal
Sylvester Comprehensive Cancer Center, University of Miami Miller School of Medicine, Miami, FL, United States

S. Chakravarthy
Illinois Institute of Technology, Chicago, IL, United States

X. Cheng
Emory University School of Medicine, Atlanta, GA, United States

C. Clément
Institut Curie, PSL Research University, CNRS, UMR3664, Equipe Labellisée Ligue contre le Cancer; Sorbonne Universités, UPMC Univ Paris 06, CNRS, UMR3664, Paris, France

P.A. Cole
Johns Hopkins School of Medicine, Baltimore, MD, United States

J. Côté
Centre Hospitalier Universitaire de Québec Research Center and Faculty of Medicine, Laval University; St-Patrick Research Group in Basic Oncology and Laval University Cancer Research Center, Quebec City, QC, Canada

J.F. Couture
Ottawa Institute of Systems Biology, University of Ottawa, Ottawa, ON, Canada

Y. Dalal
Center for Cancer Research, National Cancer Institute, National Institutes of Health, Bethesda, MD, United States

J.E. DeNizio
Perelman School of Medicine, University of Pennsylvania, Philadelphia, PA, United States

Y. Doyon
Centre Hospitalier Universitaire de Québec Research Center and Faculty of Medicine, Laval University, Quebec City, QC, Canada

P. Dyer
University of Colorado at Boulder, Boulder, CO, United States

Y. Gu
University of Colorado at Boulder, Boulder; Colorado State University, Fort Collins, CO, United States

H. Hashimoto
Emory University School of Medicine, Atlanta, GA, United States

D. Hayward
Johns Hopkins School of Medicine, Baltimore, MD, United States

H. Hoffman
The Wistar Institute, Philadelphia, PA, United States

T. Irving
Illinois Institute of Technology, Chicago, IL, United States

R.M. Kohli
Perelman School of Medicine, University of Pennsylvania, Philadelphia, PA, United States

S. Krishnan
University of Michigan, Ann Arbor, MI, United States

T.G. Kutateladze
University of Colorado School of Medicine, Aurora, CO, United States

F. Lai
Sylvester Comprehensive Cancer Center, University of Miami Miller School of Medicine, Miami, FL, United States

B. Laurent
Boston Children's Hospital; Harvard Medical School, Boston, MA, United States

M.Y. Liu
Perelman School of Medicine, University of Pennsylvania, Philadelphia, PA, United States

K. Luger
Howard Hughes Medical Institute, University of Colorado at Boulder, Boulder; Institute for Genome Architecture and Function, Colorado State University, Fort Collins, CO, United States

R.D. Makde
Center for Eukaryotic Gene Regulation, The Pennsylvania State University, University Park, PA, United States

R. Marmorstein
University of Pennsylvania; Abramson Family Cancer Research Institute, Perelman School of Medicine at the University of Pennsylvania, Philadelphia, PA, United States

F. Mattiroli
University of Colorado at Boulder, Boulder, CO, United States

C.E. McCullough
University of Pennsylvania, Philadelphia, PA, United States

R.K. McGinty
Center for Eukaryotic Gene Regulation, The Pennsylvania State University, University Park, PA, United States

C.A. Musselman
University of Iowa, Iowa City, IA, United States

U. Muthurajan
University of Colorado at Boulder, Boulder, CO, United States

G.J. Narlikar
University of California, San Francisco, CA, United States

A. Patel
Emory University School of Medicine, Atlanta, GA, United States

D. Quénet
Center for Cancer Research, National Cancer Institute, National Institutes of Health, Bethesda, MD, United States

D. Ray-Gallet
Institut Curie, PSL Research University, CNRS, UMR3664, Equipe Labellisée Ligue contre le Cancer; Sorbonne Universités, UPMC Univ Paris 06, CNRS, UMR3664, Paris, France

C. Ren
Icahn School of Medicine at Mount Sinai, New York, NY, United States

E. Sahakian
H. Lee Moffitt Cancer Center, Tampa, FL, United States

N. Sekulic
Perelman School of Medicine, University of Pennsylvania, Philadelphia, PA, United States

E. Seto
George Washington University Cancer Center, School of Medicine and Health Sciences, The George Washington University, Washington, DC, United States

Y. Shi
Boston Children's Hospital; Harvard Medical School, Boston, MA, United States

R. Shiekhattar
Sylvester Comprehensive Cancer Center, University of Miami Miller School of Medicine, Miami, FL, United States

E. Skordalakes
The Wistar Institute, Philadelphia, PA, United States

C. Steegborn
Department of Biochemistry, University of Bayreuth, Bayreuth, Germany

D. Sturgill
Center for Cancer Research, National Cancer Institute, National Institutes of Health, Bethesda, MD, United States

B. Suenkel
Department of Biochemistry, University of Bayreuth, Bayreuth, Germany

S. Tan
Center for Eukaryotic Gene Regulation, The Pennsylvania State University, University Park, PA, United States

R.C. Trievel
University of Michigan, Ann Arbor, MI, United States

I. Vassias
Institut Curie, PSL Research University, CNRS, UMR3664, Equipe Labellisée Ligue contre le Cancer; Sorbonne Universités, UPMC Univ Paris 06, CNRS, UMR3664, Paris, France

A. Villagra
George Washington University Cancer Center, School of Medicine and Health Sciences, The George Washington University, Washington, DC, United States

L. Zeng
Icahn School of Medicine at Mount Sinai, New York, NY, United States

X. Zhang
Emory University School of Medicine, Atlanta, GA, United States

C.Y. Zhou
University of California, San Francisco, CA, United States

K. Zhou
University of Colorado at Boulder, Boulder; Colorado State University, Fort Collins, CO, United States

M.-M. Zhou
Icahn School of Medicine at Mount Sinai, New York, NY, United States

PREFACE

These two volumes of *Methods of Enzymology* cover the rapidly developing field of Epigenetics. The central dogma of molecular biology, first proposed by Francis Crick in 1956, provided a framework for understanding the transfer of genetic information, which flows from DNA to RNA to protein. However, in the early 1980s, it was discovered that methylation of the DNA can change its function, hinting that genetic information transfer could be altered in other ways. In the 1990s, it was discovered that gene function could indeed be altered in many other ways and the modern field of Epigenetics was born. Epigenetics, a term first coined by Conrad Waddington in 1942 as "the branch of biology which studies the causal interactions between genes and their products, which bring the phenotype into being," is now known as the study of heritable changes in gene expression due to internal or environmental signals that results in the change of cellular function or physiological phenotype, but that is not caused by changes in the genetic information. An example of epigenetic regulation is cell differentiation, whereby cells of an organism with identical genetic information, such as cells of the nose, eyes, and hair, carry out different cellular functions.

At the heart of epigenetics is the regulation of chromatin, the packaged form of DNA. The building blocks of chromatin are nucleosome core particles containing about 146 bp of DNA wrapped around an octamer of histone proteins, two copies each of histones H2A, H2B, H3, and H4. Chromatin mediates all DNA-templated events, including DNA transcription, replication, and repair, and is regulated by many proteins and noncoding RNAs. The proteins that regulate chromatin include posttranslational modification (PTM) "writer" enzymes, "eraser" enzymes that remove these PTMs, and "reader" proteins that bind chemically modified histones or DNA. PTMs of DNA include methylation at the 5 position of cytosine and various oxidation states of 5-methyl-cytosine. PTMs of histones include acetylation, methylation, and ubiquitination on lysine residues; methylation and citrullination of arginine residues; and phosphorylation of threonine and serine residues. ATP-dependent chromatin-remodeling enzymes that function to reposition nucleosomes within chromatin; histone chaperone proteins that insert or eviction of histone variants in and out of chromatin, respectively; and noncoding RNA molecules also contribute to chromatin regulation.

Chromatin-regulatory proteins work together to mediate epigenetic regulation, with ramifications for cellular function and physiological phenotype, and over the last decade, it has become apparent that the dysfunction of epigenetic regulators can drive many diseases including metabolic and neurodegenerative disorders and various cancers. Over the last decade, we have seen significant progress on understanding the molecular mechanisms underlying epigenetic regulation leading to new insights into cellular function and physiological phenotype, the development of new technologies, and the development of small-molecule probes to study these biological processes and small epigenetic drugs that are currently in clinical trials to treat disease.

The remarkable progress in the field of epigenetic research that has occurred over the last two decades is highlighted in these two volumes of *Methods of Enzymology*, entitled Enzymes of Epigenetics. In Volume 1, Chapters 1 through 5 cover *Chromatin Structure and Histones*. This includes strategies for in vitro chromatin assembly (Chapter 1) and the assembly of protein–chromatin complexes (Chapter 2) for biochemical, biophysical, and structural studies, preparation of recombinant centromeric nucleosomes with and without nonhistone centromere proteins (Chapter 3), functional characterization of histone deposition by histone chaperones (Chapter 4), and methods to study nucleosome sliding by ATP-dependent chromatin-remodeling enzymes (Chapter 5).

In Volume 1, Chapters 6 through 13 cover *Posttranslational Histone Modification Enzymes and Complexes*. This includes in vitro activity assays for MYST histone acetyltransferases and adaptation for high-throughput inhibitor screening (Chapter 6), and the preparation, biochemical analysis, and structure determination of the classical (Chapter 7) and sirtuin (Chapter 8) histone lysine deacetylases, SET domain histone methyltransferases (Chapter 9), LSD1/KDM1A (Chapter 10), and JmjC (Chapter 12). Also included are chapters on LSD1 histone demethylase assays and inhibition (Chapter 11) and preparation and analysis of native chromatin-modifying complexes (Chapter 13).

In Volume 1, Chapters 14 and 15 cover *Histone Modification Binders* and include the preparation, biochemical analysis, and structure determination of the acetyl-lysine reader bromodomains (Chapter 14) and the readers of the methyllysine mark (Chapter 15).

In Volume 1, Chapters 16 through 20 cover *DNA Modifications and Nucleic Acid Regulators*. This includes quantification of oxidized 5-methylcytosine bases and TET enzyme activity (Chapter 16), characterization of how

DNA modifications affect DNA binding by C_2H_2 zinc finger proteins (Chapter 17), regulation of chromosome ends by the riboprotein telomerase complex (Chapter 18), detection and analysis of long noncoding RNAs (Chapter 19), and identification of centromeric RNAs involved in histone dynamics in vivo (Chapter 20).

In Volume 2, Chapters 1 through 8 cover *Epigenetic Technologies*. This includes identification and quantification of histone PTMs using high-resolution mass spectrometry (Chapter 1), substrate specificity profiling of histone-modifying enzymes using peptide microarray (Chapter 2), an open-source platform to analyze such microarrays (Chapter 3), chemical biology approaches for characterization of epigenetic regulators (Chapter 4), mapping lysine acetyltransferase–ligand interactions by activity-based capture (Chapter 5), methods for investigating histone acetylation stoichiometry and turnover rate (Chapter 6), semisynthesis of acetylated and sumoylated histone analogs (Chapter 7), and an IF-FISH approach for covisualization of gene loci and nuclear architecture in fission yeast (Chapter 8).

In Volume 2, Chapters 9 through 11 cover *Small-Molecule Epigenetic Regulators* and include the preparation and analysis of sirtuin deacetylase inhibitors (Chapter 9) and activators (Chapter 10) and methyltransferase inhibitors (Chapter 11).

In Volume 2, Chapters 12 through 15 cover *Epigenetics and Biological Connections* and include exploring the dynamic relationship between cellular metabolism and chromatin structure using SILAC-mass spec and ChIP-sequencing (Chapter 12), proteomic methods to investigate the dynamics of histone turnover in the central nervous system (Chapter 13), ChIP-seq techniques to map the epigenome of senescent cells (Chapter 14), and exploiting chromatin biology to understand immunology (Chapter 15).

I expect that these volumes will be a useful resource for investigators in the epigenetics field as well as those outside of the epigenetics field who would like to incorporate epigenetics into their own research programs.

RONEN MARMORSTEIN
Abramson Family Cancer
Research Institute, Perelman School of Medicine at the University of Pennsylvania, Philadelphia, PA, United States

PART I

Chromatin Structure and Histones

CHAPTER ONE

In Vitro Chromatin Assembly: Strategies and Quality Control

U. Muthurajan*, F. Mattiroli*, S. Bergeron*, K. Zhou*,¶, Y. Gu*,¶, S. Chakravarthy†, P. Dyer*, T. Irving†, K. Luger*,‡,§,1

*University of Colorado at Boulder, Boulder, CO, United States
†Illinois Institute of Technology, Chicago, IL, United States
‡Howard Hughes Medical Institute, University of Colorado at Boulder, Boulder, CO, United States
§Institute for Genome Architecture and Function, Colorado State University, Fort Collins, CO, United States
¶Colorado State University, Fort Collins, CO, United States
[1]Corresponding author: e-mail address: karolin.luger@colorado.edu

Contents

1. Introduction 4
2. Chromatin Reconstitution Strategies 7
   2.1 Fluorescent Labeling of the Histones and Refolding into Octamer 8
   2.2 End-Labeling of DNA 10
   2.3 Nucleosome Assembly 11
   2.4 Trinucleosome Assembly 13
   2.5 "Microscale" or Step Dilution Method for Nucleosome Reconsitution 15
3. Quality Control Strategies for In Vitro Reconstituted Chromatin 16
   3.1 Micrococcal Nuclease Digestion 16
   3.2 Analysis of MNase-Digested Products Via Bioanalyzer 2100 20
   3.3 Size-Exclusion Chromatography Coupled with Multiangle Light Scattering 24
   3.4 Analytical Ultracentrifugation 26
   3.5 Atomic Force Microscopy Analysis of Trinucleosomes 29
   3.6 Small-Angle X-ray Scattering 33
4. Summary and Conclusion 36
References 38

Abstract

Chromatin accessibility is modulated by structural transitions that provide timely access to the genetic and epigenetic information during many essential nuclear processes. These transitions are orchestrated by regulatory proteins that coordinate intricate structural modifications and signaling pathways. In vitro reconstituted chromatin samples from defined components are instrumental in defining the mechanistic details of such processes. The bottleneck to appropriate in vitro analysis is the production of high quality, and quality-controlled, chromatin substrates. In this chapter, we describe methods for in vitro chromatin reconstitution and quality control. We highlight the strengths and

Methods in Enzymology, Volume 573
ISSN 0076-6879
http://dx.doi.org/10.1016/bs.mie.2016.01.002

weaknesses of various approaches and emphasize quality control steps that ensure reconstitution of a bona fide homogenous chromatin preparation. This is essential for optimal reproducibility and reliability of ensuing experiments using chromatin substrates.

1. INTRODUCTION

In eukaryotic cells, chromatin refers to a hierarchical complex of DNA and proteins (histone and nonhistone), that implements the proper structural and functional regulation of all genetic and epigenetic information. The minimal chromatin unit is the nucleosome, formed by a core histone octamer and 147 base pairs DNA wrapped around it (Fig. 1A; Nuc147). Nucleosomes and their higher order assemblies hinder direct access to the DNA for the multitude of nuclear machineries that mediate DNA-related processes such as transcription, replication, and DNA damage repair. These machineries gain access to the packaged DNA, via tightly regulated structural transitions that expose the genetic and epigenetic information by controlling nucleosome dynamics, both spatially and temporally.

Highly defined nucleosome and chromatin samples are essential for defining the detailed mechanisms that regulate the dynamics of chromatin. For these studies, defined chromatin templates are a prerequisite. In vitro assembly can be performed at any level of complexity. We have previously described small- and large-scale preparation of high-quality nucleosomes, the minimal chromatin unit (Fig. 1A; Nuc147) to be used for structural, biochemical, and biophysical applications (Dyer et al., 2004). We have also prepared nucleosomes containing histones from species other than *Xenopus laevis*, histone variants, histone modifications, different DNA sequences, and histone point mutations (Bao, White, & Luger, 2006; Bao et al., 2004; Black et al., 2004; Brehove et al., 2015; Chakravarthy, Bao, Roberts, Tremethick, & Luger, 2004; Chakravarthy & Luger, 2006; Chatterjee et al., 2015; Clapier et al., 2008; Fierz, Kilic, Hieb, Luger, & Muir, 2012; Lu et al., 2008; Muthurajan et al., 2004; Park, Dyer, Tremethick, & Luger, 2004; Watanabe et al., 2010). Over the past few years, we have expanded our repertoire of chromatin templates assembled in vitro to include fluorescently labeled nucleosomes (either on the histones or DNA). Moreover, with the availability of highly positioned DNA sequences like the 601 DNA (Lowary & Widom, 1998), it is now possible to place linker arms of desirable lengths on either side of the minimal Nuc147

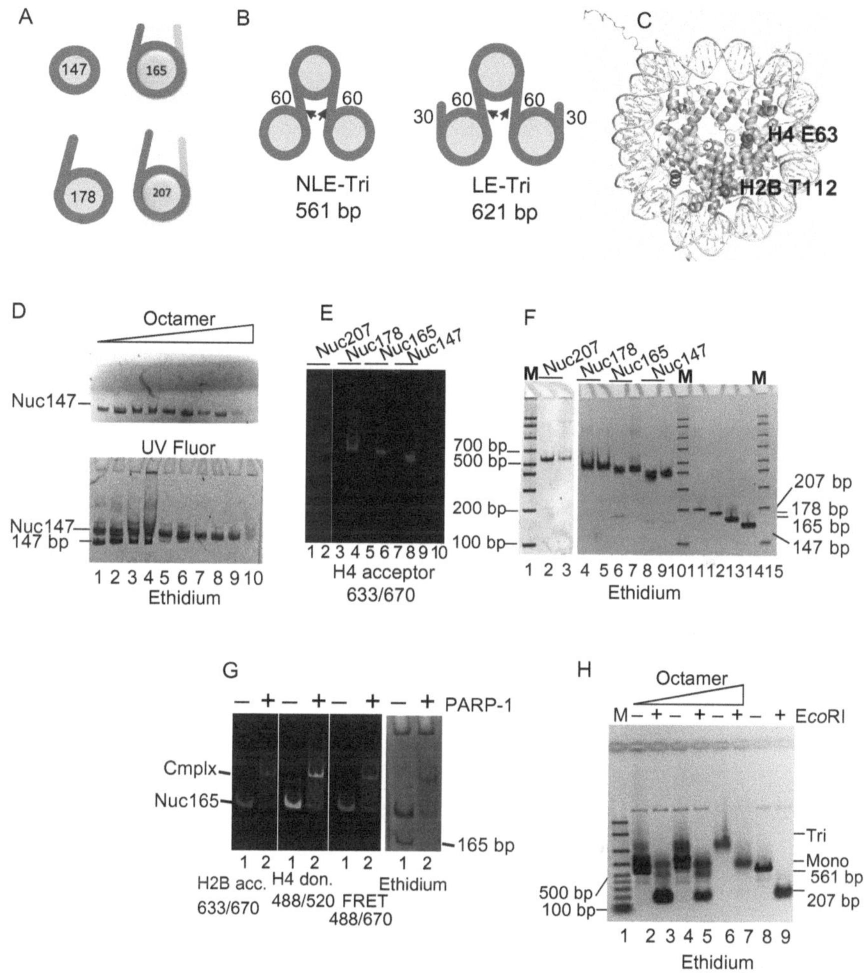

Fig. 1 Chromatin assembly via salt gradient reconstitution. (A) Schematic of mono-nucleosomes with different extranucleosomal linker lengths. (B) Schematic of the trinucleosome templates. An *Eco*RI recognition sequence is engineered in the inner linker arms of the DNA (indicated by *red arrows* (*dark gray* in the print version)). (C) Location of histone residues H4 E63 and H2B T112 used for attaching fluorescent labels (pdb 1AOI). (D) Titration of Atto647N-labeled octamer (H4E63C) onto 5S 147 bp DNA. Note free DNA and additional bands at lower octamer ratios (lanes 1–4). As more histones are titrated in, the free DNA band disappears and a more homogenous nucleosome band is observed (lanes 5–9). Upon further addition of histones, yields decrease due to sample aggregation (lane 10). (E) Fluorescently labeled nucleosomes: DNA fragments 207, 178, 165, and 147 base pairs in length, all containing the 601 positioning sequence, were assembled into nucleosomes with histone octamers labeled at H4E63C with Atto647N. Nucleosomes were analyzed by 5% native PAGE and scanned on a Typhoon imager at an emission wavelength of 670 nm. Lanes 2, 4,

(Fig. 1A). All these nucleosomes are valuable in studying dynamics of chromatin regulation in the context of mononucleosomes, where DNA extensions play a role in structure and dynamics (Clark et al., 2012; Muthurajan et al., 2014; White, Hieb, & Luger, 2016).

Increasingly, chromatin interaction with its regulatory factors is studied in a polynucleosome context. To this end, we have developed minimal higher order chromatin templates containing three nucleosomes (Fig. 1B; trinucleosome). We distinguish two types of trinucleosomes based on the availability of extranucleosomal DNA extension from the two outer nucleosomes. Trinucleosomes with extensions (linker-ended trinucleosomes—LE-Tri) provide a model for damaged chromatin with exposed DNA blunt ends required for binding proteins like PARP-1 which are known to regulate DNA repair dynamics. Trinucleosomes lacking such extensions (nonlinker-ended trinucleosomes—NLE-Tri) can therefore be used as a model for native chromatin (no exposed DNA ends) (Fig. 1B).

These chromatin templates are not trivial to prepare and they require a number of quality control steps to ensure their proper assembly, homogeneity, and stability. In this chapter, we revisit and update the salt gradient assembly method for assembling chromatin containing fluorescent labels, building on previously published articles (Dyer et al., 2004; Luger, Rechsteiner, & Richmond, 1999). We describe the use of restriction enzyme and micrococcal nuclease (MNase) digestion, native gel electrophoresis (PAGE and agarose), bioanalyzer sequencing, size-exclusion

Fig. 1—Cont'd 6, and 8 are labeled nucleosomes assembled on 207, 178, 165, and 147 bp DNA, respectively. (F) The same gel as in (E) was stained with ethidium bromide. Lanes 2 and 3 are labeled and unlabeled Nuc207, respectively; lanes 4 and 5 are labeled and unlabeled Nuc178; lanes 6 and 7 are labeled and unlabeled Nuc165; lanes 8 and 9 are labeled and unlabeled Nuc147. Lanes 11–14 are 207, 178, 165, and 147 bp DNA fragments, respectively. Note the absence of free DNA (<1%) in the nucleosome samples. Lanes marked M contain 50–2000 bp ladder. (G) EMSA using double-labeled nucleosomes (Atto647N on H2B and Alexa488 on H4) on 165 bp DNA, with and without PARP-1. Note that the donor and acceptor signal colocalizes upon PARP-1 binding (lane 2—Cmplx) indicating that the nucleosomes remain intact. (H) *Eco*RI digestion of assembled trinucleosomes. Lane 1, 50–2 k ladder; lane 2, uncut NLE-Tri; lane 3, *Eco*RI-digested NLE-Tri; lane 4, uncut NLE-Tri; lane 5, *Eco*RI-digested NLE-Tri; lane 6, uncut NLE-Tris; lane 7, *Eco*RI-digested NLE-Tris. Note that *Eco*RI digestion of trinucleosomes results in Nuc207 and/or 207 bp DNA. Note that there is no free 207 bp DNA in the *Eco*RI-digested lane for fully saturated trinucleosomes (lane 7 whereas undersaturated trinucleosomes lanes 3 and 5 contain free 207 bp DNA), indicating that all three nucleosome positioning sites are occupied. *Note that panels (E) and (F) are reused from a previous publication (Clark, Kramer, Muthurajan, & Luger, 2012).*

chromatography coupled with multiangle light scattering (SEC-MALS), analytical ultracentrifugation (AUC), atomic force microscopy (AFM), and small-angle X-ray scattering (SAXS) as tools to determine the quality and polydispersity of the chromatin templates.

2. CHROMATIN RECONSTITUTION STRATEGIES

An essential prerequisite to assembling chromatin is the availability of pure recombinant histone proteins that are devoid of posttranslational modifications and defined DNA templates. All methods described here utilize *X. laevis* histones expressed as inclusion bodies in *E. coli* and purified using previously published methods (Dyer et al., 2004). The use of *X. laevis* histones is mainly due to historical reasons; we have demonstrated that the same methods can be used to purify histones from many other species. Of all the species tested, only yeast histones are somewhat problematic, as their yields are generally lower. The purification of histones from bacteria has been described extensively (Luger, Mader, Richmond, Sargent, & Richmond, 1997; White, Suto, & Luger, 2001). In the preparation of histones, it is of utmost importance to avoid contamination with nucleic acids, as their presence will skew quantification and result in heterogeneous chromatin preparations. In our experience, this is the most prevalent reason for unsuccessful chromatin reconstitution experiments.

We generally use DNA comprised of a single or triple 601 sequence, due to their superior positioning properties. Other DNA sequences can be used with the same protocol, but the quality and stability of the final product may vary. Accuracy of positioning may also be variable, especially on longer DNA fragments. Details of the preparation of DNA for in vitro chromatin assembly can be found in Dyer et al. (2004).

Histone proteins are refolded into octamer as a first step toward assembling chromatin. Octamer is assembled by mixing each unfolded histone at equimolar amounts and refolding in 2 *M* salt. The refolded complex contains a mix of octamer, H3–H4 and H2A–H2B and excess histones. Histone octamer is purified by resolving the complexes on a size-exclusion column (Superdex 200), as described (Dyer et al., 2004). The histone octamer is not stable at ionic strength lower than 2 *M*, and thus not a physiologically relevant entity, but it is a convenient starting point for assembling chromatin in vitro because it contains the correct stoichiometry of properly refolded histone complexes. Reconstitutions can also be performed with H2A–H2B dimer and $(H3–H4)_2$ tetramer, as described earlier (Dyer et al., 2004).

2.1 Fluorescent Labeling of the Histones and Refolding into Octamer

Comprehensive in vitro biophysical and structural characterization of chromatin complexes is greatly aided by the usage of labeled components. In particular, we have widely used fluorescently labeled histones to assemble nucleosomes that were used for solution-state binding affinity measurements (Chodaparambil et al., 2014; D'Arcy et al., 2013; Dechassa, Wyns, & Luger, 2014; Dechassa et al., 2011; Fierz et al., 2012; Klein et al., 2016; Muthurajan et al., 2014; Park et al., 2004; White et al., 2016; Winkler & Luger, 2011; Winkler, Luger, & Hieb, 2012). Only one of the four histones, H3, contains a cysteine, and this can be mutated without any adverse effect on nucleosome structure (Luger et al., 1997; White et al., 2001). The absence of functional cysteine groups on histone proteins has enabled us to engineer single cysteine point mutations that can be used for the attachment of maleimide-based fluorophores in positions that do not affect the overall structure of nucleosomes.

Our preferred fluorophores are Atto647N (acceptor) and Alexa 488 or Oregon Green 488 (donor), but other compatible fluorescence resonance energy transfer (FRET) pairs can be used (Winkler et al., 2012). Histones H4 E63C and/or H2B T112C (Fig. 1C) in combination with H3 C110A are the most commonly used sites for labeling the histone octamer. Both H4- and H2B-labeling mutants can be refolded together to make double-labeled octamers containing both acceptor and donor fluorophores. The fluorescently labeled components can then be incorporated into nucleosomes or nucleosomal arrays via one of the several assembly protocols discussed previously (Chakravarthy & Luger, 2006; Dyer et al., 2004).

2.1.1 Equipment

- Dialysis tubing (6- to 8-kDa cutoff, 2.3-cm flat width)
- HiLoad 16/60 Superdex 200 HR preparation-grade gel-filtration column and AKTA protein purification system, equipped with UV detector and fraction collector (GE Healthcare)
- Nanospec or any spectrophotometer
- Standard SDS-PAGE equipment, 15% SDS-PAGE gels
- Typhoon scanner (GE Healthcare)
- ImageQuant TL for visualization and analysis

2.1.2 Buffers and Reagents

- Unfolding buffer for labeled histones: 20 m*M* Tris (hydroxymethyl)-aminomethane hydrochloride (Tris–HCl; pH 7.5), 6 *M* guanidinium hydrochloride, 20 m*M* sodium chloride (NaCl), 0.01 m*M* Tris (2-carboxyethyl)phosphine (TCEP); pH adjusted to 7 prior to addition—made fresh on the day of unfolding

- Unfolding buffer for unlabeled histones: 6 *M* guanidinium HCl, 20 m*M* Tris–HCl, pH 7.5, 5 m*M* dithiothreitol (DTT)—made fresh on the day of unfolding
- Refolding buffer: 2 *M* NaCl, 10 m*M* Tris–HCl (pH 7.5), 1 m*M* ethylenediaminetetraacetic acid (EDTA), 5 m*M* β-mercaptoethanol (BME)
- Maleimide reactive dyes (Alexa 488, Atto647N, Alexa 532, etc.)—10 m*M* stock in dimethyl sulfoxide (DMSO) or dimethyl fluoride (DMF) (Thermofisher)—freshly made or frozen stocks; thaw only once
- 10 m*M* Glutathione or DTT

2.1.3 Procedure

1. Add 2.0 mL of unfolding buffer to the lyophilized histone/s (2 mg/mL final) to be labeled, wrap in aluminum foil (TCEP is light sensitive), and dissolve gently for 45 min at room temperature using a rotator.
2. Measure histone concentration at 276 nm and add an equimolar amount of fluorophore (up to twofold excess of fluorophore will work). Incubate overnight with rotation at 4°C in the dark.
3. Quench the reaction by adding reduced 10 m*M* (final) glutathione or DTT. Incubate 1 h at 4°C.
4. Resuspend the other histone(s) with 2.0 mL of unfolding buffer and dissolve gently for 45 min at room temperature.
5. Determine the concentration of the unlabeled histone(s).
6. Mix the four histones at equimolar ratio, then bring the final protein concentration to 1 mg/mL using unfolding buffer.
7. Place the sample into dialysis bag and dialyze against cold refolding buffer at 4°C. It is essential to ensure that the dialysis bag is spinning freely; otherwise, precipitation of histones might occur. Use a volume of refolding buffer at least 100-fold greater than your sample for a thorough buffer exchange and change buffer at least three times total (4 h each with at least one overnight).
8. Concentrate dialyzed sample to 1–2 mL or 0.5–2% of the column volume to be used for purification.
9. Spin sample and inject into a Superdex 200 gel-filtration column (10/30 or 16/60) equilibrated in refolding buffer. Collect 1 mL fractions.

10. Run 15% SDS gel of eluted fractions, image gel on Typhoon to confirm labeling, then stain for protein. Determine fractions containing octamer, pool conservatively and concentrate.
11. Determine concentration of octamer and efficiency of labeling by determining moles dye per mole protein on the nanospec and freeze in liquid nitrogen by adding 0–20% (v/v) glycerol and store in −80°C.

2.1.4 Notes

1. Unfolding buffer used to unfold labeling mutants should not contain DTT or BME, however, once the labeling is quenched, BME can be used in the refolding buffer.
2. We typically dedicate columns to each fluorophore used to prevent cross-contamination. If this is not feasible, then the column will need stringent cleaning (per manufacturer's protocol) to remove the fluorophore.
3. We typically achieve 50–60% labeling efficiency.
4. If the octamer is going to be used immediately, it can be stored at 4°C where it is usually stable for several weeks, at least for nonvariant octamer.

2.2 End-Labeling of DNA

In a scenario when histone labeling is not desired (for eg, if labeling of histones disrupts functional interactions, or to confirm the presence of DNA in a complex with increased sensitivity), the DNA can also be labeled. Fluorescently labeled primers are used to amplify double-stranded labeled DNA by polymerase chain reaction (PCR) amplification. DNA can also be labeled internally, by using mutated bases at strategic positions. We will not discuss the method for internal DNA labeling here but more information can be obtained from Luo, North, and Poirier (2014).

2.2.1 Equipment

- PCR machine
- Agarose/native acrylamide gel apparatus
- GE Healthcare Typhoon for fluorescence imaging
- ImageQuant TL for visualization and analysis

2.2.2 Buffers and Reagents

- Fluorescently labeled primers (eg, IDT)
- DNA polymerase

- PCR buffer
- dNTP stock solutions
- PCR purification kit (Zymo Research)

2.2.3 Procedure

1. Set up PCR using labeled primer(s) and analyze the product on a native gel. 5–10% acrylamide can be used for smaller DNA fragments, 1% agarose is suggested for DNA larger than 500 bp.
2. Purify DNA using a commercially available kit and verify size by PAGE.
3. Verify labeling efficiency by determining moles of dye per mole DNA. Typically, almost 100% of the amplified DNA should be fluorescently labeled.
4. Scan the gel on Typhoon prior to staining with Sybr Gold or ethidium bromide. Store in amber tubes or in a light protected environment to prevent photobleaching of the fluorophore.

2.2.4 Notes

1. The PCR reactions should be set up in amber tubes to keep the samples light protected.

2.3 Nucleosome Assembly

Nucleosome assembly via salt gradient is traditionally employed to obtain large quantities of nucleosomes for crystallization and other biochemical as well as biophysical assays. With strongly positioning DNA sequences like the 601 DNA this method results in high yields of homogenous mononucleosomes. We have previously described the method for reconstituting nucleosomes in detail (Dyer et al., 2004). Here, we assume that the reader is familiar with these approaches and focus on the generation of high-quality nucleosomes containing one or two fluorescent labels.

2.3.1 Procedure

The method for preparing fluorescently labeled nucleosomes is identical to that for unlabeled nucleosomes (Dyer et al., 2004), except for the following variations:

1. The beaker containing the DNA: octamer mixtures during salt gradient reconstitution should be wrapped in aluminum foil to protect the fluorescent label from photobleaching.
2. Native PAGE analysis of fluorescently labeled nucleosomes should be carried out in the dark as well and the gel should be scanned on a

Typhoon imager at the fluorophore-specific wavelength prior to staining with ethidium bromide or SyBr Gold.

3. To determine nucleosome concentration, use 0.1 *M* NaOH to make a 1:10 dilution of the nucleosomes, with the same dilution of TCS in 0.1 *M* NaOH as blank. Read absorbance at 260 nm and calculate concentration using epsilon derived from the molecular weight of the DNA used in the reconstitution. Accurate DNA epsilons and molecular weight can be obtained by using the DNA sequence online at http://www.basic.northwestern.edu/biotools/oligocalc.html (Kibbe, 2007).
4. Nucleosomes can be concentrated and stored in TCS buffer (20 m*M* Tris, pH 7.5, 1 m*M* EDTA, 1 m*M* DTT) at 4°C for several months.

2.3.2 Notes

1. There are certain histone variants that do not refold readily to a histone octamer (eg, macroH2A; Chakravarthy et al., 2005); in such cases, the histones need to be refolded as $(H3–H4)_2$ tetramer and H2A–H2B dimer and a two-step reconstitution method is employed (Dyer et al., 2004).
2. For native PAGE, it is advisable to run a DNA ladder and an unlabeled nucleosome sample (if available), for visualization by ethidium bromide staining.

Fig. 1D shows mononucleosomes assembled on 147 bp 5S DNA using Atto647N-labeled histone octamer (H4E63C). Note that at lower ratios of octamer, there is free DNA present along with multiple nucleosome-like bands that can easily be mistaken for proper nucleosomes (lanes 1–4). At near-optimal ratios of octamer there is one discrete band of Nuc147 (lanes 5–8). As more octamer is titrated, the complex aggregates and no nucleosomes appear in the gel (lanes 9 and 10).

Fig. 1E shows mononucleosomes assembled on 207, 178, 165, and 147 bp DNA using Atto647N-labeled octamer (at H4 E63C). There is a single discrete band in each case indicating homogenous assembly. The same gel was stained with ethidium bromide (Fig. 1 F) and shows <1% free DNA in each case. These nucleosomes were successfully utilized in HIFI-FRET assays (eg, Muthurajan et al., 2014).

Fig. 1G shows nucleosomes assembled on 165 bp DNA using double-labeled octamer (Atto647N on H2B T112C; Alexa488 on H4 E63C). These nucleosomes were used in an electrophoretic mobility shift assay (EMSA) to test binding of PARP-1. They are particularly useful to ascertain the presence of histones H4 (and H3), and H2B (with H2A). As seen in the figure, the nucleosomes form a single band of slower mobility upon PARP-1 binding (lanes 2, 4, 6, and 8-Cmplx). Note that fluorescence persists in both fluorophore channels (Atto647N and Alexa488), indicating that nucleosome integrity is preserved upon PARP-1 binding. Furthermore, the presence of both labels allows us to conclude that H2B and H4 are present in the nucleosomes as well as in the supershifted complex.

2.4 Trinucleosome Assembly

In the past, the typical higher order chromatin templates used in vitro was composed of an array of 12 nucleosomes, assembled on a ~2400 bp DNA with 12 repeats of 601 DNA (Ghosh et al., 2010; Muthurajan, McBryant, Lu, Hansen, & Luger, 2011; Rogge et al., 2013). These substrates are useful for a variety of biophysical assays, but not ideal for binding studies due to the fact that they can engage in multiple binding events.

Trinucleosomes are a minimalistic model for higher order chromatin that are more amenable to biochemical and biophysical studies (Fig. 1H). The central nucleosome represents a nucleosome in a chromatin context, as it is flanked on either side by a nucleosome. Trinucleosomes are assembled using the same protocol as for mononucleosomes, however, additional steps are needed to assess their quality and homogeneity. To exclude the possibility of under-saturation of the array, a DNA restriction digestion is carried out targeting the *Eco*RI sites located in the internucleosomal DNA (Fig. 1B). Moreover, to control for oversaturation, ideally AUC needs to be performed.

2.4.1 Equipment

- Reconstitution set up described earlier
- Standard agarose gel running setup

2.4.2 Buffers and Reagents

- As for mononucleosomes
- 1 × Tris acetic acid EDTA (TAE) buffer: 40 m*M* Tris, 20 m*M* sodium acetate, 1 m*M* EDTA, pH 8
- 50% sucrose (w/v)
- Heat block or water bath set at 37°C
- 10 × NEB *Eco*RI buffer
- *Eco*RI (NEB)
- Low-EEO/Multi-Purpose/Molecular Biology Grade Agarose (Fisher BP160-100)

2.4.3 Procedure

The method for reconstituting trinucleosomes is exactly the same as that for mononucleosomes, with the following differences:

1. At every ratio of octamer titration, the number should be multiplied by three to account for three octamers on each DNA repeat. The final volume of the reconstitution can be increased to 50 μL.
2. The final concentration ofDNA in the reconstitution can be in the range of 1–5 μ*M*.
3. After the completion of TCS dialysis, set up *Eco*RI digestion for each DNA: histone octamer ratio using 1 μg of trinucleosome in a final volume of 10 μL of 1 × *Eco*RI buffer. Incubate at 37°C for 1 h.
4. While waiting for the restriction digestion, prepare a 1% agarose gel in 1 × TAE buffer.
5. After 1 h of *Eco*RI digestion, add 50% sucrose to *Eco*RI digestion to obtain a final concentration of 5% (v/v) to each tube and load on the agarose gel, run at 50 V for 2 h. Load a ladder and a 207 bp DNA marker.
6. Stain the gel in ethidium bromide for ~15 min and determine if any of the DNA: octamer ratios have 207 bp free DNA in the digested lanes. Samples with free 207 bp DNA after digestion are undersaturated, that is, they contain less than three nucleosomes.
7. The DNA: octamer ratios that do not have any free DNA, as determined in the *Eco*RI digestion step, should now be checked by AUC before moving onto large-scale reconstitutions (AUC procedure described later).

2.4.4 Notes

1. The quality of the agarose used for analyzing trinucleosomes and higher molecular weight arrays is extremely important because the arrays tend to smear in poor quality agarose.
2. It is important to run the larger complexes at 50 V, because at higher voltage the complexes might be unstable due to the heat generated.
3. Consider adding DNA-only uncut and cut samples as controls. A DNA ladder and 207 bp DNA should be included. A Nuc207 sample can also be loaded on the gel as a reference.
4. Trinucleosomes can also be assembled with fluorophore-labeled octamer as reported before (Muthurajan et al., 2014; Winkler & Luger, 2011).
5. It is possible to overassemble (have more histones than the three positioning repeats will allow), therefore, it is critical to perform both *Eco*RI digestion as well as AUC for a good quality control. *Eco*RI digestion will

look identical for correctly assembled and overassembled trinucleosomes, and the difference will be visible only in their S_{avg} values via AUC.

A representative *Eco*RI digestion of a trinucleosome sample is shown in Fig. 1H. Since the *Eco*RI digestion sites are present in the internucleosomal linker DNA, the optimal assembly ratio will contain only mononucleosome bands upon digestion (lane 7). Note that at lower octamer ratios, there is 207 bp DNA, indicative of undersaturated samples (lanes 3 and 5). Additionally, in the undersaturated trinucleosome samples, both the undigested as well as the digested samples are heterogeneous with multiple bands (lanes 2–5) due to the presence of a mix of mono-, di-, and trinucleosomes assembled on the DNA.

2.5 "Microscale" or Step Dilution Method for Nucleosome Reconsitution

We have previously described a quicker method for the small-scale reconstitution of nucleosomes using radioactively labeled DNA by simply assembling the octamer and DNA in 2 *M* salt and then diluting it with a no salt buffer (Dyer et al., 2004). This technique can be used for unlabeled DNA or labeled DNA, or labeled histones.

Fig. 2 demonstrates mononucleosome assembly via microscale/step dilution method, note that at lower octamer ratios, multiple bands are visible when free DNA is present potentially due to the presence of nonnative distribution of histones on DNA (lanes 1 and 2). At higher ratios of octamer,

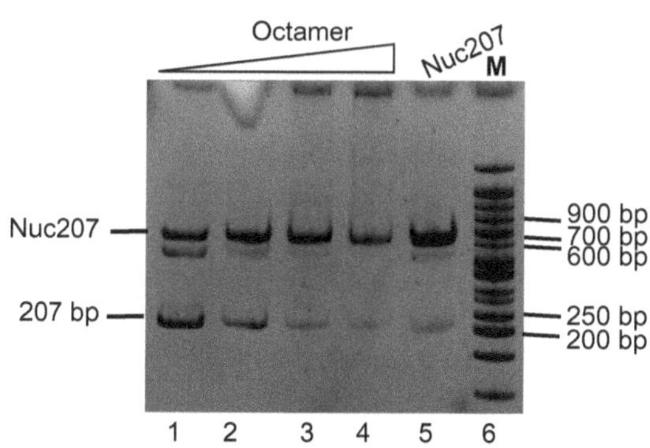

Fig. 2 "Microscale"/step dilution method for nucleosome reconstitution. Lanes 1–4: 207 bp DNA titrated with 1.2, 1.6, 2.0, 2.2-fold excess octamer, respectively; lane 5, Nuc207 control reconstituted via salt gradient; lane 6, DNA ladder.

there is a single discrete band representing Nuc207 (lanes 3 and 4) that runs similar to the salt gradient reconstituted nucleosomes (lane 5) on a native gel. Details for this method are found in Dyer et al. (2004). These nucleosomes are usually not suitable for experiments that require high-quality chromatin.

3. QUALITY CONTROL STRATEGIES FOR IN VITRO RECONSTITUTED CHROMATIN

A variety of quality control techniques are used in our lab to assess in vitro assembled chromatin. Quality control measures are geared to ascertain the contents in the reconstituted nucleosomes either visually (gel electrophoresis, AFM) or by analyzing the properties of assembled chromatin in solution (AUC, SEC-MALS, and SAXS). Traditionally, gel electrophoresis is one of the time-tested methods and is very easy to perform in any lab. Native PAGE is sensitive to shape and charge, and works particularly well with mononucleosomes. Indeed, it has been utilized in the past to distinguish different positions of the histone octamer on 146 base pairs of DNA (Dyer et al., 2004). It is particularly powerful in conjunction with fluorescently labeled histones and can be combined with MNase digestion to quantify the amount of completely assembled nucleosomes that protect the requisite 147 bp of DNA. In recent years, more advanced techniques like SEC-MALS, AUC, SAXS, and AFM have been employed. While these require access to more advanced equipment, they are uniquely positioned to answer specific questions, as detailed later. As an easy guide to the sample requirements and the type of information obtainable from each of the quality control methods and also to find training resources refer to Table 1.

3.1 Micrococcal Nuclease Digestion

MNase is an enzyme that digests DNA in regions that are not stably bound by proteins (Cuatrecasas, Edelhoch, & Anfinsen, 1967). Once the nucleosome is assembled, the DNA wrapped around the histones is protected from MNase digestion, while the linker arms are digested. This will yield a distinct protected band of 147 bp in length. This method can be used to assess the integrity and stability of nucleosomes in solution and is also used to assess the end products of salt dialysis and chaperone-mediated assembly reactions (as shown below using nucleosome assembly protein 1—Nap1).

For MNase digestion of any sample, a pilot scale experiment needs to be performed to optimize the amount of enzyme and time of digestion. Optimally, the combination is such that the entire range of digestion products up

Table 1 Quality Control Methods Listed in Increasing Order of Complexity, with Their Advantages and Disadvantages at a Glance

Method for QC	Sample Volume, Concentration	Information Obtained	Caveats	Useful Resources
Native gel electrophoresis	2–3 μL of 5 μ*M* reconstitutions	– Homogeneity – Fluorescently labeled histones and/or DNA confirm the presence of components (scanning at appropriate wavelengths)	Any complex with similar mobility can be mistaken for nucleosomes	N/A
Restriction enzyme digestion	1 μg	– Saturation of nucleosome arrays	Saturated and oversaturated nucleosome arrays appear the same after digestion with restriction enzyme	NEB website
Micrococcal nuclease digestion	0.5–1 μg	– Size of protected fragment – Quantify amount of properly assembled nucleosomes	Biased to AT sequences	NEB website
Bioanalyzer 2100	1 μL of 0.2 mg/mL DNA	– Base pair resolution of MNase-digested products – Efficiency of nucleosome assembly	There might be small variations in how the DNA fragment sizes are determined based on the DNA ladder and/or chip	Agilent website
SEC-MALS	120 μL of 0.2–0.3 mg/mL	– Polydispersity – Stability and molecular weight	Small changes in molecular weight cannot be detected	http://www.wyatt.com/training/light-scattering-university.html

Continued

Table 1 Quality Control Methods Listed in Increasing Order of Complexity, with Their Advantages and Disadvantages at a Glance—cont'd

Method for QC	Sample Volume, Concentration	Information Obtained	Caveats	Useful Resources
AUC	400 μL of 25–30 ng/μL (0.4 absorbance at 260 or 280 nm)	– Homogeneity – Shape of molecules – For a mix of species the composition can be quantified by advanced data analysis	Molecular weights are generally an estimate only, since they are largely dependent on partial specific volumes	AUC theory and operation: https://www.beckmancoulter.com/wsrportal/wsr/support/training-education/proteomelab-xl-a-xl-i/index.htm AUC workshop for data analysis: http://www.ultrascan.uthscsa.edu/workshop.php
AFM	30 μL of 1.5 ng/μL	– Visualizing the complexes – Height profiles of nucleosomes and complexes	Best for nucleosomal arrays and larger molecular weight complexes	http://www.asylumresearch.com/Events/Classes/
SAXS	150 μL of ~2 mg/mL	– Overall shape of complexes – Maximum dimension and radius of gyration	No information about which domains are interacting or other molecular details, only a shape of the complex is obtained	SAXS data collection and processing at ALS Berkeley: http://bl1231.als.lbl.gov/2013/08/2013_als_UM_workshop.php At APS Chicago: http://small-angle.aps.anl.gov/

Sample volume and concentration are listed for nucleosomes and chromatin and is meant only as a rough guideline.

to the end-point can be observed. Additionally, quantification of MNase digestion products can be performed if a reference DNA is added to all the samples. This can serve as a reference relative to which the recovery of the products can be normalized and also serves as an internal control for any extraction steps.

3.1.1 Equipment

- Standard native PAGE equipment, 5% or 10% acrylamide native gels

3.1.2 Buffers and Reagents

- MNase (NEB)
- Proteinase K (NEB)
- MinElute PCR cleanup kit (Qiagen)
- Nucleosome assembly buffer: 20 m*M* Tris–HCl (pH 7.5), 150 m*M* NaCl, 1 m*M* EDTA (pH 8.0), 1 m*M* DTT
- 10 × MNase reaction buffer: 500 m*M* Tris–HCl, pH 7.9, 50 m*M* $CaCl_2$
- 6 × DNA loading buffer (LB): 0.25% bromophenol blue, 0.25% xylene cynol, 15% (w/v) ficoll in water

3.1.3 Procedure

Nucleosomes assembled via any method and trinucleosomes can be tested by MNase digestion.

1. Mix 20 μL sample (200 n*M* DNA or nucleosomes) with 10 μL of 10 × MNase buffer, 1 μL of 100 × BSA, 30 U MNase (or appropriate amount as determined in pilot experiments; test a wide range of enzyme 0–100 U MNase) and bring the volume to 100 μL with nuclease-free water.
2. Incubate at room temperature for 10 min.
3. Quench the reaction with 25 μL of 0.5 *M* EDTA and then add 25 μg of proteinase K.
4. Incubate for 30 min at 50°C.
5. Add a reference DNA to all samples (1–5 n*M* final concentration).
6. Purify the DNA using a MinElute Qiagen commercial kit that allows elution of DNA in a small volume (10 μL) of buffer.
7. Load 10 μL onto 5% or 10% native PAGE and run in 0.2 × TBE at 150 V for 60 min.

3.1.4 Notes

1. Always include samples with no MNase or a "zero" time point along with DNA only controls.

2. Standard phenol : chloroform precipitation and ethanol extraction of the DNA can also be used to purify the final DNA products, but this step also leads to higher DNA losses and variability, and precludes size analysis and quantification by bioanalyzer (see later).
3. Note that different lengths of DNA might be protected for nucleosomes containing histone variants (Bao et al., 2004; Dechassa et al., 2011).

Fig. 3A depicts a gel representing MNase digestion of mononucleosomes assembled via salt gradient reconstitution on 207 bp DNA. Note that unassembled 207 bp DNA is completely digested (lanes 2 and 3), while the DNA–histone mix without reconstitution results in a smear of bands indicative of formation of nonnucleosomal species (lanes 4 and 5). For salt gradient assembled Nuc207, upon MNase treatment a discrete 147 bp protected band is observed (lanes 6 and 7).

Fig. 3B is a representative MNase digestion of nucleosomes assembled on 147 bp and 207 bp DNA, mediated by Nap1 (chaperone-mediated assembly). A frequent concern regarding chaperone-assembled nucleosomes is whether the assembled nucleosomes are identical to salt gradient reconstituted nucleosomes. MNase digestion shows that Nap1-assembled nucleosomes protect 147 bp of DNA for both Nuc147 (lanes 7 and 8) as well as Nuc207 (lanes 10 and 11). Note that DNA and histones mixed together in the absence of Nap1 (lanes 4 and 5) show a smear of protected bands due to absence of positioned nucleosomes.

Fig. 3C represents a typical MNase digestion pattern of trinucleosomes. The samples show three discreet digestion products, corresponding to the tri-, di-, and mononucleosome species (561 bp, ~400 bp, and ~147 bp, respectively, obtained by random cuts in the linker DNA). As the digestion proceeds for longer time, the mononucleosome protection band (147 bp) becomes more prominent, while the higher species (tri- and dinucleosomes) slowly disappear.

3.2 Analysis of MNase-Digested Products Via Bioanalyzer 2100

Gel electrophoresis is a low-resolution method of analyzing MNase-digested products, since it may not resolve small base pair differences effectively. Additionally, quantification of relative amounts of different bands is difficult. The Agilent 2100 Bioanalyzer provides an alternative method that is fast, efficient, highly sensitive, quantitative, accurate, and reproducible with minimal sample volume requirements (Falk et al., 2015). Bioanalyzers have several applications including protein analysis; here, we have used it as a tool to analyze products resulting from MNase digestion.

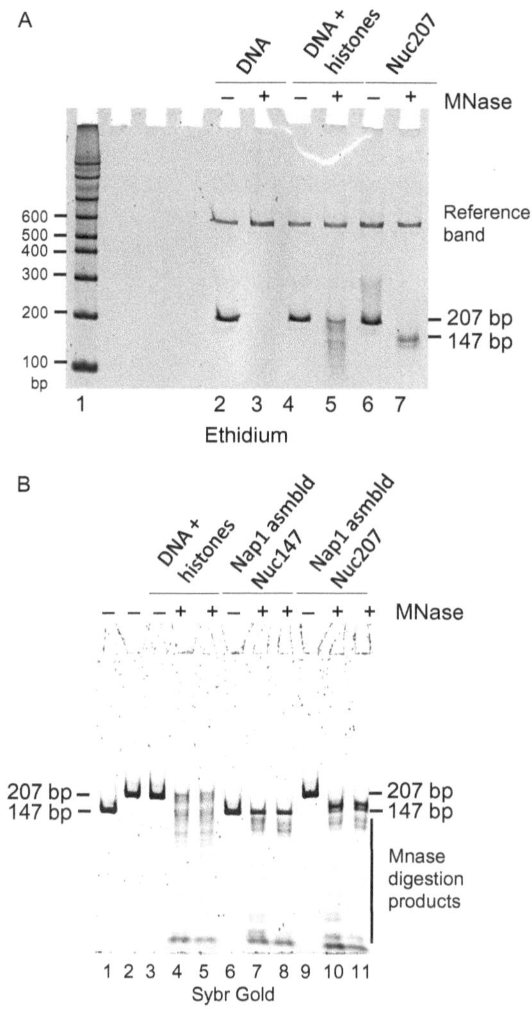

Fig. 3 MNase digestion of mono- and trinucleosomes. (A) Micrococcal nuclease digestion of Nuc207 assembled via salt gradient dialysis. Lane 1, 50–2 k ladder; lane 2, undigested 207 bp DNA; lane 3, MNase-digested 207 bp DNA; lane 4, DNA–histone mix (1:1), untreated; lane 5, DNA–histone mix, MNase treated; lane 6, salt gradient assembled Nuc207, untreated; lane 7, Nuc207 MNase treated. Note that DNA is digested completely (lane 3). DNA–histone mix results in a smear of protected bands because histones do not get positioned into nucleosomes without reconstitution (lane 5). Nuc207 upon MNase treatment gives a characteristic protected band of ~147 bp DNA (lane 7). (B) MNase digestion of chaperone-assembled nucleosomes. Lane 1, 147 bp DNA; lane 2, 207 bp DNA; lanes 3–5, 207 bp DNA + histones with 0, 0.5, and 1 U MNase; lanes 6–8, Nap1-assembled Nuc147 with 0, 0.5, and 1 U MNase; lanes 9–11, Nap1-assembled Nuc207 with 0, 0.5, and 1 U MNase. Digested samples were treated with Proteinase K and run on 5% native PAGE stained with Sybr Gold, followed by visualization on the Typhoon imager. Note that the Micrococcal nuclease protection assay indicates similar sized protected fragments (147 bp) for both Nap1-assembled nucleosomes (Nuc147 and Nuc207).

(Continued)

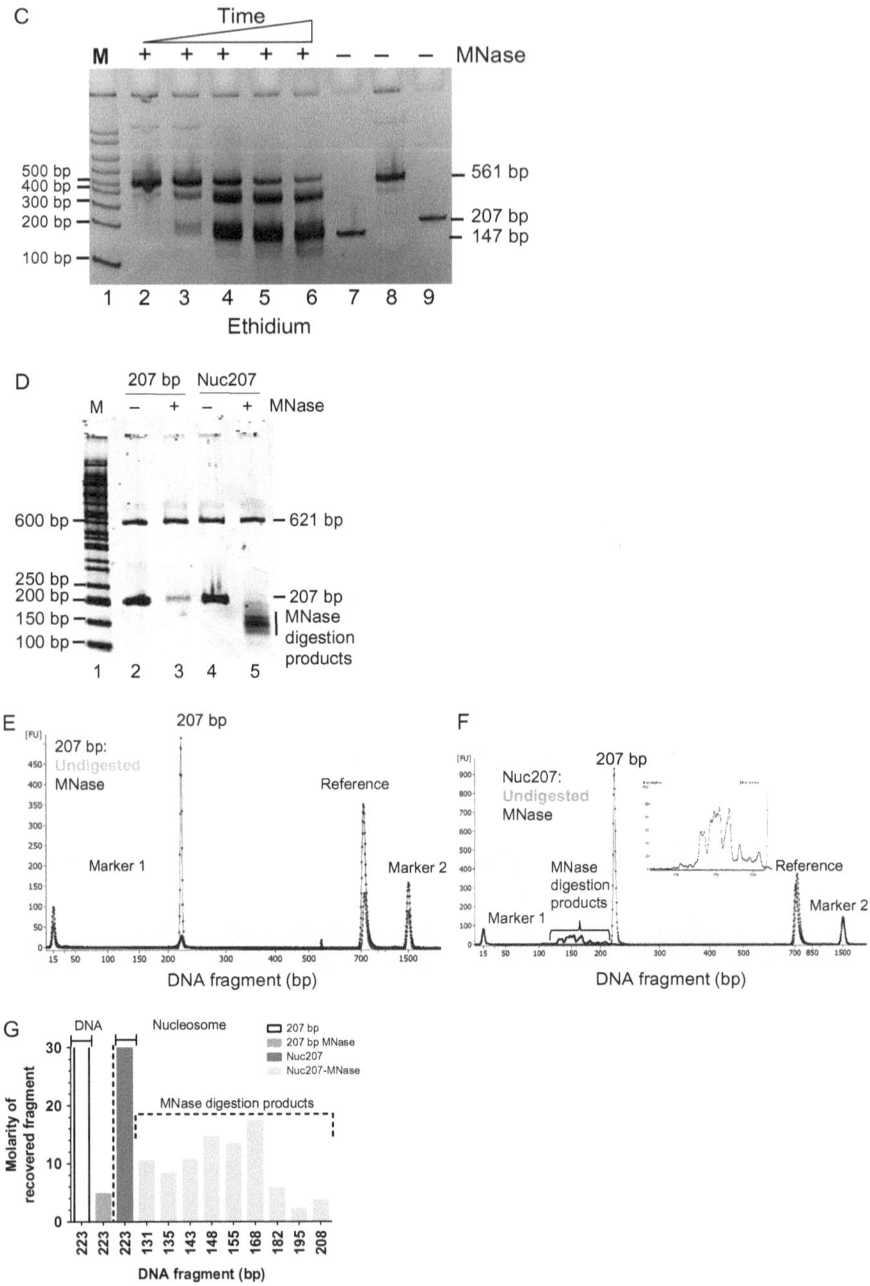

Fig. 3—Cont'd (C) Micrococcal nuclease digestion of trinucleosomes: NLE-Tri were incubated with MNase for 0, 2, 4, 6, and 8 min. Bands corresponding to 147, 414, and 561 bp were obtained indicating the presence of mono-, di-, and trinucleosome lengths. Lane 1: 50–2 k ladder; lanes 2 through 6: 0–8 min of MNase treatment; lane 7: 147 bp DNA; lane 8: 561 bp DNA; lane 9: 207 bp DNA. (D) MNase-digested and undigested 207 bp DNA and Nuc207 were analyzed by 10% native PAGE after DNA extraction. Lane 1: DNA ladder; lanes 2 and 3: Undigested and MNase-digested

3.2.1 Equipment

- Agilent Bioanalyzer 2100
- DNA 1000 chip

3.2.2 Buffers and Reagents

- Purified DNA sample (MinElute PCR cleanup kit—Qiagen)
- DNA 1000 Reagents (5067-1505 Agilent)

3.2.3 Procedure

1. Subsequent to MNase digestion, add 1–5 n*M* of a reference DNA (ideally somewhat longer than the products to be analyzed) that will function as a loading control in each lane as described earlier.
2. Purify the samples using a MinElute PCR kit and elute DNA in 10 μL.
3. 1 μL of each sample is loaded on DNA 1000 chip and analyzed on Bioanalyzer 2100. The suggested quantitative range for this chip is 0.1–50 ng/μL.
4. Data can be visualized and analyzed on the free software 2100 Expert from Agilent.
5. Quantification of the data for recovered DNA can be done using Microsoft Excel. Here we used the calculated molarity value output from the "2100 Expert" software. The reference DNA is used to normalize the amount of DNA fragment in each lane (normalized value = molarity of the DNA fragments of interest/molarity of the reference DNA). The normalized values are used for comparison between different lanes and to determine percent fragment recovered.

Fig. 3—Cont'd 207 bp DNA, respectively; lanes 4 and 5: Undigested and MNase-digested Nuc207, respectively. (E) Analysis of MNase-digested Nuc207 via Bioanalyzer 2100. Undigested 207 bp DNA (gray trace) overlaid with MNase-digested DNA (black trace). Note that 207 bp DNA is almost completely digested (samples loaded in lanes 2 and 3 in (D)). (F) Overlay of MNase digestion profile for undigested (gray trace) and MNase-treated Nuc207 (black trace). For nucleosomes, several protected bands are visible around 150 bp (samples loaded in lane 4 and 5 in DNA; see inset). (G) Bar graph of quantification data obtained from Bioanalyzer 2100. The added advantage of analyzing MNase digestion products on a Bioanalyzer is that the molarity of each DNA fragment is calculated based on the DNA markers loaded along with the samples. Hardly any fragment is recovered upon digestion of free DNA, while digestion of Nuc207 reveals several fragments in the range of 130–200 bp. The *lines* on top of the undigested DNA and Nuc207 indicate continuity to 100% on *Y*-axis.

3.2.4 Notes

1. Use filter tips and gloves to lower the level of contaminants and DNA degradation.
2. Always use DNA only controls to compare with nucleosome samples and internal standards to confirm recovery.
3. DNA 1000 chips have a separation range between 25 and 1000 bp.

Fig. 3D shows a 10% native acrylamide gel of extracted DNA after MNase digestion of 207 bp DNA (lanes 2 and 3) or Nuc207 (lanes 4 and 5). 207 bp DNA is almost completely digested, while Nuc207 has several protected bands.

Fig. 3E and F shows the Bioanalyzer results for 207 bp DNA (Fig. 3E) and Nuc207 (Fig. 3F) for the samples that are shown in Fig. 3D. The undigested (gray trace) and digested (black trace) samples are overlaid in each panel. While 207 bp DNA has no defined protection pattern, Nuc207 clearly shows a distinct protection around 150 bp, (gray trace) and inset. This method allows for a more accurate analysis of the degradation pattern at lower levels of MNase, compared to gel electrophoresis (Fig. 3D).

In addition, the output data from the Bioanalyzer allow for quantification of the DNA fragments. As shown in Fig. 3G, the molarity of individual fragments calculated by the software can be used to measure the relative amount of DNA protected during MNase digestion of Nuc207. It is apparent that while 207 bp DNA is completely digested, a variety of digestion products were recovered for Nuc207 including some that were only a few base pairs apart. This allows us to estimate the amount of nucleosomes that are formed in each sample. Importantly, this method can be used to compare the efficiency of different assembly conditions including various chaperones or reconstitution methods in that it allows us to monitor only properly formed nucleosomes. Optimization of MNase digestion conditions is an essential prerequisite. In fact, over digestion could result in an underestimation of the amount of nucleosome in the samples.

3.3 Size-Exclusion Chromatography Coupled with Multiangle Light Scattering

Multiangle light scattering (MALS) uses laser light scattering to provide information about shape, size, and polydispersity of protein samples in solution and is widely used for determining homogeneity, stoichiometry, and molecular weight of the complexes. When connected in line with a SEC column, MALS can provide information about the different species as they separate on the column. This provides a wealth of information about homogeneity, stability, and stoichiometry of the injected sample under the conditions tested. However, differences less than twofold in molecular weight or shape cannot be detected. Nevertheless, this technique is powerful for analyzing complexes of nucleosome and chromatin regulatory proteins (Clark et al., 2012; Yang, van der Woerd, Muthurajan, Hansen, & Luger, 2011). A size-exclusion column (run by an AKTA protein purification system) is connected to a MALS and refractive index (RI) detector. The SEC should be selected depending on the size of the complexes that need to be resolved.

3.3.1 Equipment

- Wyatt Instruments HELEOS and rEX detectors
- GE Healthcare AKTA protein purification system and SEC column

3.3.2 Buffer and Reagents

- 25 m*M* Tris, pH 7.5, 200 m*M* NaCl, 2 m*M* arginine

3.3.3 Procedure

1. Prepare 120 μL of nucleosome solution at about 0.2–0.5 mg/mL (use higher concentrations for smaller proteins).
2. Equilibrate the column with buffer prefiltered via 0.1 μm membrane.
3. Spin the sample at 14,000 rpm for 10 min and inject 100 μL into the SEC column.
4. Unicorn software is used to run the SEC, while the ASTRA software from Wyatt records the MALS and RI data. Refer to the manufacturer manual for details.

3.3.4 Notes

1. Unstable reducing agents such as DTT or BME, or other unstable buffer components like detergents should be avoided. TCEP should be used as the reducing agent.
2. Thoroughly dialyze or buffer exchange samples in the running buffer to avoid high background due to buffer differences.
3. It is recommended that a BSA standard (2 mg/mL) be run as a control prior to testing samples.
4. Resolution power and lack of contaminations on the SEC column are essential to minimize light scattering background signal. Refer to the manufacturer's instruction for cleaning procedures.

Nuc165 injected onto Superdex 200 elutes in a single symmetric peak and the calculated molecular weight matches the theoretical value (Fig. 4). This

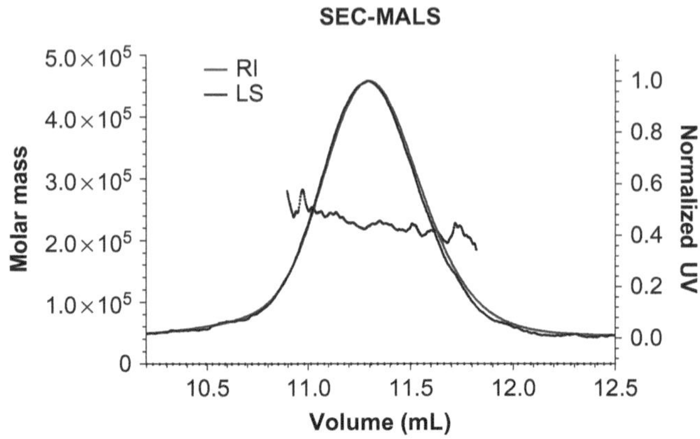

Fig. 4 Size-exclusion chromatography coupled with multiangle light scattering (SEC-MALS). A profile for mononucleosomes assembled on 165 bp DNA (Nuc165). The experimentally determined molecular weight (2.24×10^5 Da) matches closely with the theoretical molecular weight (2.1×10^5 Da) for Nuc165.

technique was applied to analyze nucleosome-MeCP2 and nucleosome-PARP-1 complexes (Clark et al., 2012; Yang et al., 2011).

3.4 Analytical Ultracentrifugation

AUC is a powerful technique that allows for particle analysis in solution, especially in conjunction with newly developed data analysis software. The rate of sedimentation of a molecule depends on both its shape and size; therefore, information pertaining to both can be obtained using this method. AUC can be used to analyze chromatin of any size (mono-, tri-, or 12-mer arrays) and their complexes, but is particularly useful for larger samples where native PAGE either is not applicable, or has insufficient resolution. Data should be collected in intensity mode rather than absorbance to reduce noise levels (Kar, Kingsbury, Lewis, Laue, & Schuck, 2000), where intensity refers to the transmitted light incident on the sample. The intensity data can easily be converted to pseudo-absorbance data using the software Ultrascan III. The partial specific volume is a key parameter required by the software to determine molecular weight and is affected by the buffer components. AUC can be combined with fluorescence detector to obtain enhanced detection sensitivity, but this special application is not discussed here.

A list of institutes that have access to this instrument worldwide is available at http://www.uslims3.uthscsa.edu/index.php. A good starting point for a novice would be to locate the nearest facility that has the equipment and expertise required to operate AUCs. Additionally, several institutes also provide paid services and can run the experiment and data analysis for the users. However, some rudimentary guidelines are provided here for the experimenters to follow. These, in combination with a previous publication (Rogge et al., 2013), would be a good starting point for researchers who wish to use AUC.

3.4.1 Equipment

- AUC cells with quartz windows
- Beckman XLA or XLI AUC
- Computer with Ultrascan III installed for data analysis and access to Laboratory Information management system (LIMS) for two-dimensional spectrum analysis and genetic algorithm coupled with Monte-Carlo analysis calculations

3.4.2 Buffers and Reagents

- AUC buffer: 20 m*M* Tris, pH 7.5, 1 m*M* EDTA, 1 m*M* DTT, or any other buffer (see notes).

3.4.3 Notes

1. Clean AUC cell components and assemble the cell (Fig. 5A and B).
2. To perform an AUC run, sample absorbance at 260 or 280 nm of ~0.3–0.4 is required for high-quality data. Prepare 400 μL of trinucleosomes (or other complexes) in the final dialysis buffer. Collect data in intensity mode, use water in reference channels (Fig. 5C–F).
3. The size of the particles to be analyzed will influence the rotor speed during data collection, generally, the smaller the size of the particle, the higher the speed needed to sediment it (eg, for mononucleosome samples use 40,000 rpm, for a trinucleosome sample use 30,000 rpm) (Fig. 5G–J).
4. Data are collected as number of scans representing the sedimentation of the sample from the top to the bottom of the cell. The number of scans should be decided such that the sample fully sediments. This value depends on the number of samples in the rotor, the molecular weight of the complexes, the speed, etc. Generally, 150 scans could be collected per sample.
5. AUC data analysis can be performed using the Ultrascan III software, following the user manual details: http://www.ultrascan2.uthscsa.edu/manual2/index.html (Brookes & Demeler, 2006; Cao & Demeler, 2005; Demeler, 2005a, b; Demeler & Van Holde, 2004).

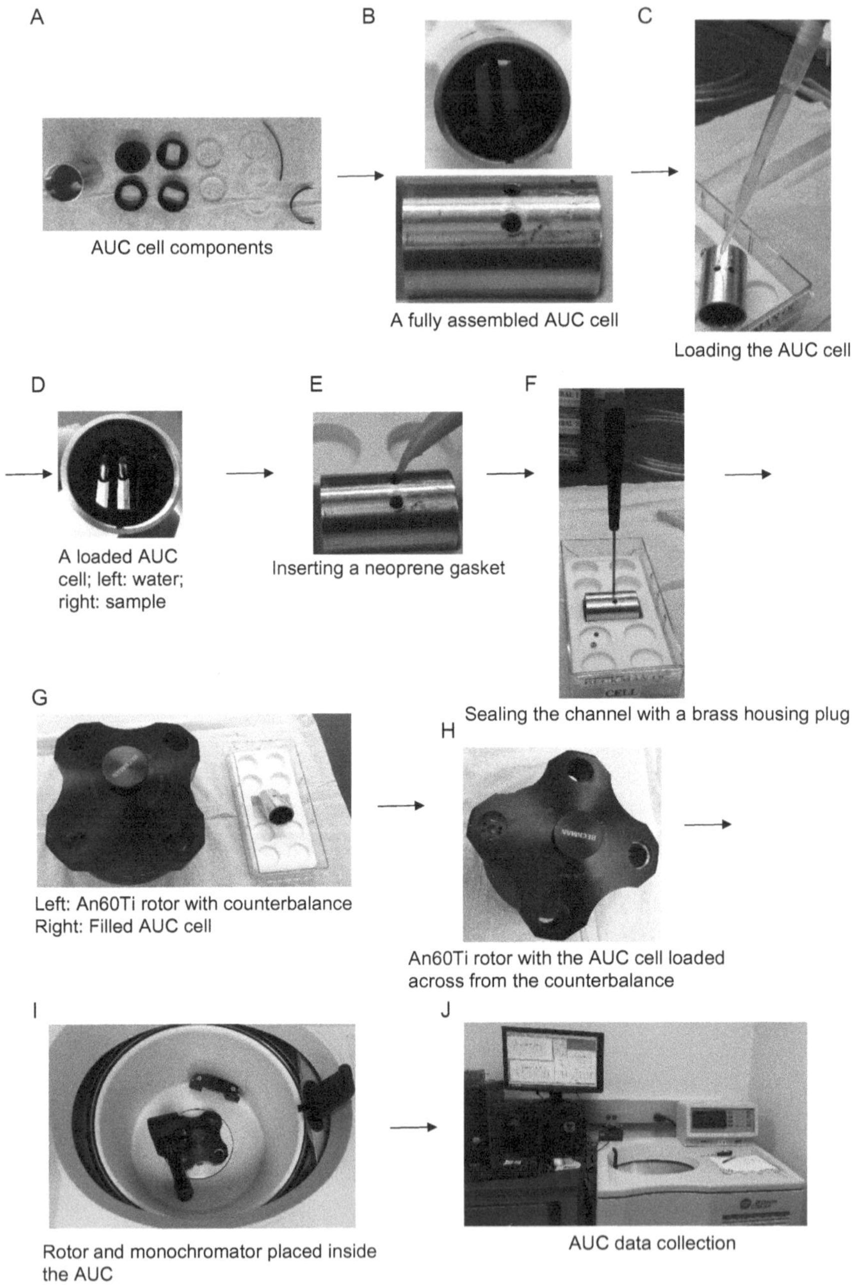

Fig. 5 See legend on opposite page.

3.4.4 Procedure

1. Any buffer is compatible with AUC, but DTT above 1 m*M* should be avoided. BME and detergents should also be avoided due to background scatter.

Fig. 6A shows an example of a nucleosome reconstituted on 165 bp of DNA (Nuc165), analyzed by sedimentation velocity. This particle sediments with an S_{avg} of ~11.2S and with a frictional ratio (f/f_0) of about 1.4 (Fig. 6A). Fully saturated trinucleosomes sediment with an S_{avg} of ~17S (Fig. 6B) and a frictional ratio of ~3 as would be expected for an extended array. As a case study, the trinucleosomes in complex with the chromatin architectural protein methyl-CpG-binding protein 2 (MeCP2) was analyzed via AUC. The MeCP2-bound trinucleosomes have a mid-point *S* value of ~20S in accordance with an increase in the molecular weight of the complex (Fig. 4C). In addition, for the MeCP2 bound trinucleosomes, the f/f_0 value dropped to ~1.9, indicating that the extended trinucleosomes form a more condensed structure upon MeCP2 binding, consistent with electron microscopy data of longer arrays (Ghosh et al., 2010).

Monodispersity of the sample can be assessed by inspecting the trend of the sedimentation curve. A drift of the curve to higher *s* values is indicative of some degree of polydispersity. For example, the nucleosome sample shown in Fig. 6A runs as a largely monodisperse sample, while the MeCP2 complex shown in Fig. 6B suggests some heterogeneity within the sample. This is also revealed in the analysis of the solutes performed by the software Ultrascan III.

3.5 Atomic Force Microscopy Analysis of Trinucleosomes

AFM allows measurement of height profiles, volume, and physical properties of macromolecules based on their capacity to deflect laser light off of a

Fig. 5 Overview of an AUC experiment and equipment. (A) Components of an AUC cell. (B) An assembled AUC cell. (C) Two sample loading apertures on the side of the cell housing. (D) The reference channel is loaded with 400 μL of water and the sample channel has 400 μL of the sample. (E and F) The channels are sealed with a neoprene gasket and a brass housing plug. (G) The correct weight is inserted into the counterbalance and placed in the rotor. (H) The sample AUC cell is placed across from the counterbalance. (I) The rotor is placed inside the centrifuge chamber and the monochromator is inserted in its slot. (J) An initial radial calibration is done at 3000 rpm to ascertain that the cells are not leaking and to calibrate the rotor. Subsequently, the centrifuge is allowed to equilibrate for an hour to allow the temperature and vacuum to equilibrate prior to starting the run. Data are collected until the sample has completely sedimented to the bottom of the cell. Data analysis is done using Ultrascan III and the associated LIMS.

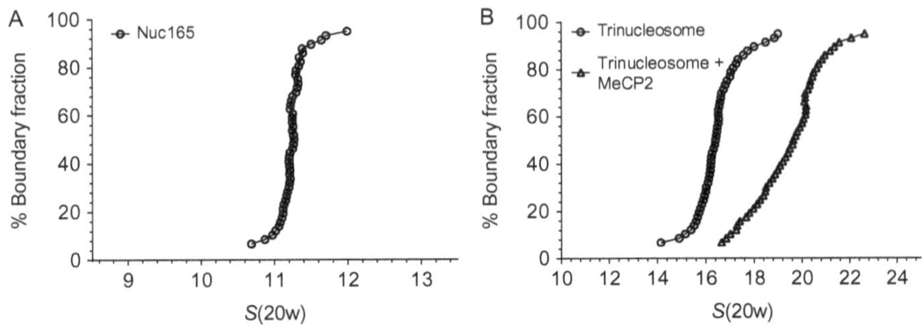

Fig. 6 Analysis of chromatin by AUC. (A) Sedimentation velocity profile for Nuc165 indicates an S_{avg} value of ~11.2S. The vertical nature of the $G(s)$ plot indicates homogeneity. (B) The sedimentation profile for trinucleosomes was overlaid with trinucleosomes in complex with a threefold excess of MeCP2. The latter showed a forward shift of S_{avg} from ~17S to ~20S indicative of MeCP2 binding to and compacting trinucleosomes.

cantilever as it scans the sample surface (Fig. 7A–E). A large number of images can be generated and should be analyzed for statistical purposes. Sample requirements are minimal for this method. We have described AFM data collection in air (dry) which is a rapid and convenient method for visualizing complexes. AFM is quite versatile and can be used to collect images in liquid either statically or in a flow cell. The greatest advantage of this technique is the direct visualization of chromatin and its complexes, it is particularly useful for chromatin arrays and larger complexes (Lohr, Bash, Wang, Yodh, & Lindsay, 2007; Lyubchenko & Shlyakhtenko, 2015; Nazarov et al., 2016).

3.5.1 Equipment

- Asylum Research MFP-3D AFM or comparable instrument
- Olympus 240TS tips or comparable brand
- Freshly cleaved mica discs (SPI Supplies) glued to glass slides (VWR) (Fig. 7F–G)
- 3-Aminopropyltriethoxysilane (APTES) (Sigma)
- Syringe filters and Syringes or Steriflips

3.5.2 Buffers and Reagents

- Sample buffer—20 m*M* Tris, pH 7.5, 1 m*M* EDTA, 1 m*M* DTT, filtered through a 0.2-μm filter

3.5.3 Procedure

1. Cleave mica using a piece of scotch tape by pressing it down on the mica disc and peeling it off repeatedly until a smooth mica surface is obtained

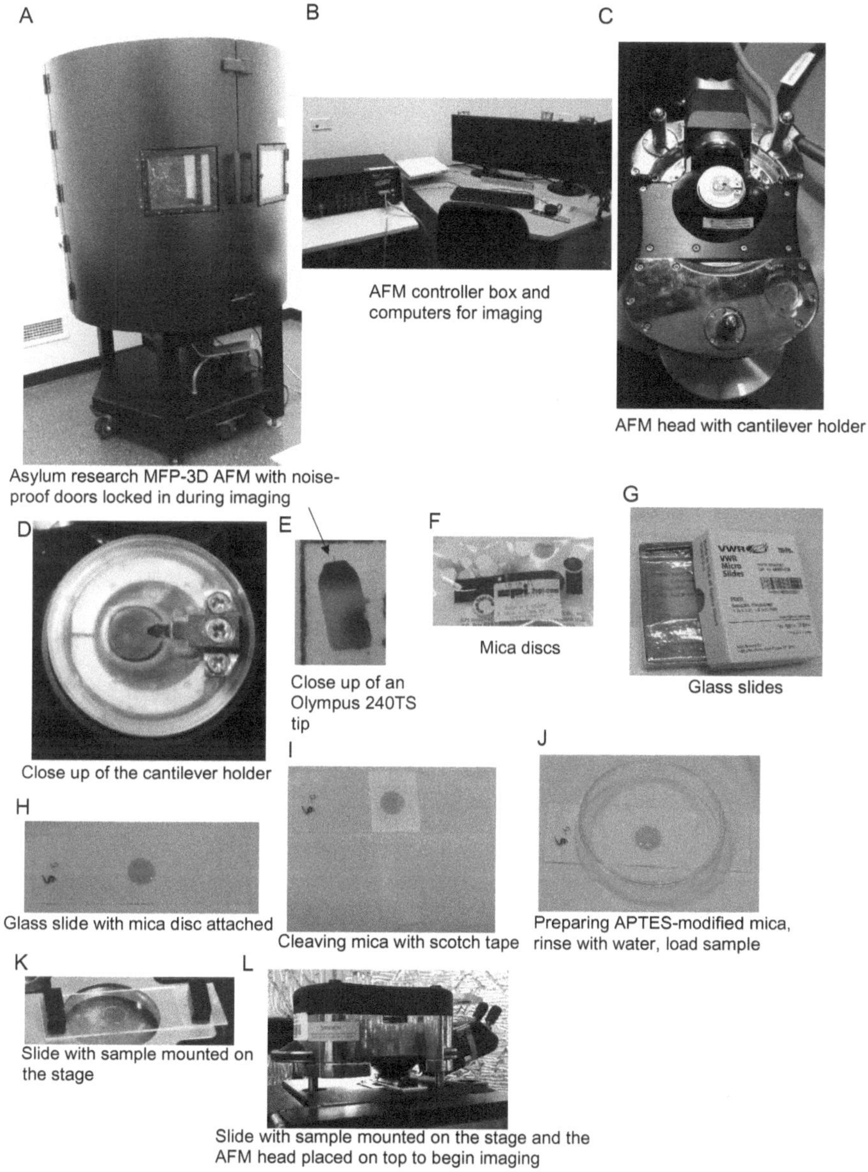

Fig. 7 Overview of an AFM experiment and equipment. (A) Asylum MFP-3D AFM. (B) Controller box and computers for data collection. (C) AFM head inverted to show cantilever holder. (D) Close up of the cantilever holder. (E) Close up of a 240TS cantilever with a tip (black arrow). (F) Mica discs are purchased separately and glued on to standard glass slides (G and H). (I) Mica is cleaved using a piece of scotch tape to obtain a smooth surface which is then modified using APTES (J). (K) The sample is then incubated on modified mica and placed on the stage. (L) The cantilever holding the tip is lowered on to the sample and images are collected.

(Fig. 7I). Modify with APTES: 1:1000 dilution in water, add 30 μL on to the mica disc, incubate covered at room temperature for 30 min, gently rinse with filtered water, and dry in a gentle stream of pure nitrogen (Fig. 7H–J).

2. Dilute trinucleosome or nucleosomal array samples to ~1.5 ng/μL, however, the dilution range might vary depending on the sample (Fig. 7J).
3. Add ~30 μL of the diluted sample on the APTES-mica surface and wait for 15 min.
4. Rinse off surface with buffer and dry in a gentle stream of air or pure nitrogen.
5. Wipe off buffer droplets from around the sample area, place slide on stage, secure with magnets.
6. Start scanning 5 × 5 μm areas. Identify areas with well-spread chromatin specimens and collect 2 × 2 μm scans. Follow these up with 1 × 1 μm scans and go to smaller areas if the instrument capability allows. At higher resolution scans (smaller area) are more sensitive and therefore likely to contain noise (Fig. 7K–L).

3.5.4 AFM Data Processing

Images are flattened and analyzed using the MFP-3D software provided by Asylum or by a comparable software. In this example, each image was divided into four quadrants and zoomed in digitally to obtain a clearer view of well-separated trinucleosomes. Multiple free hand lines were drawn through the trinucleosomes, and height profiles were obtained. For statistical analysis a large number of images (at least a few dozens) should be analyzed. Results can be analyzed by Microsoft Excel for easy visualization and relative comparison.

3.5.5 Notes

1. Mononucleosomes are too small to be analyzed by AFM and thus this method is not recommended for these samples.

In the 200 × 200 nm images obtained with trinucleosomes, the three nucleosomes held together by linker arms were clearly visible and the height profile was ~1.5–2 nm (Fig. 8A and B; right panel in B for height profile). Complexes of trinucleosomes bound to chromatin-interacting proteins, such as MeCP2 can be visualized (Fig. 8C and D). MeCP2 compacts trinucleosomes as previously shown for nucleosomal arrays (Hansen, Ghosh, & Woodcock, 2010; Ghosh et al., 2010). Some clustering of nucleosomes (more than one trinucleosome) can be clearly seen (Fig. 8D). The contrast in the amplitude and phase traces (Fig. 8C and D, middle and right

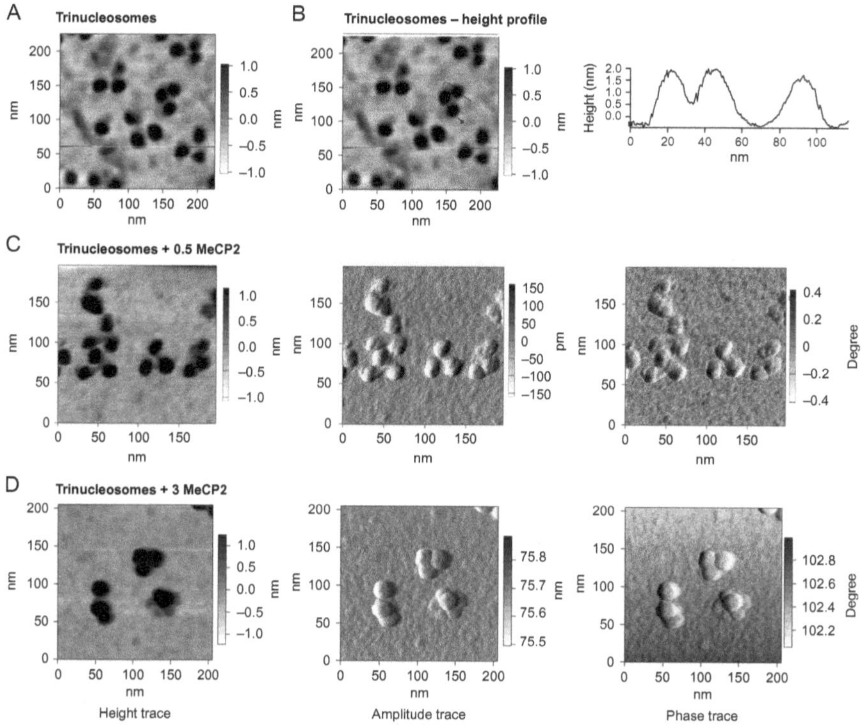

Fig. 8 Atomic force microscopy (AFM) to visualize trinucleosomes. (A) Trinucleosomes assembled by salt gradient reconstitution were imaged in air using AFM. A 250 × 250 nm scan was imaged (*left panel*). (B) A free-hand line drawn through a representative trinucleosome and the height profile of the nucleosomes (*right panel*) indicates a height of 1.5–2 nm. (C) A 200 × 200 nm scan of trinucleosomes with 0.5-fold of MeCP2 show both uncompacted and compacted trinucleosomes (height trace, *left panel*; amplitude trace, *middle panel*; amplitude trace, *right panel*). (D) A 200 × 200 nm scan of trinucleosomes with a threefold excess of MeCP2 show mostly compacted trinucleosomes (height trace, *left panel*), (amplitude trace, *middle panel*), (amplitude trace, *right panel*). Note that while linker DNA was visible in the uncompacted and partially compacted trinucleosomes (A and B) it is completely invisible in fully compacted trinucleosomes in (C).

panels) allows us to better visualize the change in structure. We have successfully employed AFM to analyze magnesium-dependent and PARP-1-induced chromatin compaction of macro-H2A containing 12-mer arrays and trinucleosomes, respectively (Muthurajan et al., 2011, 2014).

3.6 Small-Angle X-ray Scattering

SAXS of a purified sample in solution can be analyzed to infer size and shape information. SAXS data are collected at synchrotron sources equipped with a SAXS beamline (eg, ALS—SYBILS BL 12.3.1, Berkeley; APS Chicago—BioCAT, Sector 18ID; SSRL BL4-2 Stanford University, Stanford). For larger complexes such as mono- and trinucleosomes, the beamline additionally should be equipped with a mobile detector for collecting lower angle scattering. Data collection can be done either statically or continuously as the sample is being eluted off of a SEC (SEC-SAXS) (Malaby et al., 2013).

In the static mode, data are collected at multiple concentrations in replicates. Concentration and buffer-dependent sample aggregation is more apparent in data collected statically. In line SEC-SAXS is free of these pitfalls. A high-flux beam and advent of noiseless high-speed detectors allow short exposure times (thereby reducing radiation damage of samples) and virtually real-time data reduction and analysis. SAXS data can provide information about the structure, shape, and size of chromatin and its complexes with interacting proteins. A new development, time-resolved SAXS can potentially be utilized to monitor real-time conformational changes induced by chromatin-binding proteins (Graceffa et al., 2013; Kathuria et al., 2014).

3.6.1 Equipment

- Synchrotron beamline with SAXS data collection capabilities preferably with in-line SEC. All data in the present work was collected at Sector 18ID (BioCAT) located at Advanced Photon Source located in Argonne National Laboratory, Chicago
- Size-exclusion column (Superose 6 or Superdex 200)
- GE Healthcare FPLC system

3.6.2 Buffers and Reagents

- Sample buffer: 25 m*M* Tris–HCl, pH 7.5, 2 m*M* arginine, 50 m*M* NaCl and 20 m*M* Tris–HCl, pH 7.5, 1 m*M* EDTA, 1 m*M* DTT—filtered via 0.2 μm membrane.

3.6.3 Procedure

1. Write a proposal requesting beam time and wait for approval. Proposal submission procedures and deadlines can be found on beamline websites.
2. Concentrate samples to ~2 mg/mL in ~150 μL. Save an aliquot for examining the complexes by other methods (such as gel electrophoresis and AUC).
3. Prepare samples and ship them to the beamline scientist a day or two in advance. For protein stocks that are stored in −80°C, ship on dry ice. For chromatin samples, prepare a makeshift ice pack by filling Ziploc bags

with ice and place sample tubes sealed with parafilm inside a 50 mL falcon tubes.

4. Arrive at the beamline the night before to do the wet lab set up like preparation of buffers and any complexes (eg, trinucleosome-MeCP2).
5. Clean the quartz sample chamber after each run using a Hamilton syringe pump sequentially with water, bleach, water, isopropanol, and water to remove cross-contamination between runs.
6. Data reduction (radial averaging of the scattering intensities to generate $I(q)$ vs q plots), usually involves software indigenous to the beamline and should be done at the beamline. Subtract buffer from sample data.
7. All further data processing can be done with the ATSAS suite. R_g and D_{max} can be determined by Primus in ATSAS (Petoukhov et al., 2012).
8. Ab initio modeling is done using Dammin (Svergun, 1999).
9. Images of the envelopes can be generated in VMD by first generating a shell in pdb2vol (Humphrey, Dalke, & Schulten, 1996; Wriggers & Chacón, 2001; Wriggers, Milligan, & Mccammon, 1999).

3.6.4 Notes

1. Inline SEC-SAXS is better for SAXS data collection on chromatin due to fractionation and removal of aggregates prior to SAXS data collection. Even though the samples undergo dilution, it is still possible to calculate the exact concentration if the scattering plots and the UV chromatogram are synchronized.
2. It is important to keep track of sample amounts loaded onto SEC and to compare it with the relative UV signal of the eluate to determine if any aggregates were retained in the column.
3. Collect data for individual components as well as complexes.

Mononucleosomes (Nuc165—~2.5 mg/mL) prepared by salt gradient reconstitution were subjected to SAXS data collection combined with in-line SEC. An R_g of ~46.9 Å and a D_{max} of ~145 Å was determined (Fig. 9A left and middle panels). Ab initio modeling produced a low-resolution model that matched our previously published data (Fig. 9A right panel; Yang et al., 2011).

SAXS analysis of trinucleosomes confirmed previous AUC and AFM data, suggesting that in solution these are extended with an R_g of ~154 Å and a D_{max} of ~500 Å (Fig. 9B, left and middle panels). Remarkably, the three nucleosome particles were visible as spheres attached via a string of linker DNA (Fig. 9B, right panel).

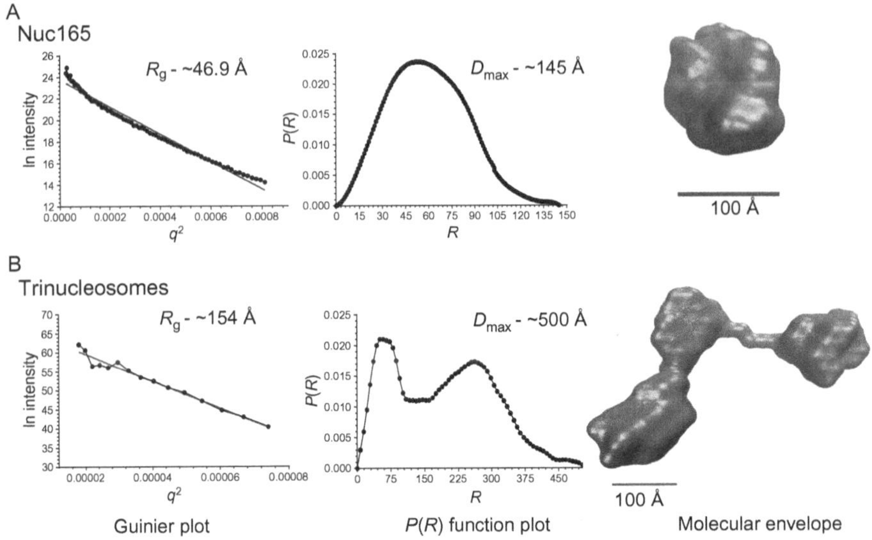

Fig. 9 Small-angle X-ray scattering (SAXS) data for in vitro assembled chromatin. (A) *Left panel*: Guinier plot for data acquired for Nuc165 indicates a radius of gyration (R_g) of 46.9 Å. *Middle panel*: $P(R)$ function plot for Nuc165 indicates a maximum dimension (D_{max}) of ~145 Å. *Right panel*: Ab initio model calculated from SAXS data for Nuc165, note the globular area for the nucleosome core. Image was generated using VMD. (B) SAXS as quality control for trinucleosomes. *Left panel*: Guinier plot for trinucleosomes indicates an R_g of ~154 Å. *Middle panel*: $P(R)$ function plot for trinucleosomes indicates a D_{max} of ~510 Å. *Right panel*: Ab initio model for trinucleosomes calculated from SAXS data.

4. SUMMARY AND CONCLUSION

For the preparation of highly defined chromatin, it is imperative to start with ingredients of highest purity. Both the DNA and the histone octamer need to be purified using stringent measures to prevent any contaminants to be carried over into the assembly reaction (such plasmid DNA for DNA preparations and dimer/tetramer/aggregates/single histones, or nucleic acid contaminants for octamer). Determining the precise concentrations of the components is also essential. Contamination of histones with nucleic acids, or inaccurate stoichiometries of components, is the most frequent cause for failed chromatin reconstitutions. This is common for any method of nucleosome assembly used to make chromatin substrates.

As a general rule of thumb, nucleosomes stored in low salt buffer such as TCS are stable at 4°C. However, such low salt concentrations may not be ideal when studying chromatin-binding proteins, which are often highly charged. In these experiments, we suggest using buffers containing salt concentration in the range of 25–250 m*M*, where the nucleosome is also stable.

Assembly of nucleosomes by salt gradient dialysis, when done properly, is an efficient way for generating homogenous nucleosome species. In fact, combined with purification over a preparatory gel electrophoresis setup these nucleosomes can be used directly for crystallization (Tan & Davey, 2011). In the cell, nucleosomes are assembled and disassembled by histone chaperones and ATP-remodelers (Hughes & Rando, 2014). It is plausible that these proteins may yield nucleosomes with subtly altered thermodynamic or structural properties.

Nucleosomes assembled by histone chaperones in vitro contain free DNA, nucleosomes, nonnucleosomal species, and histone–DNA aggregates. The presence of these contaminants is a caveat when using these samples in functional assays with chromatin regulatory proteins and likely have been the source of numerous discrepancies in the published literature. For example, many chromatin-binding proteins bind free DNA in addition to nucleosomes. The free DNA present in such preparations will compete for binding of the chromatin-interacting protein and this in turn will skew the binding data. Partially assembled nucleosomes may exhibit different binding properties compared to fully folded nucleosomes. These caveat exist for virtually every assembly method to different degrees. It is important to be aware of such issues to properly control for and then interpret the experimental data.

Rigorous quality control measures of chromatin substrates are essential to ensure reproducibility and validity of conclusions. Gel electrophoresis, while simple and therefore widely used in the past, only provides limited information and can instill a false sense of confidence in sample homogeneity. Current methods such as SEC-MALS and AUC allow one to determine the molecular weight and shape of the solutes. AFM allows visualization of the assembled chromatin, and a statistical analysis can be performed based on the height profile data if desired. SAXS performed in line with SEC allows us to collect data exclusively on a homogenous population of nucleosomes as they elute off of the SEC column and obtain shape and structural information. AUC and SAXS provide the added advantage of studying molecules in solution unhindered by any immobilization. Using a multitude of quality control methods allows us to obtain consistent high-quality chromatin substrates successfully.

REFERENCES

Bao, Y., Konesky, K., Park, Y.-J., Rosu, S., Dyer, P. N., Rangasamy, D., ... Luger, K. (2004). Nucleosomes containing the histone variant H2A.Bbd organize only 118 base pairs of DNA. *The EMBO Journal*, *23*(16), 3314–3324. http://doi.org/10.1038/sj.emboj.7600316.

Bao, Y., White, C. L., & Luger, K. (2006). Nucleosome core particles containing a poly(dA.dT) sequence element exhibit a locally distorted DNA structure. *Journal of Molecular Biology*, *361*(4), 617–624. http://doi.org/10.1016/j.jmb.2006.06.051.

Black, B. E., Foltz, D. R., Chakravarthy, S., Luger, K., Woods, V. L. J., & Cleveland, D. W. (2004). Structural determinants for generating centromeric chromatin. *Nature*, *430*(6999), 578–582. http://doi.org/10.1038/nature02766.

Brehove, M., Wang, T., North, J., Luo, Y., Dreher, S. J., Shimko, J. C., ... Poirier, M. G. (2015). Histone core phosphorylation regulates DNA accessibility. *The Journal of Biological Chemistry*, *290*(37), 22612–22621. http://doi.org/10.1074/jbc.M115.661363.

Brookes, E., & Demeler, B. (2006). Genetic algorithm optimization for obtaining accurate molecular weight distributions from sedimentation velocity experiments. In C. Wandrey & H. Cölfen (Eds.), *Analytical ultracentrifugation VIII. Progress in colloid and polymer science* (Vol. 131, pp. 78–82). Springer.

Cao, W., & Demeler, B. (2005). Modeling analytical ultracentrifugation experiments with an adaptive space-time finite element solution of the Lamm equation. *Biophysical Journal*, *89*(3), 1589–1602.

Chakravarthy, S., Bao, Y., Roberts, V. A., Tremethick, D., & Luger, K. (2004). Structural characterization of histone H2A variants. *Cold Spring Harbor Symposia on Quantitative Biology*, *69*, 227–234. http://doi.org/10.1101/sqb.2004.69.227.

Chakravarthy, S., Gundimella, S. K. Y., Caron, C., Perche, P.-Y., Pehrson, J. R., Khochbin, S., & Luger, K. (2005). Structural characterization of the histone variant macroH2A. *Molecular and Cellular Biology*, *25*(17), 7616–7624. http://doi.org/10.1128/MCB.25.17.7616-7624.2005.

Chakravarthy, S., & Luger, K. (2006). The histone variant macro-H2A preferentially forms "hybrid nucleosomes" *The Journal of Biological Chemistry*, *281*(35), 25522–25531. http://doi.org/10.1074/jbc.M602258200.

Chatterjee, N., North, J. A., Dechassa, M. L., Manohar, M., Prasad, R., Luger, K., ... Bartholomew, B. (2015). Histone Acetylation near the Nucleosome Dyad Axis Enhances Nucleosome Disassembly by RSC and SWI/SNF. *Molecular and Cellular Biology*, *35*(23), 4083–4092. http://dx.doi.org/10.1128/MCB.00441-15. PMID, 26416878. PMCID, PMC4628061. Epub 2015 Sep 28.

Chodaparambil, J. V., Pate, K. T., Hepler, M. R. D., Tsai, B. P., Muthurajan, U. M., Luger, K., ... Weis, W. I. (2014). Molecular functions of the TLE tetramerization domain in Wnt target gene repression. *The EMBO Journal*, *33*(7), 719–731. http://doi.org/10.1002/embj.201387188.

Clapier, C. R., Chakravarthy, S., Petosa, C., Fernandez-Tornero, C., Luger, K., & Muller, C. W. (2008). Structure of the Drosophila nucleosome core particle highlights evolutionary constraints on the H2A-H2B histone dimer. *Proteins*, *71*(1), 1–7. http://doi.org/10.1002/prot.21720.

Clark, N. J., Kramer, M., Muthurajan, U. M., & Luger, K. (2012). Alternative modes of binding of poly(ADP-ribose) polymerase 1 to free DNA and nucleosomes. *The Journal of Biological Chemistry*, *287*(39), 32430–32439. http://doi.org/10.1074/jbc.M112.397067.

Cuatrecasas, P., Edelhoch, H., & Anfinsen, C. B. (1967). Fluorescence studies of the interaction of nucleotides with the active site of the nuclease of Staphylococcus aureus. *Proceedings of the National Academy of Sciences of the United States of America*, *58*(5), 2043–2050.

D'Arcy, S., Martin, K. W., Panchenko, T., Chen, X., Bergeron, S., Stargell, L. A., ... Luger, K. (2013). Chaperone Nap1 shields histone surfaces used in a nucleosome

and can put H2A-H2B in an unconventional tetrameric form. *Molecular Cell, 51*(5), 662–677. http://doi.org/10.1016/j.molcel.2013.07.015.
Dechassa, M. L., Wyns, K., Li, M., Hall, M. A., Wang, M. D., & Luger, K. (2011). Structure and Scm3-mediated assembly of budding yeast centromeric nucleosomes. *Nature Communications, 2*, 313. http://doi.org/10.1038/ncomms1320.
Dechassa, M. L., Wyns, K., & Luger, K. (2014). Scm3 deposits a (Cse4-H4)2 tetramer onto DNA through a Cse4-H4 dimer intermediate. *Nucleic Acids Research, 42*(9), 5532–5542. http://doi.org/10.1093/nar/gku205.
Demeler, B. (2005a). *UltraScan—A comprehensive data analysis software package for analytical ultracentrifugation experiments.* 210–230.
Demeler, B. (2005b). Hydrodynamic methods. In H. Rashidi & L. Buehler (Eds.), *Bioinformatics basics: Applications in biological science and medicine* (2nd ed., pp. 226–255). CRC Press LLC.
Demeler, B., & Van Holde, K. E. (2004). Sedimentation velocity analysis of highly heterogeneous systems. *Analytical Biochemistry, 335*(2), 279–288.
Dyer, P. N., Edayathumangalam, R. S., White, C. L., Bao, Y., Chakravarthy, S., Muthurajan, U. M., & Luger, K. (2004). Reconstitution of nucleosome core particles from recombinant histones and DNA. *Methods in Enzymology, 375*, 23–44.
Falk, S. J., Guo, L. Y., Sekulic, N., Smoak, E. M., Mani, T., Logsdon, G. A., … Black, B. E. (2015). CENP-C reshapes and stabilizes CENP-A nucleosomes at the centromere. *Science, 348*(6235), 699–703.
Fierz, B., Kilic, S., Hieb, A. R., Luger, K., & Muir, T. W. (2012). Stability of nucleosomes containing homogenously ubiquitylated H2A and H2B prepared using semisynthesis. *Journal of the American Chemical Society, 134*(48), 19548–19551.
Ghosh, R. P., Nikitina, T., Horowitz-Scherer, R. A., Gierasch, L. M., Uversky, V. N., Hite, K., … Woodcock, C. L. (2010). Unique physical properties and interactions of the domains of methylated DNA binding protein 2. *Biochemistry, 49*(20), 4395–4410.
Graceffa, R., Nobrega, R. P., Barrea, R. A., Kathuria, S. V., Chakravarthy, S., Bilsel, O., & Irving, T. C. (2013). Sub-millisecond time-resolved SAXS using a continuous-flow mixer and X-ray microbeam. *Journal of Synchrotron Radiation, 20*(6), 820–825.
Hansen, J. C., Ghosh, R. P., & Woodcock, C. L. (2010). Binding of the Rett syndrome protein, MeCP2, to methylated and unmethylated DNA and chromatin. *IUBMB Life, 62*(10), 732–738.
Hughes, A. L., & Rando, O. J. (2014). Mechanisms underlying nucleosome positioning in vivo. *Annual Review of Biophysics, 43*, 41–63. http://doi.org/10.1146/annurev-biophys-051013-023114.
Humphrey, W., Dalke, A., & Schulten, K. (1996). VMD: Visual molecular dynamics. *Journal of Molecular Graphics, 14*(1), 33–38. 27–28.
Kar, S. R., Kingsbury, J. S., Lewis, M. S., Laue, T. M., & Schuck, P. (2000). Analysis of transport experiments using pseudo-absorbance data. *Analytical Biochemistry, 285*, 135–142.
Kathuria, S. V., Kayatekin, C., Barrea, R., Kondrashkina, E., Graceffa, R., Guo, L., … Bilsel, O. (2014). Microsecond barrier-limited chain collapse observed by time-resolved FRET and SAXS. *Journal of Molecular Biology, 426*(9), 1980–1994.
Kibbe, W. A. (2007). OligoCalc: An online oligonucleotide properties calculator. *Nucleic Acids Research, 35*(Web Server issue), W43–W46. http://doi.org/10.1093/nar/gkm234.
Klein, B. J., Muthurajan, U. M., Lalonde, M. E., Gibson, M. D., Andrews, F. H., Hepler, M., … Kutateladze, T. G. (2016). Bivalent interaction of the PZP domain of BRPF1 with the nucleosome impacts chromatin dynamics and acetylation. *Nucleic Acids Research, 44*(1), 472–484. http://doi.org/10.1093/nar/gkv1321. Epub 2015 Nov 30. PubMed PMID: 26626149; PubMed Central PMCID: PMC4705663.
Lohr, D., Bash, R., Wang, H., Yodh, J., & Lindsay, S. (2007). Using atomic force microscopy to study chromatin structure and nucleosome remodeling. *Methods (San Diego, Calif.), 41*(3), 333–341. http://doi.org/10.1016/j.ymeth.2006.08.016.

Lowary, P., & Widom, J. (1998). New DNA sequence rules for high affinity binding to histone octamer and sequence-directed nucleosome positioning. *Journal of Molecular Biology, 276*(1), 19–42.

Lu, X., Simon, M. D., Chodaparambil, J. V., Hansen, J. C., Shokat, K. M., & Luger, K. (2008). The effect of H3K79 dimethylation and H4K20 trimethylation on nucleosome and chromatin structure. *Nature Structural & Molecular Biology, 15*(10), 1122–1124. http://doi.org/10.1038/nsmb.1489.

Luger, K., Mader, A. W., Richmond, R. K., Sargent, D. F., & Richmond, T. J. (1997). Crystal structure of the nucleosome core particle at 2.8 A resolution. *Nature, 389*(6648), 251–260. http://doi.org/10.1038/38444.

Luger, K., Rechsteiner, T. J., & Richmond, T. J. (1999). Preparation of nucleosome core particle from recombinant histones. *Methods in Enzymology, 304*, 3–19.

Luo, Y., North, J. A., & Poirier, M. G. (2014). Single molecule fluorescence methodologies for investigating transcription factor binding kinetics to nucleosomes and DNA. *Methods (San Diego, Calif.), 70*(2–3), 108–118. http://doi.org/10.1016/j.ymeth.2014.09.011.

Lyubchenko, Y. L., & Shlyakhtenko, L. S. (2015). Chromatin imaging with time-lapse atomic force microscopy. *Methods in Molecular Biology (Clifton, N.J.), 1288*, 27–42. http://doi.org/10.1007/978-1-4939-2474-5_3.

Malaby, A. W., Kathuria, S. V., Nobrega, R. P., Bilsel, O., Muthurajan, U., Luger, K., ... Chakravarthy, S. (2013). Complementary techniques enhance the quality and scope of information obtained from SAXS. In *Transactions of American Crystallography Association Annual Meeting; July 20–24, 2013, Honolulu, HI.*

Muthurajan, U. M., Bao, Y., Forsberg, L. J., Edayathumangalam, R. S., Dyer, P. N., White, C. L., & Luger, K. (2004). Crystal structures of histone Sin mutant nucleosomes reveal altered protein-DNA interactions. *The EMBO Journal, 23*(2), 260–271. http://doi.org/10.1038/sj.emboj.7600046.

Muthurajan, U. M., Hepler, M. R. D., Hieb, A. R., Clark, N. J., Kramer, M., Yao, T., & Luger, K. (2014). Automodification switches PARP-1 function from chromatin architectural protein to histone chaperone. *Proceedings of the National Academy of Sciences of the United States of America, 111*(35), 12752–12757. http://doi.org/10.1073/pnas.1405005111.

Muthurajan, U. M., McBryant, S. J., Lu, X., Hansen, J. C., & Luger, K. (2011). The linker region of macroH2A promotes self-association of nucleosomal arrays. *The Journal of Biological Chemistry, 286*(27), 23852–23864. http://doi.org/10.1074/jbc.M111.244871.

Nazarov, I., Chekliarova, I., Rychkov, G., Ilatovskiy, A. V., Crane-Robinson, C., & Tomilin, A. (2016). AFM studies in diverse ionic environments of nucleosomes reconstituted on the 601 positioning sequence. *Biochimie, 121*, 5–12. http://doi.org/10.1016/j.biochi.2015.11.010. Epub 2015 Nov 14. PubMed PMID: 26586109.

Park, Y., Dyer, P. N., Tremethick, D. J., & Luger, K. (2004). A new fluorescence resonance energy transfer approach demonstrates that the histone variant H2AZ stabilizes the histone octamer within the nucleosome. *The Journal of Biological Chemistry, 279*(23), 24274–24282. http://doi.org/10.1074/jbc.M313152200.

Petoukhov, M. V., Franke, D., Shkumatov, A. V., Tria, G., Kikhney, A. G., Gajda, M., ... Svergun, D. I. (2012). New developments in the ATSAS program package for small-angle scattering data analysis. *Journal of Applied Crystallography, 45*(2), 342–350.

Rogge, R. A., Kalashnikova, A. A., Muthurajan, U. M., Porter-Goff, M. E., Luger, K., & Hansen, J. C. (2013). Assembly of nucleosomal arrays from recombinant core histones and nucleosome positioning DNA. *The Journal of Visualized Experiments* (79). http://doi.org/10.3791/50354. PubMed PMID: 24056546; PubMed Central PMCID: PMC3864267.

Svergun, D. I. (1999). Restoring low resolution structure of biological macromolecules from solution scattering using simulated annealing. *Biophysical Journal, 76*(6), 2879–2886.

Tan, S., & Davey, C. A. (2011). Nucleosome structural studies. *Current Opinion in Structural Biology*, *21*(1), 128–136. http://doi.org/10.1016/j.sbi.2010.11.006.

Watanabe, S., Resch, M., Lilyestrom, W., Clark, N., Hansen, J. C., Peterson, C., & Luger, K. (2010). Structural characterization of H3K56Q nucleosomes and nucleosomal arrays. *Biochimica et Biophysica Acta*, *1799*(5–6), 480–486. http://doi.org/10.1016/j.bbagrm.2010.01.009.

White, A. E., Hieb, A. R., & Luger, K. (2016). A quantitative investigation of linker histone interactions with nucleosomes and chromatin. *Scientific Reports*, *6*, 19122. http://doi.org/10.1038/srep19122. PubMed PMID: 26750377; PubMed Central PMCID: PMC4707517.

White, C. L., Suto, R. K., & Luger, K. (2001). Structure of the yeast nucleosome core particle reveals fundamental changes in internucleosome interactions. *The EMBO Journal*, *20*(18), 5207–5218. http://doi.org/10.1093/emboj/20.18.5207.

Winkler, D. D., & Luger, K. (2011). The histone chaperone FACT: Structural insights and mechanisms for nucleosome reorganization. *The Journal of Biological Chemistry*, *286*(21), 18369–18374. http://doi.org/10.1074/jbc.R110.180778.

Winkler, D. D., Luger, K., & Hieb, A. R. (2012). Quantifying chromatin-associated interactions: The HI-FI system. *Methods in Enzymology*, *512*, 243–274. http://doi.org/10.1016/B978-0-12-391940-3.00011-1.

Wriggers, W., & Chacón, P. (2001). Using Situs for the registration of protein structures with low-resolution bead models from X-ray solution scattering. *Journal of Applied Crystallography*, *34*(6), 773–776.

Wriggers, W., Milligan, R. A., & Mccammon, J. A. (1999). Situs: A package for docking crystal structures into low-resolution maps from electron microscopy. *Journal of Structural Biology*, *125*(2), 185–195.

Yang, C., van der Woerd, M. J., Muthurajan, U. M., Hansen, J. C., & Luger, K. (2011). Biophysical analysis and small-angle X-ray scattering-derived structures of MeCP2-nucleosome complexes. *Nucleic Acids Research*, *39*(10), 4122–4135.

CHAPTER TWO

Preparation, Crystallization, and Structure Determination of Chromatin Enzyme/Nucleosome Complexes

R.K. McGinty, R.D. Makde[1], S. Tan[2]
Center for Eukaryotic Gene Regulation, The Pennsylvania State University, University Park, PA, United States
[2]Corresponding author: e-mail address: sxt30@psu.edu

Contents

1. Introduction 44
2. Prepare Nucleosome Core Particles 44
3. Prepare Chromatin Protein 47
4. Purify Chromatin Complex 47
   4.1 Analytical Purification 47
   4.2 Preparative Purification 52
5. Concentrate Chromatin Complex 54
6. Crystallize Chromatin Complex 56
   6.1 Crystallization 56
   6.2 Fluorescent Labeling of Chromatin Protein for Cocrystallization with the Nucleosome 57
7. Postcrystallization Soaks to Improve Diffraction 61
References 63

Abstract

Structural studies of chromatin complexes composed of chromatin factors or enzymes bound to the nucleosome have been constrained by the ability to produce high-quality complexes in the amounts appropriate for biophysical studies and by the difficulty of crystallizing these complexes. We describe here procedures and approaches to prepare chromatin complexes, to crystallize chromatin complexes, and to improve diffraction properties through postcrystallization soaks. Special attention is paid to evaluating the quality of the purified chromatin complexes as well as assessing the presence of the chromatin protein or enzyme in crystals. The methods described for preparing and purifying chromatin complexes should be applicable to biochemical, biophysical, and other structural approaches including cryoelectron microscopy.

[1] Present address: High Pressure & Synchrotron Radiation Physics Division, Bhabha Atomic Research Centre, RRCAT, Indore, India.

Methods in Enzymology, Volume 573
ISSN 0076-6879
http://dx.doi.org/10.1016/bs.mie.2016.01.003

1. INTRODUCTION

As the fundamental repeating unit of chromatin, the nucleosome is a hub of genetic activity in the eukaryotic nucleus. This activity involves interactions of chromatin factors and enzymes with not just the individual histone and DNA components of the nucleosome but with the nucleosome as a unit (McGinty & Tan, 2016). Despite the fundamental importance of such interactions to eukaryotic biology, we currently possess only a limited view of the molecular details of nucleosome interactions due to the relatively few three-dimensional structures available (Armache, Garlick, Canzio, Narlikar, & Kingston, 2011; Barbera et al., 2006; Kato et al., 2013; Makde, England, Yennawar, & Tan, 2010; Maskell et al., 2015; McGinty, Henrici, & Tan, 2014; Zhou et al., 2015). Remedying this deficiency will require determining atomic structures of different classes of chromatin factors and enzymes with the nucleosome so that paradigms for chromatin interactions can be elucidated. We present here procedures for the crystal structure determination of chromatin complexes with special attention to preparing such complexes, applicable to other structural methods including cryoelectron microscopy, and postcrystallization soaks to improve crystal diffraction quality.

The basic workflow involved in the crystal structure determination of a chromatin complex is shown in Fig. 1, and each item except for structure determination is discussed here (there are few aspects of crystal structure determination which are specific to chromatin complexes). The linear form of this ideal workflow does not, of course, reflect the reality of most projects where practical issues force one to proceed in a decidedly nonlinear fashion, for example, to redesign chromatin protein or nucleosome constructs. Crystallizing chromatin complexes remains a challenging endeavor, and it is the unusual chromatin complex project that progresses smoothly through purification, concentration, crystallization, diffraction studies, and structure determination. However, the substantial biological insights afforded provide compelling justification to pursue such structural studies.

2. PREPARE NUCLEOSOME CORE PARTICLES

The preparation of recombinant nucleosome core particles (NCPs) appropriate for structural studies has been described in detail in previous publications (Luger, Rechsteiner, & Richmond, 1999a, 1999b; Luger,

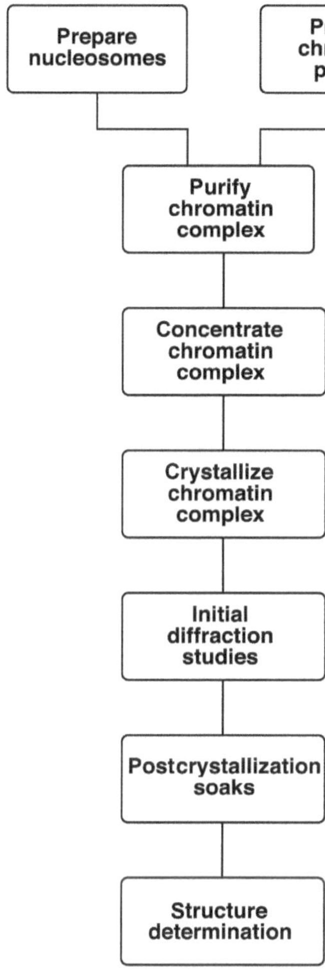

Fig. 1 Basic workflow for the preparation, crystallization, and structure determination of a chromatin complex.

Rechsteiner, Flaus, Waye, & Richmond, 1997), and these descriptions are still comprehensive and appropriate today. The reader is encouraged to follow closely the procedures described in these publications since the details can determine both the efficiency and quality of the resulting nucleosomes. We offer here some observations based on our experiences, which we hope will augment the descriptions in these publications.

1. We prefer to use anion-exchange high-performance chromatography over preparative gel electrophoresis to purify the reconstituted NCPs for the scalability and throughput advantages. While preparative

gel electrophoresis could offer higher resolution purification, we have not yet encountered situations that require such potential higher resolution.

2. We use GE Healthcare Source 15Q resin for the anion-exchange purification of reconstituted NCPs and find this resin to be at least as good as the Tosoh TSK DEAE-5PW employed in the Luger et al. publications (Luger, Rechsteiner, et al., 1997; Luger et al., 1999a, 1999b). Although we have not evaluated the GE Healthcare Mono Q resin for purifying reconstituted nucleosomes, we suspect this resin would also be appropriate based on our other comparisons of Source Q and Mono Q resins.
3. We routinely evaluate Source Q fractions of reconstituted NCPs by native polyacrylamide gel electrophoresis (PAGE), but one has to be careful in interpreting the results since the peak fractions will contain significant concentrations of salt, which promote dissociation of the NCPs during electrophoresis. For a more accurate assessment of NCP quality, samples should be dialyzed against the storage low salt buffer before electrophoresis. Otherwise, the amount of free DNA and noncanonical nucleosome species might be overestimated.
4. The choice of the nucleosomal DNA to reconstitute with histones may be dependent on whether the associated chromatin factor or enzyme has any sequence preference. In the case that the chromatin factor or enzyme has no sequence preference, one could use the symmetric human alpha satellite nucleosome positioning sequence used successfully to crystallize the NCP at 1.9 Å resolution (Davey, Sargent, Luger, Maeder, & Richmond, 2002; Richmond & Davey, 2003). However, one should be aware that preparing a symmetric nucleosome positioning sequence requires significantly more effort than preparing an asymmetric nucleosome positioning sequence (four preparative enzymatic steps vs just one). We have successfully used the asymmetric Widom 601 nucleosome positioning sequence used extensively in the chromatin field for our structural studies (Kim, Chatterjee, Jennings, Bartholomew, & Tan, 2015; Lowary & Widom, 1998; Makde et al., 2010; McGinty et al., 2014; Thåström et al., 1999), and we observe slightly higher NCP reconstitution yields using this sequence. One potential disadvantage is a lower yield of plasmid containing multiple inserts of the Widom 601 DNA sequence: typically, a 12 L preparation of pST55-16xNCP601a (a high copy number plasmid containing 16 copies of a 147 bp Widom DNA repeat) provides 50–100 mg of plasmid, which yields 20–50 mg of the 147 Widom DNA repeat. Comparable yields

for other asymmetric nucleosome positioning sequences can be twice as high.

5. The plasmid containing multiple inserts of the nucleosome positioning sequence should be propagated in the HB101 *Escherichia coli* strain for the recA genotype to reduce recombination of the relatively unstable plasmid. Even when using HB101, one may notice a small fraction of plasmid with less than the full number of inserts, but inserts appear to be lost as individual units and not fractions of an insert.
6. For crystallizing chromatin complexes, it is helpful to have minimally 3–5 mg of purified NCPs at a concentration of ~10 mg/mL.

3. PREPARE CHROMATIN PROTEIN

It is difficult to provide general instructions on how to prepare chromatin proteins since each protein is highly individual. Seeing that one will need milligram quantities of the chromatin protein, recombinant methods are often used to overexpress the individual protein or protein complex in an appropriate heterologous system. Standard approaches should be used to express and purify the protein (Burgess & Deutscher, 2009; Coligan et al., 1995). We try to prepare at least 5 mg of protein at a final concentration of >5 mg/mL in an appropriate buffer, with higher concentrations required for larger proteins and complexes and those requiring high salt for concentration. The concentration buffer conditions are optimized to reduce aggregation of the concentrated protein. We use dynamic light scattering (DLS) to assess the aggregation state (we strive for a single peak with a polydispersity of <20%). We also aim for a purity of >95%, noting that the chromatin protein does not need to be absolutely pure since further purification of the chromatin protein/nucleosome complex will be carried out (see item 4 below).

4. PURIFY CHROMATIN COMPLEX

4.1 Analytical Purification

We prefer to use size exclusion chromatography to purify the chromatin protein or enzyme in complex with the nucleosome (the complex will be referred to as the "chromatin complex") for several reasons. Size exclusion chromatography is a relatively robust and simple method, it is scalable (enabling both analytical and preparative uses), and the buffer exchange properties allow samples to be concentrated immediately after the run.

Preparative electrophoresis of the chromatin complex is an alternative that may be particularly useful if overlapping chromatography peaks of different complexes can be resolved by electrophoresis. We do note that not all chromatin complexes stable to size exclusion chromatography will be stable to electrophoresis. Selective precipitation with polyethylene glycol is yet another alternative purification method for chromatin complexes (Arnaudo et al., 2013).

In order to ensure stoichiometric preparations of chromatin complexes, we first optimize reconstitution buffer, especially salt concentration, and chromatin protein/nucleosome ratios using analytical scale reconstitutions prior to proceeding to preparative scales. An example of this optimization is described here.

1. Equilibrate a Superdex 200 column in filtered MilliQ water at a flow rate of 0.2–0.4 mL/min. We have used both Superdex 200 HR 10/30 and Superdex 200 increase 10/300 columns with similar results though Superdex increase 10/300 columns can be run at higher flow rates.
2. Prepare 300 mL of appropriate initial size exclusion chromatography buffer filtered through a 0.45-μm filter. The precise buffer composition should be selected to stabilize the chromatin complex, a decision that may require iterative optimization. We use 5 m*M* Tris–Cl pH 7.6, 35–75 m*M* NaCl, 1 m*M* DTT, and any complex-specific additives for our initial attempts. When possible, our initial choice of salt concentration is based on quantitative binding experiments using fluorescence quenching HIFI binding experiments in varying conditions (Hieb, D'Arcy, Kramer, White, & Luger, 2012; Kim et al., 2015; McGinty et al., 2014; Winkler, Luger, & Hieb, 2012).
3. Equilibrate Superdex 200 column in the same chromatography buffer.
4. Plan the recipe for the chromatin complex. For analytical reconstitutions, we typically prepare 400 μL of complex with 0.5–1.0 mg/mL NCP and 2.2–2.5 molar equivalents of a chromatin protein anticipated to bind with a 2:1 chromatin protein:NCP stoichiometry. This corresponds to a slight excess of the chromatin protein to drive saturation. In cases where a 1:1 nucleosome binding is expected, the molar equivalents are halved. Depending on the size and shape of the chromatin protein, the excess chromatin protein can often be separated from the chromatin protein/NCP complex by size exclusion chromatography.

   An example used to reconstitute the PRC1 ubiquitylation module/NCP complex is provided here.

(a) Record the concentration of the available chromatin protein and nucleosome:

PRC1 ub module prep 1	10.0 mg/mL	235 μ*M*
NCP prep 196	22.3 mg/mL	109 μ*M*

(b) Decide how much NCP you wish to use. Then calculate the amount in moles of chromatin protein you need to reconstitute the appropriate stoichiometric chromatin protein/NCP complex plus additional chromatin protein to drive formation of the chromatin complex. The final volume should be less than 500 μL.

18.0 μL	109 μ*M* NCP prep 196	(Final concentration 1 mg/mL)
20.8 μL	235 μ*M* PRC1 ub module prep 1	(2.5 molar equivalents to NCP)
361 μL	Reconstitution buffer	(10 m*M* Tris–Cl pH 7.6, 35 m*M* NaCl, 1 m*M* DTT, 10 μ*M* $ZnSO_4$)
400 μL	Total volume	

5. Prepare the chromatin complex mixture by adding the components in order (buffer, then NCP, then PRC1 ub module), usually at room temperature. Add the chromatin protein in 2–5 aliquots with at least 5 min between each addition:

361 μL	Reconstitution buffer
18.0 μL	109 μ*M* NCP prep 196
5.2 μL	235 μ*M* PRC1 ub module prep 1
	Obs: precipitation, then immediate clearing
5.2 μL	235 μ*M* PRC1 ub module prep 1
	Obs: precipitation, then immediate clearing
5.2 μL	235 μ*M* PRC1 ub module prep 1
	Obs: precipitation, then immediate clearing
5.2 μL	235 μ*M* PRC1 ub module prep 1
	Obs: precipitation, then immediate clearing

6. Purify the chromatin complex by size exclusion chromatography over the equilibrated Superdex 200 column:
   (a) Load the sample through the sample loop using a 1-mL disposable syringe.
   (b) Elute the sample using a flow rate of 0.4 mL/min (or 0.75 mL/mL for Superdex 200 increase columns) collecting 1 min = 0.4 mL fractions and monitoring both A_{260} and A_{280} if possible. The PRC1 ub module/NCP complex usually elutes between 0.42 and 0.48 column volumes, while the free PRC1 ub module protein usually elutes around 0.58 column volumes (Fig. 2A and B).
7. Analyze peak fractions by SDS-PAGE using an 18% gel.
8. The quality of complex is assessed as follows:
   (a) Examine the width and symmetry of the chromatographic peak of the chromatin complex. A front shoulder often suggests superstoichiometric nonspecific binding to the nucleosome as observed initially with the reconstitution described earlier (Fig. 2A). Meanwhile a back shoulder may suggest unsaturated complex or free nucleosome. The presence of a peak for the chromatin protein implies unbound protein resulting from too stringent of reconstitution conditions or too much protein added in the reconstitution mixture. We also plot and examine the A_{260}/A_{280} ratio that acts as a surrogate for the NCP to chromatin protein ratio (Fig. 2D). A flat A_{260}/A_{280} ratio across the chromatin complex peak indicates a homogenous and saturated complex (Fig. 2E and F). One can also use the absorbance of a fluorophore attached to the chromatin protein to further interrogate the quality of the complex based on alignment of the fluorophore absorbance and A_{260} peaks (Fig. 2F, and see more below).
   (b) Examine the stoichiometry of the chromatin protein and histones in peak fractions via SDS-PAGE. Homogeneous, saturated complexes show similar stoichiometry across the entire peak as opposed to higher levels of the chromatin protein relative to histones at the front edge of the peak vs the back edge.
9. Based on the above analysis, adjust the reconstitution buffer conditions until a high-quality complex is obtained. For the PRC1 ubiquitylation module–NCP complex described earlier, the NaCl concentration in the reconstitution buffer was systematically increased to 75 mM to eliminate the observed front shoulder indicative of superstoichiometric binding of PRC1 to the NCP (Fig. 2A and B).

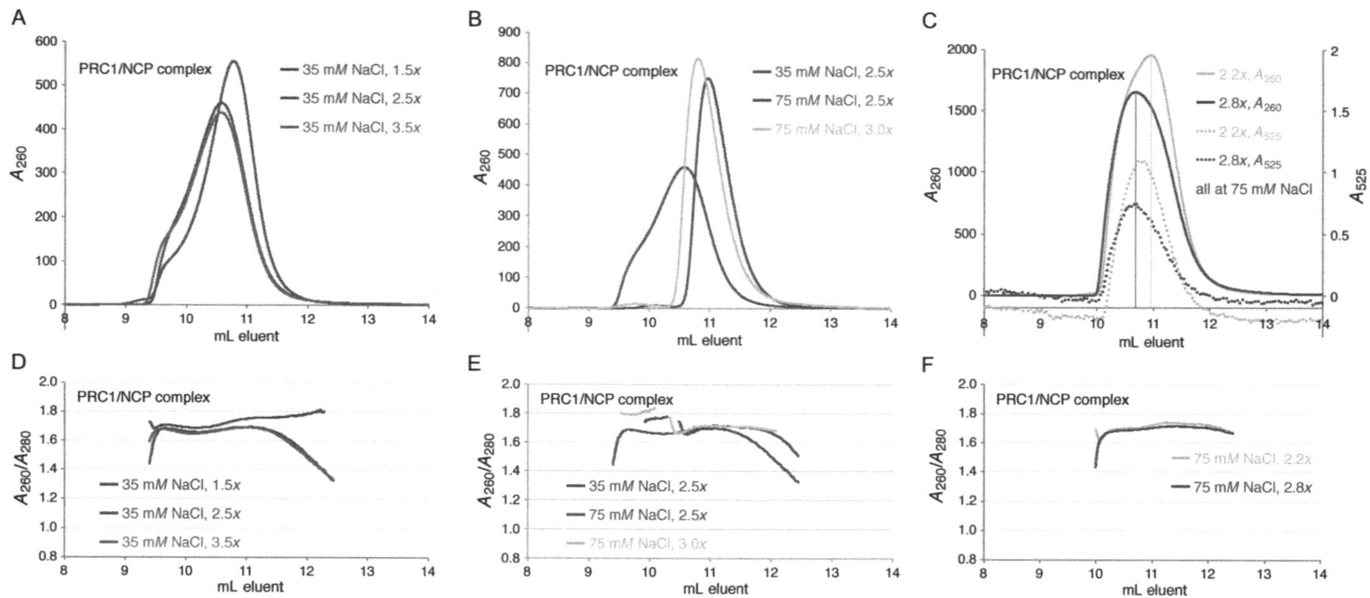

Fig. 2 Size exclusion chromatography for the purification and analysis of a reconstituted chromatin complex. (A) and (B) A_{260} chromatograms for reconstituted chromatin complexes on an analytical scale using specified salt concentrations and molar equivalents of chromatin protein relative to NCP. (C) A_{260} and A_{525} for fluorescently labeled chromatin complex reconstituted on a preparative scale. (D–F) A_{260}/A_{280} ratio for chromatograms in top panels. (See the color plate.)

10. With optimized buffer conditions established, we then optimize the ratio of chromatin protein to nucleosome by repeating analytical reconstitutions followed by size exclusion chromatography, titrating the chromatin protein into the nucleosome. We first use a broad titration series, for example, 1.1:1, 2.2:1, 3.3:1, etc., to determine preferred stoichiometry and to further rule out superstoichiometric binding by examining peak profile and retention time shift upon increasing chromatin protein (Fig. 2A and D). This broad titration series can then be refined using a more targeted titration (Fig. 2B and E). Again fluorescently labeled chromatin factor can aid in the analysis. Such optimization led us to use 2.8–3.0:1 PRC1 ub module:NCP for preparative scale reconstitutions (Fig. 2C and F). Note that at a ratio of 2.8:1 PRC1 ub module:NCP, the A_{260} peak (detecting mostly nucleosomal DNA) and the A_{525} peak (detecting the fluorescently labeled chromatin enzyme) align, whereas at a ratio of 2.2:1, the two peaks do not align (Fig. 2F).
11. If the Superdex 200 column will not be used for additional runs, wash the column in water at 0.4 mL/min and then store in 25% ethanol at 0.2 mL/min flow rate.

4.2 Preparative Purification

After optimizing the reconstitution on analytical scale, we reconstitute and purify the chromatin complex on a preparative scale. Here, we describe the procedure used in our laboratory to purify chromatin complexes by size exclusion chromatography on preparative scale for crystallization.

1. Equilibrate a Superdex 200 column in optimized buffer as described earlier for analytical scale experiments.
2. Plan the recipe for the chromatin complex. This might be the single most important step of the procedure and due care is warranted.

   An example used to reconstitute the PRC1 ub module/NCP complex is provided here.

   (a) Record the concentration of the available chromatin protein and nucleosome:

PRC1 ub module prep 3	8.54 mg/mL	200 μ*M*
NCP prep 222	12.9 mg/mL	62.9 μ*M*

(b) Decide how much NCP you wish to use. Then calculate the amount in moles of chromatin protein you need to reconstitute the appropriate stoichiometric chromatin protein/NCP complex. Determine the amount of salt contributed by the chromatin protein buffer and the amount of additional salt needed to raise the overall reconstitution salt concentration to the desired level. Finally, calculate the volume of 1 *M* Tris–Cl pH 7.6 to add an additional 25 m*M* Tris–Cl (to counteract the relatively low pH 6.0 in the NCP storage solution). The final volume should be less than 500 μL.

121 μL	62.9 μ*M* NCP prep 222	~1.55 mg
7.5 μL	1 *M* Tris–Cl pH 7.6	
1.9 μL	5 *M* NaCl	(75 m*M* final concentration)
110 μL	200 μ*M* PRC1 ub module prep 3	(2.9:1 ratio)
59.6 μL	MilliQ water	
300 μL	Total volume	

3. Prepare the chromatin complex mixture by adding the components in order (water, NCP, then Tris, then salt, then PRC1 ub module), usually at room temperature. Add the chromatin protein in 2–5 aliquots with at least 5 min between each addition and record observations. It is not uncommon to observe some precipitation upon addition of the chromatin protein that clears upon mixing, especially at lower salt concentrations.

59.6 μL	MilliQ water
121 μL	62.9 μ*M* NCP prep 222
7.5 μL	1 *M* Tris–Cl pH 7.6
1.9 μL	5 *M* NaCl
27.6 μL	200 μ*M* PRC1 ub module prep 3
	Obs: precipitation, then immediate clearing
27.6 μL	200 μ*M* PRC1 ub module prep 3
	Obs: precipitation, then immediate clearing

27.6 μL	200 μ*M* PRC1 ub module prep 3
	Obs: precipitation, then immediate clearing
27.6 μL	200 μ*M* PRC1 ub module prep 3
	Obs: precipitation, with some cloudiness remaining
300 μL	Total volume

4. Purify the chromatin complex by size exclusion chromatography over the equilibrated Superdex 200 column as described earlier for analytical reconstitutions (Fig. 2C and F).
5. Pool the chromatin complex (PRC1 ub module/NCP here) Superdex fractions (typically 3–5 fractions of 0.4 mL/fraction, or 1.2–2 mL) in Eppendorf tubes. We often add PMSF to 0.1 m*M* to minimize proteolysis of the complex.

5. CONCENTRATE CHROMATIN COMPLEX

1. Concentrate the chromatin complex pool using a Vivaspin 500, 10 kDa cutoff centrifuge concentration device to a final concentration of ~10 mg/mL. Use a microcentrifuge speed of 10K rpm or manufacturer's recommended centrifugation speed for all concentration centrifugation steps. For ill-behaved samples, consider concentrating an aliquot to first confirm that the chromatin complex can be concentrated to high concentrations without aggregation before committing the entire sample.
   - **(a)** Wash concentration device twice with size exclusion chromatography buffer and discard filtrates.
   - **(b)** Concentrate chromatin complex pool through successive spins of 400–500 μL until the final volume is less than 150 μL. 5–10 min is usually sufficient to concentrate 500 μL per spin. Substantially, longer centrifugation times may indicate aggregation of the sample in the concentration buffer.
   - **(c)** Transfer concentrated chromatin complex sample to a new Eppendorf tube.
   - **(d)** Add 500 μL of size exclusion chromatography buffer to concentration device and allow to stand for 5 min at room temperature. Then

centrifuge at 10K rpm for 5–10 min to concentrate this wash to less than 10 μL.

(e) Combine wash with concentrated chromatin complex pool. Mix gently.

(f) Centrifuge chromatin complex sample for 5 min at 20°C and then transfer supernatant to new Eppendorf tube. Avoid any particulate matter at the bottom of the tube.

(g) Analyze aggregation state of concentrated sample by DLS or similar appropriate method.

2. Quantitate the chromatin complex. We typically quantitate using the UV absorbance of the nucleic acid component of the NCP within the chromatin complex under denaturing basic conditions to dissociate the chromatin and histone proteins from the nucleosomal DNA. With some complexes, we do not observe a clear peak at 260 nm. Rather the absorbance continues to increase at lower wavelengths. In such cases, quantitation using this protocol can lead to overestimation of complex concentration. Additionally, when using chromatin proteins with significant absorption at 260 nm, this protocol will need to be adjusted accordingly.

(a) For samples with an expected concentration of 1–25 mg/mL, prepare 1:200 dilution of chromatin complex in 0.2 N NaOH: 497.5 μL 0.2 N NaOH + 2.5 μL sample.

(b) Blank UV spectrophotometer with quartz cuvette containing 0.2 N NaOH.

(c) Record UV spectrum of chromatin complex in 0.2 N NaOH from 320 to 220 nm. Record A_{320} and A_{260} of sample.

(d) Calculate concentration of chromatin complex. For example, for the PRC1 ub module/NCP complex in the size exclusion chromatography example in the previous section, the molecular weight of the PRC1 ub module/NCP was 290,234 Da (2:1 complex), and the molecular weight of the 147 bp double-stranded DNA in the complex was 97,020:

(i) adjusted A_{260} of 1:200 dilution in 0.2 N NaOH $= A_{260} - A_{320} = 0.564$

(ii) A_{260} of stock $= 200 \times 0.564 = 112.8$

(iii) concentration of DNA in stock $= 112.8 \times 40$ μg/mL $=$ 4.51 mg/mL

(iv) concentration of chromatin complex $= 4.51$ mg/mL $\times$ 290,234/97,020 $= 13.5$ mg/mL

The extinction coefficient was adjusted from that of DNA in neutral pH conditions based on empirical measurements. We have observed that this calculation can occasionally overestimate the chromatin complex concentration, so it is best to consider the calculated chromatin complex concentration as an estimate and not a firm measurement.

(e) Calculate the purification yield of the chromatin complex based on the limiting reagent when reconstituting the chromatin complex (in this example, the NCP). For example:

(i) started with 1.55 mg of NCP complex, equivalent to 7.59 nmol for the 204,921 Da NCP

(ii) ended with 100 μL of PRC1 ub module/NCP complex @ 13.5 mg/mL, or 4.65 nmol PRC1 ub module/NCP complex

(iii) purification yield = 4.65/7.59 = 61.3% (yields are typically between 50% and 80%)

3. Use the chromatin complex immediately for crystallization trials, or store complex on ice in cold room, or where appropriate, add glycerol to 20%, flash freeze in liquid nitrogen, and store at −80°C.

6. CRYSTALLIZE CHROMATIN COMPLEX

6.1 Crystallization

Standard crystallization approaches can be used to crystallize the chromatin complex. As with purification of chromatin proteins, the process of crystallizing a chromatin complex is highly variable and dependent on the properties of the individual complex. Nevertheless, basic principles and concepts will apply to the crystallization of many chromatin complexes, and we share such ideas here:

1. As with most any crystallization trial, one should survey a range of pH, salt, and precipitating conditions. Since many chromatin complexes are sensitive to salt, it may be prudent to bias toward low salt conditions and near neutral pH. These conditions can be fulfilled by commercial low ionic strength crystallization screens and screens for protein–nucleic acid complexes (Hampton Natrix Screens, Hollis Lab Protein-Nucleic Acid Complex Crystal Screen (Pryor, Wozniak, & Hollis, 2012)).

2. For initial crystallization screens, we use 96-well microbatch under oil trials using 1 μL of sample + 1 μL of the precipitating solution. Robotics can be helpful for setting up such trials or ones using even smaller volumes, but robotics are not necessary.

3. Divalent ions can play different roles in chromatin complex crystallization. Magnesium ions (Mg^{2+}) are often used in protein/nucleic acid cocrystallization. However, Mg^{2+} is also used to condense chromatin (Fredericq, Hacha, Colson, & Houssier, 1991), and NCPs are crystallized at the salting in transition balancing Mn^{2+} as a precipitating agent and KCl as a solubilizing agent (Hanson, Alexander, Harp, & Bunick, 2004; Luger, Mäder, Richmond, Sargent, & Richmond, 1997; Rhodes, Brown, & Klug, 1989). As such, use of millimolar concentrations of Mg^{2+} in the crystallization trials could favor precipitating the chromatin complex instead of controlled crystallization. It should be noted, however, that the Sir3 silencing protein was crystallized with NCPs in the presence of 10 m*M* $MgCl_2$ (Armache et al., 2011).
4. As with crystallizing any protein, the use of orthologous proteins can be a fruitful strategy. In the case of the RCC1/nucleosome complex, initial crystals were grown using human RCC1, but the crystals were hollow and produced very poor diffraction. Substituting yeast RCC1 (Srm1) produced morphologically more perfect crystals, but nonisotropic diffraction of X-rays limited to about 7 Å in the worse directions. On the other hand, Drosophila RCC1 (Bj1) complexed with the nucleosome produced crystals with both external and internal order, permitting diffraction data collection to 2.9 Å (Makde et al., 2010; Makde & Tan, 2013).

6.2 Fluorescent Labeling of Chromatin Protein for Cocrystallization with the Nucleosome

One recurrent challenge we observe in the crystallization of chromatin complexes is the undesired crystallization of subcomplexes. This is especially true when working with chromatin proteins bound to the NCP, where the nucleosome particle often crystallizes without the chromatin protein bound. This can lead to much time wasted optimizing false leads to obtain crystals that are large enough to verify their contents by SDS-PAGE. To short circuit this problem and streamline crystal screening, we use fluorescently labeled chromatin proteins during screening. This allows us to verify the presence of the chromatin protein in any crystallization hits *in situ*, following procedures developed more generally for protein crystallization (Meyer, Betzel, & Pusey, 2015; Pusey et al., 2015).

We label our chromatin proteins through the N-α amino group using 5-(and-6)-carboxyrhodamine 6G, succinimidyl ester using pH conditions that favor conjugation specifically at the N-terminus. This fluorophore is small and inexpensive and has stable spectrophotometric properties over a wide range of pHs often used in crystallization screens. The N-terminal labeling is convenient and unlikely to interfere with crystallization as terminal regions of proteins are often unstructured in macromolecular crystals. Of note, it is possible that some nonspecific labeling of lysine side chain amines may occur in this pH-controlled reaction. However, at labeling percentages under 2% of the total chromatin protein pool, any resulting heterogeneity is minimal. Moreover, if undesired side chain modification of the chromatin protein prevents nucleosome binding, the improperly labeled protein is often purified away from the chromatin complex during size exclusion chromatography and at trace levels causes little to no change of the overall chromatin protein to nucleosome ratio. This methodology can also be applied to other fluorophores and conjugation chemistries.

An example of fluorescent labeling of a chromatin protein and incorporation into a chromatin complex is shown here.

1. Prepare your chromatin protein for labeling. In order to achieve efficient labeling of the chromatin protein, reactive nucleophiles (especially primary amines) that can consume the fluorophore must be removed or minimized. Buffer exchange can be performed by dialysis, size exclusion chromatography, or through a desalting column. Additionally, if the chromatin protein is sufficiently concentrated, dilution into labeling buffer may be adequate given the low levels of labeling required. In anticipation of labeling requirements, the chromatin protein can often be concentrated in a suitable buffer for labeling.
2. We typically label 2–5 mg of chromatin protein at 1–4 mg/mL. An example of labeling the PRC1 ubiquitylation module is shown here.
   (a) Dissolve 5-(and-6)-carboxyrhodamine 6G, succinimidyl ester in DMF (dimethyl formamide) at 10 m*M*.
   (b) Decide on the quantity of chromatin protein to label and prepare labeling mixture

250 μL	235 μ*M* PRC1 ub module prep 1 (2.5 mg)
738 μL	Labeling buffer (10 m*M* HEPES pH 7.5, 300 m*M* NaCl)
12 μL	10 m*M* carboxyrhodamine succinimidyl ester
1000 μL	Total volume (2.5 mg/mL chromatin protein)

(c) Mix at room temperature for 1 h.

(d) Purify the labeled chromatin protein. We prefer to use size exclusion chromatography as it further purifies the chromatin protein and adds an extra level of quality control, but a desalting column is a suitable alternative. We perform size exclusion chromatography in a buffer optimized for concentration of the chromatin protein after clarification of the labeling reaction mixture in a tabletop microcentrifuge. We note that two successive spins are often required to remove all particulate matter.

3. Concentrate the labeled chromatin protein as described earlier for the unlabeled protein.

4. Quantitate the labeled protein and labeling efficiency. Because the fluorophore absorbs at 280 nm, the $A_{280,\mathrm{total}}$ for the labeled chromatin protein is the sum of the $A_{280,\mathrm{protein}}$ and the $A_{280,\mathrm{fluor.}}$. Therefore, the protein concentration and labeling percentage can be determined by subtracting $A_{280,\mathrm{fluor.}}$ from the $A_{280,\mathrm{total}}$. This $A_{280,\mathrm{fluor.}}$ can be calculated by measuring the absorbance of the fluorophore at its peak absorbance ($A_{532,\mathrm{fluor.}}$ in this example) and 280 nm. In our typical chromatin protein buffers, the $A_{532,\mathrm{fluor.}}/A_{280,\mathrm{fluor.}}$ ratio for carboxyrhodamine is approximately 3. We refer to this ratio as the correction factor (CF). Importantly, most chromatin proteins have little to no absorption above 300 nm. Thus, by measuring the absorbance of the labeled protein at 280 and 532 nm, we can calculate the concentration of the protein and the fluorophore in our labeled protein samples. An example for the PRC1 ub module is shown here.

a. Determine the CF for carboxyrhodamine in chromatin protein concentration buffer $= A_{532,\mathrm{fluor.}}/A_{280,\mathrm{fluor}}$. For other fluorophores replace A_{532} with A_{max}. $A_{280,\mathrm{fluor}} = A_{532,\mathrm{total}}/\mathrm{CF}$

b. Measure $A_{280,\mathrm{total}}$, $A_{532,\mathrm{total}}$, and $A_{600,\mathrm{total}}$ at appropriate dilution for labeled chromatin protein

(i) adjusted $A_{280,\mathrm{total}}$ of 1:21 dilution $= A_{280,\mathrm{total}} - A_{600,\mathrm{total}} = 0.304$

(ii) $A_{280,\mathrm{total}}$ of stock $= 21 \times 0.304 = 6.38$

(iii) adjusted $A_{532,\mathrm{total}}$ of 1:21 dilution $= A_{532,\mathrm{fluor.}} = A_{524,\mathrm{total}} - A_{600,\mathrm{total}} = 0.064$

(iv) $A_{532,\mathrm{total}}$ of stock $= 21 \times 0.064 = 1.34$

(v) [Chromatin protein] $= (A_{280,\mathrm{total}} - A_{280,\mathrm{fluor}})/\varepsilon_{\mathrm{protein}} = (6.38 - 1.34/3)/0.924\ \mathrm{mL/mg} = 6.42\ \mathrm{mg/mL} = 151\ \mu M$

(vi) Percent labeling $= (A_{532,\mathrm{total}}/\varepsilon_{\mathrm{fluor.}})/$[Chromatin protein] $= 1.34/92{,}000\ M^{-1}/0.000151\ M = 9.6\%$

5. Reconstitute the chromatin complex with the fluorescently labeled chromatin protein. We typically include 0.5–2% labeled chromatin protein in our reconstitutions. An example using PRC1 ub module/NCP is shown below.

   Prepare 0.5% labeled PRC1 ub module

95.9 μL	234.7 μ*M* unlabeled PRC1 ub module
10.2 μL	120.5 μ*M* PRC1 ub module (9.6% labeled)

   Prepare reconstitution mixture

69.9 μL	MilliQ water
124 μL	61.9 μ*M* NCP prep 201
7.5 μL	1 *M* Tris–Cl pH 7.6
2.2 μL	5 *M* NaCl
24.1 μL	0.5% labeled PRC1 ub module
	Obs: precipitation, then immediate clearing
24.1 μL	0.5% labeled PRC1 ub module
	Obs: precipitation, then immediate clearing
24.1 μL	0.5% labeled PRC1 ub module
	Obs: precipitation, then immediate clearing
24.1 μL	0.5% labeled PRC1 ub module
	Obs: precipitation, then immediate clearing
300 μL	Total volume

6. The chromatin complex can be purified, concentrated, quantitated, and used in crystallization screens as described earlier for the unlabeled complex. We record the A_{525} along with the A_{260} and A_{280} during size exclusion chromatography. This allows us to visualize the alignment of the chromatin protein (A_{525}) and NCP (A_{260}) independently to confirm saturation of the chromatin complex (Fig. 2C).
7. Observe fluorescence of crystals. We observed crystals grown with fluorescently labeled chromatin complexes using a Nikon SM7 1500 stereoscope equipped with a Nikon Intensilight C-HGFI light source, a

DsRed filter (Ex 545/30 nm, DM 570 nm, BA 620/60 nm), and a CoolSnapEZ Turbo 1394 camera.

7. POSTCRYSTALLIZATION SOAKS TO IMPROVE DIFFRACTION

Postcrystallization soaks of chromatin complex crystals appear to be particularly important for obtaining high-resolution diffraction. This is consistent with the previous observations that the order within NCP crystals could be improved by postcrystallization soaks in methylpentanediol (MPD) or other alcohols (Richmond, Finch, Rushton, Rhodes, & Klug, 1984; Struck, Klug, & Richmond, 1992). Therefore, it is highly advisable to spend significant efforts exploring postcrystallization soaking conditions once chromatin complex crystals have been grown, assuming of course that sufficient numbers of such crystals can be grown.

The underlying approach is to find a base soaking solution in which the crystal is stable, and then to gradually increase the concentration of the additive, usually a dehydration agent. We usually perform this procedure using loops to transfer crystals between drops of different compositions, but one can imagine other procedures using dialysis buttons or diffusing volatile reagents. One should also consider using dehydration equipment such as the HC-1 device developed at EMBL Grenoble, which may provide a more reproducible and direct way of manipulating the hydration state of macromolecular crystals (Bowler et al., 2015; Sanchez-Weatherby et al., 2009; Wheeler, Russi, Bowler, & Bowler, 2012).

1. Determine a crystallization base soaking solution in which the crystal is stable.

   We usually start with the precipitant used in the relevant crystallization trial and increase the precipitant concentration by 1.25 fold (though older crystallization drops may need higher precipitant concentrations). So, for example, if the precipitant contained 20 m*M* Tris–Cl pH 7.6, 50 m*M* NaAc, and 6% PEG2000-MME, we would use 20 m*M* Tris–Cl pH 7.6, 50 m*M* NaAc, and 7.5% PEG2000-MME as the trial crystallization base soaking solution.

   Transfer a small crystal into the trial crystallization base soaking solution and observe the crystal under a microscope. If the crystal dissolves, it is likely that the precipitant concentration needs to be increased, although the pH or the salt concentration might have to be manipulated as well. If the crystal cracks, it is likely the precipitant concentration needs to be decreased. The trial crystallization base soaking solution is ideally one in which the crystal is stable for at least 15 min at the appropriate temperature. We usually perform soaking experiments at room temperature, but we did observe that soaking RCC1/nucleosome crystals at 4°C was important for obtaining high-resolution diffraction (Makde & Tan, 2013).

2. Determine a soaking protocol that permits the additive to be soaked into the crystal without damaging the external morphology of the crystal.
   (a) First decide what soaking additive and what final concentration of this additive you wish to examine. Possible soaking additives are MPD, PEG400, PEG2000, and PEG550-MME. Then decide what size steps of the additive concentration to employ. For example, you might wish to soak MPD into the crystal to a final concentration of 24% in 4% steps. In this case, you would prepare the crystallization base soaking solution with 0%, 4%, 8%, 12%, 16%, 20%, 24% MPD.
   (b) Use a crystal loop to transfer a crystal into 10 μL of the crystallization base soaking solution with no added MPD. Incubate at room temperature for desired length of time (eg, 5 min). Observe the crystal under a microscope for possible changes in external morphology.
   (c) Transfer crystal into 10 μL of the crystallization base soaking solution with 4% MPD. Incubate and observe as before.
   (d) Repeat step (c) until the crystal is transferred to the final additive concentration (24% MPD in this example). If cracking or defects are observed during the process, continue with the procedure for at least two more additive additions. It is not uncommon to observe cracking or defects followed by annealing of these cracks or defects upon further soaks, reflecting repacking of the molecules in the crystal.
   (e) Where appropriate, flash cool the crystal by plunging into liquid nitrogen before storage in liquid nitrogen.

Notes:

(i) Unfortunately, there is no simple soaking protocol that will handle all cases. The appropriate protocol needs to be determined empirically. Potentially important variables include the identity of the soaking additive, the final concentration of the soaking additive, the number of steps used to reach the final concentration of the soaking additive, the method used to achieve such steps (eg, transfer via loops), the soaking temperature, and the soaking time per step.

(ii) Since the ultimate criteria of judging success of a soaking protocol are the quality of X-ray diffraction and since one will want to examine many soaking protocols, one approach is to process many crystals through the various soaking protocols and to then examine the diffraction in succession at a home or synchrotron source.

(iii) It is advisable to examine a variety of dehydration soaking agents including but not necessarily limited to MPD, PEG400, PEG2000, and PEG550-MME.

(iv) You can assess the maximum concentration for an additive by increasing the concentration of the additive until the crystals show defects that do not anneal with further additions of the additive.

(v) The soaking protocol can manipulate more than one additive, perhaps to introduce a cryoprotectant as well as a dehydrating agent. For example, you could increase both glycerol and PEG400 conditions simultaneously or in succession.

(vi) Make sure to allow enough time for the crystal to equilibrate against the soaking solution, particularly for the final soak. This may be particularly important for the reproducibility of the soaking protocol.

(vii) We have observed that postcrystallization protocols optimized for crystals harvested from fresh drops do not necessarily work with crystals harvested from older crystallization drops.

(viii) One may wish to decrease the incremental step of the dehydrating agent or additive in the soaking solutions to achieve more gradual dehydration. For example, we used 2% steps of PEG400 when dehydrating RCC1/nucleosome crystals (Makde et al., 2010; Makde & Tan, 2013).

REFERENCES

Armache, K.-J., Garlick, J. D., Canzio, D., Narlikar, G. J., & Kingston, R. E. (2011). Structural basis of silencing: Sir3 BAH domain in complex with a nucleosome at 3.0 Å resolution. *Science (New York, NY)*, *334*(6058), 977–982.

Arnaudo, N., Fernandez, I. S., McLaughlin, S. H., Peak-Chew, S. Y., Rhodes, D., & Martino, F. (2013). The N-terminal acetylation of Sir3 stabilizes its binding to the nucleosome core particle. *Nature Structural & Molecular Biology*, *20*(9), 1119–1121.

Barbera, A. J., Chodaparambil, J. V., Kelley-Clarke, B., Joukov, V., Walter, J. C., Luger, K., et al. (2006). The nucleosomal surface as a docking station for Kaposi's sarcoma herpesvirus LANA. *Science (New York, NY)*, *311*(5762), 856–861.

Bowler, M. W., Mueller, U., Weiss, M. S., Sanchez-Weatherby, J., Sorensen, T. L.-M., Thunnissen, M. M. G. M., et al. (2015). Automation and experience of controlled crystal dehydration: Results from the European Synchrotron HC1 Collaboration. *Crystal Growth & Design*, *15*(3), 1043–1054.

Burgess, R. R., & Deutscher, M. P. (Eds.), (2009). *Vol 463: Guide to Protein Purification: Vol. 463. Methods in enzymology* (pp. 1–851): Amsterdam: Elsevier.

Coligan, J. E., Dunn, B. M., Speicher, D. W., & Wingfield, Paul T. (Eds.), (1995). *Current protocols in protein science*. New York: Wiley.

Davey, C. A., Sargent, D. F., Luger, K., Maeder, A. W., & Richmond, T. J. (2002). Solvent mediated interactions in the structure of the nucleosome core particle at 1.9 a resolution. *Journal of Molecular Biology, 319*(5), 1097–1113.

Fredericq, E., Hacha, R., Colson, P., & Houssier, C. (1991). Condensation and precipitation of chromatin by multivalent cations. *Journal of Biomolecular Structure & Dynamics, 8*(4), 847–865.

Hanson, B. L., Alexander, C., Harp, J. M., & Bunick, G. J. (2004). Preparation and crystallization of nucleosome core particle. *Methods in Enzymology, 375*, 44–62.

Hieb, A. R., D'Arcy, S., Kramer, M. A., White, A. E., & Luger, K. (2012). Fluorescence strategies for high-throughput quantification of protein interactions. *Nucleic Acids Research, 40*(5).

Kato, H., Jiang, J., Zhou, B.-R., Rozendaal, M., Feng, H., Ghirlando, R., et al. (2013). A conserved mechanism for centromeric nucleosome recognition by centromere protein CENP-C. *Science (New York, NY), 340*(6136), 1110–1113.

Kim, S.-A., Chatterjee, N., Jennings, M. J., Bartholomew, B., & Tan, S. (2015). Extranucleosomal DNA enhances the activity of the LSD1/CoREST histone demethylase complex. *Nucleic Acids Research, 43*(10), 4868–4880.

Lowary, P. T., & Widom, J. (1998). New DNA sequence rules for high affinity binding to histone octamer and sequence-directed nucleosome positioning. *Journal of Molecular Biology, 276*(1), 19–42.

Luger, K., Mäder, A. W., Richmond, R. K., Sargent, D. F., & Richmond, T. J. (1997a). Crystal structure of the nucleosome core particle at 2.8 A resolution. *Nature, 389*(6648), 251–260.

Luger, K., Rechsteiner, T. J., Flaus, A. J., Waye, M. M., & Richmond, T. J. (1997b). Characterization of nucleosome core particles containing histone proteins made in bacteria. *Journal of Molecular Biology, 272*(3), 301–311.

Luger, K., Rechsteiner, T. J., & Richmond, T. J. (1999a). Expression and purification of recombinant histones and nucleosome reconstitution. *Methods in Molecular Biology (Clifton, NJ), 119*, 1–16.

Luger, K., Rechsteiner, T. J., & Richmond, T. J. (1999b). Preparation of nucleosome core particle from recombinant histones. *Methods in Enzymology, 304*, 3–19.

Makde, R. D., England, J. R., Yennawar, H. P., & Tan, S. (2010). Structure of RCC1 chromatin factor bound to the nucleosome core particle. *Nature, 467*(7315), 562–566.

Makde, R. D., & Tan, S. (2013). Strategies for crystallizing a chromatin protein in complex with the nucleosome core particle. *Analytical Biochemistry, 442*, 138–145.

Maskell, D. P., Renault, L., Serrao, E., Lesbats, P., Matadeen, R., Hare, S., et al. (2015). Structural basis for retroviral integration into nucleosomes. *Nature, 523*(7560), 366–369.

McGinty, R. K., Henrici, R. C., & Tan, S. (2014). Crystal structure of the PRC1 ubiquitylation module bound to the nucleosome. *Nature, 514*(7524), 591–596.

McGinty, R. K., & Tan, S. (2016). Recognition of the nucleosome by chromatin factors and enzymes. *Current Opinion in Structural Biology, 37*, 54–61.

Meyer, A., Betzel, C., & Pusey, M. (2015). Latest methods of fluorescence-based protein crystal identification. *Acta Crystallographica. Section F, Structural Biology and Crystallization Communications, 71*(Pt 2), 121–131.

Pryor, E. E., Wozniak, D. J., & Hollis, T. (2012). Crystallization of Pseudomonas aeruginosa AmrZ protein: Development of a comprehensive method for obtaining and optimization of protein-DNA crystals. *Acta Crystallographica Section F, Structural Biology and Crystallization Communications, 68*(Pt 8), 985–993. http://doi.org/10.1107/S1744309112025316.

Pusey, M., Barcena, J., Morris, M., Singhal, A., Yuan, Q., & Ng, J. (2015). Trace fluorescent labeling for protein crystallization. *Acta Crystallographica. Section F, Structural Biology and Crystallization Communications*, *71*(Pt 7), 806–814.

Rhodes, D., Brown, R. S., & Klug, A. (1989). Crystallization of nucleosome core particles. *Methods in Enzymology*, *170*, 420–428.

Richmond, T. J., & Davey, C. A. (2003). The structure of DNA in the nucleosome core. *Nature*, *423*(6936), 145–150.

Richmond, T. J., Finch, J. T., Rushton, B., Rhodes, D., & Klug, A. (1984). Structure of the nucleosome core particle at 7 A resolution. *Nature*, *311*(5986), 532–537.

Sanchez-Weatherby, J., Bowler, M. W., Huet, J., Gobbo, A., Felisaz, F., Lavault, B., et al. (2009). Improving diffraction by humidity control: A novel device compatible with X-ray beamlines. *Acta Crystallographica. Section D, Biological Crystallography*, *65*(Pt 12), 1237–1246.

Struck, M. M., Klug, A., & Richmond, T. J. (1992). Comparison of X-ray structures of the nucleosome core particle in two different hydration states. *Journal of Molecular Biology*, *224*(1), 253–264.

Thåström, A., Lowary, P. T., Widlund, H. R., Cao, H., Kubista, M., & Widom, J. (1999). Sequence motifs and free energies of selected natural and non-natural nucleosome positioning DNA sequences. *Journal of Molecular Biology*, *288*(2), 213–229.

Wheeler, M. J., Russi, S., Bowler, M. G., & Bowler, M. W. (2012). Measurement of the equilibrium relative humidity for common precipitant concentrations: Facilitating controlled dehydration experiments. *Acta Crystallographica. Section F, Structural Biology and Crystallization Communications*, *68*(Pt 1), 111–114.

Winkler, D. D., Luger, K., & Hieb, A. R. (2012). Quantifying chromatin-associated interactions: The HI-FI system. *Methods in Enzymology*, *512*, 243–274.

Zhou, B.-R., Jiang, J., Feng, H., Ghirlando, R., Xiao, T. S., & Bai, Y. (2015). Structural mechanisms of nucleosome recognition by linker histones. *Molecular Cell*, *59*(4), 628–638. http://doi.org/10.1016/j.molcel.2015.06.025.

CHAPTER THREE

Preparation of Recombinant Centromeric Nucleosomes and Formation of Complexes with Nonhistone Centromere Proteins

N. Sekulic[1], B.E. Black[2]
Perelman School of Medicine, University of Pennsylvania, Philadelphia, PA, United States
[2]Corresponding author: e-mail address: blackbe@mail.med.upenn.edu

Contents

1. Introduction 68
2. Preparation of Histones for CENP-A Nucleosome Assembly 69
   2.1 Expression of His-H2A and H2B Monomers and (CENP-A/H4)$_2$ Heterotetramers 69
   2.2 Purification of (CENP-A/H4)$_2$ Heterotetramer 71
   2.3 Purification of His-H2A, H2A, and H2B Monomers 74
   2.4 Histone H2A/H2B Refolding 77
3. Preparation of DNA for CENP-A Nucleosome Assembly 79
   3.1 Preparation of DNA by PCR 79
   3.2 Preparation of DNA from plasmid excision 83
4. CENP-A Nucleosome Assembly 84
   4.1 Reconstitution of CENP-A Nucleosomes 85
   4.2 Checking Nucleosome Quality by Native PAGE 86
   4.3 Nucleosome Repositioning by Thermal Shifting 88
   4.4 Purification of Nucleosomes by Preparative Native PAGE 88
5. Complex of CENP-A Nucleosome with CENP-C 90
   5.1 Expression and Purification of CENP-C 90
   5.2 Determining Stoichiometry of CENP-A Nucleosome and CENP-C 93
Acknowledgments 95
References 95

Abstract

Centromeres are present on each chromosome to direct proper segregation during cell division. The understanding of how the histone H3 variant, CENP-A, epigenetically marks the location of the centromere on the chromosome has been advanced, in part, through the study of histone complexes, nucleosomes, and nucleosomal complexes

[1] Current address: Biotechnology Centre of Oslo and Department of Chemistry, University of Oslo, Oslo, Norway.

Methods in Enzymology, Volume 573
ISSN 0076-6879
http://dx.doi.org/10.1016/bs.mie.2016.01.014

with nonhistone centromere proteins. In this chapter, we describe the preparation of recombinant versions of these complexes. The methodology is firmly rooted in classic nucleosome reconstitution methods, but we highlight the aspects of the preparations that diverge from those used for the methods established with canonical histones. We also provide a method for producing PCR-amplified nucleosomal DNA sequences in milligram quantities that is particularly useful for studies where multiple sequences and/or chemical modifications are desired. Lastly, we describe our approach to assemble and analyze a complex between the recombinant human CENP-A nucleosome and one of its binding partners, CENP-C.

1. INTRODUCTION

The centromere is the chromosomal locus that is essential for proper segregation of sister chromatids during cell division (McKinley & Cheeseman, 2016; Sekulic & Black, 2012). In mitosis, a protein megacomplex—the kinetochore—forms at centromeres and assures proper attachment of duplicated chromosomes to microtubule filaments that guide each sister chromatid to the two daughter cells. Centromeres in many species, including humans, form on repetitive and AT-rich DNA. While this form of DNA is found at all normal human centromeres, it is paradoxically neither necessary nor sufficient for specifying centromere location. Multiple lines of evidence have focused attention on specialized nucleosomes containing the histone H3 variant, CENP-A, as an epigenetic mark that specifies the location of the centromere. Once CENP-A seeds a new centromere, a constitutive centromere associated network (CCAN) of proteins are recruited to the centromere. Centromere complexes built from their component parts, starting with the CENP-A nucleosome, have been essential to develop our current understanding of the nature of centromeric chromatin (Black, Brock, Bédard, Woods, & Cleveland, 2007; Carroll, Silva, Godek, Jansen, & Straight, 2009; Guse, Carroll, Moree, Fuller, & Straight, 2011; Sekulic, Bassett, Rogers, & Black, 2010; Tachiwana et al., 2011). Here, we describe the preparation of these nucleosomes. The general scheme builds solidly on the foundation provided by classic nucleosome reconstitution approaches (Dyer et al., 2004; Luger, Rechsteiner & Richmond, 1999), but there are several specific variations that make it important to specify in this chapter how we generate highly pure CENP-A-containing nucleosomes in sufficient quantities for a variety of biochemical and biophysical experiments. The nucleosomes are formed as core particles (ie, ~147 bp DNA) or as nucleosomes with longer DNA (eg, ~200 bp) and with native centromere sequences or synthetic sequences (eg, the Widom "601" sequence;

Lowary & Widom, 1998). Finally, we detail our approach to forming a complex with CENP-C, one of the two CCAN components known to directly contact CENP-A nucleosomes at the centromere (Carroll, Milks, & Straight, 2010; Falk et al., 2015, 2016; Kato et al., 2013).

2. PREPARATION OF HISTONES FOR CENP-A NUCLEOSOME ASSEMBLY

2.1 Expression of His-H2A and H2B Monomers and $(CENP\text{-}A/H4)_2$ Heterotetramers

Histones H2A and H2B are expressed as monomers in inclusion bodies. $(CENP\text{-}A/H4)_2$ is expressed as a soluble heterotetramer (Fig. 1).

2.1.1 Equipment

Shaking incubator (New Brunswick Scientific Innova 44R Incubator/Shaker), Fernbach flasks (2800 mL), Centrifuge bottles (1 L) (Nalgene Cat # 3141-1006), Refrigerated centrifuge, Sorvall RC 6 Plus (Thermo Fisher) with SLC-4000 rotor (Thermo Fisher).

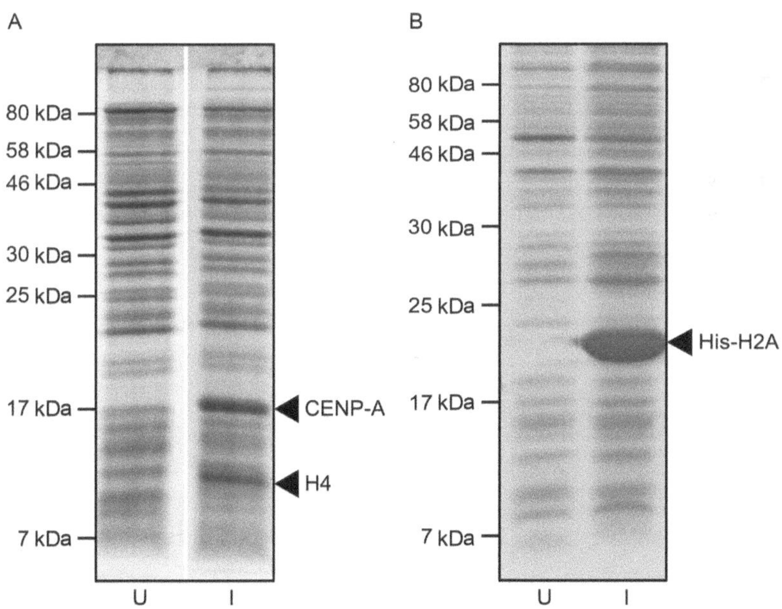

Fig. 1 15% SDS-PAGE analysis of protein inductions for $(CENP\text{-}A/H4)_2$ (A) and His-H2A (B). In each panel, "uninduced sample" (U) is shown on left and "induced sample" (I) is shown on right. Bands corresponding to induced proteins are indicated with *black triangles*.

2.1.2 Reagents

pET-12 plasmid with human H2B (Ampicillin [Amp] resistance); pET-12 plasmid with a bicistronic expression cassette encoding human CENP-A and H4 (Amp resistance); pET-30b+ with human H2A (Kanamycin [Kan] resistance); BL21 [DE3] pLysS (Promega) strain of bacteria (pLysS encodes lysozyme and the chloramphenicol [Cam] resistance marker); 2 × YT broth (dissolve in dH_20: 16 g Bacto Tryptone, 10 g Bacto Yeast Extract, 5 g NaCl per liter); Ampicillin (Amp) (Cellgro Cat# 61-238-RH); Chloramphenicol (Cam) (Sigma-Aldrich Cat # C0378); Kanamycin (Kan) (Sigma-Aldrich Cat # K1377); Phenylmethylsulfonyl fluoride (PMSF) (Sigma-Aldrich Cat # P7626); β-mecaptoehanol (BME) (Sigma-Aldrich Cat # M6250); Aprotinin from bovine lung (Sigma-Aldrich Cat# A6106); Leupeptin (Sigma-Aldrich Cat # L2884); Pepstatin (Sigma-Aldrich Cat # P5318); Isopropyl β-D-1 thiogalactopyranoside (IPTG) (Denville Cat # C18280).

2.1.3 Buffers

Wash Buffer: 50 m*M* Tris–Cl pH 7.5 at 4°C, 100 m*M* NaCl, 1 m*M* EDTA. Immediately before use supplement with protease inhibitors (1 m*M* PMSF, 0.01–0.02 TIU/mL aprotinin, 2.5 μg/mL leupeptin, 2.5 μg/mL pepstatin, and 7 m*M* BME) and filter.

2.1.4 Protocol

Day 1

1. Transform one of the histone-encoding plasmids into BL21 [DE3] pLysS cells (follow standard transformation protocol for pLysS cells) and plate onto LB + Amp + Cam for H2B and CENP-A/H4 plasmids or LB + Kan + Cam for His-H2A plasmid.

Day 2

2. Start one or more 50 mL overnight cultures in 250 mL flasks in an orbital shaker at 37°C. Use 2 × YT media supplemented with appropriate antibiotic (final concentration of 50 mg/L of Amp or Kan and 34 mg/L Cam).
3. Prepare 2–6 L of 2 × YT media for the next day and prewarm at 37°C overnight.

Day 3

4. Add Amp or Kan (final concentration of 50 mg/L) and Cam (final concentration of 34 mg/L) to each 1 L of 2 × YT in 2.8 L Fernbach flasks.

5. Inoculate each 1 L of media with 15–20 mL of the combined overnight cultures (if more then one 50 mL culture was grown for a particular plasmid) and shake at 37°C in an orbital shaker monitoring the OD_{600}.
6. Grow cells to an OD_{600} of 0.6–0.7. This should take approximately 1.5–2 h.
7. Induce cells with 1 m*M* IPTG (final concentration).
8. Grow cells for ~4 h.
9. Harvest cells in precooled 1 L centrifuge bottles. Spin at 4°C, 4000 rpm for 20 min in a precooled SLC-4000 rotor.
10. Remove the supernatant and resuspend each pellet separately in 25 mL of Wash Buffer supplemented with 1 m*M* PMSF, 7 m*M* BME, 2.5 μg/mL Leupeptin, 2.5 μg/mL Pepstatin, and 0.01–0.02 TIU/mL Aprotinin. Transfer resuspended cells from each of the 1 L cultures to a 50 mL conical tube.
11. Freeze resuspended cells and store at −80°C.

2.2 Purification of $(CENP\text{-}A/H4)_2$ Heterotetramer

$(CENP\text{-}A/H4)_2$ heterotetramers are expressed off a bicistronic construct where a single mRNA encodes both proteins (Black et al., 2004). Upon expression, they form a heterotetramer in bacterial cells. The heterotetramer is bound to bacterial nucleic acid, so hydroxyapatite can be used to bind under conditions permissive to nucleic acid interactions (ie, 80 m*M* phosphate), and then $(CENP\text{-}A/H4)_2$ is eluted by high monovalent salt (NaCl), leaving the nuclei acid bound on the column. A cation exchange column further purifies $(CENP\text{-}A/H4)_2$ and effectively concentrates the complex. This method generally yields 2–4 mg of pure soluble tetramer (Fig. 2B) per litre of grown culture.

2.2.1 Equipment

Wheaton Dounce-Type Stainless Steel homogenizer (15 mL capacity); Sonicator, Branson Sonifier 450 with 3/4″ disruptor horn with solid tip; Refrigerated centrifuge, Sorvall RC 6 Plus (Thermo Fisher); Fixed angle rotor, SS-34 (Thermo Fisher); 50 mL polycarbonate tubes (Nalgene Cat #3117-9500); XK-50/100 Column housing (GE healthcare Cat # 28-9889-65); Peristaltic pump, ÄKTA prime (GE Healthcare); Hydroxyapatite Bio-gel HT gel (Bio-Rad Cat # 1300151); Dialysis tubing, SnakeSkin MWCO 3.5 kDa (Thermo Fisher Cat # 68305); Cation exchange column, HiTrap SP FF 1 mL column (GE healthcare Cat # 29051324)

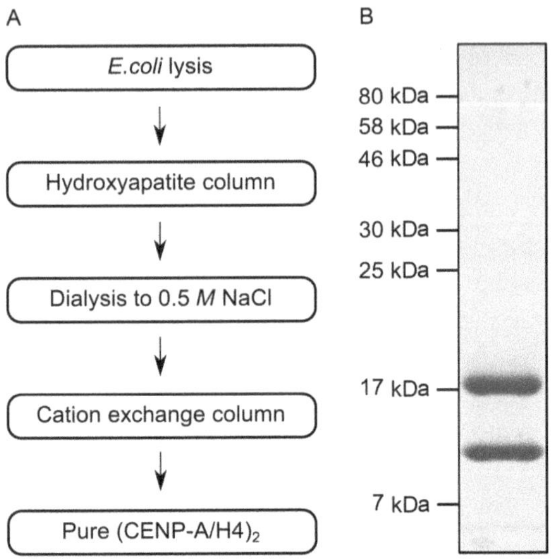

Fig. 2 Purification of the $(CENP\text{-}A/H4)_2$ heterotetramer. (A) Scheme showing steps involved in purification. (B) 15% SDS-PAGE of the purified $(CENP\text{-}A/H4)_2$ heterotetramer.

2.2.2 Reagents

PMSF, BME, Aprotinin from bovine lung, Leupeptin, Pepstatin

2.2.3 Buffers

All buffers should be filtered and degassed prior to use.

Hydroxyapatite Start Buffer (*HT Start Buffer*): 80 m*M* sodium phosphate pH 6.9, 300 m*M* NaCl. Add fresh: 2.5 μg/mL Leupeptin, 2.5 μg/mL Pepstatin, 0.01–0.02 TIU/mL Aprotinin, 1 m*M* PMSF, 14 m*M* BME.

Hydroxyapatite Wash Buffer (*HT Wash Buffer*): 80 m*M* sodium phosphate pH 6.9, 500 m*M* NaCl. Add fresh protease inhibitors: 2.5 ug/mL leupeptin, 2.5 μg/mL pepstatin, 0.01–0.02 TIU/mL aprotinin, 1 m*M* PMSF, 14 m*M* BME.

Hydroxyapatite Elution Buffer (*HT Elution Buffer*): 80 m*M* sodium phosphate pH 6.9, 2.5 *M* NaCl. Add fresh: 1 m*M* PMSF, 14 m*M* BME.

Cation Exchange Start Buffer: 20 m*M* sodium phosphate pH 6.9, 500 m*M* NaCl. Add fresh: 1 m*M* PMSF, 14 m*M* BME.

Cation Exchange Elution Buffer: 20 m*M* sodium phosphate pH 6.9, 2 *M* NaCl. Add fresh: 1 m*M* PMSF, 14 m*M* BME.

2.2.4 Protocol

1. Thaw 25 mL of bacterial suspension frozen in step #11, Section 2.1.4. Homogenize resuspended cells in 10 mL batches in an ice-cold stainless

steel homogenizer (~5 full passes until mixture passes through dounce smoothly to indicate homogenization has occurred). Add HT Start Buffer to 40 mL and centrifuge in polycarbonate tubes in SS-34 rotor at 26,000 × *g* for 15 min. Discard supernatant and resuspend the cell pellet in 35 mL HT Start Buffer by pipetting up and down, and repeat centrifugation steps. Repeat for total of three washes.

The $(CENP\text{-}A/H4)_2$ heterotetramer remains with cellular nucleic acid in the pellet prior to sonication (see next step). By washing the pellet at this step, we are removing many bacterial contaminants.

2. Resuspend pellet in 30 mL HT Start Buffer supplemented with protease inhibitors. Sonicate in 50 mL conical tube for 30 pulses (pulse duration of 0.2 s, 1 pulse/s) at 80% amplitude. Repeat twice with a 2 min cooling period in ice/water slurry for a total of 90 pulses. Spin in polycarbonate tubes in an SS-34 rotor at 26,000 × *g* for 15 min and collect supernatant.
3. Prepare hydroxyapatite slurry by washing it with HT Start Buffer without protease inhibitors (3 ×; each with 300 mL). Allow slurry to settle for 10 min before removing excess buffer and fines—cloudy upper level at the top of the bed of slurry. Pack washed slurry in XK-50/100 column housing following manufactures instructions. Allow 2–3 cm to settle under gravity, then open the column outlet and allow the gel to pack under gravity flow. When the bed is stable, close the column outlet and adjust upper adaptor so that plunger is 1 cm from the bed. Connect the column to ÄKTA prime system preequilibrated with HT Start Buffer supplemented with protease inhibitors and equilibrate column with one and a half column volume of HT Start Buffer supplemented with protease inhibitors.

   Use 2 mL of resin per 1 mL of sonicated extract. This ratio is based on the limited DNA binding capacity of the resin and the high DNA content in the sonicated extract.
4. Centrifuge sample from step #2 again at 26,000 × *g* for 15 min to reclarify and immediately apply it to the column using peristaltic pump with a flow rate of 1 mL/min.
5. Wash column with 1 column volume of HT Start Buffer and 3 column volumes of the HT Wash Buffer at 3 mL/min.
6. Elute $(CENP\text{-}A/H4)_2$ with HT elution buffer at 1 mL/min.
7. Pool $(CENP\text{-}A/H4)_2$ fractions and dialyze overnight using dialysis tubing to Cation Exchange Start Buffer but make sure to match final NaCl concentration to the Start Buffer concentration (500 m*M* NaCl) using

formula $C_1V_1 + C_2V_2 = C_3V_3$, where C_1V_1 are NaCl concentration and volume of pulled fractions, C_2V_2 are NaCl concentrations and volume (use 2 L) of dialysis buffer, and C_3V_3 are final concentration and combined volume of dialysis buffer + pulled fractions.

Excess salt in the sample could interfere with the next step of purification.

8. Clarify sample by centrifuging in polycarbonate tubes using SS-34 rotor at 26,000 × *g* for 15 min and then apply to a 1 mL HiTrap SP FF preequilibrated with Cation Exchange Start Buffer.
9. Wash the column with the same buffer for 10 column volumes, then step increase salt to 900 m*M* (20% Cation Exchange Elution Buffer) and wash for another 20 column volumes.
10. Finally, elute pure protein with 2 *M* salt (100% Cation Exchange Buffer).
11. Pool fractions containing $(CENP\text{-}A/H4)_2$. Add glycerol to 50% and store at −20°C.

2.3 Purification of His-H2A, H2A, and H2B Monomers

Human H2B is expressed from a vector with Ampicillin resistance and its purification from inclusion bodies follows the protocol for recombinant *Xenopus* H2B outlined by Richmond and colleagues (please refer to Luger et al., 1999). Human H2A is expressed from a pET30b + vector with Kanamycin resistance. This generates a fusion protein with tags on the N-terminus (a tandem 6 × His tag and an S-tag) followed by a PreScission protease cutting sequence (LEVLFQ↓GPLG; the GPLG moiety is all that remains on the final H2B polypeptide after cleavage). The N-terminal tags on human H2A are required for the highest expression levels into inclusion bodies (Fig. 1B). For generating H2A protein with the tags present, follow the same general protocol as for H2B. Tag removal by PreScission protease requires nondenaturing conditions, so following the initial gel filtration step in urea, the samples are dialyzed out of urea and into PreScission protease Buffer for overnight cutting and then back in the urea based buffer for final cation exchange column purification. Typical yields of the human histones are ~40 mg/L of culture for cleaved H2A and ~20 mg/L of culture for H2B.

2.3.1 Equipment

Wheaton Dounce-Type Stainless Steel homogenizer (15 mL capacity); Sonicator, Branson Sonifier 450 with 3/4″ disruptor horn with solid tip; Refrigirated centrifuge, Sorvall RC 6 Plus (Thermo Fisher); Fixed angle rotor, SS-34 (Thermo Fisher); 50 mL polycarbonate tubes (Nalgene Cat # 3117-9500); Microcentrifuge, Eppendorf 5424 (Thermo Fisher); ÄKTA purifier (GE Healthcare); Gel filtration column, High Prep 16/60 Sephacryl S-200 HR column (GE Healthcare Cat # 17-1166-01); Cation Exchange Column, HiTrap SP FF 5 mL (GE Healthcare Cat # 17-5054-01); Lyophilizer

2.3.2 Reagents

PMSF, Triton X-100, Dithiothreitol (DTT), Urea, AG501-X8 (D) deionizing resin (BioRad Cat # 142-6425), Guanidinium hydrochloride (GuHCl), PreScission protease (GE Healthcare Cat # 27-0843-01)

2.3.3 Buffers

Note: Urea containing solutions should always be freshly prepared and used (within 24 h) as solution of urea may develop significant concentration of reactive cyanate ions upon standing. Urea stock solutions should always be deionized with deionizing resin before filtering. Spent resin will change color from blue to yellow after absorbing ions. Urea solutions should not be refrigerated due to decreased solubility at low temperature.

All buffers must be filtered and degassed for 45 min prior to their use on the column.

Wash Buffer: 50 m*M* Tris–HCl pH 7.5 at 4°C, 100 m*M* NaCl, 1 m*M* EDTA

Unfolding Buffer: 7 *M* guanidinium hydrochloride, 20 m*M* Tris–HCl pH 7.5 at 20°C, 10 m*M* DTT

Gel Filtration Buffer: 7 *M* deionized urea, 20 m*M* sodium acetate pH 5.2, 1 *M* NaCl, 5 m*M* BME, 1 m*M* EDTA

PreScission Protease Buffer: 50 m*M* Tris–HCl pH 7.5 at 20°C, 150 m*M* NaCl, 5 m*M* BME

Cation Exchange Start Buffer: 7 *M* deionized urea, 20 m*M* sodium acetate pH 5.2, 0.2 *M* NaCl, 5 m*M* BME, 1 m*M* EDTA

Cation Exchange Elution Buffer: 7 *M* deionized urea, 20 m*M* sodium acetate pH 5.2, 0.6 *M* NaCl, 5 m*M* BME, 1 m*M* EDTA

Dialysis Buffer: 10 m*M* Tris–HCl pH 7.0, 2 m*M* BME

2.3.4 Protocol

1. Thaw batches of 25 mL of the appropriate cell suspension from step #11 from Section 2.1.4, at room temperature.

2. Dounce homogenize cells at 4°C with five strokes in 10 mL batches and sonicate in 25 mL batches. See details in steps 1 and 2 from Section 2.2.4.
3. Centrifuge immediately for 20 min at 4°C at 26,000 × *g*.
   The pellet contains inclusion bodies of the histone proteins.
4. Resuspend each pellet with 20 mL of Wash Buffer supplemented with 1% (v/v) Triton X-100. Spin for 10 min at 26,000 × *g*, 4°C. Repeat this step to completely wash the inclusion bodies.
5. Combine two pellets in one 50 mL conical tube by resuspending in Wash Buffer (ie, containing no Triton) and then centrifuge for 10 min at 26,000 × *g*, 4°C.
6. Soak the pellet for 30 min in 350 μL DMSO at room temperature.
7. Slowly add 10 mL of Unfolding Buffer and resuspend pellet at room temperature for 1 h with gentle stirring and pipetting up and down.
8. Centrifuge for 10 min at 26,000 × *g* at 20°C. Carefully remove the supernatant containing the solubilized histone, avoiding the pellet.
9. Reextract the pellet to remove any remaining histone with 2 mL Unfolding Buffer (adjust this volume, if necessary, depending on the size of the remaining pellet). Transfer to eppendorf tube and centrifuge in microcentrifuge at 20,000 × *g* for 10 min.
10. Inject sample on the High Prep 16/60 Sephacryl S-200 HR column that has been preequilibrated with Gel Filtration Buffer.
    This step separates the histone from DNA and larger protein/DNA aggregates. DNA is the first to elute from the column. Some of the early histone-containing fractions may contain high levels of contaminating DNA that would compromise subsequent purification steps. So it is very important to minimize the amount of DNA in the final prep by assuring A_{260}:A_{280} is <1 for each of the fractions to be pooled in this step.
11. Analyze peak fractions by 15% SDS-PAGE. Combine those containing histone and where A_{260}:A_{280} is <1.
    For untagged H2B and H2A containing the N-terminal tags, skip to step #16. For removal of His and S-tag of H2A, it is necessary to change buffer from denaturing urea buffer to PreScission Protease Buffer.
12. Dialyze protein solution into PreScission Protease Buffer. Dialyze in two serial steps of 2 L each; the first for 2–3 h at room temperature and the second overnight at 4°C.
13. Add PreScission protease according to the user manual to the protein solution (200 U protease:10 mg of target protein). Let the solution incubate overnight with very gentle agitation on an orbital shaker at 4°C.

14. Monitor the extent of cleavage by 15% SDS-PAGE and proceed to next step when protein is completely cut.
    His-H2A runs around ~19 kDa and cleaved H2A runs ~14 kDa.
15. Dialyze protein solution in Cation Exchange Start Buffer matching the final 0.5 *M* NaCl concentration of the buffer as outline in step #7, Section 2.2.4. Sample should dialyze for at least 3 h prior to loading onto the cation exchange column.
16. Clarify the sample by centrifuging in polycarbonate tubes in SS-34 rotor at 26,000 × *g* for 15 min.
17. Inject sample on a 5 mL cation exchange SP FF column preequilibrated with 5 column volumes of Cation Start Buffer.
18. Wash the column with 10 volumes of Cation Start Buffer. Elute the histone by gradually increasing Cation Exchange Elution Buffer to 100% over 20 column volumes.
19. Use 15% SDS-PAGE to identify fractions containing pure histone. Pool fractions together and dialyze extensively against Dialysis Buffer. Dialyze 3 × against 3 L each. The final dialysis step should be >8 h (eg, overnight).
    Extensive dialysis is needed to remove urea.
20. Lyophilize in aliquots each containing ~4 mg histone.

2.4 Histone H2A/H2B Refolding

Monomeric histones purified from inclusion bodies are paired with their partner histone (H2A and H2B to make H2A/H2B heterodimers and H3 and H4 to make $[H3/H4]_2$ heterotetramers). Histones are combined in denaturing conditions and then slowly transferred to high salt, urea-free buffer for refolding. CENP-A is copurified in a soluble heterotetramer with its binding partner H4, so it does not require a refolding step.

2.4.1 Equipment

Glass Pasteur pipette; Refrigerated centrifuge, Sorvall RC 6 Plus (Thermo Fisher); Fixed angle rotor, SS-34 (Thermo Fisher); 50 mL polycarbonate tubes (Nalgene Cat # 3117-9500); Microcentrifuge, Eppendorf 5424 (Thermo Fisher); Dialysis tubing, SnakeSkin MWCO 7kD (Thermo Fisher Cat # 38700); Concentrators, Amicon Ultra 4 mL centrifugal filters 10 NMWL (Millipore Cat # UFC801008); ÄKTA purifier (GE Healthcare); Gel filtration column, High Load 16/600 Superdex S-200 PG column (GE Healthcare, Cat # 28989335)

2.4.2 Buffers

Unfolding Buffer: 6 *M* GuHCl, 20 m*M* Tris–HCl pH 7.5 at 20°C, 5 m*M* DTT

Refolding Buffer: 2 *M* NaCl, 10 m*M* Tris–HCl pH 7.5 at 20°C, 1 m*M* EDTA, 5 m*M* BME

2.4.3 Protocol

Day 1

1. Calculate the volume of Unfolding Buffer to be added to each sample of lyophilized histone protein partners so as to reach a concentration of ~2 mg/mL. Add calculated volume to lyophilized histone by gently pipetting up and down using a glass Pasteur pipette. Allow material to dissolve for 1 h at room temperature.

   If using multiple tubes of a particular histone, combine into one tube at this time.
2. Determine the molar concentration of each histone spectroscopically (A_{280}) using exact molecular extinction coefficient and exact molecular weight (see Table 1 for values for human histones H2A and H2B).
3. Determine which of the histone partners has the smaller molar abundance that will limit the size of the refolding reaction. Calculate the appropriate volume of each histone partner to mix at a precise equimolar ratio. Combine the histones and adjust the final volume with Unfolding Buffer to reach 1 mg/mL total histone concentration.

   It is critical to be very precise with combining histones in equimolar amounts since a molar excess of one binding partner will result in precipitation that can compromise the entire refolding step.
4. Dialyze histones using dialysis tubing with three changes against 2 L of Refolding Buffer. The first dialysis should be at room temperature for about 1 h, the second should start at the room temperature for 30 min and then be moved to 4°C (so the proteins cool slowly) for 3–4 h, and the third dialysis should go overnight at 4°C.

   If precipitate is visible during any of these steps, harvest the samples and centrifuge at 26,000 × g to remove precipitate. Collect supernatant and continue dialyzing.

Table 1 Molecular Weight and Extinction Coefficients for Histones Cleaved H2A and H2B used in CENP-A Nucleosome Assembly

Histone	MW (D)	ε (Per cm/*M*)
Cleaved H2A	14,419.97	3840
H2B	13,906.12	6400

Day 2

5. Clarify the dialyzed solution to remove any precipitated product (polycarbonate tubes, SS-34 rotor, 26,000 × *g*, 4°C, 15 min).
6. Concentrate mixture using Amicon Ultra 4 centrifugal filter unit. Spin at 1000 × *g* with pipetting up and down every 7 min until total volume reaches ~1.5 mL.
7. Centrifuge at 26,000 × *g* for 15 min (polycarbonate tubes, SS-34 rotor, 4°C).
8. Inject the sample onto the High Load Superdex 16/600 S-200 PG gel filtration column that is already preequilibrated with Refolding Buffer. Run the column at 1 mL/min with Refolding Buffer monitoring for complex elution at A_{280}. H2A/H2B heterodimers elute near their predicted Stokes' radius (28.4 kDa).
9. Identify fractions with equimolar amounts of histone partners by 15% SDS-PAGE and combine them.
10. Concentrate fractions to >3.0 mg/mL. Add glycerol to a final concentration of 50% and gently mix. Store at −20°C.

3. PREPARATION OF DNA FOR CENP-A NUCLEOSOME ASSEMBLY

We use two different methods for generation of DNA for CENP-A nucleosome assembly. One is based on a PCR reaction from a plasmid that has a single copy of the nucleosome template DNA, and the other one is based on a mega prep of a plasmid containing multiple tandem copies of the nucleosome template DNA sequence as described previously (Dyer et al., 2004; Fig. 3). We typically use a shallow salt elution gradient during anion exchange chromatography to separate nucleosome template DNA from primers (critical for PCR-based method) or from plasmid backbone (critical for the plasmid prep based method; see Fig. 4).

3.1 Preparation of DNA by PCR

We use a PCR method for generating up to 400 μg of DNA per 96-well plate (Falk et al., 2015). In this method, DNA is amplified from a single plasmid borne copy of the nucleosome DNA sequence. The PCR product is purified away from primers and other contaminants in the reaction mix by anion exchange chromatography. Because it does not require initial

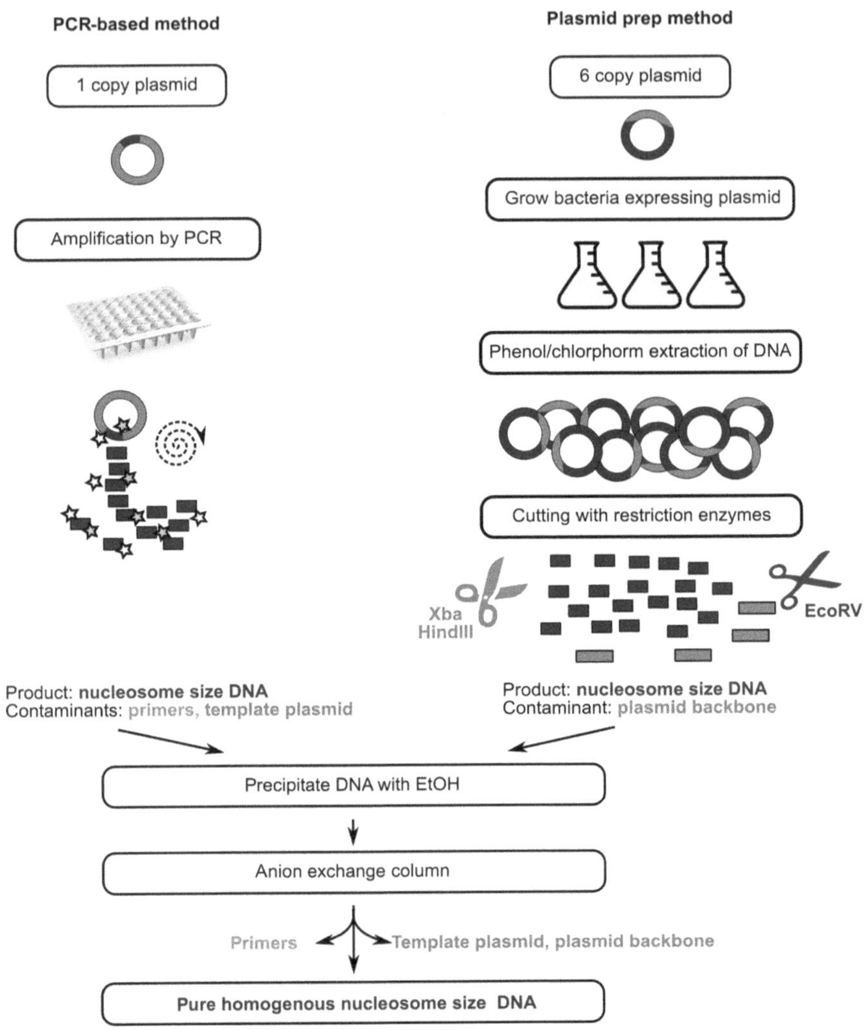

Fig. 3 Overview of DNA preparation for CENP-A nucleosome assembly by PCR method and by mega plasmid prep method. (See the color plate.)

generation of a plasmid with tandem copies of the nucleosome template DNA sequence, this method is particularly useful in studies requiring modest quantities of several different templates (eg, either sequence variations or the primer-mediated incorporation of fluorescent or nonfluorescent chemical moieties). Optimal PCR conditions (eg, annealing temperature) should be determined for each pair of primers and nucleosomal DNA sequence before scaling to the multiwell format.

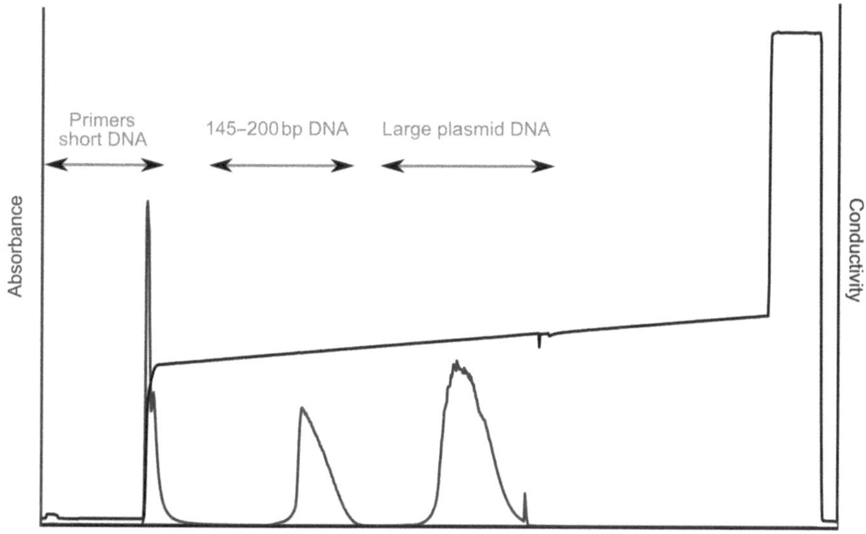

Fig. 4 Chromatogram of separation of nucleosome template DNA from major contaminants.

3.1.1 Equipment

PCR thermocycler, Eppendorf MasterCyclerPro (Eppendorf); Refrigerated tabletop centrifuge, Eppendorf 5810R with swinging bucket rotor and adapters for 50 mL tubes, max speed 3220 × *g* (Fisher); Resource Q anion exchange column, for 1 mL: prepacked Resource Q 1 mL (GE Healthcare Cat # 17-1177-01), for 20 mL: housing HR 16/10 column (GE Healthcare Cat # 19-7403-01) + Resource 15Q resin (GE Healthcare Cat # 17-0947-01).

3.1.2 Reagents

PCR primers (100 pmol/μL), *Taq* DNA polymerase (NEB Cat # M0267), Deoxynucleotide set (Sigma-Aldrich Cat # DNTP100), sodium acetate, ice-cold 100% ethanol

3.1.3 Buffers

10× PCR Buffer: 200 m*M* Tris–HCl pH 8.8 at 25°C, 100 m*M* ammonium sulfate, 100 m*M* KCl, 20 m*M* magnesium sulfate, 1% (v/v) Triton-100

25× dNTPs: 5 m*M* dATP, 5 m*M* dCTP, 5 m*M* dGTP, 5 m*M* dTTP

3 M Sodium acetate pH 5.2

ResourceQ Buffer A: 20 m*M* Tris–HCl pH 7.0 at 23°C

ResourceQ Buffer B: 20 m*M* Tris–HCl pH 7.0 at 23°C, 2 *M* NaCl
TE (10, 0.1)—10 m*M* Tris–HCl pH 8 at 23°C, 0.1 m*M* EDTA

3.1.4 Protocol

1. Determine the optimal annealing temperature for primer/plasmid combination by varying temperature in 1–2°C steps from 50 to 70°C for 50 μL reactions. Check on 1% agarose gel for presence of single band corresponding to expected DNA product. For a 50 μL PCR reaction use recipe outlined in Table 2 and start PCR using program outlined in Table 3. The very short extension time is because of the short length of nucleosomal DNA; the extension is probably finished by the time the thermocycler heats up to the extension temperature.
2. Scale up for multiwell format and perform 50 μL reactions in each well of 96-well plates.
3. Pool the product from one plate and check for purity on the 1% agarose gel. If you scale to multiple 96-well plates, combine all product at this point.
4. Product of up to three plates (~12 mL) can be combined in one 50 mL conical tube for ethanol precipitation. Ethanol precipitate DNA by addition of 1/10 volume of 3 *M* sodium acetate, pH 5.2, and 2.5 volumes of 100% ice-cold ethanol. For the best yield leave precipitated DNA at −20°C overnight. Centrifuge in a tabletop centrifuge (Eppendorf 5810R with swinging bucket rotor and adapters for 50 mL tubes) at highest speed (3220 × *g*) for 30 min at 4°C.

Table 2 Quantities and Reagents for PCR Reaction for Production of 145–200 bp DNA Sequences for Nucleosome Reconstitutions

Volume (μL)	Component	Final Concentration
0.5	Forward primer	1 pmol/μL
0.5	Reverse primer	1 pmol/μL
5	10 × PCR buffer	1 ×
2	25 × dNTPs	200 μ*M*
1	DNA template (25 ng)	0.5 μg/μL
0.25	*Taq* DNA polymerase	
40.75	Nuclease-free water	
50	Total	

Table 3 PCR Cycling Program for Production of 145–200 bp DNA Sequences for Nucleosome Reconstitutions

Temperature (°C)	Time	Step
95	5 min	Initial melt
95	30 s	32× cycles
50–70	25 s	
72	5 s	
72	5 min	Final extension

5. Wash the pellet with 10 mL of 70% ethanol to remove salt. Centrifuge as in previous step for 20 min at 4°C. Remove the supernatant and let pellet dry on air for ~20 min.
6. Resuspend DNA in 5–10 mL of ResourceQ Buffer A by gently pippeting up and down and load on the preequilibrated Resource Q column.

   For smaller preps (1–3 plates) use 1 mL column and for larger preps use 20 mL.
7. Elute insert and plasmid by running a linear gradient between 25% and 45% ResourceQ Buffer B over 40 column volumes (Fig. 4). Clean the column with 100% ResourceQ Buffer B.

   Primers elutes at 25% B while 145–200 bp DNA elutes during the gradient. The gradient needs to be optimized for each DNA sequence. Run fractions on the 1% agarose gel to confirm size and purity of the DNA.
8. Pool the fractions containing the nucleosome size DNA, ethanol precipitate DNA as in step #4, and wash as in step #5.
9. Resuspend DNA in TE (10, 0.1) so to have final concentration above 5 mg/mL and store at −20°C.

3.2 Preparation of DNA from plasmid excision

We have generated a plasmid with six identical repeats of α-satellite DNA (Falk et al., 2015). Our construct was based on α-satellite sequence from the human X chromosome (Yang, Hansen, Oishi, Ryder, & Hamkalo, 1982). This natural α-satellite sequence was used by Bunick and colleagues to design a palindromic 145 bp template for nucleosomes that would aid in crystallization (Harp et al., 1996). The palindromic template was used to obtain the first atomic resolution X-ray crystal structure of the nucleosome

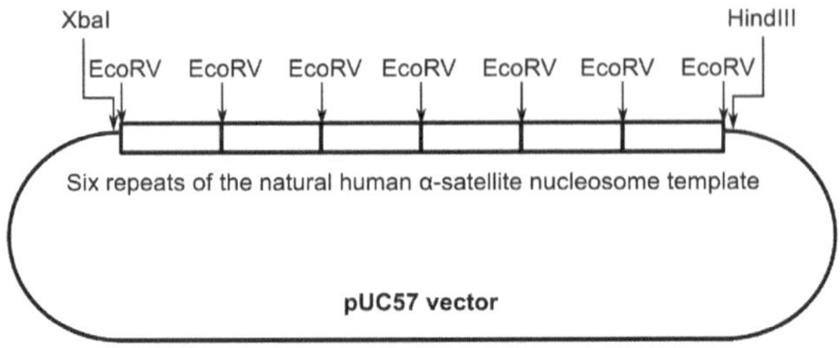

Fig. 5 Diagram of plasmid with six copies of α-satellite nucleosome template DNA.

(Luger, Mäder, Richmond, Sargent, & Richmond, 1997). Nucleosomes (canonical nucleosomes with H3 or centromere-specific CENP-A-containing nucleosomes) are phased on the 171 bp monomer repeat of human α-satellite on native centromeres (Hasson et al., 2013), and the preferred site maps precisely to the one used as the basis for the earlier crystallographic studies. Reconstituted CENP-A nucleosomes preferentially position themselves on this 145 bp segment when present in a longer (195 bp) template fragment (Falk et al., 2015). For production, we use a pUC57 plasmid containing six identical 145 or 147 bp long DNA repeats. The entire array is flanked by XbaI and HindIII restriction sites. The sequence of each repeat is: ATC(A)AATATCCACCTGCAGATTCTAC CAAAAGTGTATTTGGAAACTGCTCCATCAAAAGGCATGTTCA GCTCTGTGAGTGAAACTCCATCATCACAAAGAATATTCTGAG AATGCTTCCGTTTGCCTTTTATATGAACTTCCT(C)GAT. Nucleotides in the brackets are present only in the 147 bp construct. Each repeat is separated by an EcoRV restriction site, so that resulting DNA after EcoRV digestion has blunt ends (Fig. 5).

We follow the same general protocol for plasmid growth, isolation, and EcoRV digestion as Dyer et al. (2004). We use anion chromatography for the final separation of plasmid and insert, as described, in Section 3.1.4, steps #4–9.

4. CENP-A NUCLEOSOME ASSEMBLY

Heterodimers of H2A/H2B, heterotetramers of $(CENP\text{-}A/H4)_2$, and DNA are dialyzed from high salt to low salt to allow for nucleosome formation. Dialysis can be done in a stepwise manner as was described in Hayes and Lee (1997), but we typically use the gradient dialysis method for canonical nucleosome reconstitution described by Dyer et al. (2004) and Luger et al. (1999). We describe it here so that it complements the other parts of our protocol and to specify how we perform CENP-A nucleosome reconstitutions.

4.1 Reconstitution of CENP-A Nucleosomes

4.1.1 Equipment

Two channel peristaltic pump; Dialysis equipment: for volumes <50 μL, Slide-A-lyzer MINI dialysis device 7 kDa MWCO, 0.1 mL (Thermo Fisher Cat # 69560); for volumes 50–200 μL, D-tube dialyzer MINI 6–8 kDa MWCO (Millipore Cat # 71504); for volumes 200–700 μL, D-tube dialyzer MIDI 6–8 kDa MWCO (Millipore Cat # 71507); for volumes 0.7–3 mL, Slide-A-lyser cassette 10 kDa MWCO (Thermo Fisher Cat # 66380); for volumes 3–10 mL, Slide-A-lyzer cassette 10 kDa MWCO (Thermo Fisher Cat # 66810); for volumes >10 mL, SnakeSkin 7 kDa MWCO (Thermo Fisher Cat # 38700)

4.1.2 Buffers

Reconstitution Buffer—high salt (*Rb-HS*): 10 m*M* Tris–HCl pH 7.5 at 4°C, 2 *M* NaCl, 1 m*M* EDTA, 1 m*M* DTT

Reconstitution Buffer—low salt (*Rb-LS*): 10 m*M* Tris–HCl pH 7.5 at 4°C, 0.25 *M* NaCl, 1 m*M* EDTA, 1 m*M* DTT

Reconstitution Buffer—no salt (*Rb-NS*): 20 m*M* Tris–HCl pH 7.5 at 4°C, 1 m*M* EDTA, 1 m*M* DTT

4.1.3 Protocol

Day 1

1. Calculate volumes of (CENP-A/H4)$_2$ heterotetramer, H2A/H2B heterodimer, and DNA that have equimolar amounts of molecules.

   Histones must be present precisely at a 1 heterotetramer:2 heterodimers molar ratio, because any excess of one over the other will cause precipitation that can compromise the entire reconstitution reaction. DNA can be in slight excess (eg, 1.1-fold molar excess). Excess DNA can be removed by running a preparative native PAGE (see Section 4.4).

2. Setup the reconstitution mixture by combining in the following order of addition: DNA, 4 *M* NaCl (use a volume that will result in the concentration of NaCl of 2 *M* in the starting reconstitution mixture prior to dialysis), and histones. Add Rb-HS to achieve a final DNA

concentration of ~0.7 mg/mL. This is the optimal concentration for nucleosome reconstitution.

3. Incubate on ice for 30 min. Transfer to dialysis tubing and begin gradient dialysis with a flow rate of 1.5 mL/min overnight for 16 h using a two channel peristaltic pump as described in Luger et al. (1999a).

   In brief, the nucleosome reconstitution mixture is dialyzed against 500 mL of Rb-HS in a beaker with constant stirring, at the beginning of the experiment. Tubing is inserted in the beaker that takes the buffer gradually and disposes it into a waste container. The buffer volume is maintained by adding the Rb-LS into the reconstitution beaker. In this way, the amount of salt is gradually decreased over time without a net change of the volume of buffer in the dialysis beaker. As the [salt] decreases, the histones associate with the DNA in stepwise manner (heterotetramers first and heterodimers second), forming nucleosomes.

Day 2

4. In the morning, ~500 mL of the Rb-LS buffer should be still left in the original cylinder. Transfer this buffer to a new beaker and continue dialyzing the reconstitution against the solution for another 3 h. Watch for the possible precipitation at this step and if it occurs remove it by spinning (26,000 × *g* for 15 min, 4°C) and continue dialysis.
5. Dialyze three times against 500 mL of Rb-NS buffer.

4.2 Checking Nucleosome Quality by Native PAGE

The quality of nucleosome reconstitutions is very sensitive to molar ratio of histone dimers, tetramers and DNA, that is in turn dependent on the purity of these components varying from batch to batch. High quality of nucleosomes (ie, homogenous in terms of histone stoichiometry and nucleosome positioning) can be key for downstream experiments. For instance, since biophysical investigations (X-ray crystallography, small angle scattering, analytical ultracentrifugation, etc.) require a monodispersed and homogeneous solution of nucleosomes, it is important to assess the quality of nucleosome by native PAGE before proceeding with experiments.

If excess histones are present during nucleosome reconstitution, then higher order aggregates will form. If DNA is in excess, then the nucleosome preparation will be contaminated with free DNA. Also, depending on the nature of DNA sequence, nucleosomes can assemble onto DNA in more than one position, resulting in a heterogeneous population of differentially positioned nucleosomes.

Native PAGE is used to separate protein–DNA complexes based on size, charge, and shape. Higher order aggregates will travel slower than nucleosomes and free DNA will travel the faster. Free histones will not enter the gel because of their positive charge. Also, nucleosomes assembled at different positions along the DNA will have different shape and exposed charge so they will travel differently than properly positioned nucleosomes on the native gel.

We use the method of thermal shifting (Dyer et al., 2004) to reinforce single nucleosome position on DNA (see Section 4.3) and preparative gel electrophoresis (see Section 4.4) to separate nucleosomes from higher order structures and free DNA. In this way, we typically obtain monodisperse homogeneous samples suitable for biophysical studies.

4.2.1 Equipment

Mini protean tetra cell casting module (Bio-Rad Cat # 165-8022); PAGE equipment

4.2.2 Reagents

2% Bisacrylamide (Bis) (BioRad Cat # 161-0142), 40% Acrylamide (Acryl) (BioRad Cat # 161-0140), Tris base (MW 121.14), Boric acid, Ethylenediaminetetraacetic acid (EDTA), Ammonium persulfate (APS), Tetramethylethylenediamine (TEMED), Ethidium bromide (EtBr), Coomassie Brilliant Blue, Sucrose

4.2.3 Buffers

30:0.5 Acryl:Bis (12 mL) (Keep protected from light at 4°C and do not use if mixture is more than 2 weeks old) : 3 mL 2% Bis; 9 mL 40% Acryl

5 × TBE (1 L) (The stock from which one makes the 0.2× TBE used in running the gels): 54 g Tris base (MW 121.14); 27.5 g Boric Acid; 20 mL EDTA (0.5 *M* stock)

5% native gel (10 mL): 1.667 mL Acryl:Bis (30:0.5); 400 μL 5 × TBE; 7.87 mL H_2O; 50 μL 10% APS; 5 μL TEMED

20% (w/v) sucrose solution

4.2.4 Protocol

1. Use mini protean tetra cell casting module to pack native gels 0.75 or 1 mm in thickness. Prerun them at 4°C, 150 V for 60 min using 0.2 × TBE as running buffer.

   This is necessary to remove ions from the gel that could perturb the mobility of the sample.

2. Load reconstituted nucleosome samples supplemented with 20% sucrose for final concentration of 5% sucrose and run at 4°C, 150 V for 60 min. Optimal time needs to be adjusted for different DNA lengths and compositions. For optimal results do not load more than 1 μg of DNA.
3. Carefully remove the gel from glass plates and stain in EtBr solution for 5 min. Wash with deionized water and image with UV light.

   EtBr intercalates DNA so both, protein/DNA complexes and free DNA are visualized by this method. Nucleosomes strongly protect DNA from EtBr intercalation, so it understains nucleosomal DNA relative to free DNA.
4. Stain the gel with Coomassie Brilliant Blue to visualize the protein components of the complex.

4.3 Nucleosome Repositioning by Thermal Shifting

Nucleosome reconstitutions for many DNA templates (including native α-satellite DNA sequences) often result in an initial heterogeneous positioning of histones. Since the heterotetramer binds to DNA first as [NaCl] is gradually lowered during reconstitution, nucleosome formation itself will tend to trap the position where it initially binds. We use thermal shifting to generate uniformly positioned nucleosomes (Dyer et al., 2004). At elevated temperatures, histone–DNA interactions are weakened, allowing the histone octamer to slide along the DNA until they find their energetically most favorable position on DNA. The length of the nucleosome template and DNA sequence play an important role in determination of time and temperature for thermal shifting. For CENP-A nucleosomes on either 195 bp or 147 bp α-satellite sequence we shift at 55°C for 2 h. Homogeneous nucleosomes travel as a single band by native PAGE.

4.4 Purification of Nucleosomes by Preparative Native PAGE

Preparative native PAGE uses the same principles of macromolecular separation as analytical native PAGE. The gel is packed in 28 or 37 mm diameter cylinder that is cooled by circulating buffer. As the samples elute from the end of the gel they are collected in Rb-NS buffer (see Section 4.1.). This method is very effective in separation of nucleosomes from higher order aggregates and free DNA. By varying gel length, thickness, and composition (Acryl:Bis ratio) it can be optimized for removal of differentially positioned or otherwise altered nucleosomes.

4.4.1 Equipment

PrepCell Model 491 (Bio-Rad Cat # 1702926); Programmable power supply EC2060P (Thermo Fisher); Peristaltic pump; UV monitor; Fraction collector; Chart recorder; Pasteur pipette; Concentrators: Amicon Ultra 4 mL centrifugal filters 50 NMWL (Millipore Cat # UFC805008).

4.4.2 Buffers

Running buffer: 0.2 × TBE buffer; Elution buffer: Rb-NS; 20% (w/v) sucrose; saturated *n*-butanol.

4.4.3 Protocol

1. Assemble gel-casting apparatus according to the manual for the PrepCell 491.
2. For a 28 mm diameter gel, prepare 20–30 mL of 5% native gel mix. This will result in 5–7 cm long gel. For a 37 mm diameter gel, prepare 40–50 mL of gel mixture for the same gel length.

   A longer PrepCell gel (ie, 7 cm) will perform better in separating species that have similar mobility in the gel (for example, differentially positioned nucleosomes). Use 37 mm diameter PrepCell for preparations larger than 2 mg.
3. Pour gel mixture carefully along cooling tube, avoiding bubble formation. Top with 1 mL of saturated *n*-butanol to assure even surface and let the gel sit and polymerize overnight at 4°C while cooling with circulating running buffer (0.2 × TBE).
4. Remove *n*-butanol by extensive washing with water.
5. Assemble PrepCell apparatus according to the manual using 12 or 6 MWCO dialysis membrane to prevent escape of nucleosomes. Use 0.2 × TBE as running/cooling buffer and Rb-NS as elution buffer.
6. Connect to peristaltic pump, UV monitor, fraction collector and chart recorder.
7. Prerun PrepCell at 10 W constant power for 90 min.
8. Concentrate heat shifted nucleosomes to ~600 μL, add sucrose to final 5% (w/v) and load on top of the gel with glass Pasteur pipette.
9. Run at constant power of 10 W, elute at 1 mL/min and collect 2 mL fractions.
10. Identify fractions containing homogenous nucleosomes, pool them together and concentrate to at least 1 mg/mL. Nucleosomes store well at 4°C at such concentrations.

5. COMPLEX OF CENP-A NUCLEOSOME WITH CENP-C

5.1 Expression and Purification of CENP-C

CENP-C protein is expressed in *E. coli* as a GST fusion protein. It is affinity purified on a glutathione column. GST is cleaved by PreScission protease, and then cation exchange is used to separate GST from CENP-C. Gel filtration is used as a final step of purification (Fig. 6). Typically we obtain 2–3 mg of CENP-C per liter of cell culture.

5.1.1 Equipment

Shaking incubator (New Brunswick Scientific Innova 44R Incubator/Shaker); Fernbach flask (2800 mL); Centrifuge bottles (1 L) (Nalgene Cat # 3141-1006); Refrigerated centrifuge, Sorvall RC 6 Plus (Thermo Fisher); SLC-4000 rotor (Thermo-Fisher); Wheaton Dounce-Type Stainless Steel homogenizer (15 mL capacity); Sonicator, Branson Sonifier 450 with 3/4″ disruptor horn with solid tip; Tabletop centrifuge (Eppendorf 5810R with swinging bucket rotor and adapters for 50 mL tubes); Snake-Skin 7kD MWCO (Thermo Fisher Cat # 38700); Glutathione column, GSTrap FF 5 mL (GE Healthcare Cat # 17-5131-01); Cation exchange

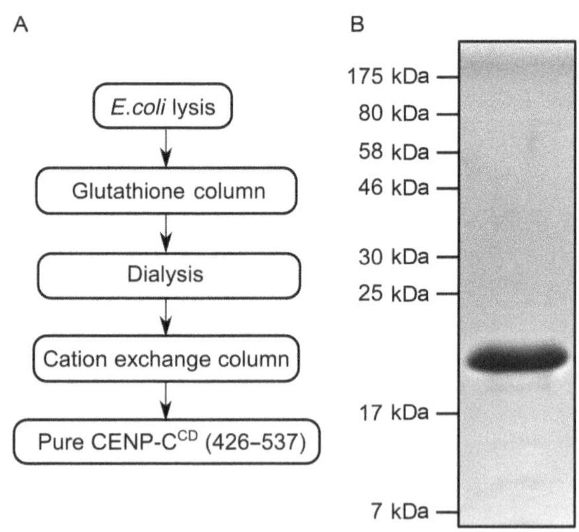

Fig. 6 Purification of CENP-C. (A) Overview of purification steps. (B) 15% SDS-PAGE of purified CENP-C^{CD}.

column, HiTrap SP Sepharose FF 5 mL (GE Healthcare Cat # 17-5157-01); Gel filtration column, High Load 16/600 Superdex S-200 PG column (GE Healthcare Cat # 28989335); Concentrators, Amicon Ultra 4 mL centrifugal filters 10 NMWL (Millipore Cat # UFC801008).

5.1.2 Reagents

pGEX-6P plasmid (GE Healthcare) with human CENP-C construct (a.a. 426–537) (Amp resistance), BL21 [DE3] pLysS (Promega) strain of bacteria, 2 × YT broth, Amp, Cam, PMSF, BME, Aprotinin from bovine lung, Leupeptin, Pepstatin, IPTG, L-Glutathione reduced (Sigma-Aldrich Cat # G4251), PreScission Protease, liquid N_2

5.1.3 Buffers

Wash/GST Equilibration Buffer: 20 m*M* sodium phosphate, pH 7.5; 100 m*M* NaCl; 0.1% Triton; 0.1% Tween; 1 m*M* EDTA—immediately before use supplement with protease inhibitors (1 m*M* PMSF, 0.01–0.02 TIU/mL aprotinin, 2.5 μg/mL leupeptin, 2.5 μg/mL pepstatin and 7 m*M* BME) and filter.

GST Wash Buffer: 20 mM sodium phosphate buffer, pH 7.5; 500 m*M* NaCl

GST Elution Buffer: 10 m*M* L-glutathione in GST equilibration buffer

Cation Exchange Loading Buffer: 20 m*M* Tris–HCl, pH 7.5 at 4°C; 100 m*M* NaCl; 0.5 m*M* EDTA; 1 m*M* DTT

Cation Exchange Elution Buffer: 20 m*M* Tris–HCl, pH 7.5 at 4°C; 2 *M* NaCl; 0.5 m*M* EDTA; 1 m*M* DTT

Gel Filtration Buffer: 20 m*M* Tris–HCl, pH 7.5 at 4°C; 200 m*M* NaCl; 0.5 m*M* EDTA; 1 m*M* DTT

5.1.4 Protocol

Day 1

1. Transform CENP-C construct into BL21 [DE3] pLysS cells (follow standard transformation protocol for pLysS cells) and plate onto LB + Amp + Cam plates.

Day 2

1. Prepare 50 mL overnight cultures in 2 × YT media supplemented with antibiotics (final concentration 50 mg/L of Amp and 34 mg/L Cam) in the 250 mL flasks. Grow overnight at 37°C on orbital shaker.
2. Prepare 4 L of 2 × YT media for the next day and let it prewarm at 37°C overnight.

Day 3

3. Add Amp (final 50 mg/L) and Cam (final 34 mg/L) to each 1 L flask of 2× YT.
4. Inoculate each 1 L flask with 15–20 mL of the overnight culture.
5. Grow cells to an OD_{600} of 0.6 in orbital shaker at 37°C. Reduce temperature to 22°C and induce with 0.1 m*M* IPTG (final concentration).
6. Grow cells at 22°C overnight.
7. Harvest cells in precooled 1 L buckets. Centrifuge at 4°C, 4000 rpm for 15 min in a precooled SLC-4000 rotor.
8. Resuspend each pellet separately in 25 mL of Wash/GST Equilibration Buffer supplemented with PMSF, BME, Leupeptin, Pepstatin, and Aprotinin and transfer to a 50 mL conical (1 L of cells per conical).
9. Freeze pellets at −80°C. Pellets can be stored at this moment or purification can be continued to next step.
10. Thaw bacterial suspension. Homogenize cells and sonicate as in steps #1 and #2 in Section 2.2.4. Centrifuge at 26,000 × *g* for 15 min and collect supernatant.
11. Load supernatant onto a GSTrap FF 5 mL column that was preequilibrated with GST Equilibration Buffer at 1 mL/min, wash at 5 mL/min with 20 column volumes of GST Wash Buffer, and elute at 1 mL/min with 10 m*M* glutathione in Equilibration Buffer.
12. Identify fractions that contain the eluted protein by 12% SDS-PAGE.
13. Pool fractions with the protein and cleave with PreScission Protease overnight at 4°C while dialyzing against Cation Exchange Loading Buffer.
14. Run 15% SDS-PAGE to assess cleavage reaction.
15. Clarify by centrifuging sample at 26,000 × *g* for 15 min.
16. Equilibrate a HiTrap SP FF 5 mL column with 10 column volumes of Cation Exchange Loading Buffer at a rate of 5 mL/min.
17. Load the cut protein at 1 mL/min and wash with 20 column volumes of Cation Exchange Loading buffer at 5 mL/min.

    GST is not binding column at this pH and is eluting in flow through.
18. Elute the protein with linear gradient going from 0% to 100% Cation Exchange Elution Buffer over 40 column volumes at 5 mL/min.
19. Check for protein separation and purity by 15% SDS-PAGE.
20. Combine fractions with pure CENP-C and concentrate using 10 kDa MWCO concentrators until the total volume <2 mL.
21. Equilibrate a High Load 16/600 Superdex S-200 PG column with 1.2 column volumes of gel filtration buffer at 1 mL/min.

22. Inject the protein sample and run the column at 1 mL/min collecting 2 mL fractions.
23. Identify the fractions with pure protein by 15% SDS-PAGE and combine them.
24. Concentrate protein to ∼4 mg/mL. The protein is stable at 4°C for several days. For longer storage add glycerol to 20%, aliquot and snap freeze in liquid N_2. Store at −80°C.

5.2 Determining Stoichiometry of CENP-A Nucleosome and CENP-C

Centromeric proteins could bind CENP-A nucleosome core particle in different sotichiometric ratios. Two molecules of CENP-C^{CD} (426–537) stably bind to one CENP-A nucleosome core particle (Falk et al., 2015), and we have used this ratio to generate sufficient amounts of the complex for biophysical experimentation.

5.2.1 Reagents

5% native PAGE, CENP-A nucleosomes, purified CENP-C^{CD} (426–537), CENP-C gel filtration buffer, EtBr, Coomassie Brilliant Blue, scalpel

5.2.2 Protocol

1. Calculate the moles of CENP-A present in 1 μg nucleosomes (varies with DNA template used). Calculate the amount of CENP-C to add at varying molar ratios to the CENP-A nucleosomes.

   One microgram of nucleosomes usually gives a strong signal for native PAGE and provides enough material for the subsequent step of the 2D analysis of the complex to measure the composition and stoichiometry of the nucleosomes using SDS-PAGE.
   (see step #5).
2. Aliquot calculated CENP-C amounts in eppendorf tubes. Adjust the volume in all tubes with CENP-C gel filtration buffer to equate to the one with the largest volume. In this way, the volume is standardized across the titration curve.
3. Add 1 × moles of CENP-A nucleosomes to each of the eppendorf tubes that already contain different amounts of CENP-C, gently mix and leave at room temperature for 20–30 min.

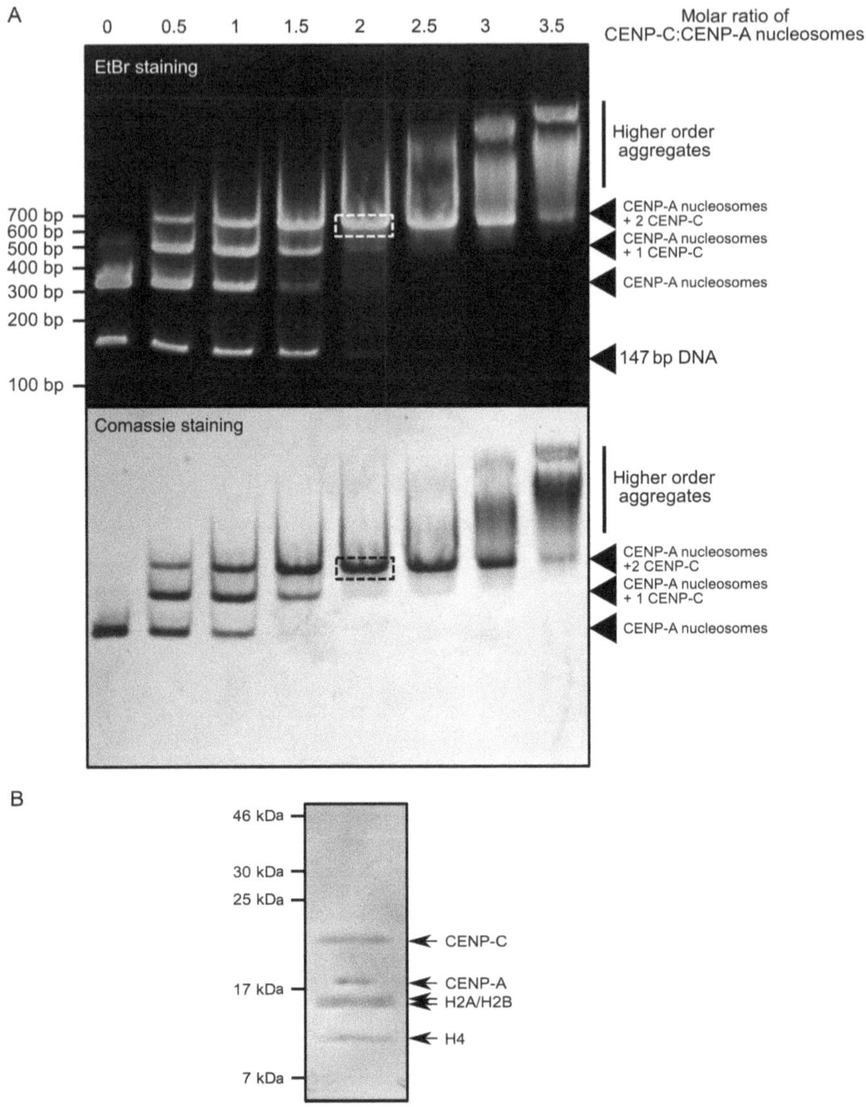

Fig. 7 Complex of CENP-C with CENP-A nucleosomes. (A) 5% native gel of titration of CENP-A nucleosomes with CENP-C with the indicated molar ratios. The same gel is shown when stained with EtBr (*upper part*) and with Coomassie stain (*lower part*). (B) 15% SDS-PAGE of single native PAGE band excised from the sample generated with a 2× molar ratio of CENP-C to CENP-A nucleosomes (labeled with *square* in the (A)).

4. Add 20% sucrose to final 5% (w/v) to each of the tubes and run the entire reaction for each titration step on a single lane of 5% native PAGE. Stain with EtBr followed by Coomassie Brilliant Blue as detailed earlier (Section 4.2; Fig. 7).

5. Excise bands containing both DNA and protein and mince them in very small pieces using scalpel, add 15 μL of 2 × SDS loading buffer, heat for 10 min at 100°C and load both supernatant and gel pieces on 15% SDS-PAGE.

   SDS gel will separate proteins based on MW, so components of the complex and their relative amounts can be identified. For instance, if the relative intensities of the bands are similar, then this indicates that there are two molecules of CENP-C per nucleosome (that itself has two molecules of each of the histones).
6. Once conditions that yield a desired stoichiometry are established, the reactions can be scaled up for biophysical studies.

   Shifting ratio slightly towards excess of CENP-C (CENP-C: nucleosome 2.2:1) can aid in saturation of the available binding sites on all nucleosomes present in the sample.

ACKNOWLEDGMENTS

We thank K. Luger (Univ. of Colorado), D. Cleveland (UCSD), and A. Straight (Stanford Univ.) for advice and plasmids and our former and present UPenn labmates D. Rogers, S. Wood-Scott, L. Guo, and S. Falk for contributing to the development of these methods. This work was supported by NIH grant GM082989 to B.E.B.

REFERENCES

Black, B. E., Foltz, D. R., Chakravarthy, S., Luger, K., Woods, V. L., & Cleveland, D. W. (2004). Structural determinants for generating centromeric chromatin. *Nature, 430*(6999), 578–582. http://doi.org/10.1038/nature02766.

Black, B. E., Brock, M. A., Bédard, S., Woods, V. L., & Cleveland, D. W. (2007). An epigenetic mark generated by the incorporation of CENP-A into centromeric nucleosomes. *Proceedings of the National Academy of Sciences of the United States of America, 104*(12), 5008–5013. http://doi.org/10.1073/pnas.0700390104.

Carroll, C. W., Milks, K. J., & Straight, A. F. (2010). Dual recognition of CENP-A nucleosomes is required for centromere assembly. *The Journal of Cell Biology, 189*(7), 1143–1155. http://doi.org/10.1083/jcb.201001013.

Carroll, C. W., Silva, M. C. C., Godek, K. M., Jansen, L. E. T., & Straight, A. F. (2009). Centromere assembly requires the direct recognition of CENP-A nucleosomes by CENP-N. *Nature Cell Biology, 11*(7), 896–902. http://doi.org/10.1038/ncb1899.

Dyer, P. N., Edayathumangalam, R. S., White, C. L., Bao, Y., Chakravarthy, S., Muthurajan, U. M., & Luger, K. (2004). Reconstitution of nucleosome core particles from recombinant histones and DNA. *Methods in Enzymology, 375*, 23–44.

Falk, S. J., Guo, L. Y., Sekulic, N., Smoak, E. M., Mani, T., Logsdon, G. A., … Black, B. E. (2015). Chromosomes. CENP-C reshapes and stabilizes CENP-A nucleosomes at the centromere. *Science (New York, N.Y.), 348*(6235), 699–703. http://doi.org/10.1126/science.1259308.

Falk, S. J., Lee, J., Sekulic, N., Sennett, M. A., Lee, T. H., & Black, B. E. (2016). CENP-C directs a structural transition of CENP-A nucleosomes mainly through sliding of DNA gyres. *Nature Structural & Molecular Biology, 23*(3), 204–208. http://doi.org/10.1038/nsmb.3175.

Guse, A., Carroll, C. W., Moree, B., Fuller, C. J., & Straight, A. F. (2011). In vitro centromere and kinetochore assembly on defined chromatin templates. *Nature*, *477*(7364), 354–358. http://doi.org/10.1038/nature10379.

Harp, J. M., Uberbacher, E. C., Roberson, A. E., Palmer, E. L., Gewiess, A., & Bunick, G. J. (1996). X-ray diffraction analysis of crystals containing twofold symmetric nucleosome core particles. *Acta Crystallographica. Section D, Biological Crystallography*, *52*(Pt. 2), 283–288. http://doi.org/10.1107/S0907444995009139.

Hasson, D., Panchenko, T., Salimian, K. J., Salman, M. U., Sekulic, N., Alonso, A., … Black, B. E. (2013). The octamer is the major form of CENP-A nucleosomes at human centromeres. *Nature Structural & Molecular Biology*, *20*(6), 687–695. http://doi.org/10.1038/nsmb.2562.

Hayes, J. J., & Lee, K. M. (1997). In vitro reconstitution and analysis of mononucleosomes containing defined DNAs and proteins. *Methods (San Diego, Calif.)*, *12*(1), 2–9. http://doi.org/10.1006/meth.1997.0441.

Kato, H., Jiang, J., Zhou, B.-R., Rozendaal, M., Feng, H., Ghirlando, R., … Bai, Y. (2013). A conserved mechanism for centromeric nucleosome recognition by centromere protein CENP-C. *Science (New York, N.Y.)*, *340*(6136), 1110–1113. http://doi.org/10.1126/science.1235532.

Lowary, P. T., & Widom, J. (1998). New DNA sequence rules for high affinity binding to histone octamer and sequence-directed nucleosome positioning. *Journal of Molecular Biology*, *276*(1), 19–42. http://doi.org/10.1006/jmbi.1997.1494.

Luger, K., Mäder, A. W., Richmond, R. K., Sargent, D. F., & Richmond, T. J. (1997). Crystal structure of the nucleosome core particle at 2.8 A resolution. *Nature*, *389*(6648), 251–260. http://doi.org/10.1038/38444.

Luger, K., Rechsteiner, T. J., & Richmond, T. J. (1999). Preparation of nucleosome core particle from recombinant histones. *Methods in Enzymology*, *304*, 3–19.

McKinley, K. L., & Cheeseman, I. M. (2016). The molecular basis for centromere identity and function. *Nature Reviews. Molecular Cell Biology*, *17*(1), 16–29. http://doi.org/10.1038/nrm.2015.5.

Sekulic, N., Bassett, E. A., Rogers, D. J., & Black, B. E. (2010). The structure of (CENP-A-H4)(2) reveals physical features that mark centromeres. *Nature*, *467*(7313), 347–351. http://doi.org/10.1038/nature09323.

Sekulic, N., & Black, B. E. (2012). Molecular underpinnings of centromere identity and maintenance. *Trends in Biochemical Sciences*, *37*(6), 220–229. http://doi.org/10.1016/j.tibs.2012.01.003.

Tachiwana, H., Kagawa, W., Shiga, T., Osakabe, A., Miya, Y., Saito, K., … Kurumizaka, H. (2011). Crystal structure of the human centromeric nucleosome containing CENP-A. *Nature*, *476*(7359), 232–235. http://doi.org/10.1038/nature10258.

Yang, T. P., Hansen, S. K., Oishi, K. K., Ryder, O. A., & Hamkalo, B. A. (1982). Characterization of a cloned repetitive DNA sequence concentrated on the human X chromosome. *Proceedings of the National Academy of Sciences of the United States of America*, *79*(21), 6593–6597.

CHAPTER FOUR

Functional Characterization of Histone Chaperones Using SNAP-Tag-Based Imaging to Assess De Novo Histone Deposition

C. Clément[*,†], I. Vassias[*,†], D. Ray-Gallet[*,†], G. Almouzni[*,†,1]

[*]Institut Curie, PSL Research University, CNRS, UMR3664, Equipe Labellisée Ligue contre le Cancer, Paris, France

[†]Sorbonne Universités, UPMC Univ Paris 06, CNRS, UMR3664, Paris, France

[1]Corresponding author: e-mail address: genevieve.almouzni@curie.fr

Contents

1. Introduction 98
2. The SNAP-Tag System to Visualize in vivo New Histone Incorporation into Chromatin: A De Novo Histone Deposition Assay in Cells 100
   2.1 Establishment of Cell Lines Expressing SNAP-Tagged Histones 100
   2.2 Protocol for the "Quench-Chase-Pulse" Assay 103
3. De Novo Histone Deposition Assay Coupled with Histone Chaperone Depletion and Labeling of Replicating Cells 106
4. De Novo Histone Deposition Assay Coupled with Histone Chaperone Depletion in Response to UV-DNA Damage 108
5. Imaging and Quantifying SNAP Labeling 111
6. Interpreting the De Novo Histone Deposition Assay Coupled with Histone Chaperone Depletion 113
7. Conclusion and Perspectives 113

Acknowledgments 115

References 115

Abstract

Histone chaperones—key actors in the dynamic organization of chromatin—interact with the various histone variants to ensure their transfer in and out of chromatin. In vitro chromatin assembly assays and isolation of protein complexes using tagged histone variants provided first clues concerning their binding specificities and mode of action. Here, we describe an in vivo method using SNAP-tag-based imaging to assess the de novo deposition of histones and the role of histone chaperones. This method exploits cells expressing SNAP-tagged histones combined with individual cell imaging

Methods in Enzymology, Volume 573
ISSN 0076-6879
http://dx.doi.org/10.1016/bs.mie.2016.04.004

to visualize directly de novo histone deposition in vivo. We show how, by combining this method with siRNA-based depletion, we could assess the function of two distinct histone chaperones. For this, we provide the details of the method as applied in two examples to characterize the function of the histone chaperones CAF-1 and HIRA. In both cases, we document the impact of their depletion on the de novo deposition of the histone variants H3.1 and H3.3, first in a normal context and second in response to DNA damage. We discuss how this cellular assay offers means to define in a systematic manner the function of any chosen chaperone with respect to the deposition of a given histone variant.

1. INTRODUCTION

Distinct regions of the genome are marked at the level of chromatin by specific histone variants and their modifications. Variants of histone H3, for example, include the replicative form H3.1, the replacement form H3.3, and the centromeric variant CenH3 (CENP-A) (Szenker, Ray-Gallet, & Almouzni, 2011). Despite their high sequence conservation—H3.1 and H3.3 differ by only five amino acids—these variants participate in marking specific regions of the genome, such as zones of active transcription, DNA repair, or the centromere (Talbert & Henikoff, 2010). Histone chaperones are key actors of histone variant management (De Koning, Corpet, Haber, & Almouzni, 2007; Gurard-Levin, Quivy, & Almouzni, 2014). They can direct histone incorporation into chromatin at the appropriate time in the cell cycle, for example, during DNA replication when a supply of new histones is required at the time of genome duplication, or during DNA damage. Providing the proper histone variant at the right place and the right time is of major importance for cellular fate and development (Santoro & Dulac, 2015; Szenker, Lacoste, & Almouzni, 2012). Indeed, improper histone variant distribution and unbalanced situations directly impact genome function and have been associated with several pathological situations (Filipescu, Muller, & Almouzni, 2014). Thus, characterizing the precise role of histone chaperones in the incorporation of histone variants into chromatin is of critical importance. To this goal, we have developed sensitive and specific assays to monitor histone variant deposition in vivo.

Following the discovery of the first histone chaperones (Laskey, Honda, Mills, & Finch, 1978; Smith & Stillman, 1989), in vitro chromatin assembly assays using crude cell extracts (Gaillard, Roche, & Almouzni, 1999; Ray-Gallet & Almouzni, 2004) revealed two distinct major pathways for histone deposition (Kaufman & Almouzni, 2006). These two pathways allow histone deposition either (i) coupled to DNA synthesis or (ii) independent of DNA synthesis. Cellular fractionation and complementation assays enabled to assign a key role to chromatin assembly factor 1 (CAF-1) in the DNA synthesis-coupled deposition (Stillman, 1986). Biochemical characterization of the complex identified three subunits, namely p48, p60, and p150 (Smith & Stillman, 1989). This complex is important not only during DNA replication but also during nucleotide excision repair (NER), a DNA repair pathway that requires DNA synthesis (Gaillard et al., 1996). The use of cell lines stably expressing epitope-tagged H3 variants combined with biochemical purification of these variants established that CAF-1 was selective for H3.1 (Tagami, Ray-Gallet, Almouzni, & Nakatani, 2004). To date, no other H3.1-dedicated chaperone has been described. Details concerning its function are provided in De Koning et al. (2007) and Gurard-Levin et al. (2014). Concerning DNA synthesis-independent histone deposition, the importance of the chaperone histone regulator A (HIRA) was first established using in vitro assays (Ray-Gallet et al., 2002). Given that chromatin disruption and dynamics are also important during transactions that do not involve DNA synthesis, for example during transcription, this pathway also engaged active research. Specifically, experiments in Xenopus egg extracts depleted for HIRA demonstrated that HIRA is critical for chromatin assembly in the absence of DNA synthesis. The HIRA complex—which contains HIRA, Ubinuclein 1 (UBN1), and calcineurin-binding protein 1 (CABIN1)—is one of the H3.3-selective histone chaperones (Tagami et al., 2004). Further details can be found in Gurard-Levin et al. (2014).

While these in vitro assays proved critical to define distinct nucleosome assembly pathways and highlight key histone chaperones (Ray-Gallet & Almouzni, 2004), further progress was hampered by the lack of method to assess the function of histone chaperones in vivo at a cellular level. The advent of the SNAP-tag system (Keppler et al., 2003) offered a unique tool to specifically visualize new histone variants and to monitor their de novo deposition at a cellular level (Jansen, Black, Foltz, & Cleveland, 2007; Ray-Gallet et al., 2011). By tagging individual histone variants, and following their fate within cells at specific times during the cell cycle or during particular genomic events, it provides a powerful method to characterize the function of histone chaperones in vivo. Importantly, with this technique, it is possible, with single-cell resolution, to detect variations within a heterogeneous cell population, variations that are otherwise averaged with more global approaches. Furthermore, at the subcellular level, it is possible to distinguish particular patterns for histone deposition within individual nuclei. In addition, when coupled with quantitative microscopy approaches to measure histone deposition in a population of cells, one can accurately scale various types of impact on histone deposition.

Here, we describe the use of the SNAP-tag system combined with siRNA depletion of individual histone chaperones to evaluate their role in the deposition of the histone variants H3.1 and H3.3. We document how this system was critical to unravel the in vivo role of histone chaperones in (i) the global deposition of histones (Ray-Gallet et al., 2011), (ii) cell cycle-dependent histone deposition (Ray-Gallet et al., 2011), and (iii) histone deposition in response to DNA damage (Adam, Polo, & Almouzni, 2013).

2. THE SNAP-TAG SYSTEM TO VISUALIZE IN VIVO NEW HISTONE INCORPORATION INTO CHROMATIN: A DE NOVO HISTONE DEPOSITION ASSAY IN CELLS

The SNAP-tag is a mutant of the human DNA repair protein O^6-alkylguanine-DNA alkyltransferase that specifically and covalently binds to benzylguanine, in an irreversible manner (Keppler et al., 2003). It can be fused to a protein of interest and this epitope-tagged protein can be expressed in cells. Importantly, cell-permeable fluorescent or nonfluorescent substrates (derivatives of benzylguanine) that react with the SNAP-tag are available (Fig. 1A and Table 1). This enables, in vivo, the specific visualization of newly synthesized proteins and the use in "quench-chase-pulse" assays (Fig. 1B) (Bodor, Rodriguez, Moreno, & Jansen, 2012): the "quench" step irreversibly labels all preexisting SNAP-tagged proteins with a nonfluorescent dye; the "chase" period allows for new SNAP-tagged proteins—which are not quenched—to be synthesized; and finally, the "pulse" step labels these new proteins with a fluorescent dye. To visualize newly synthesized molecules specifically, we performed this "quench-chase-pulse" assay using cell lines expressing SNAP-tagged H3.1 or H3.3. To concentrate on new histones that are chromatin-incorporated, a further step has to be performed via a triton extraction enabling the elimination of soluble histones.

2.1 Establishment of Cell Lines Expressing SNAP-Tagged Histones

The first SNAP-tagged cell line was devised to visualize the deposition of the centromeric histone variant, CENP-A (Jansen et al., 2007). To establish HeLa cell lines stably expressing either H3.1 or H3.3 tagged with one

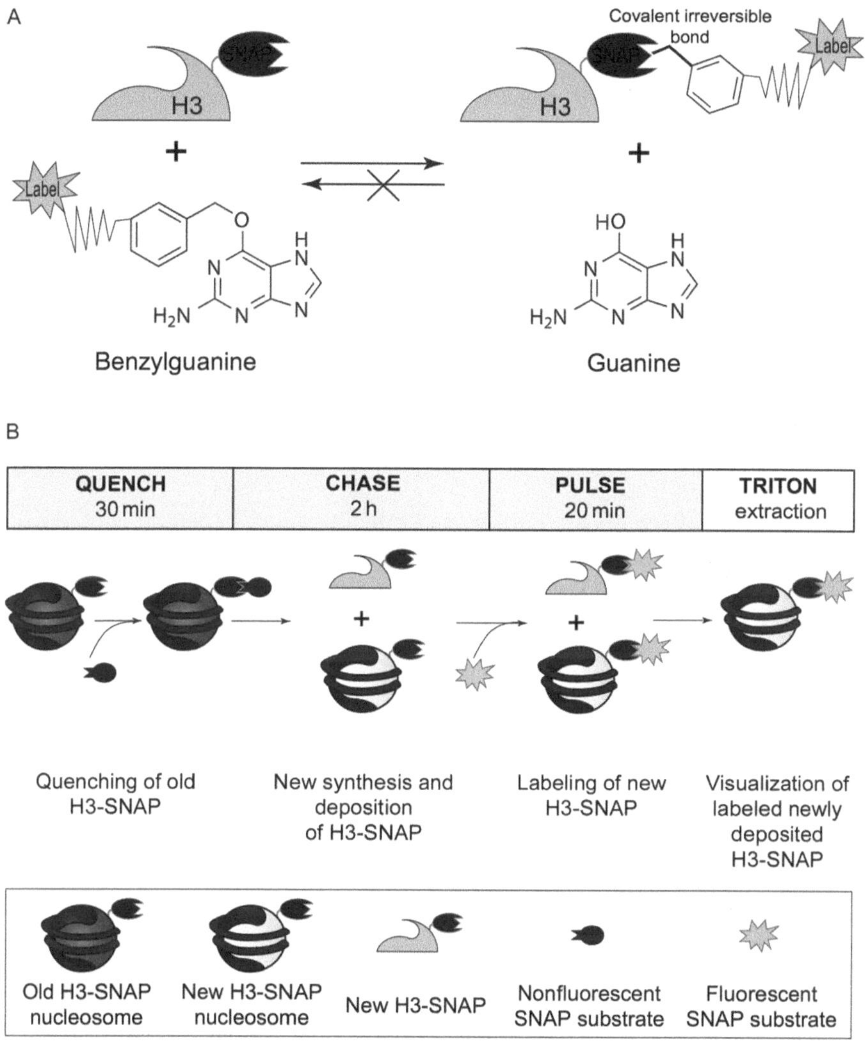

Fig. 1 SNAP-based labeling principle and de novo histone H3 deposition assay. (A) Histone H3 is fused to the SNAP-tag, a 20-kDa mutant of the DNA repair protein O^6-alkylguanine-DNA alkyltransferase that reacts specifically and rapidly with benzylguanine derivatives, leading to irreversible covalent labeling of the SNAP-tag with a synthetic probe. Many SNAP-tag substrates are cell permeable, permitting labeling of histone H3 in live cells (see Table 1). (B) Our de novo histone H3 deposition assay in stably expressing H3.1- or H3.3-SNAP cells is based on a "quench-chase-pulse" experiment. The "quench" step labels old (preexisting) H3-SNAP with a nonfluorescent SNAP-tag substrate, the "chase" step allows the synthesis and deposition of new H3-SNAP, and the "pulse" step labels only the new H3-SNAP with a red fluorescent SNAP-tag substrate (TMR). A triton extraction before fixation of the cells eliminates soluble histones in order to enrich for the nonsoluble labeled H3-SNAP that contain newly deposited histones stably bound to chromatin.

Table 1 List of Commercially Available Cell-Permeable Fluorescent Substrates Used to Label SNAP-Tag Fusion Proteins Inside Living Cells (Purchased from New England Biolabs)

	Excitation (nm)	Emission (nm)
SNAP-Cell® 360	357	437
SNAP-Cell® 505-Star	504	532
SNAP-Cell® Fluorescein	500	532
SNAP-Cell® Oregon Green®	490	514
SNAP-Cell® TMR-Star	554	580
SNAP-Cell® 430	421	444–485
SNAP-Cell® 647-SiR	645	661

SNAP-tag and three Human influenza hemagglutinin (HA) epitopes, derivatives of this approach have been used (Jansen et al., 2007; Ray-Gallet et al., 2011). Indeed, the H3.1 or H3.3 open-reading frames (ORFs) amplified from complementary DNA by PCR were cloned between the KpnI/XhoI sites of pCENP-A-SNAP-3XHA (Jansen et al., 2007), replacing the CENP-A gene. The resulting H3.1-SNAP-3xHA or H3.3-SNAP-3xHA ORFs were subcloned into pBABE and used for retroviral production and delivery into HeLa cells as described previously (Shah et al., 2004). After selection with Blasticidin S (5 mg/mL, Calbiochem), monoclonal lines were isolated by flow cytometry. Most importantly for our assays, we verified that the selected clones showed low expression levels of tagged H3.1 and H3.3 as compared to the endogenous counterparts to be able to use them as tracers (Ray-Gallet et al., 2011). Considering that the presence of the tag could potentially interfere with the functionality of the protein and impact cell properties, we also verified that they did not significantly alter cell cycle progression and general cellular properties. In addition, we also generated independently two other stable U2OS cell lines expressing H3.1-SNAP or H3.3-SNAP by cloning the genes into a pSNAPm vector (N9172S, NEB), transfection and G418 selection (400 μg/mL) of single-cell clones (Dunleavy, Almouzni, & Karpen, 2011). Considering recent advances in genome editing, the use of technologies such as the CRISPR/Cas9 system is now an attractive option that we would recommend to tag the endogenous proteins of interest directly at their natural site. The use of distinct cell lines and independent methods can be most useful to verify the generality of the properties observed in a particular context.

2.2 Protocol for the "Quench-Chase-Pulse" Assay

2.2.1 Cell Culture Conditions

Grow cells expressing the SNAP-tagged histone on glass coverslips (12 mm diameter, 0.17 mm thickness) in Dulbecco's modified Eagle's medium, 10% newborn calf serum, penicillin, and streptomycin (complete medium). Incubate cells in a humidified 37°C incubator with 5% CO_2 until they reach the desired confluence (70–80%). Labeling and washing volumes are given for one coverslip in a 24-well plate.

2.2.2 Quench

- Replace the medium with 200 μL of complete medium with 10 μ*M* SNAP-Cell Block (S9106S, New England Biolabs)
- Incubate at 37°C, 5% CO_2 for 30 min
- Wash cells twice with 1 mL of PBS
- Replace PBS with 1 mL of complete medium and incubate at 37°C, 5% CO_2 for 30 min to allow the excess SNAP-Cell Block to diffuse out
- Wash cells twice with 1 mL of PBS

2.2.3 Chase

- Replace PBS with 1 mL of complete medium
- Incubate at 37°C, 5% CO_2 for an appropriate time of "chase," to allow new SNAP-tagged histones to be synthesized and incorporated into chromatin. We typically use a "chase" period of 2 h

2.2.4 Pulse

- Replace medium with 200 μL of complete medium with 2 μ*M* fluorescent SNAP-Cell TMR-Star (tetramethylrhodamine, "TMR" in Figs. 2–4) (S9105S, New England Biolabs). Note that other SNAP-tag fluorescent substrates are available (see Table 1)
- Incubate at 37°C, 5% CO_2 for 20 min
- Wash cells twice with 1 mL of PBS
- Replace PBS with 1 mL of complete medium
- Incubate at 37°C, 5% CO_2 for 30 min, to allow the excess SNAP-Cell TMR-Star to diffuse out
- Wash cells twice with 1 mL of PBS

2.2.5 Triton Extraction

In this step, the goal is to try and eliminate soluble proteins and enrich for chromatin-incorporated histones

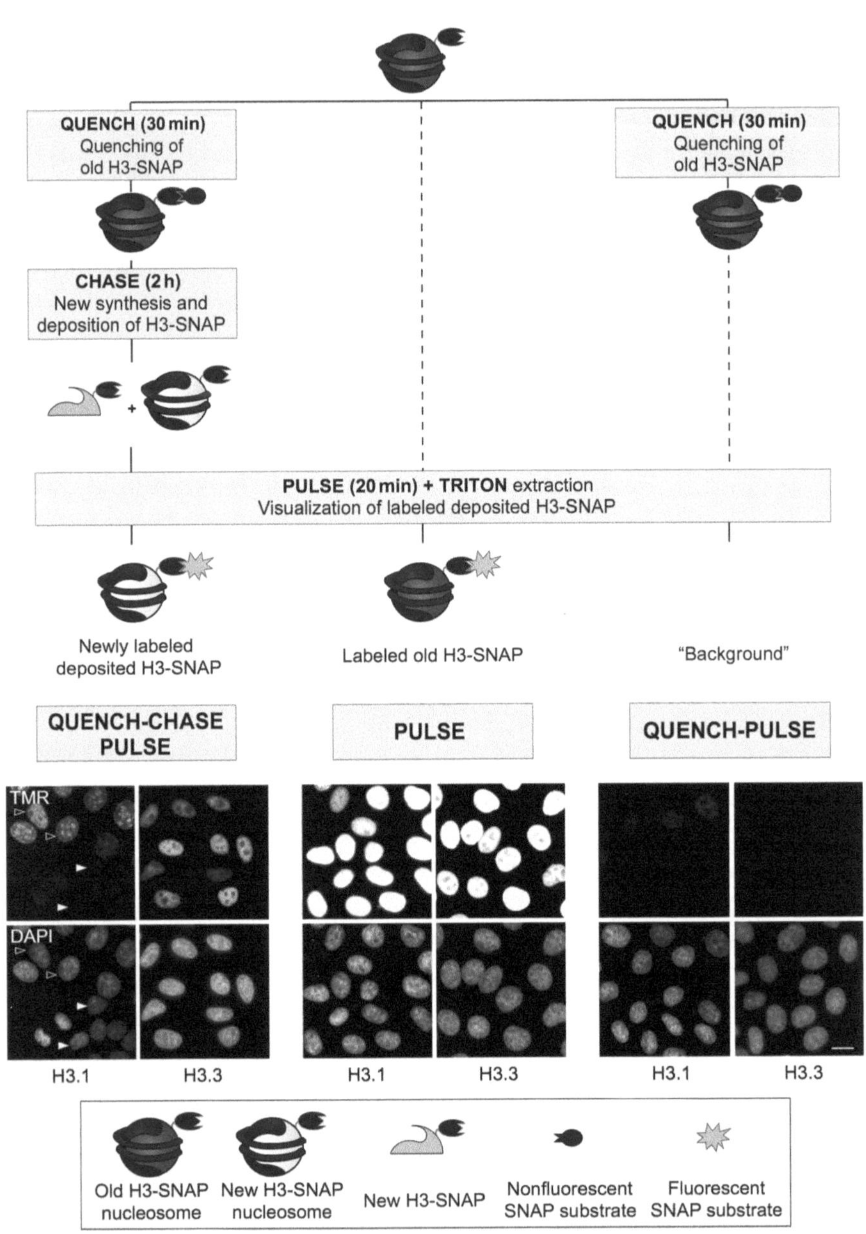

Fig. 2 Validation of the de novo histone H3 deposition assay. Schemes and cell images for the de novo histone H3 deposition assay "quench-chase-pulse" and two control experiments "pulse" and "quench-pulse" in HeLa cell lines stably expressing H3.1- or H3.3-SNAP. The "quench-chase-pulse" assay visualizes the newly deposited histones H3-SNAP (TMR): while all the cells are labeled in the H3.3-SNAP cell line, only the cells that are in S phase are fluorescent in the H3.1-SNAP cell line (see *arrows* and Fig. 3.

- Wash cells with 1 mL of cytoskeleton (CSK) buffer (10 m*M* Pipes-KOH, pH 7.0, 100 m*M* NaCl, 300 m*M* sucrose, 3 m*M* $MgCl_2$)
- Replace the buffer with 200 μL of CSK with 0.5% Triton
- Incubate at room temperature for 5 min
- Wash cells twice gently with 1 mL of CSK (the second wash should be 5 min)
- Fix cells in 2% paraformaldehyde at room temperature for 20 min
- Wash cells three times with 1 mL of PBS

2.2.6 Cell Imaging

Prepare coverslips on slides with mounting medium containing DAPI (4′,6-diaminido-2-phenylindole) or process for immunostaining beforehand. Visualize cells by microscopy and acquire images for SNAP-labeling quantification (see later).

2.2.7 Controls in the "Quench-Chase-Pulse" Assay

Two other combinations of labeling are necessary along with the "quench-chase-pulse" (Fig. 2): a "pulse" to label all preexisting histones, and a "quench-pulse" with a "quench" immediately followed by a "pulse" to monitor the efficiency of the quenching under our experimental conditions. As expected, the amount of newly synthesized and deposited histones observed after a "quench-chase-pulse" is lower than the amount of all preexisting histones after a "pulse" (Fig. 2, bottom). The low signal in the "quench-pulse" condition confirms the efficiency of the "quench" step and thus defines a minimal background signal for our assays (Fig. 2, bottom). Therefore, with the SNAP-tag system, under our experimental conditions, we have at hand a useful tool to study new histone variant deposition. Using cell lines expressing H3.1- and H3.3-SNAP, described earlier, and combining the "quench-chase-pulse" assay with siRNA depletion of histone

---

Dark arrows indicate cells in S phase, white arrows indicate cells that are not in S phase). The "pulse" assay visualizes the old (preexisting) H3-SNAP (TMR): all the cells in both H3.1 and H3.3-SNAP cell lines are labeled and show a strong fluorescence signal. The "quench-pulse" assay with a "pulse" just after the "quench" step (without a chase period) allows to control the efficiency of the quenching: no cells should be labeled (TMR) as shown in the images for both H3.1- and H3.3-SNAP cell lines. Note that the image acquisition time is the same for the three assays, explaining the overexposed image for the TMR staining in the "pulse" as compared to the "quench-chase-pulse." Scale bar, 10 μm. *Adapted from Ray-Gallet, D., Woolfe, A., Vassias, I., Pellentz, C., Lacoste, N., Puri, A., … Almouzni, G. (2011). Dynamics of histone H3 deposition in vivo reveal a nucleosome gap-filling mechanism for H3.3 to maintain chromatin integrity.* Molecular Cell, *44(6), 928–941.*

chaperones, we investigated the role of several histone chaperones in the de novo deposition of these two variants (Ray-Gallet et al., 2011).

3. DE NOVO HISTONE DEPOSITION ASSAY COUPLED WITH HISTONE CHAPERONE DEPLETION AND LABELING OF REPLICATING CELLS

Based on previous work on the histone variants H3.1 and H3.3, we further investigated the role of the histone chaperone CAF-1 on histone H3 deposition in vivo. To assess how the depletion of CAF-1 p60 by siRNA would affect H3.1/H3.3 de novo deposition, we used a modified "quench-chase-pulse" experiment in H3.1- and H3.3-SNAP HeLa cells (Fig. 3A). We performed siRNA transfections using Lipofectamine RNAimax (13778-150, Invitrogen) 48 h prior to the "quench-chase-pulse." To determine whether H3.1- and H3.3-SNAP de novo depositions are cell cycle-specific, S phase cells were labeled in the presence of 10 μM of EdU (5-ethynyl-2′-deoxyuridine) in the medium during the "pulse" step. Colabeling of SNAP-tagged histones (with TMR for example) and EdU (with Click chemistry, for example (C10337, Invitrogen)) allows the visualization of new H3.1- and H3.3-SNAP in cells undergoing replication. In Fig. 3B, new H3.1 was only deposited in replicating cells (EdU positive), at replication sites as shown by the strong colocalization between the TMR signal and the EdU staining, while non replicating cells (EdU negative) showed no labeling. In contrast, all cells displayed TMR signal for H3.3 without any colocalization between H3.3 and replication foci, indicating that this variant is deposited throughout the whole cell cycle with a broad distribution. This may seem in contrast with data from ChIP experiments that show enrichment at gene bodies, promoters and regulatory elements (Goldberg et al., 2010; Ray-Gallet et al., 2011), but can be explained if one considers that incorporation of H3.3 is broad and random at sites of unexpected dynamics, while systematic at sites of transcription. Since ChIP data obtained on populations of cells provide an average, they will only reveal changes that occur at the same place in all cells. This highlights the complementary nature of the two techniques. Notably, upon depletion of CAF-1 p60, TMR signal for H3.1 was reduced by 50%, whereas TMR signal for H3.3 was not affected. These findings establish that de novo deposition of the variant H3.1, but not H3.3, occurs at replication sites in S phase and is mediated by the histone chaperone CAF-1 in vivo. We also showed that depletion of the histone chaperone HIRA—and not CAF-1—affects de

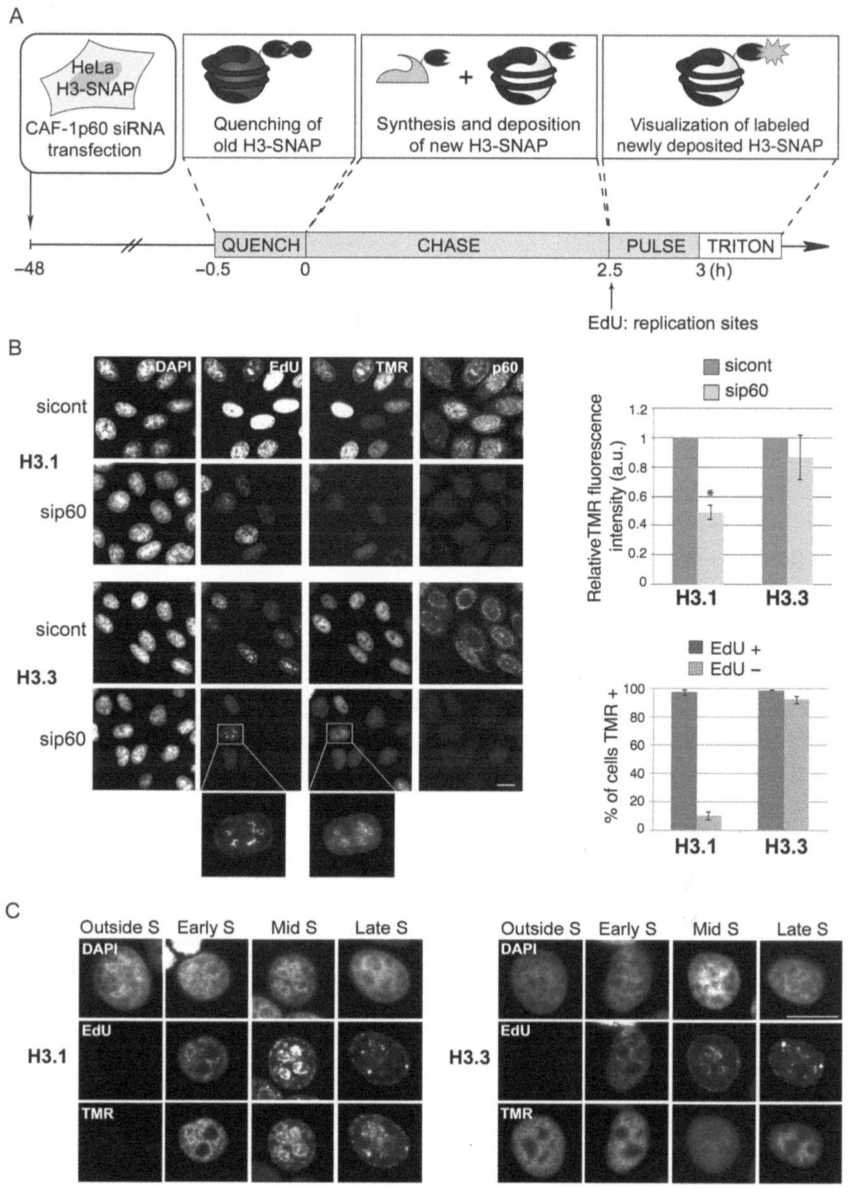

Fig. 3 De novo histone H3 deposition assay coupled with histone chaperone CAF-1 depletion and EdU staining. (A) Scheme of the de novo histone H3 deposition assay coupled with histone chaperone depletion and replication sites labeling in H3.1- and H3.3-SNAP HeLa cells. The cells are transfected with CAF-1 p60 or control siRNA before labeling of new H3-SNAP with TMR (by the "quench-chase-pulse" experiment) and replication sites with EdU. (B) (*Left*) Fluorescent microscopy visualization of new deposited

(Continued)

novo deposition of the variant H3.3 (Ray-Gallet et al., 2011). Importantly, the EdU staining reveals specific S phase patterns that can be analyzed individually (early, mid and late S phase are characteristic) (Fig. 3C). Thus, one can also refine the analysis at the level of individual nuclei by studying particular patterns. Furthermore, this assay revealed that, upon CAF-1 depletion, about 20% of cells displayed HIRA-dependent deposition of H3.3 at replication sites, suggesting a compensatory mechanism. This uncovers an alternative mode of HIRA-dependent H3.3 deposition at any region where nonnucleosomal/naked DNA is accessible, that could differ from one cell to the next and thus would not have been picked up in analyzing genome-wide H3.3 enrichment profiles. This alternative mode would function as a nucleosome gap-filling mechanism compensating for the decreased H3.1 deposition. This study advanced our understanding of the role of histone chaperones in the de novo deposition of specific histone variants in vivo during the cell cycle under normal conditions. We next wished to investigate histone dynamics after DNA damage and determine which histone chaperones could be involved and when.

4. DE NOVO HISTONE DEPOSITION ASSAY COUPLED WITH HISTONE CHAPERONE DEPLETION IN RESPONSE TO UV-DNA DAMAGE

Repair of UV damage in DNA entails specific histone dynamics in the context of chromatin (Polo & Almouzni, 2015). These can be depicted in

Fig. 3—Cont'd H3.1- or H3.3-SNAP (TMR) and replication sites (EdU) after siRNA transfection against CAF-1 p60 or control. CAF-1 p60 is detected by immunofluorescence. The insets represent enlarged images of one selected cell. Scale bar, 10 μm. (*Right*) The upper graph shows the sip60/sicont ratio corresponding to TMR fluorescence intensity for new deposited H3.1- and H3.3-SNAP. The lower graph represents the percentage of TMR-labeled H3.1- and H3.3-SNAP cells negative (EdU −) or positive (EdU +) for EdU staining. The statistical significance of the difference between siRNA and sicont—as represented by the "*" when the difference is significant—was analyzed using a Mann and Whitney test. Error bars indicate standard deviation in at least three experiments. (C) Fluorescent microscopy visualization of the different patterns of replication sites (EdU) during early, mid, or late S phase and of new incorporated H3.1 or H3.3 (TMR) after in vivo labeling by a "quench-chase-pulse" experiment. *Adapted from Ray-Gallet, D., Woolfe, A., Vassias, I., Pellentz, C., Lacoste, N., Puri, A., … Almouzni, G. (2011). Dynamics of histone H3 deposition in vivo reveal a nucleosome gap-filling mechanism for H3.3 to maintain chromatin integrity.* Molecular Cell, *44(6), 928–941.*

the "prime-repair-restore" model whereby, first, very early and at the site of the lesion, particular chromatin dynamics are critical to mark where the repair process can take place. Then, rearrangements enable the repair machinery to gain access to DNA lesions. After removal of the damage, further histone deposition occurs to fully restore chromatin organization. We used the "quench-chase-pulse" assay with the H3.3-SNAP U2OS cell line to assess the dynamics of this variant when inflicting local UV damage with UVC (see later). This assay, coupled with histone chaperone depletion, enabled us to identify the HIRA-mediated deposition of H3.3 as an early step to prime chromatin (Adam et al., 2013). This is critical to enable transcription recovery after repair is complete.

The protocol is essentially as described earlier, with the exception of the addition of a local UVC irradiation after the "chase" period. We irradiate the cells through micropore filters to induce DNA damage at discrete regions in the nucleus. A schematic representation of this assay is presented in Fig. 4A. The UVC irradiation protocol, under conditions described later, produces predominantly pyrimidine dimers that are repaired by the NER pathway as verified in Adam et al. (2013).

Protocol for UVC Irradiation

- Use a UVC lamp at 254 nm and monitor the fluence rate with a VLX-3W dosimeter (both from Vilber Lourmat)
- After the "chase" step, transfer one coverslip to a 6-cm plate without medium
- Cover the coverslip by a microfilter (isopore membrane filter with 5 m pores from Millipore)
- Irradiate the cells. We usually use a UV dose of 150 J/m^2
- Remove the filter by adding PBS into the plate
- Transfer back the coverslip into the plate containing the complete culture medium
- Perform the "pulse," triton extraction and fixation steps as described above
- Perform an immunodetection of the repair factor XPA (Xeroderma Pigmentosum, complementation group A) to visualize the sites of DNA damage when imaging cells on the microscope

For more details of the "quench-chase-pulse" assay coupled with local UV irradiation, see Adam, Dabin, Bai, and Polo (2015).

This assay combined with depletion of the HIRA and UBN1 subunits enabled us to assess the role of the HIRA complex in chromatin dynamics

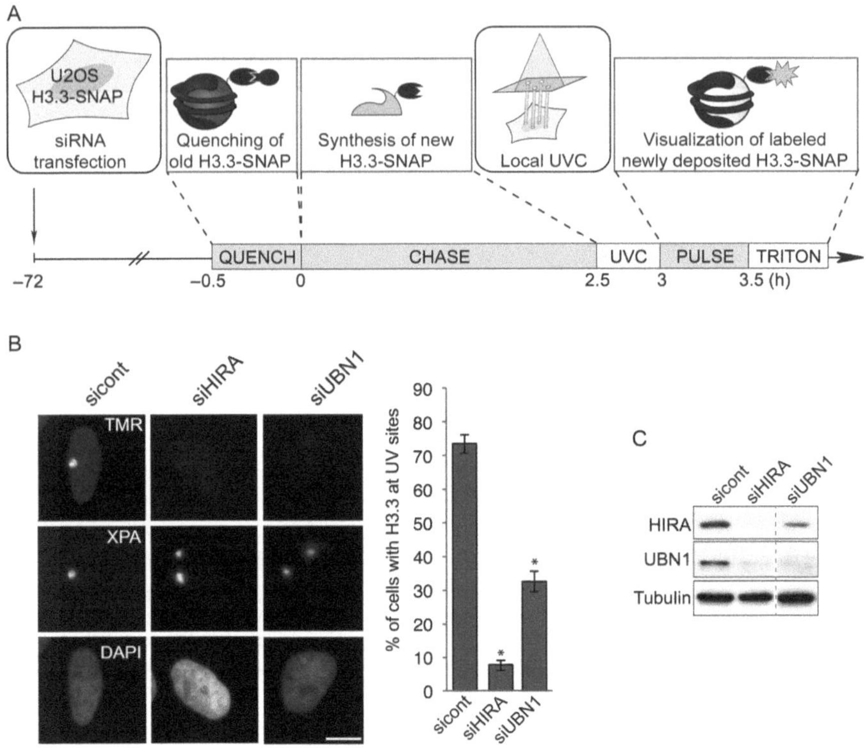

Fig. 4 De novo histone H3.3 deposition assay coupled with histone chaperone depletion in response to DNA damage. (A) Scheme of the de novo histone deposition assay in response to UVC irradiation coupled with histone chaperone depletion for monitoring accumulation of newly synthesized H3.3-SNAP at damage sites in H3.3-SNAP U2OS cells. Cells are transfected with histone chaperone or control siRNA before local UV irradiation and labeling of new H3.3-SNAP with TMR (by the "quench-chase-pulse" experiment). (B) (*Left*) Fluorescent microscopy visualization of newly deposited H3.3-SNAP (TMR) accumulating at UV-damaged sites after siRNA transfection against HIRA, UBN1, or control. Immunofluorescence labeling of XPA repair protein allows the visualization of damaged sites. Scale bar, 10 μm. (*Right*) The graph represents the percentage of cells with accumulation of new deposited H3.3-SNAP (TMR) at UV sites. The statistical significance of the difference between siRNA and sicont—as represented by the "*" when the difference is significant—was analyzed using *t*-test or one-way ANOVA followed by Tukey posttests performed with Prism 5 software. Error bars indicate standard deviation in at least three experiments. (C) SiRNA efficiencies are controlled on the Western blots. *Adapted from Adam, S., Polo, S. E., & Almouzni, G. (2013). Transcription recovery after DNA damage requires chromatin priming by the H3.3 histone chaperone HIRA.* Cell, 155*(1), 94–106.*

following DNA damage. We carried out the experiments in U2OS cells stably expressing the histone variant H3.3-SNAP. In control cells, H3.3 accumulated at the site of local DNA damage, as defined by the early repair factor XPA immunodetection. In contrast, in cells treated with HIRA and UBN1 siRNA, H3.3 failed to accumulate at XPA-positive sites (Fig. 4B). The depletion efficiencies verified by Western blot for both the HIRA and UBN1 subunits showed that the siRNA had effectively functioned (Fig. 4C). To quantify the defects in histone deposition, we determined the percentage of cells with new H3.3 at UV damage sites in control, HIRA, or UBN1 siRNA treated cells. This percentage decreased in a statistically significant manner (see later for the statistical analysis) for both HIRA and UBN1 siRNA treated cells. The most dramatic effect occurred upon depletion of the HIRA subunit (Fig. 4B). This is consistent with the central role of the HIRA subunit in maintaining the stability of the complex, as shown by the reduced level of UBN1 in HIRA-depleted cells. From the data obtained in this assay, we could conclude that the histone chaperone HIRA was essential for the deposition of the H3.3 histone variant in response to UVC irradiation (Adam et al., 2013).

5. IMAGING AND QUANTIFYING SNAP LABELING

For fluorescence image acquisition, we recommend using an epifluorescence microscope equipped with a 63× objective (we currently use a AxioImager Zeiss Z1 microscope). For consistency, it is essential to carry out image acquisition with identical parameters for all different experimental conditions and to set exposure times carefully for the entire duration of the experiment to avoid camera sensor saturation. For a first approach, 2D images are usually sufficient and recommended to help avoid bleaching. Adaptations are however possible for 3D, but would require specific settings, not detailed here.

Since TMR labeling usually gives rise to higher fluorescence background compared to classical antibody detection, we recommend extensive washing of the fluorescent substrate after the "pulse" step as mentioned above (after two quick washes, an extended wash for 30 min at 37°C 5% CO_2 is followed by two additional quick washes). TMR background level is calculated for all experiments and subtracted from all acquisitions to properly quantify specific fluorescent labeling. For each TMR image, this is achieved using the ImageJ software, by measuring the mean gray value from an area containing no nucleus in the field. We define this value as the background and subtract it from the entire image. We then perform our quantitative analysis with these background-subtracted images.

To quantify TMR fluorescence intensity in the acquired images, first, the DAPI staining allows the detection of all nuclei and the generation of a mask corresponding to the nuclei. This mask is applied to the TMR fluorescent images, enabling the evaluation of the average TMR fluorescence intensity—represented by the mean gray value (gray value to area ratio)—in each nucleus. This method can easily be automated in ImageJ by a macro allowing quantification of hundreds of nuclei in a fast and effective way. For each condition, we analyze at least a hundred nuclei. We calculate the average TMR fluorescence intensity for each condition and normalize to the control condition. We perform at least three independent experiments for each condition and standard statistical tests (Mann and Whitney test, t.test, or one-way ANOVA followed by Tukey posttests performed with Prism 5 software) are applied to the data.

For H3.3-SNAP labeling, we evaluate the mean gray value in all nuclei, as more than 95% of cells are TMR positive in "quench-chase-pulse" experiments (Ray-Gallet et al., 2011). Moreover, the pattern of labeling is rather homogenous all over the nuclei, with the exception of nucleoli, which display lower signal. In contrast, for H3.1-SNAP labeling, only replicating cells are labeled with newly incorporated H3.1 (Ray-Gallet et al., 2011). Therefore, we costain cells with EdU to detect cells in S phase. After quantification of TMR and EdU fluorescence, we select EdU positive cells and determine TMR fluorescence levels only in these cells. Thus, more cells are needed initially to reach the threshold of a hundred positive cells. Other cell cycle markers can be used effectively (Boyarchuk, Filipescu, Vassias, Cantaloube, & Almouzni, 2014).

To quantify variation in de novo histone deposition in response to local UVC damage, we use a different approach. Instead of exploring the impact of the chaperone depletion on normal incorporation, we count the number of cells with an accumulation of H3.3-SNAP labeling at damaged sites (assessed by XPA immunodetection) (Fig. 4B). Fluorescence quantification is also possible in this case, using XPA staining as a mask instead of the DAPI staining. For example, this can be done to evaluate more subtle impact on the de novo histone deposition coupled with DNA repair.

6. INTERPRETING THE DE NOVO HISTONE DEPOSITION ASSAY COUPLED WITH HISTONE CHAPERONE DEPLETION

Drawing conclusions based on this approach requires careful evaluation of the following points, especially to ascertain at which level the impact occurred. It is critical to assess whether any effect occurred at the expression level of the SNAP-tagged histone variant. Indeed, a decrease or an increase of the soluble pool of the SNAP-tagged histone variant, by mass action, could ultimately impinge on the efficiency of its de novo deposition. Thus, we systematically monitor variations in histone pools in soluble, cytosolic, and nuclear extracts from both control and chaperone-depleted cells (Martini, Roche, Marheineke, Verreault, & Almouzni, 1998). By Western blot, we carefully evaluate if any significant changes in the soluble pool of SNAP-tagged histones have occurred. It is also important to assess whether the cell cycle or a chromatin-dependent process such as transcription or replication have been affected. Indeed, changes in these processes could potentially impact histone deposition without interfering directly with the incorporation mechanism. Thus, it is essential, when interpreting the results of the "quench-chase-pulse" assay, to consider not only the deposition event in itself, but the whole assembly line from histone synthesis, to incorporation into chromatin, eviction, and degradation, since all these aspects could potentially influence the results.

In addition to the global fluorescence quantification, analysis of the specific labeling pattern of the newly deposited histone variants is also very informative. In particular, the depletion may affect the site of the histone variant deposition, or the time, without necessarily affecting significantly its efficiency. The method can be further extended to investigate the de novo deposition of a histone variant in specific genomic regions. In this case, specific markers (using antibody recognizing a protein specific to the regions, or DNA FISH, or TALEN, etc.) to covisualize the regions of interest will be needed along with the visualization of new histones.

7. CONCLUSION AND PERSPECTIVES

The approach using the SNAP-tag system, pioneered for centromeric variant CENP-A incorporation in late mitosis (Jansen et al., 2007), proved to be instrumental to define the role of the CENP-A dedicated chaperone

HJURP (Dunleavy et al., 2009; Foltz et al., 2009) at this particular time of the cell cycle (Muller et al., 2014; Tachiwana et al., 2015). Here, we describe how it can be extended as a powerful assay to systematically test the contribution of histone chaperones to the de novo deposition of histone variants throughout the different phases of the cell cycle. This can also be achieved after inducing a specific local perturbation in the genome such as DNA damage. We illustrated the two cases with the dynamics of the histone variants H3.1 and H3.3. Importantly, beyond H3.1 and H3.3, we have also used this system effectively for the variants H2A and H2A.Z to explore their distinct dynamics, at euchromatic and heterochromatic loci such as pericentric heterochromatin (Boyarchuk et al., 2014). While these studies used the SNAP-tag system, the CLIP-tag—a modified version of the SNAP-tag—offers an interesting alternative and can be combined with the SNAP-tag system to allow the labeling of two different SNAP and CLIP fusion proteins simultaneously (Gautier et al., 2008). This can be most useful to compare the behavior of distinct tagged proteins and the relative impact of interference with one chaperone or another.

To date, most studies have focused on de novo deposition of histone variants. The system offers interesting avenues to explore the fate of old histones and their recycling during genomic transactions such as DNA replication, repair, or transcription, in order to elucidate, in each case, the exact role of histone chaperones in vivo (Probst, Dunleavy, & Almouzni, 2009). These approaches could help to follow both new and old material, which can be critical to determine how parental information is passed during replication or cell division. How these events function within a given cell lineage to ensure propagation of a given state, and how they may be altered during asymmetric division during a planned differentiation program (Filipescu et al., 2014) is a fascinating question. Considering the recent data highlighting the importance of CAF-1 and its role as a barrier to cell reprogramming (Cheloufi et al., 2015; Hatanaka et al., 2015; Ishiuchi et al., 2015), it is tempting to further explore how this relates to changes in histone variant dynamics. Perturbing histone chaperones—and through them the tight coordination of histone recycling and de novo deposition—provides a unique perspective to address it. In vivo monitoring of old and new histones using the SNAP-tag system, combined with advances in microscopy, along with new genome editing techniques to knockout the chaperones, should further our knowledge of the mechanisms by which the histone chaperone network mediates chromatin assembly at different stages of cellular life.

ACKNOWLEDGMENTS

We are grateful to Jean Gautier and Lisa Prendergast for critical reading of the manuscript. This work was supported by la Ligue Nationale contre le Cancer (Equipe labellisée Ligue), the European Commission Network of Excellence EpiGeneSys (HEALTH-F4-2010-257082), ERC Advanced Grant 2009-AdG_20090506 "Eccentric," the European Commission large-scale integrating project FP7_HEALTH-2010-259743 "MODHEP," ANR-11-LABX-0044_DEEP and ANR-10-IDEX-0001-02 PSL, ANR "CHAPINHIB" ANR-12-BSV5-0022-02, ANR "Epicure" ANR-14-CE16-0009, ANR "CELLECTCHIP" ANR-14-CE10-0013, and Aviesan-ITMO cancer project "Epigenomics of breast cancer."

REFERENCES

Adam, S., Dabin, J., Bai, S. K., & Polo, S. E. (2015). Imaging local deposition of newly synthesized histones in UVC-damaged chromatin. *Methods in Molecular Biology*, *1288*, 337–347.

Adam, S., Polo, S. E., & Almouzni, G. (2013). Transcription recovery after DNA damage requires chromatin priming by the H3.3 histone chaperone HIRA. *Cell*, *155*(1), 94–106.

Bodor, D. L., Rodriguez, M. G., Moreno, N., & Jansen, L. E. (2012). Analysis of protein turnover by quantitative SNAP-based pulse-chase imaging. *Current Protocols in Cell Biology*. Chapter 8, Unit 8.8.

Boyarchuk, E., Filipescu, D., Vassias, I., Cantaloube, S., & Almouzni, G. (2014). The histone variant composition of centromeres is controlled by the pericentric heterochromatin state during the cell cycle. *Journal of Cell Science*, *127*(Pt. 15), 3347–3359.

Cheloufi, S., Elling, U., Hopfgartner, B., Jung, Y. L., Murn, J., Ninova, M., … Hochedlinger, K. (2015). The histone chaperone CAF-1 safeguards somatic cell identity. *Nature*, *528*, 218–224.

De Koning, L., Corpet, A., Haber, J. E., & Almouzni, G. (2007). Histone chaperones: An escort network regulating histone traffic. *Nature Structural & Molecular Biology*, *14*(11), 997–1007.

Dunleavy, E. M., Almouzni, G., & Karpen, G. H. (2011). H3.3 is deposited at centromeres in S phase as a placeholder for newly assembled CENP-A in G(1) phase. *Nucleus*, *2*(2), 146–157.

Dunleavy, E. M., Roche, D., Tagami, H., Lacoste, N., Ray-Gallet, D., Nakamura, Y., … Almouzni-Pettinotti, G. (2009). HJURP is a cell-cycle-dependent maintenance and deposition factor of CENP-A at centromeres. *Cell*, *137*(3), 485–497.

Filipescu, D., Muller, S., & Almouzni, G. (2014). Histone H3 variants and their chaperones during development and disease: Contributing to epigenetic control. *Annual Review of Cell and Developmental Biology*, *30*, 615–646.

Foltz, D. R., Jansen, L. E., Bailey, A. O., Yates, J. R., 3rd., Bassett, E. A., Wood, S., … Cleveland, D. W. (2009). Centromere-specific assembly of CENP-A nucleosomes is mediated by HJURP. *Cell*, *137*(3), 472–484.

Gaillard, P. H., Martini, E. M., Kaufman, P. D., Stillman, B., Moustacchi, E., & Almouzni, G. (1996). Chromatin assembly coupled to DNA repair: A new role for chromatin assembly factor I. *Cell*, *86*(6), 887–896.

Gaillard, P. H., Roche, D., & Almouzni, G. (1999). Nucleotide excision repair coupled to chromatin assembly. *Methods in Molecular Biology*, *119*, 231–243.

Gautier, A., Juillerat, A., Heinis, C., Correa, I. R., Jr., Kindermann, M., Beaufils, F., & Johnsson, K. (2008). An engineered protein tag for multiprotein labeling in living cells. *Chemistry & Biology*, *15*(2), 128–136.

Goldberg, A. D., Banaszynski, L. A., Noh, K. M., Lewis, P. W., Elsaesser, S. J., Stadler, S., … Allis, C. D. (2010). Distinct factors control histone variant H3.3 localization at specific genomic regions. *Cell, 140*(5), 678–691.

Gurard-Levin, Z. A., Quivy, J. P., & Almouzni, G. (2014). Histone chaperones: Assisting histone traffic and nucleosome dynamics. *Annual Review of Biochemistry, 83*, 487–517.

Hatanaka, Y., Inoue, K., Oikawa, M., Kamimura, S., Ogonuki, N., Kodama, E. N., … Ogura, A. (2015). Histone chaperone CAF-1 mediates repressive histone modifications to protect preimplantation mouse embryos from endogenous retrotransposons. *Proceedings of the National Academy of Sciences of the United States of America, 112*(47), 14641–14646.

Ishiuchi, T., Enriquez-Gasca, R., Mizutani, E., Boskovic, A., Ziegler-Birling, C., Rodriguez-Terrones, D., … Torres-Padilla, M. E. (2015). Early embryonic-like cells are induced by downregulating replication-dependent chromatin assembly. *Nature Structural & Molecular Biology, 22*, 662–671.

Jansen, L. E., Black, B. E., Foltz, D. R., & Cleveland, D. W. (2007). Propagation of centromeric chromatin requires exit from mitosis. *The Journal of Cell Biology, 176*(6), 795–805.

Kaufman, P. D., & Almouzni, G. (2006). Chromatin assembly. In M. L. De Pamphilis (Ed.), *DNA replication in eucaryotic cells* (pp. 121–140): Cold Spring Harbor: Cold Spring Harbor Press.

Keppler, A., Gendreizig, S., Gronemeyer, T., Pick, H., Vogel, H., & Johnsson, K. (2003). A general method for the covalent labeling of fusion proteins with small molecules in vivo. *Nature Biotechnology, 21*(1), 86–89.

Laskey, R. A., Honda, B. M., Mills, A. D., & Finch, J. T. (1978). Nucleosomes are assembled by an acidic protein which binds histones and transfers them to DNA. *Nature, 275*(5679), 416–420.

Martini, E., Roche, D. M., Marheineke, K., Verreault, A., & Almouzni, G. (1998). Recruitment of phosphorylated chromatin assembly factor 1 to chromatin after UV irradiation of human cells. *The Journal of Cell Biology, 143*(3), 563–575.

Muller, S., Montes de Oca, R., Lacoste, N., Dingli, F., Loew, D., & Almouzni, G. (2014). Phosphorylation and DNA binding of HJURP determine its centromeric recruitment and function in CenH3(CENP-A) loading. *Cell Reports, 8*(1), 190–203.

Polo, S. E., & Almouzni, G. (2015). Chromatin dynamics after DNA damage: The legacy of the access-repair-restore model. *DNA Repair (Amst), 36*, 114–121.

Probst, A. V., Dunleavy, E., & Almouzni, G. (2009). Epigenetic inheritance during the cell cycle. *Nature Reviews. Molecular Cell Biology, 10*(3), 192–206.

Ray-Gallet, D., & Almouzni, G. (2004). DNA synthesis-dependent and -independent chromatin assembly pathways in Xenopus egg extracts. *Methods in Enzymology, 375*, 117–131.

Ray-Gallet, D., Quivy, J. P., Scamps, C., Martini, E. M., Lipinski, M., & Almouzni, G. (2002). HIRA is critical for a nucleosome assembly pathway independent of DNA synthesis. *Molecular Cell, 9*(5), 1091–1100.

Ray-Gallet, D., Woolfe, A., Vassias, I., Pellentz, C., Lacoste, N., Puri, A., … Almouzni, G. (2011). Dynamics of histone H3 deposition in vivo reveal a nucleosome gap-filling mechanism for H3.3 to maintain chromatin integrity. *Molecular Cell, 44*(6), 928–941.

Santoro, S. W., & Dulac, C. (2015). Histone variants and cellular plasticity. *Trends in Genetics, 31*(9), 516–527.

Shah, J. V., Botvinick, E., Bonday, Z., Furnari, F., Berns, M., & Cleveland, D. W. (2004). Dynamics of centromere and kinetochore proteins: Implications for checkpoint signaling and silencing. *Current Biology, 14*(11), 942–952.

Smith, S., & Stillman, B. (1989). Purification and characterization of CAF-I, a human cell factor required for chromatin assembly during DNA replication in vitro. *Cell, 58*(1), 15–25.

Stillman, B. (1986). Chromatin assembly during SV40 DNA replication in vitro. *Cell*, *45*(4), 555–565.
Szenker, E., Lacoste, N., & Almouzni, G. (2012). A developmental requirement for HIRA-dependent H3.3 deposition revealed at gastrulation in Xenopus. *Cell Reports*, *1*(6), 730–740.
Szenker, E., Ray-Gallet, D., & Almouzni, G. (2011). The double face of the histone variant H3.3. *Cell Research*, *21*(3), 421–434.
Tachiwana, H., Muller, S., Blumer, J., Klare, K., Musacchio, A., & Almouzni, G. (2015). HJURP involvement in de novo CenH3(CENP-A) and CENP-C recruitment. *Cell Reports*, *11*(1), 22–32.
Tagami, H., Ray-Gallet, D., Almouzni, G., & Nakatani, Y. (2004). Histone H3.1 and H3.3 complexes mediate nucleosome assembly pathways dependent or independent of DNA synthesis. *Cell*, *116*(1), 51–61.
Talbert, P. B., & Henikoff, S. (2010). Histone variants—Ancient wrap artists of the epigenome. *Nature Reviews. Molecular Cell Biology*, *11*(4), 264–275.

CHAPTER FIVE

Analysis of Nucleosome Sliding by ATP-Dependent Chromatin Remodeling Enzymes

C.Y. Zhou, G.J. Narlikar[1]

University of California, San Francisco, CA, United States

[1]Corresponding author: e-mail address: geeta.narlikar@ucsf.edu

Contents

1. Introduction 119
2. Reconstituting Nucleosomes for FRET-Based Method 121
   2.1 Histone Labeling 121
   2.2 Visualization of Labeling on an SDS-PAGE Gel 123
   2.3 DNA Labeling 124
   2.4 Octamer and Nucleosome Assembly 125
   2.5 Purification of Nucleosomes from Free DNA by Glycerol Gradient 125
   2.6 Quantification of Nucleosomes by Gel 127
3. ATP-Dependent Chromatin Remodeling Assays 128
   3.1 Gel-Based Remodeling Assay 129
   3.2 FRET-Based Remodeling Assay 131
4. Data Analysis 134
5. Adaptation to Other Types of Remodeling Reactions 134
Acknowledgments 134
References 135

Abstract

ATP-dependent chromatin remodeling complexes carry out diverse transformations of chromatin. Understanding their mechanisms requires assays that can monitor the kinetics or chromatin remodeling. In this chapter, we describe complimentary native gel-based and FRET-based methods for assaying the kinetics of ATP-driven nucleosome sliding. These methods can be readily adapted to investigate other types of nucleosomal transformations carried out by chromatin remodeling ATPases.

1. INTRODUCTION

ATP-dependent chromatin remodeling complexes play critical roles in regulating chromatin structure (Clapier & Cairns, 2009; Narlikar,

Methods in Enzymology, Volume 573
ISSN 0076-6879
http://dx.doi.org/10.1016/bs.mie.2016.01.015

Sundaramoorthy, & Owen-Hughes, 2013). These complexes consist of a core ATPase subunit and additional non-ATPase subunits. To date four major classes of remodeling complexes have been described based on homologies within the core ATPase subunit: the SWI/SNF class, the ISWI class, the CHD class, and the INO80 class. Each class of enzymes catalyzes a different type of chromatin remodeling reaction. For example, the ISWI class mainly slides nucleosomes, while the SWI/SNF class can evict the histone octamer and exchange histones in addition to nucleosome sliding. In this chapter, we describe two methods to assay ATP-driven nucleosome sliding. Both of these methods can be easily adapted, as described later, to investigate nucleosome disassembly and histone exchange. We describe experimental methods with the human SNF2h enzyme, which are part of the ISWI family of remodelers. SNF2h plays the role of generating evenly spaced nucleosomes found in heterochromatin (Fyodorov, Blower, Karpen, & Kadonaga, 2004). The nucleosome spacing activity of SNF2h can be captured in a simpler mononucleosome-based system in which SNF2h moves an end-positioned nucleosome toward the center of the DNA (Fan, He, Kingston, & Narlikar, 2003; Hamiche, Sandaltzopoulos, Gdula, & Wu, 1999; Langst, Bonte, Corona, & Becker, 1999).

This ATP-dependent nucleosome sliding reaction consists of many different kinetic steps. A major challenge in using kinetics to investigate the mechanism of nucleosome remodeling is understanding the physical basis of the kinetic rate constants that are being measured. Here we describe two different remodeling assays under single-turnover conditions (SNF2h in excess of nucleosomes): a FRET-based assay and a gel-based assay. The two assays visualize different steps of the remodeling reaction. In the FRET-based assay, the octamer is labeled with a Cy5 dye on histone H3, while the end of the DNA is labeled with a Cy3 dye (Yang, Madrid, Sevastopoulos, & Narlikar, 2006). An end-positioned nucleosome starts at high FRET. As the remodeling enzyme slides the octamer away from the end of the DNA, the distance between the two FRET probes increases and FRET decreases. We can monitor this decrease in FRET as a function over time, and fit the trace by a first-order exponential decay to determine the rate constants. It is important to keep in mind that the rate constants derived from fitting the kinetic trace reports on the rate-limiting step of remodeling. It is also important to keep in mind that this assay is limited by the Förster radius of the FRET probes. In the FRET assay, the range of detection is only within movement of the first 10 bp. In contrast, in the gel-based assay, larger nucleosome movements (>10 bp) are more easily visualized.

The different nucleosomal species are separated based on the position of the histone octamer along the DNA, as centered nucleosomes run more slowly on a native polyacrylamide gel than end-positioned nucleosomes. The gel-based assay is less easily quantifiable than the FRET-based assay and is not as sensitive to small changes in movement of the histone octamer.

2. RECONSTITUTING NUCLEOSOMES FOR FRET-BASED METHOD

We use a method derived from a previously described method that uses bacterially expressed *Xenopus laevis* histones (Dyer et al., 2004; Luger, Rechsteiner, & Richmond, 1999). The FRET-labeled nucleosomes are assembled using the 601 nucleosome positioning sequence and have a Cy5 label at position 120 of histone H2A and a Cy3 label on the short end of the nucleosomal DNA (Fig. 1). The 601 sequence is an artificial sequence obtained from in vitro selection to bind strongly to histone octamer during nucleosome reconstitutions (Lowary & Widom, 1998). We use a histone H2A with a cysteine point mutation at position 120 to allow for chemical coupling of the Cy5 dye. In addition, we mutate the single endogenous cysteine in Xenopus H3 (H3C110A). The histones are purified as described previously and stored as lyophilized powders (Luger et al., 1999). The DNA label is incorporated using a Cy3-labeled primer.

2.1 Histone Labeling

The procedure begins with reducing histone H2A completely with TCEP, a reducing agent that does not contain any –SH groups that may interfere with the subsequent labeling reaction. After reduction, the histone is labeled with Cy5-maleimide. After quenching the reaction and getting rid of the free dye, the labeling reaction is analyzed on a SDS-PAGE gel.

Fig. 1 Cartoon depicting FRET labeling scheme on nucleosome.

Notes

- The histone labeling procedure is performed at *room temperature*.
- Cy5 is light sensitive! Limit light exposure as much as possible during all steps involving Cy5 by covering samples with foil and using lids on ice buckets.

Solutions and reagents

- Lyophilized H2A-120C (1 mg)
- Buffer for labeling (final pH ~7.3)
  - 20 m*M* Tris, pH 7.0
  - 7 *M* Guanidinium HCl
  - 5 m*M* EDTA
- Cy5-maleimide (1 mg aliquots, lyophilized, and stored in −20°C)
- TCEP (0.5 *M*, pH 7)
- 2-Mercapto-ethanol (also called BME, 14.3 *M*)
- DMF (dimethylformamide)

Procedure

1. Take out a tube of H2A-120C (1 mg/tube). Dilute up to 0.1 m*M* with labeling buffer (MW of H2A: 13,960 g/mol) at room temperature. Take a Nanodrop reading of the histone, and redetermine its concentration. We will use this information later during the octamer assembly.
2. Add TCEP to the histone to a final concentration of 0.7 m*M* (stock is 0.5 *M*). Let this sit for 2 h in the dark, at room temperature. NOTE: Set aside 2 μL of the sample to run on the gel later as an unlabeled sample.
3. Dissolve the Cy5-maleimide in DMF to a final concentration of 100 m*M* (MW of Cy5-maleimide: 641.24 g/mol). Add the dye to the histone to a final concentration of 0.6 m*M* (six-fold excess dye).
4. Cover reactions with foil, let sit in the dark, covered with foil, for 3 h at room temperature.
5. Add BME to a final concentration of 80 m*M*.
6. Get rid of free Cy5 dye by concentrating and filtering through a microcon-10 (in a 15 mL conical). Load half of the sample into each microcon and spin in a centrifuge at 3800 × *g* for 10 or 15 min at a time until 1/3 of the original volume is remaining. Every time you check on the centrifuge spin, invert the conical to mix. Dilute back down to the original volume with labeling buffer. Transfer to another filter. Save the flow through.
7. Repeat step 6 at least three more times, using a new filter each time. It is important to get rid of as much free Cy5 as possible because Cy5 is very hydrophobic, so having excess Cy5 in solution may result in improper octamer assembly.

2.2 Visualization of Labeling on an SDS-PAGE Gel (Fig. 2)

Solutions and reagents

- 15% SDS-PAGE gel
- 4 × SDS loading dye (200 m*M* Tris, pH 6.8, 400 m*M* DTT, 8% SDS, 0.4% bromophenol blue, 40% glycerol)

Procedure

1. Set up a 15% SDS-PAGE gel.
2. For each sample, combine 2 μL of the sample and 98 μL of 1 × SDS loading dye. Boil for 10 min at 95°C. The Guanidinium in the samples will make them goopy and difficult to load on the gel. If this is a problem, add 100 μL more lX loading dye, and load twice as much. Also, it helps to load the samples while they are still very hot.

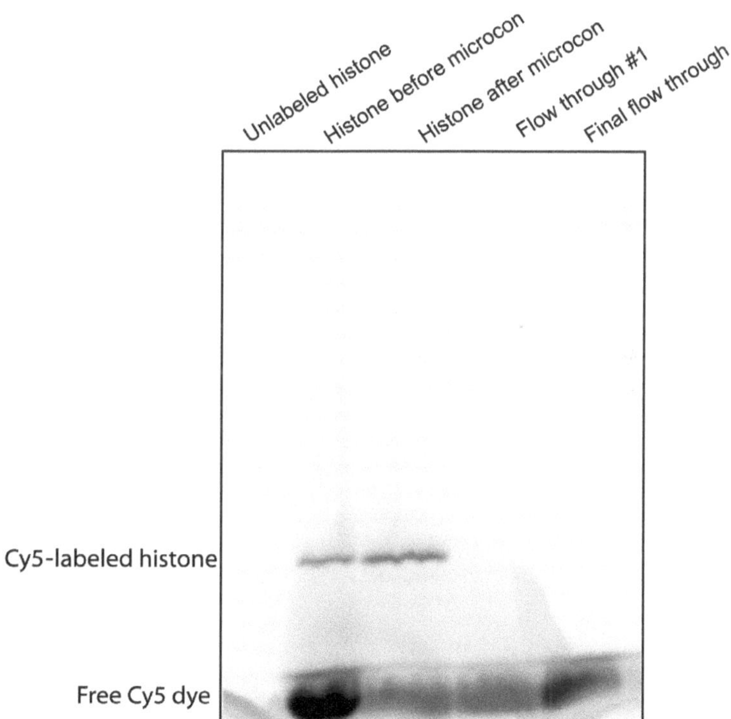

Fig. 2 SDS-PAGE gel of histone labeling. The gel was scanned on a Typhoon (GE Healthcare) using the Cy5 channel. The Cy5-labeled histone and the free Cy5 dye can be seen. The gel shows that the labeling worked, there was little to no loss of histone during the microcon steps, and most of the free dye has been removed.

3. Run the gel until the bromophenol blue dye is at the bottom (about 1 h). Then stop the gel.
4. Remove the glass plate without spacers and scan in a LAS500 or on a GE Typhoon in the Cy5 channel.

2.3 DNA Labeling

The nucleosomal DNA is labeled using a PCR primer labeled at the 5′ end with Cy3 (Integrated DNA Technologies). A large-scale 10 mL PCR is carried out in a 96-well plate using Taq polymerase, a template plasmid containing the 601 sequence, a Cy3-labeled forward primer, and an unlabeled reverse primer. For the method described here the DNA contains the 147 bp 601 sequence and an additional 60 bp from the plasmid backbone.

Solutions and reagents

- 10 × PCR buffer
- Cy3-labeled forward primer
- Reverse primer
- dNTPs
- Template containing 601 sequence
- Taq polymerase (5 U/λ)
- Native gel-loading buffer: 20% glycerol, 20 m*M* Tris, pH 7.5
- 3 *M* Sodium acetate pH 5.0
- 200 Proof ethanol
- Large 5% acrylamide gel (dimensions: 13 in. length, 6 in. height)

Procedure

Carry out the PCR reaction in a 96-well plate.

1. Transfer the PCR product into a 40 mL Oakridge tube.
2. Add 1/10 volume 3 *M* sodium acetate, pH 5.0, 3 × volume 200 proof ETOH (100% ethanol).
3. Incubate at –20°C overnight (or 3 h if time is limiting).
4. Spin down at 14,000 rpm (25,000 × *g*) for 20 min at 4°C in Oakridge tube.
5. Remove supernatant carefully using a plastic pipette and allow for the pellet to dry.
6. Redissolve pellet in ~400 μL native gel-loading buffer.
7. Load onto large 5% acrylamide gel with bucket well.
8. Cover the gel with foil and leave open only a small column on the edge of the large well. View DNA in this column band under short range UV light (using a handheld UV lamp in dark room) and cut out location of band. Then turn off UV and extend the cut to the rest of the well and

discard the UV exposed piece. This is to avoid using DNA that may have been damaged by the UV.

9. Using a clean razor blade, cut out gel band and plunge through a 10 mL syringe to disintegrate matrix.
10. Collect gel in a conical tube and add ~10 mL TE.
11. Rock overnight at room temperature.
12. Pour gel + TE mixture over a 50 mL conical-top filter unit, attach to vacuum, and collect all flow through.
13. Transfer flow through to a 40 mL Oakridge tube and repeat steps 2–5.
14. Dissolve pellet into ~200–500 μL TE and store at −20°C.

2.4 Octamer and Nucleosome Assembly

Histone octamers are assembled by mixing the labeled histone H2A in labeling buffer with other histones dissolved in refolding buffer (7 *M* Guanidinium HCl, 20 m*M* Tris, pH 7.5, 10 m*M* DTT), as described previously. H2A, H2B, H3, and H4 are mixed in a molar ratio of 1.2:1.2:1:1. The excess H2A and H2B ensures complete octamer formation and limits formation of hexamer (an octamer missing an H2A–H2B dimer), a species that is difficult to separate from the fully formed octamer on the sizing column. To approximate the amount of the labeled H2A sample, we assume a 10% loss during the labeling reaction. Octamers are assembled and purified as described previously (Dyer et al., 2004; Luger et al., 1999). Nucleosomes are assembled from the Cy5-labeled octamers and the Cy3-labeled DNA using detailed procedures described previously (Dyer et al., 2004; Luger et al., 1999). The scale at which the assembly is carried out uses 50–100 μg of DNA.

2.5 Purification of Nucleosomes from Free DNA by Glycerol Gradient

The goal of these steps is to purify the assembled mononucleosomes from free DNA and free histones. We describe a glycerol gradient-based method that is routinely used to purify this scale of nucleosome assembly. The gradient can be poured using a variety of methods. We describe how to pour it using the Gradient Master (BioComp).

Solutions and reagents

- Glycerol gradient buffer—10%: 20 m*M* Tris, pH 7.5, 1 m*M* EDTA, 10% glycerol, 0.1% NP40
- Glycerol gradient buffer—30%: same as 10% glycerol gradient buffer except with 30% glycerol
- 5% Acrylamide, 0.5 × TBE gel

Procedure

1. Prechill a Ti55 rotor and rotor buckets the night before doing the gradient at 4 degrees.
2. Concentrate nucleosomes to between 50 and 100 μL using a microcon-100. Transfer the concentrated sample to a fresh tube. Using a P200, determine the exact volume of the sample. Write that volume on the outside of the tube. It is best to underestimate the volume. If you think the volume is 70 μL, it is best to record it as 68 μL.
3. Turn on the ultracentrifuge and start the vacuum.
4. Pour the glycerol gradient:
   - **a.** Using the metal tool, draw a line all the way around the sample tube to designate the halfway point. Label the tube with sample number if necessary. Place the labeled tube into the sample holder.
   - **b.** Attach the metal needles to plastic 10 mL disposable syringes.
   - **c.** Designate one syringe for the 10% glycerol solution and the other syringe to the 30% glycerol buffer. Do not mix up the syringes!
   - **d.** Rinse each syringe with about 5 mL of the designated buffer.
   - **e.** Use the needled syringe to fill the sample tube halfway with 10% glycerol. To account for the volume of the needle, fill up to slightly above the marked halfway point.
   - **f.** Fill the bottom half of the sample tube with 30% glycerol buffer. Plunge the needle to the bottom of the sample tube and start filling the sample tube with 30% glycerol from the bottom. You may notice some mixing between the two solutions at first, but you should soon notice a separation between the two. Make sure you observe that point of separation the entire time you are filling up the sample with 30% glycerol buffer.
   - **g.** Once the halfway point has been reached, take out the syringe carefully but swiftly.
   - **h.** Place caps on the sample tubes, with the tiny hole in the bottom of the cap going on last to avoid incorporating a bubble. Some sample will come out when you do this.
   - **i.** Carefully move the sample holder onto the magnetic stand of the gradient mixer. Run the gradient using the following parameters: short tubes, glycerol 10–30% v/v, 1:05 min, 86 degrees, 121 rpm).
   - **j.** Once the gradient is done, carefully take the sample holder off of the magnetic stand. From this point forward, you must be very careful when handling the gradient. Any slight mixing or disturbance of the concentration gradient will affect separation of free DNA from nucleosomes.

5. Remove the caps from the sample tubes. Carefully get rid of any bubbles that are on the surface of the gradient. Place the sample tubes into their holders in the rotor buckets.
6. Before loading the gradient with your nucleosome sample, take out an amount from the top of the gradient that corresponds to the volume of your sample.
7. Carefully load the sample by touching the pipette tip to the very surface and side of the gradient, and slowly pushing sample through. Avoid incorporating a bubble at the very end.
8. Twist the caps onto the buckets, make sure that they are completely screwed on.
9. Carefully bring the sample rack containing buckets with samples inside, as well as the Ti55 rotor to the ultracentrifuge. Make sure to walk slowly as to not disturb the gradient!
10. Spin at 35,000 rpm for 16 h.
11. Collect gradient with butterfly needles, eluting into a 96-well plate. Place 3–4 drops per well. Cover the plate with a foil seal and number every other well on the foil.
12. Run every other fraction collected on a 5% 0.5 × TBE native gel. Load directly by pushing the pipette tip through the foil seal. The solution you are loading will be clear, so it is important to prenumber the lanes on the outside of the gel using a sharpie, according to the sample you are loading.
13. Run the gel at 80 V for 30 min, or until the yellow dye reaches the very bottom of the gel.
14. Take the gel out carefully with a razor blade and transfer to a small box containing 30–50 mL of 0.5 × TBE running buffer. NOTE: the gel will be very thin and fragile!
15. Add 3 μL of 10,000 × SYBR gold. Stain while rocking for 10 min. Destain for 5 min with water.
16. Image the gel (Fig. 3).
17. Choose which fractions to collect, pool fractions and concentrate in microcon-100, to about 50–100 μL.
18. Store at 4°C, covered in foil.

2.6 Quantification of Nucleosomes by Gel

After pooling and concentrating the relevant fractions from the glycerol gradient, the nucleosomes are quantified after running on a 1 × TBE, 5% acrylamide gel using the DNA that was used for the assembly to generate

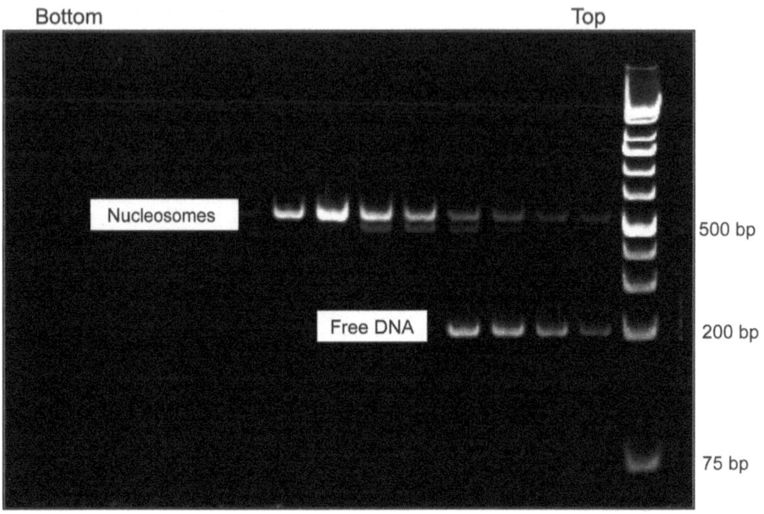

Fig. 3 Native gel of fractions from glycerol gradient. Fractions are collected using a blood collection needle, such that the bottom of the gradient elutes first. A DNA size marker is loaded in the rightmost lane. Nucleosomes are well separated from free DNA, and fractions containing pure nucleosomes are pooled and concentrated.

a standard curve. The nucleosomes and DNA are stained with ethidium bromide and the gel scanned using a GE Typhoon imager. For the quantification, it is important to take into account a 2.5-fold quenching of the ethidium fluorescence signal when the DNA is wrapped around histones.

3. ATP-DEPENDENT CHROMATIN REMODELING ASSAYS

General guidelines for all experiments

- To thaw the remodeling enzyme, warm the outside of the tube between your hands until you see it is almost completely thawed, and then immediately put onto ice. Once thawed, the enzyme is stable on ice for a few hours. Do not refreeze. The experiments described below use the human SNF2h enzyme, which can be obtained in high purity from expression in either Sf9 insect cells or *Escherichia coli* cells (Fan et al., 2003; Leonard & Narlikar, 2015).
- All light sensitive materials must be kept under foil or under ice bucket lids.
- Mix everything very carefully and thoroughly at every step. Do a quick spin of the reaction tubes when everything has been added to make sure everything is at the bottom of the tube.

- The reactions described below are carried out under single-turnover conditions in which [SNF2h] > [nucleosomes]. These conditions allow for fitting of the entire time course by a first-order rate equations as described in Section 4. However the protocols can be easily adapted to multiple turnover conditions [SNF2h] < [nucleosomes]. In this case initial rates are measured in which less than 10% of the nucleosomes are remodeled.

3.1 Gel-Based Remodeling Assay

This assay is adapted from methods originally developed to assay nucleosome sliding by Drosophila ISWI complexes (Hamiche et al., 1999; Langst et al., 1999). In this assay, we first mix enzyme and ATP in a standard reaction buffer. After incubating at room temperature for 5 min, we start the reaction by adding nucleosomes. At each desired time point, we will take out a small portion of the reaction and quickly combine it with a stop solution consisting of ADP and stop plasmid. The stop plasmid can generally be any ~3 kb plasmid vector. The ADP inhibits the active site of the enzyme and the plasmid chases the enzyme off of the nucleosome. Once the reaction is complete, the time points are run on a 5% native gel, stain with SYBR gold, and visualized on a gel imager. The protocol described below is for a specific concentration of SNF2h and nucleosomes. These concentrations can be varied.

Final conditions

100 n*M* SNF2h
2 m*M* ATP-Mg
3 m*M* free Mg
20 n*M* nucleosomes
0.02% NP-40
70 m*M* KCl
~12 m*M* HEPES, pH 7.5
11% Glycerol
Room temperature

Stock solutions and buffers

SNF2h storage buffer: 25 m*M* HEPES, pH 7.5, 210 m*M* KCl, 20% glycerol
Nucleosome buffer with NP-40
30 m*M* $MgCl_2$
30 m*M* ATP-Mg
Stop solution: 0.7 mg/mL stop plasmid, 30 m*M* ADP, 20% glycerol

Reaction Setup (for a Single Reaction)	
Enzyme (3 ×, 300 n*M*)	13.4 μL
$MgCl_2$ (30 m*M*)	4 μL
ATP-Mg (30 m*M*)	2.6 μL
Nucleosomes (5 ×, 100 n*M*)	8 μL add to start
H_2O	12
Total	40 μL

Procedure

Before beginning, pour a 0.5 × TBE 5% acrylamide gel with a 10 well comb. Label the wells and set up the gel box with buffer.

1. Dilute SNF2h to 3 × (600 n*M*) desired concentration in SNF2h storage buffer. Aliquots are stable all day on ice. It is preferred that aliquots are not refrozen and reused.
2. Dilute nucleosomes to 5 × (100 n*M*) desired concentration in nucleosome buffer with NP-40.
3. Aliquot the stop solution into eight small PCR tubes, one for each time point. 5 μL/time point.
   Suggestions for time points: 30 s, 1, 2, 3, 4, 5, 10, and 15 min.
4. Combine $MgCl_2$, water, and ATP-Mg.
5. Add enzyme, mix well, and do a quick spin of the sample.
6. Incubate at room temperature for 5 min.
7. At the same time, pipette 16 μL nucleosomes into a separate tube and incubate at room temperature for 5 min.
8. After incubating, start the reaction by adding 16 μL of nucleosomes with a P20. At the same time, press the "start" button on the timer.
9. Quickly discard the tip of the P20 and change to a preloaded P200 (set at 25 μL) to mix the reaction.
10. At each time point, take out 5 μL of the reaction and mix with the pre-aliquoted stop solution. Aim for the first time point to be at 30 s.
11. Take time points up to 10 min. Make sure to record your actual time points as precisely as possible.
12. Load the entire time point (10 μL) into each lane of the gel.
13. Run the gel at 80 V for 1 h, or until the xylene cyanol (light blue) dye almost reaches the bottom of the gel.

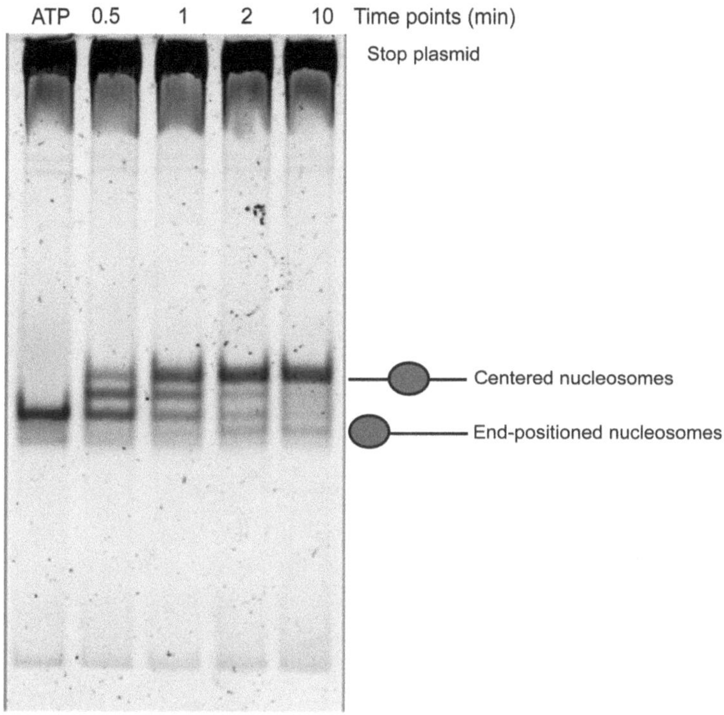

Fig. 4 Gel remodeling assay. After starting the reaction, 5 μL time points are taken and quenched with equal volume of stop solution. Samples at each time point are then run on a 5% native gel. Nucleosomes start end positioned, but quickly become centered as the remodeling reaction progresses.

14. Stain with SYBR gold for 30 min, using 2 × as much as normal. (Use at 1000 × instead of 10,000 ×.)
15. Visualize the gel (Fig. 4).

3.2 FRET-Based Remodeling Assay

This assay is carried out in the same buffer conditions as for the gel remodeling assay. The only difference is that the FRET remodeling assay will require a larger volume, to fill the space in the cuvette. The nucleosomes will start in a high FRET state and will have a high Cy5 fluorescence signal when excited with the Cy3 excitation maximum wavelength. As the reaction proceeds, the signal from Cy5 diminishes. The fluorescence measurements can be carried out in any fluorometer capable of measuring Cy3 and Cy5 fluorescence and with the sensitivity to measure the fluorescence at a concentration of ~10 n*M* Dye. We use an ISS K2 fluorometer.

Final conditions

100 n*M* SNF2h
2 m*M* ATP-Mg
3 m*M* free Mg
10 n*M* nucs
0.02% NP-40
70 m*M* KCl
~12 m*M* HEPES, pH 7.5
11% Glycerol
25°C

Stock solutions and buffers

SNF2h storage buffer (25 m*M* HEPES, pH 7.5, 210 m*M* KCl, 20% glycerol)
Nucleosome buffer with NP-40 (standard)
30 m*M* $MgCl_2$
30 m*M* ATP-Mg

Reaction Setup (for a Single Reaction)	
Enzyme (3 ×)	26.7 μL
$MgCl_2$ (30 m*M*)	8 μL
ATP-Mg (30 m*M*)	5.33 μL
Nucleosomes (5 ×)	16 μL add to start
H_2O	24 μL
Total	80 μL

Procedure

1. Determine the excitation maximum of the donor fluorescent label (Cy3) in your buffer conditions by doing an excitation scan of the fluorescently labeled nucleosomes alone in reaction buffer without SNF2h. Typically for Cy3, the excitation maximum is around 550 nm. Then using the excitation maximum, do an emission scan that spans 550–700 nm. This will provide the emission maxima for the direct excitation of Cy3 (typically around 565 nm) and for the FRET-based excitation of Cy5 (typically around 670 nm).
2. Dilute SNF2h to 3 × desired concentration in SNF2h storage buffer.

3. Dilute nucleosomes to 5 × desired concentration in nucleosome buffer with NP-40. Make new dilutions every couple hours, as nucleosomes may fall apart at low concentrations.
4. Combine $MgCl_2$, water, and ATP-Mg (or Master Mix).
5. Add enzyme, mix well, and do a quick spin of the sample.
6. Pipette sample into cuvette. Put into sample holder in the fluorometer. Incubate for 5 min.
7. At the same time, pipette 16 μL nucleosomes into a separate tube and incubate at room temperature for 5 min.
8. Meanwhile, initiate a kinetic run on the fluorometer in which the excitation wavelength is set to the excitation maximum of Cy3 and the emission wavelength to that of the emission maximum of Cy5 as determined in step 1. After the kinetic run has begun, pause acquisition.
9. While paused, open the fluorometer lid. Set a lab timer for 10 s.
10. Dial a P200 pipette to 60 μL and pull up the 16 μL nucleosomes (try to keep liquid close to the tip of the pipette to avoid introducing air bubbles).
11. Start the timer and place pipette tip down into cuvette. When timer reaches zero, start mixing, taking care to avoid bubbles. Mix four to five times.
12. When done mixing, close the fluorometer lid and restart acquisition, taking note of how many seconds have elapsed on the timer from the time of mixing to restarting of acquisition. (Typical lag time is 6–9 s.)
13. Stop acquisition when fluorescence change has plateaued (Fig. 5).
14. Clean cuvette by rinsing with water, ethanol, then water again. Dry completely before the next run.

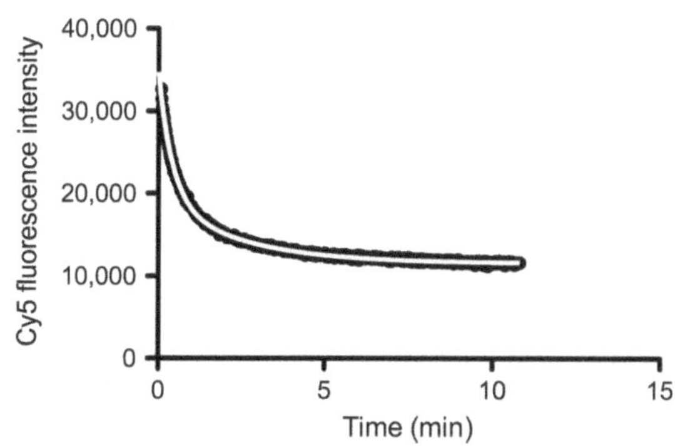

Fig. 5 FRET remodeling assay. As remodeling progresses, FRET decreases and Cy5 intensity decreases. The data are fit to a double exponential to derive two rates constants; one each from the fast and slow phases of the reaction.

4. DATA ANALYSIS

For the gel-based assay, we use ImageQuant (GE) or ImageJ software to quantify the fraction of the substrate band (end-positioned nucleosomes) relative to all the nucleosomal bands (substrate + products). Plotting the fraction substrate [(substrate)/(substrate + products)] as a function of time typically yields a plot that can be fit by a single or double exponential to obtain one or two rate constants, respectively. Programs such as Prism or MATLAB can be used to generate the fits. Similarly the data obtained by the FRET measurements can be fit by single or multiple exponentials (Fig. 5). In the case of the FRET data, additional variables are the fluorescence values for the starting material, any intermediates and the end product. In addition for the FRET data, the lag time that is recorded (step 12 in Section 3.2) has to be added to each of the recorded time points. The fit of data to multiple exponentials may indicate the presence of intermediates in the remodeling reaction or the presence of two populations of nucleosomes, which react with different rates.

5. ADAPTATION TO OTHER TYPES OF REMODELING REACTIONS

Both the assays described here can be modified to investigate the kinetics of either octamer eviction or histone exchange. Naked DNA migrates faster in a native 5% polyacrylamide gel than nucleosomal DNA. This property allows visualization of octamer eviction or octamer transfer as a function of time using a native gel-based assay (Lorch, Zhang, & Kornberg, 1999). FRET can be used to study octamer eviction by assaying for loss of FRET between labeled H3 and labeled DNA under conditions where the native gel assay reveals that the octamer is evicted rather than slid (Rowe & Narlikar, 2010). Similarly the kinetics of incorporating H2A/H2B into a nucleosome can be measured by labeling the H2A/H2B with one FRET dye and labeling the H3 or H4 present in the nucleosomes with the other FRET dye (Rowe & Narlikar, 2010). The H2A/H2B incorporation then can be observed by following the gain in FRET as a function of time.

ACKNOWLEDGMENTS

We thank J.L. for providing the raw data for Fig. 4. This work is supported by Grant R01 GM073767 to G.J.N. from the NIH.

REFERENCES

Clapier, C. R., & Cairns, B. R. (2009). The biology of chromatin remodeling complexes. *Annual Review of Biochemistry*, *78*, 273–304.

Dyer, P. N., Edayathumangalam, R. S., White, C. L., Bao, Y., Chakravarthy, S., Muthurajan, U. M., et al. (2004). Reconstitution of nucleosome core particles from recombinant histones and DNA. *Methods in Enzymology*, *375*, 23–44.

Fan, H. Y., He, X., Kingston, R. E., & Narlikar, G. J. (2003). Distinct strategies to make nucleosomal DNA accessible. *Molecular Cell*, *11*, 1311–1322.

Fyodorov, D. V., Blower, M. D., Karpen, G. H., & Kadonaga, J. T. (2004). Acf1 confers unique activities to ACF/CHRAC and promotes the formation rather than disruption of chromatin in vivo. *Genes & Development*, *18*, 170–183.

Hamiche, A., Sandaltzopoulos, R., Gdula, D. A., & Wu, C. (1999). ATP-dependent histone octamer sliding mediated by the chromatin remodeling complex NURF. *Cell*, *97*, 833–842.

Langst, G., Bonte, E. J., Corona, D. F., & Becker, P. B. (1999). Nucleosome movement by CHRAC and ISWI without disruption or trans-displacement of the histone octamer. *Cell*, *97*, 843–852.

Leonard, J. D., & Narlikar, G. J. (2015). A nucleotide-driven switch regulates flanking DNA length sensing by a dimeric chromatin remodeler. *Molecular Cell*, *57*, 850–859.

Lorch, Y., Zhang, M., & Kornberg, R. D. (1999). Histone octamer transfer by a chromatin-remodeling complex. *Cell*, *96*, 389–392.

Lowary, P. T., & Widom, J. (1998). New DNA sequence rules for high affinity binding to histone octamer and sequence-directed nucleosome positioning. *Journal of Molecular Biology*, *276*, 19–42.

Luger, K., Rechsteiner, T. J., & Richmond, T. J. (1999). Preparation of nucleosome core particle from recombinant histones. *Methods in Enzymology*, *304*, 3–19.

Narlikar, G. J., Sundaramoorthy, R., & Owen-Hughes, T. (2013). Mechanisms and functions of ATP-dependent chromatin-remodeling enzymes. *Cell*, *154*, 490–503.

Rowe, C. E., & Narlikar, G. J. (2010). The ATP-dependent remodeler RSC transfers histone dimers and octamers through the rapid formation of an unstable encounter intermediate. *Biochemistry*, *49*, 9882–9890.

Yang, J. G., Madrid, T. S., Sevastopoulos, E., & Narlikar, G. J. (2006). The chromatin-remodeling enzyme ACF is an ATP-dependent DNA length sensor that regulates nucleosome spacing. *Nature Structural & Molecular Biology*, *13*, 1078–1083.

PART II

Posttranslational Histone Modification Enzymes and Complexes

CHAPTER SIX

In Vitro Activity Assays for MYST Histone Acetyltransferases and Adaptation for High-Throughput Inhibitor Screening

C.E. McCullough*, R. Marmorstein*,†,1

*University of Pennsylvania, Philadelphia, PA, United States

†Abramson Family Cancer Research Institute, Perelman School of Medicine at the University of Pennsylvania, Philadelphia, PA, United States

[1]Corresponding author: e-mail address: marmor@mail.med.upenn.edu

Contents

1. Introduction 140
2. Preparation of Recombinant MYST Proteins 142
3. Kinetic Analysis 144
   3.1 Primer on Steady-State Kinetic Analysis 144
   3.2 Specific Kinetic Properties of MYST Acetyltransferases 145
   3.3 Controlling for Chemical Acetylation 146
   3.4 Kinetic Assay Using Radioactive Acetyl-CoA 147
4. Thermofluor Assay to Investigate Cofactor and Inhibitor Binding 147
5. Inhibitor Screening 149
   5.1 General Considerations 149
   5.2 Buffer Considerations 151
   5.3 Assays 152
   5.4 Thermofluor Binding Assay 154
   5.5 Eliminating Promiscuous Inhibitors 155
   5.6 Pan-Assay Interference Compounds 155
   5.7 Colloidal Aggregators 156
6. Prospects and Conclusions 157

Acknowledgments 158

References 158

Abstract

Lysine acetylation is a posttranslational modification that is carried out by acetyltransferases. The MYST proteins form the largest and most diverse family of acetyltransferases, which regulate gene expression, DNA repair, and cell cycle homeostasis, among other activities, by acetylating both histone and nonhistone proteins. This chapter will describe methods for the preparation and biochemical characterization of

Methods in Enzymology, Volume 573
ISSN 0076-6879
http://dx.doi.org/10.1016/bs.mie.2016.01.016

MYST family acetyltransferases, including protocols for the preparation of recombinant protein, enzyme assays for measuring steady-state parameters, and binding assays to measure cofactor and inhibitor binding. We also provide details on adapting these assays for high-throughput screening for small molecule MYST inhibitors. This chapter seeks to prepare researchers for some hurdles that they may encounter when studying the MYST proteins so that there may be better opportunity to plan appropriate controls and obtain high-quality data.

1. INTRODUCTION

Lysine acetylation is a posttranslational modification that is conserved from yeast to humans and is correlated with numerous regulatory functions in both the nucleus and cytoplasm of the cell. Of the different HAT (histone acetyltransferase) families that have been characterized to date, the MYST proteins form the largest and most diverse family (Yang, 2004). MYST family proteins regulate gene expression, DNA repair, cell cycle homeostasis, and other activities (Sapountzi & Cote, 2011) by acetylating both histone and nonhistone proteins (Cereseto et al., 2005; Iizuka & Stillman, 1999; Lin et al., 2009; Mellert & McMahon, 2009; Sapountzi & Cote, 2011).

Although MYST proteins share a structurally conserved acetyl coenzyme A (acetyl CoA) core-binding region with other HATs, they employ several unique structural and catalytic properties within their conserved HAT domain. The MYST HAT domains contain N- and C-terminal regions that flank the core-binding region that are structurally distinct from other HAT families. These include an N-terminal C_2HC zinc-binding region and C-terminal helix-turn-helix motifs required for chromatin regulatory activities (Lafon, Chang, Scott, Jacobson, & Pillus, 2007; Yan, Barlev, Haley, Berger, & Marmorstein, 2000). For catalysis, the MYST proteins have been shown to employ a ping-pong catalytic mechanism (Yan et al., 2000), although one study was more consistent with a ternary complex mechanism (Berndsen, Albaugh, Tan, & Denu, 2007). In the ping-pong mechanism, the general base glutamate first deprotonates the active site cysteine so that the acetyl group from acetyl CoA can be transferred to the cysteine. The glutamate then protonates the leaving cofactor and deprotonates the substrate lysine so that the cysteine can transfer the acetyl group to the lysine. Finally, both the glutamate and cysteine may react with water to return to their initial state (Fig. 1) (Yan et al., 2000). More recently, MYST catalytic activity was also found to require active site lysine autoacetylation for cognate substrate acetylation (Yuan et al., 2012).

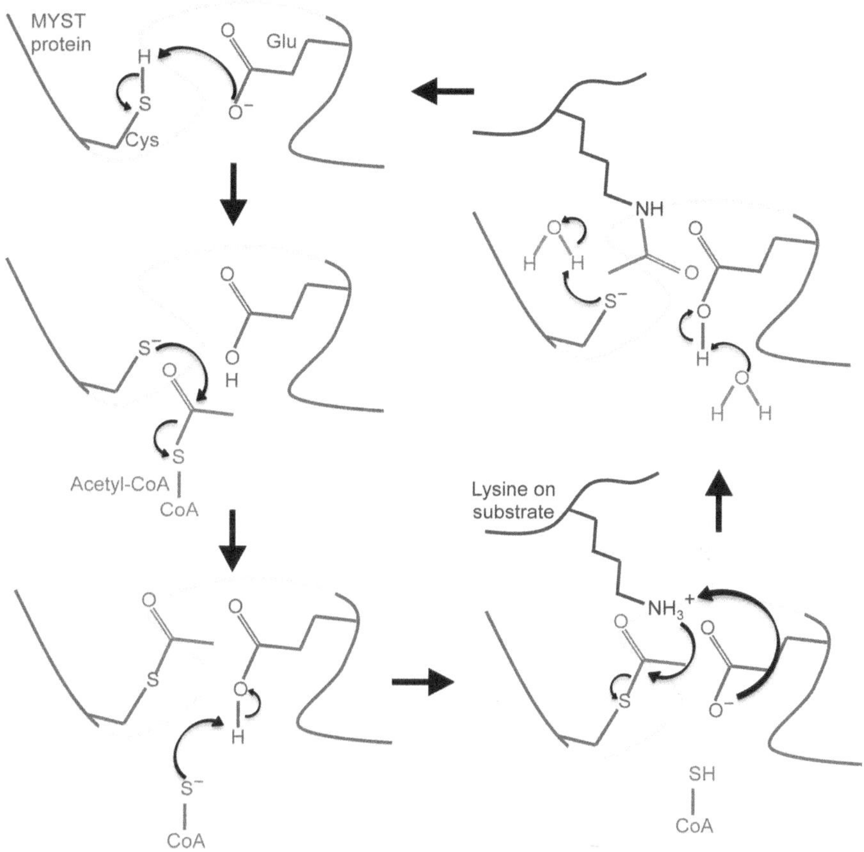

Fig. 1 "Ping-pong" catalytic mechanism employed by MYST family acetyltransferases. A glutamate in the active site acts as the general base to first deprotonate the active site cysteine. Next the acetyl group from acetyl CoA is transferred to the cysteine. The glutamate then protonates the leaving cofactor. The glutamate is then able to deprotonate the substrate lysine so that the cysteine can transfer the acetyl group to the lysine. Finally both the glutamate and cysteine react with water to return to their initial state. (See the color plate.)

Given the diverse regulatory roles, it is not surprising that aberrant protein acetylation or acetyltransferase function within the MYST family is correlated with several human diseases. For example, the HAT domain of the MYST protein MOZ (monocytic leukemia zinc-finger protein) can form translocation products with the CBP (CREB-binding protein) HAT in a subset of acute myeloid leukemias (Kitabayashi et al., 2001). Tip60 is linked to the onset of Alzheimer's disease (Baek et al., 2002; Cao & Sudhof, 2001, 2004; Kinoshita, Whelan, Berezovska, & Hyman, 2002) and is downregulated in lung and colon cancer (Lleonart et al., 2006), but upregulated in epithelial tumors (Hobbs et al., 2006). One study has reported that loss of acetylation on the MOF (males absent on the first) target lysine 16 of histone H4 is common in many human malignancies (Fraga et al., 2005). Additionally, mutated MYST proteins can become oncogenic (Lafon et al., 2007).

Because of their connections to disease, there have been efforts to develop inhibitors of MYST family acetyltransferases. The development of a bisubstrate CoA-peptide MYST inhibitor has been reported. While this inhibitor is more potent than the natural product HAT inhibitors anacardic acid and curcumin, it exhibited only micromolar IC_{50} values and poor selectivity for the MYST proteins vs other families of acetyltransferases (Wu, Xie, Wu, Zhang, & Zheng, 2009). Virtual screening to identify small molecule inhibitors of Tip60 led to the identification of compounds with better potency than curcumin; however, these inhibitors still exhibit poor potency with IC_{50} values in the mid-micromolar range and do not exhibit selectivity between Tip60 and other HAT families (Wu et al., 2011). Because all the current MYST inhibitors exhibit weak selectivity and/or poor pharmacokinetic properties, it is important to identify and develop new and more potent MYST HAT inhibitors that can be used both as probes to further investigate the function of MYST HATs and as lead compounds for drug development.

This chapter will describe methods for the preparation and biochemical characterization of MYST family acetyltransferases, with a particular focus on human MOF (hMOF). This chapter will include protocols on the preparation of recombinant protein, enzyme assays for measuring steady-state parameters, and a thermofluor assay to investigate cofactor and inhibitor binding. We will also present details on adapting these assays for high-throughput screening for small molecule MYST inhibitors. We place a particular focus on how MYST family proteins may need to be treated differently from other related enzymes.

2. PREPARATION OF RECOMBINANT MYST PROTEINS

We generally follow previously described methods for purifying the MYST proteins (hMOF specifically) (Yuan et al., 2012), but with some modifications. To study MYST activity in vitro we purify the HAT domain of hMOF, residues 174–449, with an N-terminal 6 × His tag separated by a TEV protease recognition site. The plasmid is transformed into *Escherichia coli* BL21(DE3) codon plus RIL (Stratagene) cells and the protein is overexpressed in TB media by induction with 1 m*M* IPTG and grown at 15°C overnight after the OD reaches 0.8–1.5. The cells are harvested and lysed by sonication in lysis buffer containing 50 m*M* HEPES (pH 7.5), 0.5 *M* NaCl, 2 m*M* β-mercaptoethanol, 5 m*M* imidazole, 5% glycerol, 0.1% CHAPS, and 0.1 m*M* phenylmethylsulfonyl fluoride. The lysate is cleared by centrifugation at 28,000 × *g* for 30 min at 4°C. The supernatant is then loaded onto nickel-charged NTA resin (Qiagen) that has been equilibrated with lysis buffer. The resin is then washed with 10 column volumes of lysis buffer containing 25 m*M* imidazole and then eluted with lysis buffer containing 250 m*M* imidazole buffer in 10-mL fractions. Because we found that the 6 × His tag does not interfere with hMOF activity or crystallization, we do not cleave the tag. Instead we proceed directly from the affinity column to size exclusion chromatography, skipping ion exchange gel chromatography due to hMOF sensitivity to low salt conditions. The hMOF is concentrated and injected onto an FPLC using a HiLoad Superdex 75 16/60 gel filtration column using a gel filtration buffer composed of 20 m*M* HEPES (pH 7.5), 0.5 *M* NaCl. We then concentrate the hMOF protein to ~50 μ*M*, aliquot to ~10 μL, flash-freeze using liquid nitrogen, and store at −80°C. Once we thaw an aliquot of hMOF we discard the remainder of that aliquot as multiple flash freezing has a negative impact on enzymatic activity. In our experience hMOF expresses and purifies fairly well, and 1 L of cells will yield approximately 5 mg of protein (Fig. 2). In the case of hMOZ and yESA1, we found that

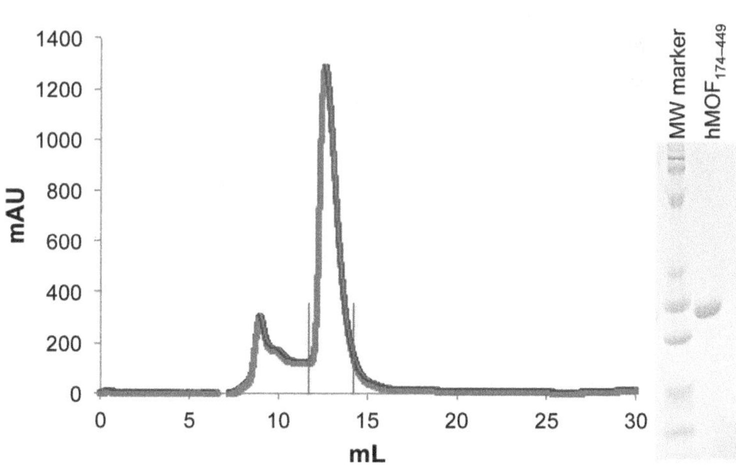

Fig. 2 *Left*: Chromatogram from size exclusion chromatography of $hMOF_{174–449}$ using a Superdex 75 column. The collected peak is indicated with *vertical lines*. *Right*: SDS-PAGE gel showing purified $hMOF_{174–449}$.

the proteins express better in LB rather than TB, but still express about 10 × less protein than hMOF. Both MOZ and ESA1 tolerate ion exchange well, and achieve a higher level of activity if sizing buffer/storage buffer is at a low pH and contains reducing agent. We use 20 m*M* sodium citrate, pH 5.5, 100 m*M* NaCl, and 1 m*M* DTT.

3. KINETIC ANALYSIS

3.1 Primer on Steady-State Kinetic Analysis

The steady-state kinetic parameters of a protein include the Michaelis constant (K_m), which is the concentration where the reaction rate of the enzyme is at one-half of its maximal velocity (V_{max}) and is related to the affinity of an enzyme for its substrate, and the catalytic constant (k_{cat}), which is the rate that an enzyme turns over its substrates (Fig. 3).

To determine the kinetic parameters for a given enzyme, very carefully conducted activity assays must be performed. The basic steps to follow to determine these parameters for a bisubstrate system such as the MYST family are to hold one substrate at a saturating concentration (~10 × K_m) while the other substrate is varied. About half of the concentrations of the varied substrate should be below K_m and the other half above K_m. Generally a minimum of eight concentrations should be used (Brooks et al., 2004). Results are then graphed as substrate concentrations vs reaction rate to obtain a Michaelis–Menten curve. For a thorough overview of how to perform these experiments see Brooks et al. (2004). Also, when performing these experiments it is important to keep the following in mind:

Maintain steady state: A kinetics reaction must be run under steady-state conditions, meaning that the reaction is occurring at a linear rate and less

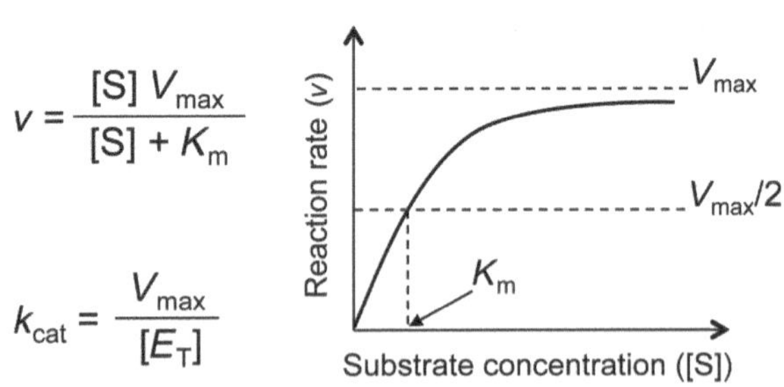

Fig. 3 Michaelis–Menten equation, k_{cat} equation, and graphical representation of the Michaelis–Menten equation where v = reaction rate, [S] = substrate concentration, V_{max} = maximum reaction velocity, K_m = Michaelis constant, and [E_T] = total enzyme concentration.

than 10% of the substrate is consumed. These parameters ensure that substrate concentration ([S]) can be assumed to remain constant for a given point. To find steady-state conditions a reaction progress curve should be obtained before starting kinetic assays. See Brooks et al. (2004) for an explanation of how to run a reaction progress experiment.

[Enzyme]⋘[substrate]: Enzyme concentration must be significantly lower than substrate concentration (generally at least 100 × lower). This allows for enzyme concentration to be equated to zero and excluded from the Michaelis–Menten equation.

Graph as a function of reaction rate: To obtain k_{cat} from a graph of Michaelis–Menten data, assay output signal must be converted to reaction rate. To do so, a standard curve should be generated so that output can be converted to a known amount of substrate conversion. For example, in our radioactivity assays the readout is in CPM (counts per minute) based on how much ^{14}C-acetyl is transferred from the acetyl CoA to H4. To generate our standard curve we measure a known number of moles of ^{14}C-acetyl CoA so that CPM can be related to molar transfer of the acetyl group in a linear equation. That value is then divided by reaction volume to obtain a reaction rate with units of $M\,s^{-1}$.

3.2 Specific Kinetic Properties of MYST Acetyltransferases

The MYST family of acetyltransferases exhibit high substrate K_m values. For example, the K_m for the yESA1 HAT domain is ~920 μ*M* for histone H3 (Yan et al., 2000). An important consideration when studying the MYST family is that a high K_m for the substrate peptide causes saturating conditions to be extremely high. For example in the case of yESA1, 9.2 m*M* of H3 would be 10 × K_m and thus saturating. To accommodate this MYST family characteristic, we have mainly used radioactivity-based assays to measure enzyme activity. Radioactive readouts tend to be quite sensitive and are not typically limited by binding capacities. For example if we require the attachment of the peptide substrate to a bead for a fluorescence-based assay, the binding capacity of the bead might be less than saturating for the peptide and we might then not be able to achieve maximum velocity (V_{max}). Furthermore, when using a synthetic peptide as the substrate, using such a high concentration may lead to trifluoroacetic acid (TFA) contamination that could destabilize the enzyme. According to Genscript's website, synthetic peptides may contain 10–45% TFA salt (Genscript TFA Removal Service). This amount of TFA could be damaging to the enzyme of interest, leading to a much lower V_{max} in assays where the peptide is the saturating component, or a decrease in activity as peptide concentration increases. To avoid this issue, it may be important to remove the TFA for kinetic studies. If that option is not available, then using more stringent buffering conditions or using slightly less than $10 \times K_m$ as an assumed saturating condition may be the best solution.

3.3 Controlling for Chemical Acetylation

Another hurdle that we have encountered in our lab is chemical acetylation of substrates (Olia et al., 2015). Because the acetyl group on acetyl CoA is a fairly labile bond it is possible for the acetyl transfer to occur in the absence of enzyme, especially with high concentrations of acetyl CoA in the presence of basic peptides. To control for this background chemical acetylation, it is imperative to include a control experiment that contains substrate and acetyl CoA but no enzyme. This can be especially important when testing very low activity levels such as when studying catalytic mutants. The chemical acetylation should then be subtracted from assay readouts to obtain a more accurate value for enzyme activity. For an example see Fig. 4.

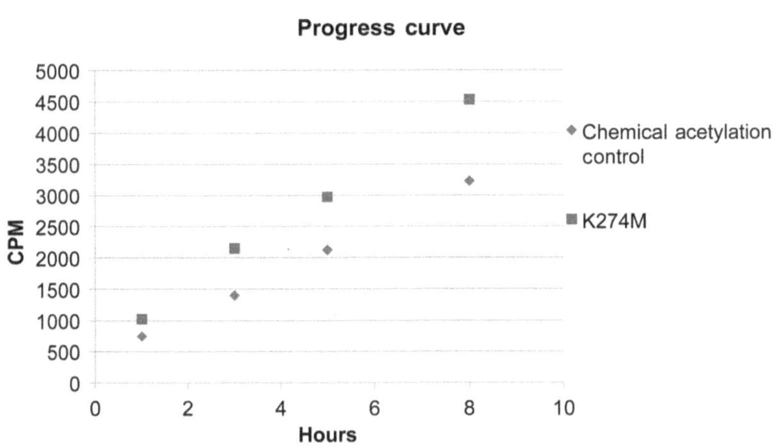

Fig. 4 Progress curve showing the enzymatic activity of the hMOF$_{174–449}$ K274M mutant over time as compared to a chemical acetylation control, which contains only the substrate and acetyl CoA, but no enzyme. The acetylation that occurs in the presence of the mutant enzyme is just slightly greater than the chemical acetylation observed in the absence of enzyme, indicating that this mutant has lost much of its catalytic activity. Note, however, that the activity levels of the mutant enzyme could be misleading without the chemical acetylation control.

3.4 Kinetic Assay Using Radioactive Acetyl-CoA

To perform kinetic experiments using a radioactivity-based assay we first mix hMOF at 50 n*M* with H4 peptide in PCR tubes in a buffer containing 40 m*M* Tris (pH 8.0), 100 m*M* NaCl, 800 μ*M* Cys, and 7.5 μ*M* bovine serum albumin (BSA). Depending on which substrate is tested, H4 will either be at varied concentrations or saturating. We begin the assay by adding ^{14}C-acetyl CoA (which will also be varied or saturating depending on which substrate is being examined) in a final volume of 50 μL. Because these assays are so sensitive and we want each experiment to run for exactly the same amount of time, we space out each reaction by 30 s (that is, begin reaction one at time 0:00, two at 0:30, three at 1:00, etc.). We allow the reaction to proceed for 8 min, which we have previously determined to be within the linear range for the enzyme. To stop the reaction, we spot 20 μL each of the reaction onto two P81 filter papers, which will bind the positively charged H4 peptide. This step, again, is done in 30-s intervals so that each reaction proceeds for exactly 8 min (that is, spot reaction one at time 8:00, two at 8:30, three at 9:00, etc.). The filter papers are then washed 3 × in 20 m*M* HEPES (pH 7.5). Because water is insoluble in some scintillation fluids and might interfere with signal readout, we dip each filter paper in acetone and lay each one out to dry on a paper towel. We then put each filter paper in a scintillation vial, add scintillation fluid, and read our results on a scintillation counter where CPM will correlate with the amount of ^{14}C-acetyl that has been transferred from the acetyl CoA to the H4 peptide and thus enzyme activity. We can then use this data to prepare Michaelis–Menten curves for the enzyme (Fig. 5).

4. THERMOFLUOR ASSAY TO INVESTIGATE COFACTOR AND INHIBITOR BINDING

A thermofluor assay measures the melting temperature of a protein (Pantoliano, Rhind, & Salemme, 2000). If the protein is stabilized by a cofactor or ligand it will have a higher melting temperature. For example, apo hMOF HAT domain has a T_m of about 37°C; however, upon addition of 100 μ*M* CoA, acetyl CoA, or acetonyl CoA the T_m increases to approximately 47°C, 50°C, and 52°C, respectively. Interestingly, however, we do not observe a strong thermal shift of hMOF in the presence of H4 peptide (Fig. 6). The thermofluor assay uses SYPRO® Orange, a dye that fluoresces in a hydrophobic environment. Thus if SYPRO® Orange is present as a

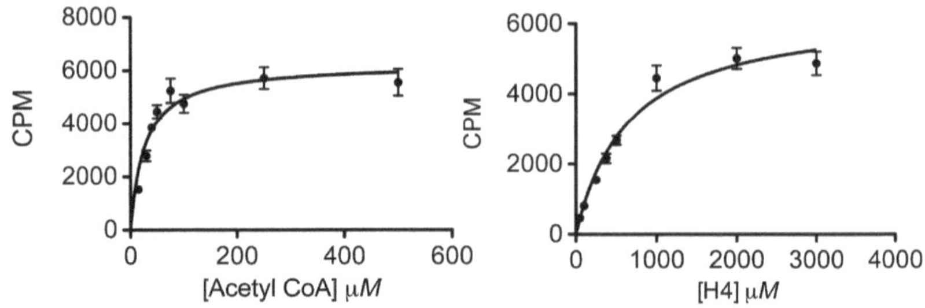

K_m for Acetyl CoA	26.72 ± 5.1 μ*M*
K_m for $H4_{1-19}$	648.8 ± 90.35 μ*M*
k_{cat}	2.88 s^{-1}

Fig. 5 Michaelis–Menten curves and calculated steady-state parameters for $hMOF_{174-449}$. Note that these values vary somewhat from the aforementioned K_m values for hMOF of 50 μ*M* and 400 μ*M* for acetyl CoA and H4 peptide, respectively, likely due to the difference in enzyme preparation and buffer conditions. To calculate the k_{cat} for this data, V_{max} was converted from CPM to *M* s^{-1} using a standard curve (data not shown) and divided by the total enzyme concentration of 50 n*M*.

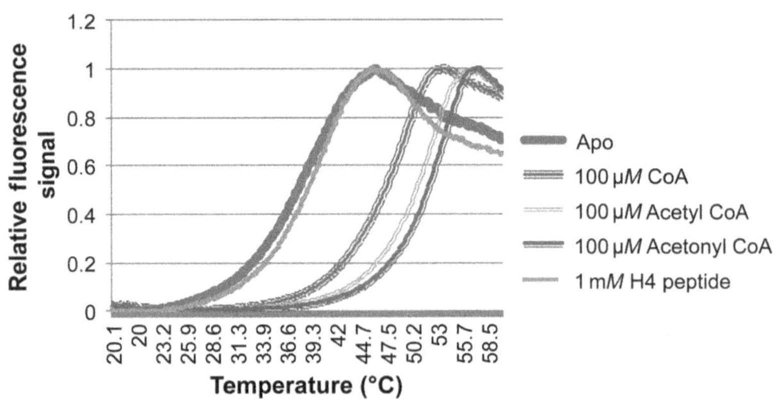

Fig. 6 Thermofluor assay as a readout of hMOF ligand binding. Data for apo hMOF, and hMOF in the presence of 1 m*M* H4 peptide, and 100 μ*M* CoA, acetyl CoA, and acetonyl CoA, respectively. T_m is the inflection point for each curve: ~37°C for apo hMOF, ~47°C for hMOF with CoA, ~50°C for hMOF with acetyl CoA, ~52°C for hMOF with acetonyl CoA, and ~38°C for hMOF with H4 peptide. (See the color plate.)

protein is melting it will bind the exposed hydrophobic patches and will fluoresce with increasing intensity as a protein unfolds. This data can then be plotted to produce a melting curve. All parameters for this assay can be determined empirically and may vary from protein to protein, but in general approximately 1 μ*M* of protein (15 μL) is incubated with 100 μ*M* of compound (1 μL) followed by addition of a 1:300 dilution of a 5000 × concentrate of SYPRO® Orange (4 μL) to a 20 μL final volume. Each sample is prepared in a minimum of triplicate and read using RT-PCR measuring from 20°C to 95°C.

There are some caveats to this assay. First, because of the nature of SYPRO® Orange fluorescence in a hydrophobic environment, detergent in the buffer can interfere with signal. In our experience, baseline fluorescence of SYPRO® Orange in the presence of detergent (0.05% Triton-X) is significantly increased and completely masks any melting data from the protein. Decreasing to 0.01% Triton-X shows some improvement but still complicates the melting curve of the protein. It is also possible that the detergent may directly affect protein unfolding properties. If detergent is absolutely required, then other dye options would need to be explored to employ this assay. Additionally, if another protein is present in the sample (a contaminant or BSA, for example) that protein will melt also, producing its own melting curve in tandem with the protein of interest and thus complicating the data.

5. INHIBITOR SCREENING

5.1 General Considerations

There are certain parameters that need to be considered for all high-throughput screening campaigns. These are reviewed in Williams and Scott (2009) and Inglese et al. (2007). This chapter seeks to touch on a subset of key considerations that we encountered when developing a MYST inhibitor screening assay and should not be treated as a comprehensive screening guide. Briefly, one should consider what assay is going to be most appropriate for the system being tested and what buffer conditions are appropriate for that assay. Typically, an additive of some kind will be required to terminate all of the reactions in HTS simultaneously; for our assays we have used 3.7% HCl for this purpose. Also, a control inhibitor is often required to monitor screening quality statistics. Acetonyl-CoA is a nonhydrolyzable acetyl CoA analog that has served us well as a control inhibitor against MYST proteins. An orthogonal secondary assay must also be planned to counter-screen any hits from the primary assay to eliminate any false positives that may have interfered with the detection system of the primary assay. There are also a few standard statistics for HTS that should be performed before screening compound libraries. These are discussed in Coma, Herranz, and Martin

$$Z' = 1 - \frac{(3\sigma_s + 3\sigma_b)}{|\mu_s - \mu_b|}$$

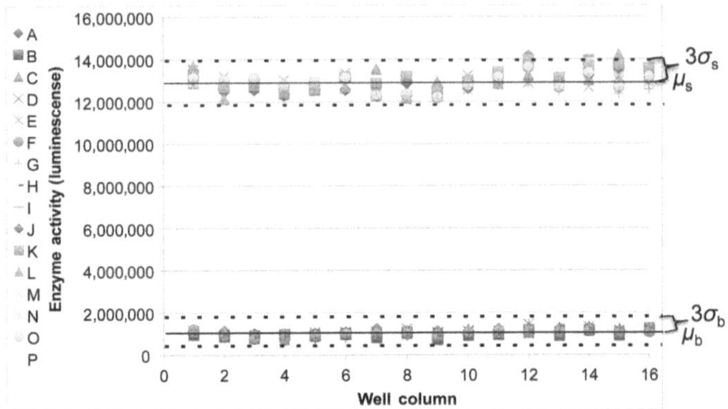

Fig. 7 Equation and sample data for determining assay *Z*-factor. Z' = *Z*-factor, σ_s = standard deviation of the positive signal, σ_b = standard deviation of the background signal, μ_s = mean of the positive signal, and μ_b = mean of the background signal. Each point in the graph represents the readout for a single reaction within a 384-well plate.

(2009). One important validation statistic is the *Z*-factor, a parameter which measures the screening window of an assay to assess assay quality (Zhang, Chung, & Oldenburg, 1999) (Fig. 7).

The *Z*-factor can be any value from 0 to 1 where assay quality increases with increasing *Z*-factor, which would indicate a large signal window with low signal variability. To obtain a *Z*-factor, a control plate should be run in the assay where half of the reactions are the controls for positive signal (enzyme with no inhibitor present/DMSO only) and the other half of the reactions are controls for background signal (either containing no enzyme, a catalytically dead mutant of the enzyme, or WT enzyme plus a saturating amount of control inhibitor, depending on what is available). The means and standard deviations of the results can then be calculated and put into the equation shown in Fig. 7 to calculate the *Z*-factor. Generally the minimum acceptable *Z*-factor value is 0.5, with higher values being more desirable.

In general, when screening for inhibitors each substrate concentration should be set at K_m within a reaction. This allows the assay to be sensitive to competitive inhibitors while maintaining robust signal (Williams & Scott, 2009). However, this can pose a problem for MYST family proteins, which tend to have a very high K_m for their substrates. It must be determined if enough of the substrate can be obtained to perform an entire screening campaign, which often involves a minimum of thousands of reactions. Furthermore, if the assay requires immobilization of the substrate, then the binding surface must have a high enough capacity to reach substrate K_m. It may be reasonable to use concentrations below K_m of the substrate if the K_m is too high to reasonably achieve in the screen. However in this case, it is important to consider that lowering the substrate concentration will slow the reaction rate. This may mean that your reaction will have to be run for a longer amount of time to obtain a detectable signal-to-noise ratio. In this case, it is important that the enzyme be stable enough to survive the duration of the assay. Additionally, lowering the concentration of one substrate in a bisubstrate system causes the screen to have a greater bias for finding inhibitors that are competitive with that substrate as it could then be easier to compete off the enzyme.

5.2 Buffer Considerations

It is important to remember that when transitioning from low-throughput to high-throughput assays there will potentially be a transition to more forms of solution transfer which may translate into a greater variety of contact surfaces for the enzyme/reaction mixture (tubing, transfer pins, etc.). In our experience working with the MYST protein hMOF, we have found that while we have had limited difficulty with protein stickiness in low throughput, we lose a significant portion of the protein through the course of transfer steps. To alleviate the effects of a much lower signal than expected when transitioning from low throughput to high throughput, we tested the addition of BSA to our assay buffer, finding that it increased our signal by about five times.

Cysteine reactive compounds can lead to a high number of false hits in a screening campaign. In the case of MYST HATs, which have been demonstrated to employ an active site cysteine for acetyltransferase activity (Yan et al., 2000), this can be a particular problem. It may be appropriate to add some amount of reducing agent to the reaction buffer to limit the amount of inhibition due to compounds reacting with the active site cysteine. This, however, can lead to more problems as it has been shown that redox cycling, or production of peroxide caused by strong reducing agent interactions with some compounds, can also lead to a high false hit rate (Johnston et al., 2008). To circumvent this issue, we have included in some of our assays free cysteine in an attempt to outcompete cysteine reactive compounds.

Small amounts of nonionic detergent can also help to decrease a false hit rate (Feng et al., 2007). This is especially true when screening MYST family acetyltransferases where screening at a higher compound concentration may lead to colloidal aggregation, which will be discussed in more detail in Section 5.5.

5.3 Assays

In the case of the MYST acetyltransferases, we have developed a radioactivity-based assay and an ELISA to use in high throughput. We have also developed a thermofluor assay to be used as a secondary orthogonal assay in low throughput.

5.3.1 Radioactivity-Based Assay

Because of the sensitivity and robustness of radioactivity assays, we decided to adapt the radioactivity-based P81 filter paper assay to be used in high throughput as a primary assay. The assay is performed by dispensing hMOF (or inactive mutant hMOF E350Q as a negative control) into each well of a 384-well plate followed by the addition of compound and then an H4 peptide/^{14}C-acetyl CoA solution to final concentrations of 50 n*M* hMOF, 50 μ*M* ^{14}C- acetyl CoA, and 400 μ*M* H4 peptide (both substrates are used at $\sim K_m$) in 40 m*M* Tris (pH 8.0), 100 m*M* NaCl, 800 μ*M* Cys, and 7.5 μ*M* BSA. The reaction is allowed to proceed for 1 h and then 3.7% HCl is added to each well to stop the reaction. The reaction is then transferred from the 384-well plate to four 96-well P81 paper-embedded mesh bottom plates. A vacuum manifold is used to pull excess liquid from the wells. Next, each well is washed with 10 m*M* HEPES (pH 7.5), followed by addition of acetone, which is also vacuumed off. The plate bottoms are then sealed and scintillation fluid is added to each well. The plates are then sealed and read on a TopCount plate reader. The *Z*-factor for this assay has remained >0.6 for all optimized trials and we have confirmed the ability of the assay to detect inhibition by acetonyl CoA.

5.3.2 ELISA

We also developed an ELISA, which is performed in automation by first binding H4-GST to a 384-well glutathione-coated plate. We have found that hMOF acetylates H4-GST almost as well as H4 peptide, whereas GST-H4 barely becomes acetylated at all. 50 n*M* enzyme followed by test compounds is added to the plate followed by 50 μ*M* acetyl CoA. The reaction is allowed to react for 3 h to obtain a suitable signal window. Primary anti-H4AcK antibody is added and allowed to bind for 30 min. Next, the secondary antibody conjugated to horseradish peroxidase (HRP) is added and allowed to bind for 30 min. Chemiluminescent substrate is added and is oxidized by HRP to produce light that can be detected on a plate reader. The plate is washed between each step. The *Z*-factor has remained reproducibly above 0.6.

This assay provides a good example of how requiring a binding surface may limit how much substrate can be included in the assay, thus slowing reaction time. In the case of this assay, because screening could take multiple hours, the stability of the assay was tested as a function of time. Another option is to perform the full reaction in a separate plate, keeping the H4-GST concentration at K_m and then transferring the full reaction to the glutathione plate second. This, however, adds an extra step, so it is important to weigh the pros and cons that adding the extra step would introduce to the assay and the screening campaign.

Another important aspect of this assay to note is that not all primary antibodies developed to recognize the acetylated substrate will perform equally well. In the case of this assay, multiple antibodies were characterized via a dot blot as well as screening within the assay to ensure we could achieve the highest signal-to-noise ratio with the best performing antibody. In our experience, Histone H4 tetra-acetyl antibody (pAb) (Cat# 39179, 39228) and Histone H4ac (pan-acetyl) antibody (pAb) (Cat# 39243, 39244) from Active Motif have performed relatively well in our assay, giving signal windows of approximately 10 × and 4 ×, respectively.

5.3.3 Other Assays

There are a number of other assays that we have considered or used that despite their successes with other systems, we have not yet been able to optimize to work effectively with the MYST family of acetyltransferases. Some methods such as Alpha (Amplified Luminescent Proximity Homogeneous Assay) and LANCE® TR-FRET (time resolution-fluorescence resonance energy transfer) assays rely on both substrate binding to beads as well as antibody recognition of the substrate. Briefly, for these assays there are two beads, one donor bead and one acceptor bead. For our purposes, one bead is conjugated to H4 peptide and the other bead is conjugated to an antibody that recognizes acetylated H4K16. Thus, when hMOF acetylates the H4 peptide on one bead, the antibody on the other bead will bind the acetylated peptide, bringing the donor and acceptor beads in close proximity so that a signal can be detected. In the case of the Alphascreen® when the beads are in close proximity and illuminated with 680 nm light, singlet oxygen is transferred from the donor bead to the acceptor bead, which will react to produce

520–620 nm light (PerkinElmer Life and Analytical Sciences, 2011). In the case of the LANCE® TR-FRET assay, when the beads are in close proximity irradiated europium chelate molecules from the donor beads transfer energy to the acceptor bead which contain a dye that converts that energy to light which emits at 665 nm (Legault, Roby, Beaudet, & Rouleau, 2006). These assays can be particularly tricky to use with MYST proteins because, as mentioned before, substrate peptide K_m tends to be very high, meaning that the binding capacity of the beads would also need to be quite high. If the beads are unable to bind a high level of the H4 peptide either the level of peptide will have to be decreased, slowing reaction rate, or the assays will display what is known as a "hooking effect" where the free peptide outcompetes the peptide that is bound to the bead to artifactually lower signal (PerkinElmer Life and Analytical Sciences, 2011). Furthermore, the success of the assay depends on the antibody-conjugated bead being specific for a given substrate, and it could take excessive amounts of time and money to test several antibody-conjugated beads to find the best one for optimum signal, if multiple antibodies are even available.

Another assay that has been troublesome for us due to antibody recognition of substrate is fluorescence polarization. To perform this assay we ordered the H4 peptide conjugated to an FITC dye. To our surprise, none of the antibodies we tested were able to recognize the acetylated peptide in the presence of the FITC tag. We have also tried to use a CoA detection assay. Briefly this assay uses a sulfur reactive dye that reacts with CoA after acetyl CoA is cleaved in the enzymatic assay (Bulfer, McQuade, Larsen, & Trievel, 2011; Trievel, Li, & Marmorstein, 2000). This assay has been ineffective with the MYST proteins for multiple reasons. First, there are eight cysteines in our construct of the hMOF HAT domain. Despite the low levels of enzyme used in activity assays (50 n*M*), the apparent interaction between these cysteines and the dye appears to cause high background signal. Furthermore, the sensitivity of this assay has not been robust enough to detect the generally low levels of activity of MYST proteins.

5.4 Thermofluor Binding Assay

We have taken advantage of the thermofluor assay as a secondary orthogonal screen to perform in low throughput to validate compounds identified in the primary assay. We have also begun efforts to translate this assay from low throughput to high throughput for fragment screening. We currently perform the high-throughput assay by adding 15 μL of 10 μ*M* hMOF in 20 m*M*

HEPES, pH 7.5, 300 mM NaCl to each well, followed by 1 μL of 20 mM fragment compound to a final concentration of 1 mM compound (or 100 μM CoA as a positive control). After allowing the compounds to incubate with the protein for 30 min, 4 μL of a 1:300 dilution of 5000 × SYPRO® Orange dye is added to each well. The plate is then sealed and read on a real-time qPCR instrument. For this assay, we have had some difficulty with deviations in control T_m and obtaining quality screening statistics. Further optimization will be required to determine if this assay will be suitable for high-throughput screening of MYST proteins.

5.5 Eliminating Promiscuous Inhibitors

After completing a screening campaign, it is important to focus on the highest quality hits that have the potential to be developed into probes or therapeutics. In general, a good inhibitor should display reproducible dose-dependent inhibition, have a Hill slope of approximately 1.0, bind reversibly to the protein (although covalent inhibitors can be beneficial under certain circumstances, as discussed in chapter 8 of Copeland (2005)), and also be selective for the protein of interest. When a compound identified through screening does not meet these criteria, it becomes important to scrutinize it carefully to identify and eliminate compounds that are disrupting enzymatic activity through nonspecific interactions. For example, MYST family acetyltransferases were found to be particularly sensitive toward cysteine reactive compounds, mercury-containing compounds, and copper-containing compounds. As mentioned earlier, MYST proteins have an active site cysteine responsible for acetyl transfer; thus, cysteine reactive compounds may nonspecifically form disulfide bonds to inhibit the protein. Similarly, mercury-containing compounds are also able to interact nonspecifically with the active site cysteine. Additionally, in our experience, copper-containing compounds are especially active against the MYST family proteins. We hypothesize that this occurs because the copper interferes with the active site cysteine, or otherwise may be displacing the zinc in the zinc finger or both. These three types of compounds are poor candidates for further optimization.

5.6 Pan-Assay Interference Compounds

Pan-assay interference compounds (PAINS) are a class of compounds that mimic genuine inhibitors by interfering with assay readouts or undergoing nonspecific reactions with proteins. About half of PAINS known today fall into just 16 structural categories (Baell & Walters, 2014), and learning these structures can save a lot of time and effort. However, because about half of PAINS do not fit into these categories it is also prudent to check the literature for other reports of a selected compound hitting unrelated proteins as this could be a red flag (Baell & Walters, 2014).

In our experience with MYST proteins, it is especially important to verify the identity and purity of the identified hits in the library. Compounds within a library can break down over time into reactive products (Baell & Walters, 2014). Through the course of our assays, we have found that the MYST protein hMOF had numerous hits where mass spectrometry revealed a plethora of compounds were present in the sample when there should have only been one. This leads us to believe that MYST proteins may be particularly sensitive to these sorts of reactive breakdown products.

Testing for the reversibility of a compound can be a fairly straightforward way to determine if it is interacting with the protein of interest through desirable interactions. If a compound is specifically binding a region of a protein in a dose-dependent manner, then the enzyme should regain activity when the compound is diluted (Copeland, 2005). Briefly, to perform this experiment (in the context of the MYST protein, hMOF), 5 μM enzyme is first incubated with $10\times IC_{50}$ of a compound for $\sim$30 min. The solution is then diluted $100\times$ such that the enzyme concentration is at 50 nM and the compound is at $0.1\times IC_{50}$, where little to no activity should be expected from an inhibitor. A reaction is then started by adding H4 peptide and acetyl CoA at K_m and time points are taken every 5 min for an hour. As a positive control for reversibility, we measure enzyme activity with no inhibitor added (DMSO control) as well as enzyme activity with the reversible inhibitor acetonyl CoA. As a negative control for reversibility, we use merbromin, a mercury-containing compound that interacts with the active site cysteine. If a compound reversibly binds to the enzyme then the activity of the enzyme should increase with time aligning with the positive controls, and if a compound is irreversible, the enzyme will remain inhibited and little activity will be seen over time, aligning with the negative control (Fig. 8). Compounds that are found to be irreversible inhibitors may be nonspecifically modifying the protein of interest and are much less likely to serve as good candidates for optimization.

5.7 Colloidal Aggregators

Colloidal aggregators are compounds that aggregate and sequester a protein to limit its activity (McGovern, Helfand, Feng, & Shoichet, 2003). This can be a particular problem when screening MYST proteins as compounds tend

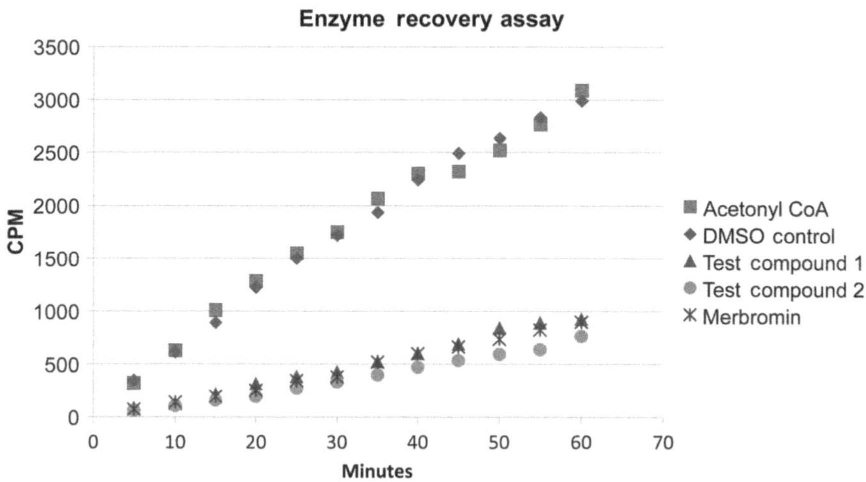

Fig. 8 Sample data for an enzyme recovery assay with a MYST protein. Note that both of the test compounds display irreversible inhibition, as they inhibit enzyme activity after dilution, aligning with the control irreversible inhibitor merbromin, a mercury-containing compound that interacts with the active site cysteine. Also note that acetonyl CoA, a reversible inhibitor, aligns very closely to the DMSO control. (See the color plate.)

to be screened at higher inhibitor concentrations (low micromolar range) where they can have a higher tendency to aggregate. In general, if a compound is reversibly interacting with an enzyme in a 1:1 stoichiometry then the Hill slope for that compound with the enzyme should be ~1 (Feng et al., 2007), and if the Hill slope is higher than 1, this may indicate that the compound is acting through colloidal aggregate inhibition. Furthermore, colloidal aggregates can be solubilized by the addition of small amounts of nonionic detergents (~0.01% Triton-X) (McGovern et al., 2003). If running an activity assay in the presence of detergent causes the inhibitory activity of a compound to be lost or significantly decreased, this may be a sign that the compound is acting via aggregation rather than by a specific interaction. To avoid initial identification of colloidal aggregators in a primary screen it can be helpful to include small amounts of nonionic detergent in the assay buffer. However, if detergent interferes with the assay readout of the primary screen, it is important to then include detergent in secondary screens to eliminate compounds acting through colloidal aggregation.

6. PROSPECTS AND CONCLUSIONS

The MYST family proteins remain one of the more understudied families of the acetyltransferases. This is likely, at least in part, due to their weak affinity for their substrates, which can make them especially challenging to study. Furthermore, the MYST proteins appear to be only weakly inhibited by validated inhibitors, and significantly sensitive to nonspecific inhibitors further complicating efforts to further probe their activity and find new inhibitors. This chapter seeks to prepare researchers for some hurdles that they may encounter when studying the MYST proteins so that there may be better opportunity to plan stringent controls and obtain high-quality data.

ACKNOWLEDGMENTS

This work was supported by the National Institutes of Health grants R01 GM060293, R01 GM098910, and P01 AG031862 to R.M., and T32 GM008275 to C.E.M. We acknowledge support of the molecular screening facility at the Wistar Institute (NIH P30 CA010815) and the DNA Sequencing Facility at the Perelman School of Medicine, University of Pennsylvania (NIH P30 CA016520).

REFERENCES

Baek, S. H., Ohgi, K. A., Rose, D. W., Koo, E. H., Glass, C. K., & Rosenfeld, M. G. (2002). Exchange of N-CoR corepressor and Tip60 coactivator complexes links gene expression by NF-kappaB and beta-amyloid precursor protein. *Cell*, *110*(1), 55–67. Retrieved from http://www.ncbi.nlm.nih.gov/pubmed/12150997.

Baell, J., & Walters, M. A. (2014). Chemistry: Chemical con artists foil drug discovery. *Nature*, *513*(7519), 481–483. http://dx.doi.org/10.1038/513481a.

Berndsen, C. E., Albaugh, B. N., Tan, S., & Denu, J. M. (2007). Catalytic mechanism of a MYST family histone acetyltransferase. *Biochemistry*, *46*(3), 623–629. http://dx.doi.org/10.1021/bi602513x.

Brooks, H. B., Geeganage, S., Kahl, S. D., Montrose, C., Sittampalam, S., Smith, M. C., & Weidner, J. R. (2004). Basics of enzymatic assays for HTS. In G. S. Sittampalam, N. P. Coussens, H. Nelson, M. Arkin, D. Auld, & C. Austinet al.*Assay guidance manual*. Bethesda, MD: Eli Lilly & Company and the National Center for Advancing Translational Sciences.

Bulfer, S. L., McQuade, T. J., Larsen, M. J., & Trievel, R. C. (2011). Application of a high-throughput fluorescent acetyltransferase assay to identify inhibitors of homocitrate synthase. *Analytical Biochemistry*, *410*(1), 133–140. http://dx.doi.org/10.1016/j.ab.2010.11.004.

Cao, X., & Sudhof, T. C. (2001). A transcriptionally [correction of transcriptively] active complex of APP with Fe65 and histone acetyltransferase Tip60. *Science*, *293*(5527), 115–120. http://dx.doi.org/10.1126/science.1058783.

Cao, X., & Sudhof, T. C. (2004). Dissection of amyloid-beta precursor protein-dependent transcriptional transactivation. *Journal of Biological Chemistry*, *279*(23), 24601–24611. http://dx.doi.org/10.1074/jbc.M402248200.

Cereseto, A., Manganaro, L., Gutierrez, M. I., Terreni, M., Fittipaldi, A., Lusic, M., … Giacca, M. (2005). Acetylation of HIV-1 integrase by p300 regulates viral integration. *EMBO Journal*, *24*(17), 3070–3081. http://dx.doi.org/10.1038/sj.emboj.7600770.

Coma, I., Herranz, J., & Martin, J. (2009). Statistics and decision making in high-throughput screening. *Methods in Molecular Biology*, *565*, 69–106. http://dx.doi.org/10.1007/978-1-60327-258-2_4.

Copeland, R. A. (2005). Evaluation of enzyme inhibitors in drug discovery. A guide for medicinal chemists and pharmacologists. *Methods of Biochemical Analysis*, *46*, 1–265. Retrieved from http://www.ncbi.nlm.nih.gov/pubmed/16350889.

Feng, B. Y., Simeonov, A., Jadhav, A., Babaoglu, K., Inglese, J., Shoichet, B. K., & Austin, C. P. (2007). A high-throughput screen for aggregation-based inhibition in a large compound library. *Journal of Medicinal Chemistry*, *50*(10), 2385–2390. http://dx.doi.org/10.1021/jm061317y.

Fraga, M. F., Ballestar, E., Villar-Garea, A., Boix-Chornet, M., Espada, J., Schotta, G., … Esteller, M. (2005). Loss of acetylation at Lys16 and trimethylation at Lys20 of histone H4 is a common hallmark of human cancer. *Nature Genetics*, *37*(4), 391–400. http://dx.doi.org/10.1038/ng1531.

Genscript TFA Removal Service. Retrieved from http://www.genscript.com/tfa_removal_service.html.

Hobbs, C. A., Wei, G., DeFeo, K., Paul, B., Hayes, C. S., & Gilmour, S. K. (2006). Tip60 protein isoforms and altered function in skin and tumors that overexpress ornithine decarboxylase. *Cancer Research*, *66*(16), 8116–8122. http://dx.doi.org/10.1158/0008-5472.CAN-06-0359.

Iizuka, M., & Stillman, B. (1999). Histone acetyltransferase HBO1 interacts with the ORC1 subunit of the human initiator protein. *Journal of Biological Chemistry*, *274*(33), 23027–23034. Retrieved from http://www.ncbi.nlm.nih.gov/pubmed/10438470.

Inglese, J., Johnson, R. L., Simeonov, A., Xia, M., Zheng, W., Austin, C. P., & Auld, D. S. (2007). High-throughput screening assays for the identification of chemical probes. *Nature Chemical Biology*, *3*(8), 466–479. http://dx.doi.org/10.1038/nchembio.2007.17.

Johnston, P. A., Soares, K. M., Shinde, S. N., Foster, C. A., Shun, T. Y., Takyi, H. K., … Lazo, J. S. (2008). Development of a 384-well colorimetric assay to quantify hydrogen peroxide generated by the redox cycling of compounds in the presence of reducing agents. *Assay and Drug Development Technologies*, *6*(4), 505–518. http://dx.doi.org/10.1089/adt.2008.151.

Kinoshita, A., Whelan, C. M., Berezovska, O., & Hyman, B. T. (2002). The gamma secretase-generated carboxyl-terminal domain of the amyloid precursor protein induces apoptosis via Tip60 in H4 cells. *Journal of Biological Chemistry*, *277*(32), 28530–28536. http://dx.doi.org/10.1074/jbc.M203372200.

Kitabayashi, I., Aikawa, Y., Yokoyama, A., Hosoda, F., Nagai, M., Kakazu, N., … Ohki, M. (2001). Fusion of MOZ and p300 histone acetyltransferases in acute monocytic leukemia with a t(8;22)(p11;q13) chromosome translocation. *Leukemia*, *15*(1), 89–94. Retrieved from http://www.ncbi.nlm.nih.gov/pubmed/11243405.

Lafon, A., Chang, C. S., Scott, E. M., Jacobson, S. J., & Pillus, L. (2007). MYST opportunities for growth control: Yeast genes illuminate human cancer gene functions. *Oncogene*, *26*(37), 5373–5384. http://dx.doi.org/10.1038/sj.onc.1210606.

Legault, M., Roby, P., Beaudet, L., & Rouleau, N. (2006). *Comparison of LANCE ultra TR-FRET to PerkinElmer's classical LANCE TR-FRET platform for kinase applications*. Shelton, CT: PerkinElmer Life and Analytical Sciences Application Note.

Lin, Y.-Y., Lu, J.-Y., Zhang, J., Walter, W., Dang, W., Wan, J., … Zhu, H. (2009). Protein acetylation microarray reveals that NuA4 controls key metabolic target regulating gluconeogenesis. *Cell*, *136*(6), 1073–1084. http://dx.doi.org/10.1016/j.cell.2009.01.033.

Lleonart, M., Vidal, F., Gallardo, D., Diaz-Fuertes, M., Rojo, F., Cuatrecasas, M., … Ramon y Cajal, S. (2006). New p53 related genes in human tumors: Significant downregulation in colon and lung carcinomas. *Oncology Reports*, *16*(3), 603–608. Retrieved from http://www.ncbi.nlm.nih.gov/pubmed/16865262.

McGovern, S. L., Helfand, B. T., Feng, B., & Shoichet, B. K. (2003). A specific mechanism of nonspecific inhibition. *Journal of Medicinal Chemistry*, *46*(20), 4265–4272. http://dx.doi.org/10.1021/jm030266r.

Mellert, H. S., & McMahon, S. B. (2009). hMOF, a KAT(8) with many lives. *Molecular Cell, 36*(2), 174–175. http://dx.doi.org/10.1016/j.molcel.2009.10.005.

Olia, A. S., Barker, K., McCullough, C. E., Tang, H. Y., Speicher, D. W., Qiu, J., … Marmorstein, R. (2015). Nonenzymatic protein acetylation detected by NAPPA protein arrays. *ACS Chemical Biology, 10*, 2034–2047.

Pantoliano, M. W., Rhind, A. W., & Salemme, F. R. (2000). Microplate thermal shift assay for ligand development and multi-variable protein chemistry optimization. US Patent No. 6,020,141.

PerkinElmer Life and Analytical Sciences. (2011). User's guide to alpha assays protein: Protein Interactions.

Sapountzi, V., & Cote, J. (2011). MYST-family histone acetyltransferases: Beyond chromatin. *Cellular and Molecular Life Sciences, 68*(7), 1147–1156. http://dx.doi.org/10.1007/s00018-010-0599-9.

Trievel, R. C., Li, F.-Y., & Marmorstein, R. (2000). Application of a fluorescent histone acetyltransferase assay to probe the substrate specificity of the human p300/CBP-associated factor. *Analytical Biochemistry, 287*(2), 319–328. http://dx.doi.org/10.1006/abio.2000.4855.

Williams, K. P., & Scott, J. E. (2009). Enzyme assay design for high-throughput screening. *Methods in Molecular Biology, 565*, 107–126. http://dx.doi.org/10.1007/978-1-60327-258-2_5.

Wu, J., Wang, J., Li, M., Yang, Y., Wang, B., & Zheng, Y. G. (2011). Small molecule inhibitors of histone acetyltransferase Tip60. *Bioorganic Chemistry, 39*(1), 53–58. http://dx.doi.org/10.1016/j.bioorg.2010.11.003.

Wu, J., Xie, N., Wu, Z., Zhang, Y., & Zheng, Y. G. (2009). Bisubstrate inhibitors of the MYST HATs Esa1 and Tip60. *Bioorganic & Medicinal Chemistry, 17*(3), 1381–1386. http://dx.doi.org/10.1016/j.bmc.2008.12.014.

Yan, Y., Barlev, N. A., Haley, R. H., Berger, S. L., & Marmorstein, R. (2000). Crystal structure of yeast Esa1 suggests a unified mechanism for catalysis and substrate binding by histone acetyltransferases. *Molecular Cell, 6*(5), 1195–1205. Retrieved from http://www.ncbi.nlm.nih.gov/pubmed/11106757.

Yang, X. J. (2004). The diverse superfamily of lysine acetyltransferases and their roles in leukemia and other diseases. *Nucleic Acids Research, 32*(3), 959–976. http://dx.doi.org/10.1093/nar/gkh252.

Yuan, H., Rossetto, D., Mellert, H., Dang, W., Srinivasan, M., Johnson, J., … Marmorstein, R. (2012). MYST protein acetyltransferase activity requires active site lysine autoacetylation. *EMBO Journal, 31*(1), 58–70. http://dx.doi.org/10.1038/emboj.2011.382.

Zhang, J. H., Chung, T. D., & Oldenburg, K. R. (1999). A simple statistical parameter for use in evaluation and validation of high throughput screening assays. *Journal of Biomolecular Screening, 4*(2), 67–73. Retrieved from http://www.ncbi.nlm.nih.gov/pubmed/10838414.

CHAPTER SEVEN

Preparation and Biochemical Analysis of Classical Histone Deacetylases

A. Villagra*, E. Sahakian†, E. Seto*[,1]

*George Washington University Cancer Center, School of Medicine and Health Sciences, The George Washington University, Washington, DC, United States
†H. Lee Moffitt Cancer Center, Tampa, FL, United States
[1]Corresponding author: e-mail address: seto@gwu.edu

Contents

1. Introduction 162
2. Generation of HDAC Stable Cell Lines Using Lentivirus 164
 2.1 Initial Titration for Selecting Antibiotic Resistance 164
 2.2 Establishment of Monoclonal Populations 166
3. Conventional HDAC Activity Assay 167
 3.1 Labeling of H4 Peptide with [^{3}H] for HDAC Assays 168
 3.2 HDAC Assay 169
4. Assessment of Selective Inhibition of HDACs Using HDAC Knockout Cells 169
 4.1 Quantification of Deacetylase Activity in Multiplate Readers 171
5. HDAC Chromatin Immunoprecipitation Assay 174
 5.1 Cross-Linking 174
 5.2 Chromatin Preparation 175
 5.3 Sample Sonication 176
 5.4 Shear Sample 176
 5.5 Chromatin Immunoprecipitation 177
 5.6 Sample Wash and Elution 177
6. Identification of Specific HDAC Substrates 179
Acknowledgments 180
References 180

Abstract

Histone deacetylase assays were first developed in the 1970s, and subsequently refined in the 1990s with the cloning of HDAC enzymes. Most of these early assays, relying on traditional in vitro chemical methodologies, are still applicable today. More recently, however, cell-based HDAC assays that measure HDAC activities in physiological conditions are emerging. Also, there is a continuing development of assays that can measure an isolated HDAC in the absence of other HDAC activities. This chapter reviews some of the older established methods for assaying HDAC activities, as well as introduces more recently developed nontraditional assays.

Methods in Enzymology, Volume 573
ISSN 0076-6879
http://dx.doi.org/10.1016/bs.mie.2016.04.005

ABBREVIATIONS

APCs antigen-presenting cells
ChIP chromatin immunoprecipitation
CPM counts per minute
EDTA ethylenediaminetetraacetic acid
FBS fetal bovine serum
HAT histone acetyltransferase
HCl hydrochloric acid
HDAC histone deacetylase
HEPES 4-(2-hydroxyethyl)-1-piperazineethanesulfonic acid
HPLC high-performance liquid chromatography
IP immunoprecipitation
IPLB immunoprecipitation lysis buffer
KCl potassium chloride
LiCl lithium chloride
LPS lipopolysaccharide
M molar
mg milligram
μg microgram
$MgCl_2$ magnesium chloride
mL milliliter
m*M* millimolar
NaCl sodium chloride
PBS phosphate-buffered saline
qRT-PCR quantitative real-time polymerase chain reaction
RCF (*g*) relative centrifugal force
rpm revolutions per minute
TE Tris-EDTA buffer
μL microliter

1. INTRODUCTION

Lysine acetylation at the N-terminal of histones was initially described as a transcriptional regulatory mechanism in cells. In a nonmodified state, the highly positive N-terminal ends of histones interact with DNA, generating an obstacle for the binding of transcription factors and, perhaps, the recruitment of proteins necessary to read the "writing pattern" on nucleosomes to exert transcription. Acetylation of histones potentially neutralizes these positive charges promoting a relaxed nucleosome conformation, which allow the binding of transcription factors and other proteins. These acetyl modifications are introduced by a heterogeneous group of proteins named histone acetyltransferases (HATs), most of them exist in multiprotein complexes that can be selectively recruited to DNA upon exogenous or endogenous cellular stimuli (Lee & Workman, 2007). Contrary to the action of HATs, acetyl groups can be removed by another group of proteins collectively known as histone deacetylases (HDACs). The 18 potential HDACs identified in humans are divided in two families and four classes. The classical HDAC family of zinc-dependent enzymes is composed by classes I, II, and IV, and the class III NAD^+-dependent enzymes belonging to the sirtuin family of HDACs. This chapter will deal with only the classical HDACs. The class I HDACs (HDAC1, 2, 3, and 8) are most closely related to the yeast deacetylase RPD3, and the class II HDACs are subdivided into class IIa (HDAC4, 5, 7, and 9) and class IIb (HDAC6 and 10). Both subclasses share homology with the yeast deacetylase HDA1. Finally, the last HDAC discovered, HDAC11, is in its own class IV and does not share homology with either RPD3 or HDA1 yeast deacetylases.

The role of HDACs was initially thought to be limited to their effects on histones. Later studies, however, revealed that HDACs encompass more complex regulatory functions dependent on their tissue expression, cellular compartment distribution, and stage of cellular differentiation (Glozak, Sengupta, Zhang, & Seto, 2005; Minucci & Pelicci, 2006). Although major advances have been made in understanding the role of specific HDACs in cell proliferation and survival, their role in the regulation of many other biological processes requires more intensive investigations. Additionally, HDAC enzymes, originally described as histone modifiers, were later demonstrated to modify a variety of other proteins unrelated to chromatin. The expanded role of HDACs over nonhistone substrates has been explored in numerous areas, including the modulation of proteins related to cell cycle, apoptosis, immune regulation, oncogenesis, metabolism, cellular differentiation, etc. The specific action and selectivity of individual HDACs over most of these biological processes is still not completely understood and, therefore, requires development of new and improved current HDAC assays.

This chapter discusses various functional assays to evaluate the activity of HDACs on histone and nonhistone targets. In addition to discussions of commonly used, established techniques, we present protocols of new techniques aimed toward dissecting functional characterization of specific HDACs, including use of comparative models such as parallel assays with HDAC knockout and/or knockdown cell lines.

2. GENERATION OF HDAC STABLE CELL LINES USING LENTIVIRUS

There are many ways to produce HDACs in vitro and in vivo for the purpose of assaying HDAC activities. In our hands, we have found that most HDACs expressed in stable mammalian cell lines are suitable for most HDAC assays. Generation of stable cell lines is a well-established technique and has been previously described in many publications. Among the methods described, lentiviral particles offer many advantages when compared to other methodologies, including those using plasmid vectors. Perhaps the most significant benefits are lentivirus' exceptional ability to infect both replicating and nonreplicating cells and, more importantly, their capacity of genomic integration, which provides stable expression of the transduced gene. We have successfully used this method to generate a large number of cells expressing various HDACs and, as an example, present here the generation of a Flag-tagged HDAC11 stable cell line.

This protocol is conducted over a span of several days. Although the general instructions are standardized to work with adherent cells, the protocol can be modified to work with suspension cells. The lentiviral particles overexpressing Flag-HDAC11 were reported previously (Cheng, Lienlaf, Perez-Villarroel, et al., 2014), and one of the models of choice is the mouse macrophage RAW264.7 cells. Lentiviral particles are commercially available already packaged by different suppliers; however, this procedure can be added as a preliminary step to the protocol (Cribbs, Kennedy, Gregory, & Brennan, 2013; Wang & McManus, 2009).

We suggest using a control empty vector for each condition to ensure specificity of the target gene overexpression. Once a lentiviral stock with a suitable titer is obtained, transduce the lentiviral particles into the RAW264.7 cell line. We recommend using a wide range of MOIs (multiplicity of infection). MOI equals of number of viral particles per cell, for example, a MOI of 1 = infection with one viral genome per cell.

2.1 Initial Titration for Selecting Antibiotic Resistance

The vector carrying the transgene of interest must have an additional antibiotic resistance gene in order to facilitate the selection of positive transduced cells. In this particular case, we used a vector carrying puromycin resistance. Therefore, a preliminary step to test the minimal concentration of this antibiotic needed to kill the cells is necessary. Briefly, we seeded a 96-well plate with 1×10^4 RAW264.7 cells per well and tested, in triplicate, the following concentrations of puromycin; 0 (untreated), 0.1, 0.5, 1.0, 2.5, 5.0, 7.5, 10.0, 12.5, 15.0, 17.5, and 20.0 μg/mL. Change media every 48 h, keeping the aforementioned concentration of puromycin in each well. Usually, cellular death will be observed after 5–7 days of culture. The minimal concentration of puromycin needed to kill 100% of the cells will be the concentration to use in the selection of stable cell lines.

Day 1

1. Seed 1×10^4 cells per well in a 96-well plate using 120 μL of RPMI supplemented with 10% fetal bovine serum (FBS) and incubate overnight at 37°C in a humidified incubator in an atmosphere of 5% CO_2 to allow complete adherence. Avoidance of antibiotics and fungicides is highly suggested. This recommended amount of cells will result in approximately 75% confluence. Some variation and further optimizations may be required when working with other cell types. We also suggest using all wells in the plate; this will avoid any loss of media due to evaporation.

Day 2

2. Replace the media of each well with 120 μL of hexadimethrine bromide 2 mg/mL (Sigma™ Cat. # H9268). Incubate for 5 min. Some primary cells are sensitive to hexadimethrine bromide.
3. We suggest creating eight different triplicate dilutions of the lentiviral particles as follows: 0.0, 0.01, 0.05, 0.1, 0.5, 1.0, 5.0, and 10.0 MOI. At this point half of the plate will be in use, 3×8 for your target gene and 3×8 for your control vector. However, the entire plate must be filled with media to avoid evaporation. Incubate overnight at 37°C in a humidified incubator in an atmosphere of 5% CO_2. Some specific cell lines may require shorter periods of incubation.

Day 3

4. Replace media with 120 μL of RPMI supplemented with 10% FBS and incubate for 48 h at 37°C in a humidified incubator in an atmosphere of 5% CO_2.

Day 5

5. Replace media with 120 μL of RPMI supplemented with 10% FBS and 5.0 μg/mL of puromycin (as determined in the preliminary step). Incubate at 37°C in a humidified incubator in an atmosphere of 5% CO_2 for 48 h. Carefully wash the cells with phosphate-buffered

saline (PBS) and replace media with the selecting antibiotic every 48 h. During this period you will observe massive cellular death across the entire plate, and after 6–8 days you will observe single colonies in some wells. In general, wells containing more than one colony should be marked and avoided for expansion.

2.2 Establishment of Monoclonal Populations

6. The following expansion step can be done in several ways. We prefer to expand wells with single colonies in the same plate until an eye-sight cluster is observed. Using a micropipette tip, scrape the cluster and transfer to a new 12-well plate; continue culturing the cells using the same conditions as described earlier. Test the expression of the protein after cellular confluence is achieved. Fig. 1 shows the expression of Flag-HDAC11 in two different monoclonal populations isolated using this method.

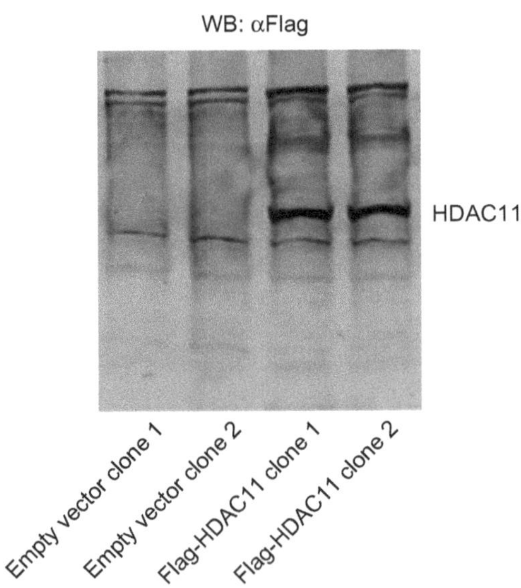

Fig. 1 Evaluation of Flag-HDAC11 in murine RAW264.7 cells. RAW264.7 cells were transduced with lentiviral particles containing Flag-HDAC11 or empty vector. The expression of Flag-HDAC11 expression was evaluated after the establishment of monoclonal populations for each condition.

3. CONVENTIONAL HDAC ACTIVITY ASSAY

Several protocols have been described for the preparation of substrates for HDAC activity assays. One of the first procedures and the most commonly used for many years, involved incubation of immature chicken erythrocytes, calf thymus nuclei, or HeLa cells with ^{3}H- or ^{14}C-acetate followed by the isolation of radiolabeled, hyperacetylated histones (Carmen, Rundlett, & Grunstein, 1996; Hendzel, Delcuve, & Davie, 1991; Inoue & Fujimoto, 1969; Sun, Spencer, Chen, Li, & Davie, 2003). The main advantage in using biologically acetylated substrates is that they contain physiologically relevant, naturally acetylated sites. An alternative to in vivo labeling of histones is to use purified recombinant HAT to add ^{3}H-acetate onto purified histones in vitro (Wade, Jones, Vermaak, & Wolffe, 1999). A benefit to using this method is that the histone substrates can be labeled with very high-specific activity. However, this method requires the preparation of high-quality HAT. Finally, a third method involves the chemical acetylation of histones with ^{3}H- or ^{14}C-acetic anhydride. This last method offers the advantage of obtaining very high-specific activity substrates, although it does present the problem of introducing nonspecific acetyl groups onto lysine residues.

The Guarente laboratory analyzed Sir2 deacetylase activity by high-pressure liquid chromatography (HPLC) using unlabeled, acetylated peptides corresponding to the N-termini of histones H3 or H4 (Imai, Armstrong, Kaeberlein, & Guarente, 2000). Later, a fluorogenic HDAC assay was developed that is well suited for high-throughput activity screening (Wegener, Wirsching, Riester, & Schwienhorst, 2003), including non-radioactive HDAC fluorescent activity assay and HDAC colorimetric assay kits that are commercially available.

In this section, we provide a straightforward protocol that uses a labeled peptide corresponding to the N-terminal of histone H4 as a substrate. This protocol offers a quick and very convenient approach for measuring HDAC activity, even though it is a method that could introduce substrates with nonnaturally acetylated sites. The original procedure was pioneered in the Schreiber lab (Taunton, Hassig, & Schreiber, 1996) and was modified in our laboratory. It is important to note that many HDACs are also capable of deacetylating nonhistone cellular proteins. Consequently, this protocol can be applied to studies of nonhistone proteins by synthesizing and labeling peptides corresponding to the nonhistone substrate of interest.

The following protocols were originally published in 2004 (Rezai-Zadeh et al., 2004) and repeated here for completeness.

3.1 Labeling of H4 Peptide with [^{3}H] for HDAC Assays

1. Synthesize or purchase a H4 peptide (SGRGKGGKGLGKGGAKR HRKVLR) corresponding to residues 2–24 of histone H4 with a free amine at the N-terminus and an amide at the C-terminus. Synthetic H4 peptide should be purified by HPLC to greater than 90% purity.
2. For radiolabeling of the synthetic H4 peptide, add 5 mCi of [^{3}H]acetic acid (2–5 Ci/mmol in ethanol) to 0.4 mg of the H4 peptide. Add 10 μL of freshly prepared BOP solution (0.24 *M* BOP and 0.2 *M* triethylamine in acetonitrile) to the peptide/label mix and rock gently on a rocker at room temperature overnight.
3. Use a Microcon-SCX or similar spin column to purify the labeled peptide. Prewash column with 500 μL of 10 m*M* hydrochloric acid (HCl) in methanol once, and then with 500 μL of 10 m*M* HCl in 10% methanol. Spin down wash solutions from the column and discard.
4. Load 250 μL of the labeling mixture onto the prewashed column and spin column at 1200 × *g* for 1 min.
5. Wash column twice with 500 μL of 10 m*M* HCl in 10% methanol. Invert column and place in a new collection tube.
6. To elute the labeled peptide, apply 50 μL of 3 N HCl in 50% isopropanol to the column and spin at 14,000 × *g* for 15 s. This step can be repeated once to ensure complete elution of the labeled peptide.
7. In a fume hood, dry the labeled peptide with the cap of the collection tube open or in a SpeedVac. Add 500 μL of dd H_2O to dissolve the peptide. Aliquot dissolved radioactive peptide and store at –80°C. The following equation is used to estimate the purity of the radiolabeled peptide:

$$\frac{\text{Purified CPM}}{\text{Ethylacetate} - \text{extractable CPM}},$$

where purified counts per minute (CPM) refers to the CPM of the final purified peptide and the ethyl acetate-extractable CPM refers to the CPM of the same volume of the purified peptide diluted in dd H_2O and extracted by ethyl acetate (see later). A successful acetylation reaction should yield a value not smaller than 1000.

3.2 HDAC Assay

1. For each reaction, the following reagents are mixed in a microcentrifuge tube: 40 μL of 5 × HDAC buffer (50 m*M* Tris–HCl [pH 8.0], 750 m*M* NaCl, and 50% glycerol), 20,000 cpm [^{3}H]acetyl histone H4 peptide, the HDAC enzyme, and dd H_2O to a total volume of 200 μL.
2. Incubate reactions at room temperature overnight.
3. Stop the reaction by the addition of 50 μL of stop solution (1 *M* HCl and 0.4 *M* acetic acid).
4. Extract released [^{3}H]acetate by adding 400 μL of ethyl acetate to the stop reaction. Vortex the mixture briefly, and centrifuge at 14,000 × *g* for 3 min at room temperature to separate the phases.
5. Transfer 200 μL of the organic phase to a scintillation vial and measure CPM.

4. ASSESSMENT OF SELECTIVE INHIBITION OF HDACs USING HDAC KNOCKOUT CELLS

In the last two decades, HDACs have been discovered to be involved in numerous normal and abnormal physiological activities. For this reason, there is now a growing need to develop novel assays to study the enzymatic activities of HDACs under physiological and pathological conditions. Standard assays such as analysis of immunopurified products were not adequate, and newer techniques discussed here will aim toward addressing this need.

As previously mentioned, there is a growing interest to study the effect of individual HDACs on specific substrates. This knowledge will, in turn, increase the understanding of the regulatory capabilities of single HDACs on diverse biological processes. Currently, there is an urgent ongoing effort to design and produce new small molecules to inhibit HDAC enzymatic activity in a more selective fashion. However, the final validation of these potential ultraselective inhibitors requires specific tools and assays to accurately identify their selectivity and potency based on individual HDACs. Most of the commercially available assays rely on the evaluation of purified HDACs. However, in most cases these nonphysiological conditions do not mirror the actual effects encountered in cells. For example, it has been demonstrated that purified HDACs do not retain their full enzymatic activity, and in some situations they may even completely lose their activity. In order to avoid these problems, our group has developed a specific test to evaluate the selectivity of ultraselective HDAC inhibitors. This technique uses a comparative evaluation of total deacetylase activity in cells isolated from HDAC knockout and wild-type (WT) mice. For example, to evaluate the selectivity of a putative ultraselective HDAC6 inhibitor we used cells isolated from HDAC6–/– as well as WT mice. We prefer to isolate total T-cells from the mouse as they do not require extensive preparative procedures.

For the purposes of this chapter, we have only discussed T-cell isolation; however, this assay can be used to isolate other cells such as B-cells and myeloid cells. This protocol is an adaptation of the EasySep™ mouse T-cell isolation system from StemCell™ Technologies (Cat. # 19851). A number of steps have been modified to suit the particular experimental procedure described in this section, with the goal of using the isolated cells for subsequent HDAC assays. Briefly, the system contains a cocktail of monoclonal antibodies that bind cell surface antigens and positively label non-T-cells. A magnetic bead is conjugated to these antibodies and these cells are removed using a magnetic device. Left over cells in the suspension cocktail will contain all the T-cells (a negative selection technique). Generally, T-cell isolation is done via negative selection to avoid activation via CD3 signaling.

1. Splenocytes are extracted by mechanical dissociation of the spleen with 10 mL of RPMI media.
2. Cells are resuspended in 15 mL of PBS
3. Centrifuge cells at 1250 rpm for 5 min at room temperature.
4. Suction off supernatant.
5. Add 1 mL of PBS and thoroughly reconstitute splenocytes by pipetting two to three times. Add 9 mL to bring the total volume to 10 mL.
6. Repeat steps 3–5.
7. Count live splenocytes and reconstitute in PBS at 1×10^8 cells/mL in a 5 mL clear flow cytometry tube.
8. Add normal rat serum at 50 μL/mL of cells (blocking step).
9. Add Mouse T-Cell Isolation Cocktail at 25 μL/mL of cells. Mix well and incubate at room temperature for 10 min.
10. Vortex Streptavidin RapidSpheres for 30 s.
11. Add Streptavidin particles at 75 μL/mL of cells. Mix well and incubate at room temperature for 2.5 min.
12. Bring the suspension up to a total volume of 2.5 mL ($<2 \times 10^8$ cells), 5 mL ($<4 \times 10^8$ cells), or 10 mL ($4–8 \times 10^8$ cells) with PBS and pipette up and down gently 2–3 times.
13. Place the tube (without cap) in the magnet for 2.5 min at room temperature.

14. Leave the tube in the magnet for 5 min; pour the liquid, which contains the T-cells, off into a 15 mL conical tube.
15. Proceed with the analysis of deacetylase activity in a multimode plate reader.

4.1 Quantification of Deacetylase Activity in Multiplate Readers

This protocol requires a 2 days procedure (Day 1: Seed cells, Day 2: Treat with compounds of interest and then analyze in a multimode plate reader). For this particular procedure we used the Cytation™ system from Biotek™.

Day 1

1. Using a 96-well plate, in a 86 μL volume plate purified HDAC6–/– and WT at 10,000 cells/well.
2. Centrifuge plate at 200 × *g* for 3 min to pull all the liquid to the bottom of the well in order to ensure none is remaining on the well sides.
3. Incubate plate overnight at 37°C 5% CO_2 in cell incubator.

Day 2

Make all compound stocks for Tubastatin A—a selective HDAC6 inhibitor (Selleckchem Cat. # S8049) and a selective class I HDAC, Entinostat (MS-275) (Selleckhem Cat. # S8049) prepared as shown in Table 1.

1. For each of the compound dilution tubes add as follows (Table 1A):
   - In row A add 15 μL per well of tube 1
   - In row B add 15 μL per well of tube 2
   - In row C add 15 μL per well of tube 3
   - In row D add 15 μL per well of tube 4
   - In row E add 15 μL per well of tube 5
   - In row F add 15 μL per well of tube 6
   - In row G add 15 μL per well of tube 7
   - In row H add 15 μL per well of tube 8
2. Mix at low speed in an orbital shaker for 30 s.
3. Thaw and prepare the HDAC-Glo™ (Promega, Cat. # G6422) media as indicated in Table 1B.
4. Add 15 μL of HDAC-Glo media mix to each well.
5. Centrifuge plate 200 g for 3 min to pull all liquid into well so that none remains on the well sides.

Begin reading immediately on the Cytation multimode plate reader (assay protocol runs for 1 h and 15 min). Fig. 2 shows T-cells isolated from HDAC6 knockout and WT mice that were treated with either Entinostat

Table 1 HDAC-GLO Deacetylase Analysis. (A) Descriptive Calculation and Explanation of Compound Dilutions to Achieve a 100 nm–25 μm Treatment Range. Sample Plate Layout (96-Well Format) for Each Compound in Triplicates [Boxes 1, 2, 3 for Tubastatin A (Tuba A) and 4, 5, 6 for Entinostat]. (B) Descriptive Calculations for HDAC-GLO Media Mix Preparation

(A) Compound stock: Make enough for 10 wells for each compound (Note: for each compound prepare the stock in tube 1 of the 8-well strip). Stock concentration for each compound should be adjusted to 25 μ*M*

Add media to tubes 2–8 (as calculated in 2nd and 5th columns) and to dilute the stock as indicated

	Tuba A			**Entinostat**	
1	25 μ*M* stock		1	25 μ*M* stock	
2	90 μL media	60 μL from tube 1	2	90 μL media	60 μL from tube 1
3	75 μL media	75 μL from tube 2	3	75 μL media	75 μL from tube 2
4	75 μL media	75 μL from tube 3	4	75 μL media	75 μL from tube 3
5	135 μL media	15 μL from tube 2	5	135 μL media	15 μL from tube 2
6	135 μL media	15 μL from tube 3	6	135 μL media	15 μL from tube 3
7	135 μL media	15 μL from tube 5	7	135 μL media	15 μL from tube 5
8	DMSO	2.5 μL DMSO in 147.5 μL media	8	Triton X	1.5 μL of Triton X in 148.5 μL media

(B) Final 96-Well Plate Layout (in Triplicates)

	Tuba A			**Entinostat**			**Additional Compounds**			**Additional Compounds**		
	1	2	3	4	5	6	7	8	9	10	11	12
A		25,000			25,000			25,000			25,000	
B		10,000			10,000			10,000			10,000	
C		5000			5000			5000			5000	
D		2500			2500			2500			2500	
E		1000			1000			1000			1000	
F		500			500			500			500	
G		100			100			100			100	
H	DMSO control (Neg)			Triton X control (Pos)			Untreated control					

Preparation of HDAC-GLO media mix. 1-Premix HDAC-GLO substrate + reagent (included in the Promega kit) and aliquot into 1500 μL volume.
To add 5 μL of developer (included in Promega kit) in 1500 μL premix HDAC-GLO solution.

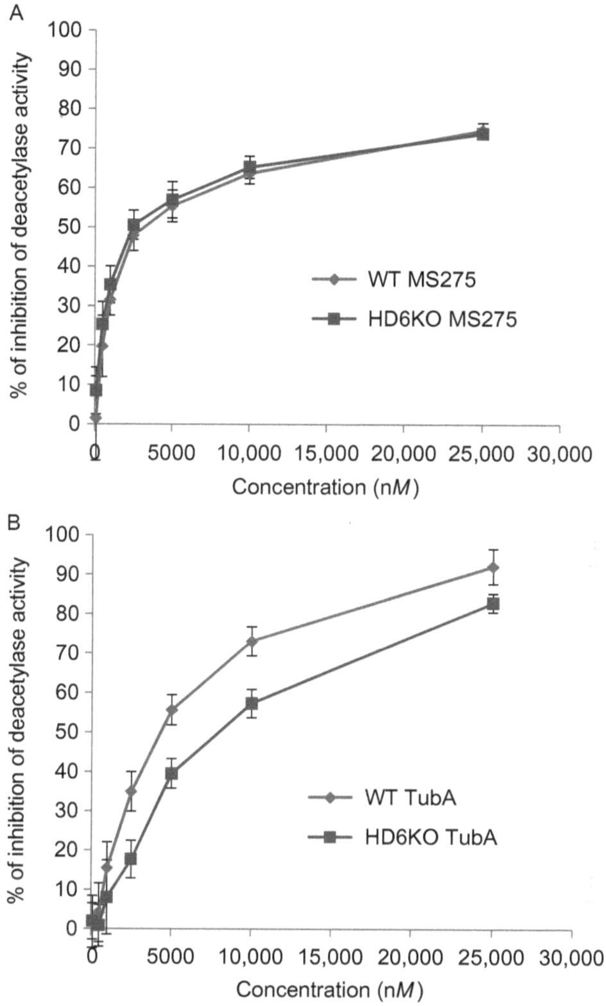

Fig. 2 Quantification of deacetylase activity using HDAC-GLO assay. T-cells isolated from either HDAC6 knockout or wild-type mice were treated with varying concentrations of Entinostat (A) or Tubastatin A (B) and deacetylase activity was measured by the addition of HDAC-GLO reagent and detected by Cytation multimode plate reader. (See the color plate.)

or Tubastatin A. As seen in Fig. 2A, the percentage of deacetylation does not change between the samples because Entinostat does not inhibit HDAC6. However, Fig. 2B indicated that Tubastatin A has selectivity toward HDAC6 since a higher level of deacetylase activity is seen in the WT T-cells when compared to the HDAC6 knockout cells. Methods for identifying deacetylated proteins are described in Section 6.

5. HDAC CHROMATIN IMMUNOPRECIPITATION ASSAY

Related to deacetylase enzymatic activity assays, another important functional assay to study HDACs is chromatin immunoprecipitation (ChIP). Although one shared characteristic among all HDACs is the lack of classical sequence-specific DNA-binding domains, it is still possible to examine the recruitment of HDACs to DNA via ChIP. HDACs physically interact tightly with histones and with proteins that do contain DNA-binding domains. Here, we describe a protocol to identify the recruitment of HDAC11 to the IL-10 promoter in antigen-presenting cells (APCs). A detailed protocol for the isolation of these cells can be found in previous publications from our group.

5.1 Cross-Linking

1. Approximately 20×10^6 APC cells from C57BL/6 and HDAC11 knockout mice per condition will be optimal.
2. Cells are collected in 50-mL conical tubes and washed twice with 10 mL 1 × PBS.
3. Remove PBS by centrifuging the samples at 1500 × *g* for 5 min at room temperature.
4. Gently reconstitute cells in 20 mL of cross-linking buffer (2.7 mL 37% formaldehyde added into 97.3 mL 1 × PBS, pH 7.2, to reach a final concentration of 1% formaldehyde in a final volume of 100 mL). Set up the sample tubes on a tube-rocker or orbital rotator and incubate for 10 min at room temperature. Both chromatin shearing efficiency and immunoprecipitation (IP) efficiency can be affected by the formaldehyde concentration and cross-linking time. In this protocol, we are using 1% formaldehyde and 10 min incubation time for RAW264.7 cells (mouse APC cell line). Cross-linking time and formaldehyde percentage should be optimized for other cell types (Lee, Johnstone, & Young, 2006; Nelson, Denisenko, & Bomsztyk, 2006).
5. Centrifuge samples at 1500 × *g* for 5 min at room temperature and remove the supernatant
6. Gently reconstituted cells in 20 mL of stop buffer (10 mL 1.25 *M* glycine added into 90 mL 1 × PBS, pH 7.2, to reach a 0.125 *M* glycine final concentration in a 100 mL final volume). Freshly prepare cross-linking buffer and stop buffer before each experiment, and add protease

inhibitor cocktail tablets immediately before use. Incubate for 10 min on a tube-rocker at room temperature.

7. Centrifuge the samples at $1500 \times g$ for 5 min at 4°C. Discard the supernatant.
8. Reconstitute cells in 20 mL of PBS, gently invert tubes for several times, and spin at $1500 \times g$ for 5 min at 4°C. Discard the supernatant.
9. Repeat step 8 if needed. Aliquot cells before centrifugation.
10. Proceed with chromatin preparation. Cell pellets from step 9 can be stored at –80°C until further use.

5.2 Chromatin Preparation

11. Remove sample tubes containing frozen cell pellets from –80°C and allow them to thaw on ice.
12. Prepare Dounce homogenizer (7 mL), set on ice.
13. Reconstitute each pellet of 20×10^6 cells in 2 mL of lysis buffer (50 m*M* 4-(2-hydroxyethyl)-1-piperazineethanesulfonic acid (HEPES), pH 7.8, 3 m*M* $MgCl_2$, 20 m*M* KCl, 0.25% Triton X-100, 0.5% IGEPAL). Gently pipette up and down until completely homogenized. Incubate on ice for 10 min.
14. Transfer cells to the Dounce homogenizer and apply 25 strokes using the loose pestle. Transfer the nuclei to a new 15-mL conical tube and centrifuge at $1500 \times g$ for 5 min at 4°C, then discard supernatant. The Dounce homogenizer has to be set on ice at all times. It should be rinsed with cold PBS three to five times between different samples. The number of strokes may vary and should be optimized for different cell types or cell lines used. In order to optimize for a particular cell line or type, take an aliquot of cells and stain with Tripan Blue and check under the microscope. Adjust stroke times until over 90% of the cells are stained but have kept round nuclei. Decrease the number of strokes if broken nuclei are observed.
15. Reconstitute nuclei pellet with 2 mL of wash buffer. Gently pipette up and down until completely homogenized. Incubate on ice for 10 min.
16. Pellet the nuclei by centrifuge at $1500 \times g$ for 5 min at 4°C. Discard the supernatant.
17. Add 600 μL sonication buffer to each tube, gently pipette up and down until homogenized. Divide the sample by transferring 300 μL nuclei sample to each 1.5 mL Eppendorf tube and proceed to Section 5.3.

5.3 Sample Sonication

18. For this specific protocol we used a Bioruptor™ XL from Diagenode. However, any bath sonicator with timing programming will be suitable. The water bath was precooled to 4°C with crushed ice before each sonication cycle and a thin layer of crushed ice was kept in the water bath for each cycle to avoid rapid temperature increase during sonication.

19. For each sonication cycle, use a program with eight pulses of 30 s followed by 30 s of resting time at 20 kHz frequency (300 W). To obtain 250–500 base pair DNA fragments for the mouse peritoneal neutrophils, approximately eight cycles are required for each sample. However, this parameter needs to be adjusted for each cell type. Centrifuge the samples at 16,000 × *g* for 10 min at 4°C. Transfer the supernatant to a new 1.5-mL Eppendorf tube (approximately 600 μL of supernatant). Sonication efficiency varies according to different cell types and equipment. It is recommended to optimize this step by a pilot experiment using a number of different sonication cycles on different cell types or cell lines. For some types of sonicators, it is recommended to check the water bath temperature and replace ice after each cycle, or the water bath could be kept on a heat adjust platform to maintain a 4°C environment.

20. Take an aliquot (20 μL) from each sample and proceed to Section 5.4. The rest of the samples can either be stored at –80°C or continue with IP.

5.4 Shear Sample

21. To each tube, add 20 μL sample aliquot, 47.2 μL H_2O, and 2.8 μL of 5 *M* NaCl in order to reach a final concentration of 200 m*M*.

22. Incubate at 65°C for 4 h or overnight.

23. Add RNAse A to a final concentration of 20 μg/mL to each sample aliquot; incubate at 37°C for 30 min.

24. Add Proteinase K to a final concentration of 100 μg/mL to each sample aliquot; incubate at 42°C for 2 h.

25. Add 200 μL of phenol/chloroform/isoamyl alcohol, pH 6.7, to each sample aliquot, vortex for 15 s and spin down at 16,000 × *g* for 5 min.

26. Transfer 15 μL of the supernatant from the top layer; add 3 μL of 5 × DNA loading dye with light xylene cyanol only.

27. Run samples on 1.5% agarose gel at 40 V for 1.5 h.

5.5 Chromatin Immunoprecipitation

28. For the preclearance add 30 μL of Protein A Agarose beads to each sample, incubate on a tube rotator for 1 h at 4°C.
29. Centrifuge the samples at 6000 × *g* for 5 min at 4°C.
30. Transfer the supernatant to a new 1.5-mL Eppendorf tube. Take an aliquot of 50 μL from each sample as input. Input samples can be stored at −80°C before reverse cross-linking.
31. Dilute 50 μL of each sample after preclearance with 450 μL dilution buffer I (50 m*M* HEPES, pH 7.8, 140 m*M* NaCl, 1 m*M* ethylenediaminetetraacetic acid (EDTA), pH 8.0, 1% Triton X-100), to a final volume of 500 μL, one dilution per antibody, into a 1.5-mL Eppendorf tube.
32. During the first dilution of each sample, add a mixture of 5 μL (5 μg) Rabbit-anti-HDAC11 antibody from Sigma™ and 20 μL (4 μg) Rabbit-anti-HDAC11 antibody from BioVision™. In the second dilution tube of each sample, add 9 μg normal Rabbit IgG-B.
33. Set the sample tubes on a tube rotator overnight, at 4°C, and cover the tubes with aluminum foil to avoid light exposure. Add 50 μL of Protein A Agarose beads to each sample; incubate on a tube rotator for 4 h at 4°C. We recommend using a mixture of two different HDAC11 antibodies that have been confirmed with good IP efficiency on macrophages (Cheng et al., 2014). It is recommended to optimize IP antibody use for each different cell type or cell line used.

5.6 Sample Wash and Elution

34. After IP, centrifuge the samples at 6000 × *g* for 5 min at 4°C and discard the supernatant.
35. Wash the pellet with 1 mL of each buffer in the following order:
    - Twice with dilution buffer I.
    - Twice with dilution buffer II (50 m*M* HEPES, pH 7.8, 500 m*M* NaCl, 1 m*M* EDTA, pH 8.0, 1% Triton X-100).
    - Twice with lithium chloride (LiCl) buffer (20 m*M* Tris–HCl, pH 8.0, 250 m*M* LiCl, 1 m*M* EDTA, pH 8.0, 0.5% Triton X-100).
    - Twice with Tris-EDTA buffer (TE) buffer (10 m*M* Tris–HCl, pH 8.0, 1 m*M* EDTA, pH 8.0).

    For each wash, centrifuge at 6000 × *g* for 5 min at 4°C.
36. Discard the supernatant from the last wash. Add 100 μL of elution buffer to each sample and incubate at 65°C for 15 min. The immunocomplexes will be in the soluble fraction.

37. Centrifuge the samples at maximum speed for 20 s. Transfer supernatant to a new 1.5-mL Eppendorf tube.

38. Remove the input sample tubes from –80°C and thaw on ice. Add 50 μL of elution buffer (50 m*M* Tris–HCl, pH 8.0, 1 m*M* EDTA, pH 8.0, 1% sodium dodecyl sulfate) to each input tube to final volume of 100 μL.

39. Add 4.16 μL 5 *M* NaCl (200 m*M* final concentration) to each IP and input sample tubes. Reverse cross-linking at 65°C for 6 h or overnight.

40. Add RNAse A to a final concentration of 20 μg/mL to each sample and incubate at 37°C for 30 min.

41. Add Proteinase K to a final concentration of 100 μg/mL to each sample and incubate at 42°C for 2 h.

42. Use Qiagen™ DNA purification kit to purify each DNA sample.

To obtain a higher amount of DNA, researchers can elute DNA twice from each column; use 50 μL elution buffer each time to obtain a higher yield of DNA.

43. Use 2 μL DNA for each qRT-PCR reaction. This is done in triplicate for each reaction, and the purified DNA must be kept in TE buffer, pH 8.0.

It is recommended that the primer efficiency and melting temperature be determined before running PCR on the ChIP samples.

This protocol has been used by our lab and collaborators (Cheng et al., 2014; Villagra, Cheng, Wang, et al., 2009). A prototypical experimental result is shown in Fig. 3, where we evaluated the recruitment of HDAC11 to the IL-10 promoter after stimulation of mouse macrophages with lipopolysaccharide (LPS).

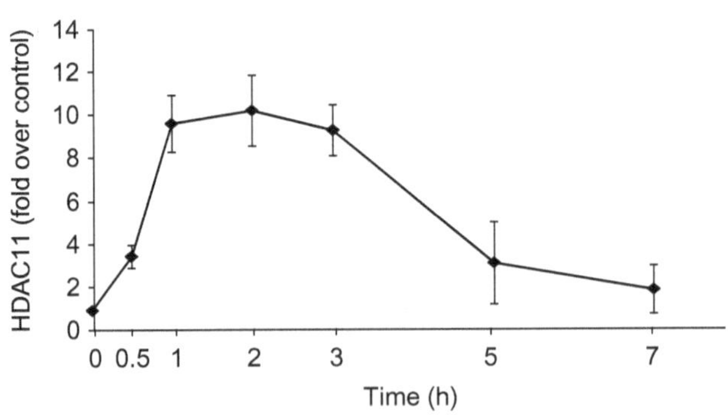

Fig. 3 HDAC11 is recruited to the IL-10 gene promoter. Macrophages were treated with LPS (1.0 μg/mL), and then harvested at baseline (time 0) or at 0.5, 1, 2, 3, 5, and 7 h after treatment. Cells were then subjected to ChIP analysis using antibodies against HDAC11. Quantitative real-time PCR analysis was performed in the region between −87 and −7 of the IL-10 gene promoter.

6. IDENTIFICATION OF SPECIFIC HDAC SUBSTRATES

Identification of specific protein substrates for particular HDACs can be achieved by several methods. Among them, the characterization of acetyl residues by mass spectrometry has proven to be a reliable approach, in particular, when the discovery of unknown substrates is the main goal (Inoue & Fujimoto, 1969; Joshi, Greco, Guise, et al., 2013). Nevertheless, this technique may not be the first choice when analyzing a limited number of pre-identified proteins. In that situation, the identification of acetyl proteins using variant-specific antibodies will provide a faster and an inexpensive approach. Unfortunately, antibodies targeting specific acetylated proteins are only available for a small number of targets, making this option not always feasible. A third option, also inexpensive and faster than mass spectrometry is a two-step assay involving the IP of all acetylated proteins and then an immunoblot against the protein(s) of interest. In this section, we describe a protocol for the identification of HDAC6 protein substrates comparing cellular extracts from HDAC6 knockdown (HDAC6KD) and control human WM164 melanoma cells.

1. Approximately 1×10^6 from HDAC6KD and control WM164 human melanoma cells are needed as starting material.
2. Cells were collected in 50-mL conical tubes and washed twice with 10 mL 1 × PBS.
3. Remove the PBS by centrifuging the samples at $1500 \times g$ for 5 min at 4°C.
4. Gently reconstitute cells in 400 μL of immunoprecipitation lysis buffer (IPLB) (50 m*M* Tris [pH 8.0], 150 m*M* NaCl, 0.5% igepal, 5 m*M* $MgCl_2$, 10% glycerol) containing 1 × protease inhibitor cocktail (Roche). Centrifuge at $10,000 \times g$ for 15 min at 4°C.
5. Add 100 μL of protein G-agarose (Millipore #16-266) and incubate for 2 h at 4°C in an orbital rotator. Centrifuge at $6000 \times g$ for 5 min at 4°C. Save 50 μL of the supernatant as an input sample.
6. Add 350 μL of IPLB to the precleared sample (700 μL total). Add 5 μg of anti-acetyl lysine antibody (Millipore #05-515) and incubate overnight at 4°C in an orbital rotator.
7. Add anti-acetyl lysine antibody, clone 4G12, agarose conjugate beads (Millipore #16-272) and incubate for 4 h at 4°C in an orbital rotator. Centrifuge at $6000 \times g$ for 5 min at 4°C.
8. Remove supernatant and wash twice with 700 μL of cold IPLB.

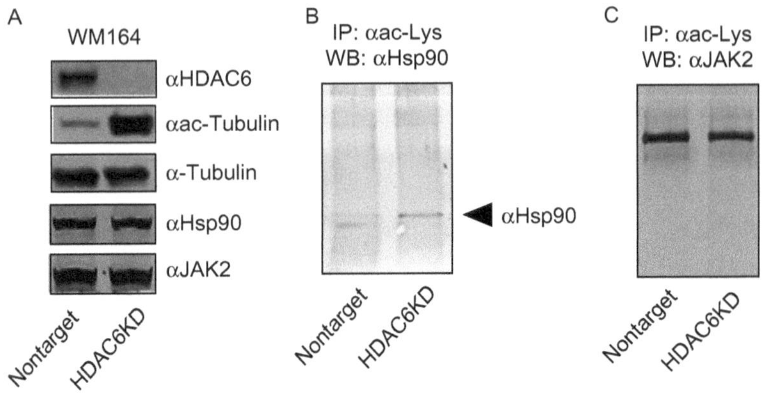

Fig. 4 (A) Nontarget and HDAC6KD WM164 melanoma cells were subjected to immunoblot using antibodies against HDAC6, acetylated α-tubulin, α-tubulin, Hsp90, and JAK2. Total acetylated proteins were immunoprecipitated from nontarget and HDAC6KD WM164 cells. The immunoprecipitated fraction was then assayed for the presence of Hsp90 (B) and JAK2 (C).

9. Add 6 × loading buffer and proceed to analyze in parallel the immunoprecipitated fractions from HDAC6KD and control cells (nontarget) by immunoblotting against Hsp90 (Fig. 4). In parallel, we have shown Jak2 as control, a protein not being deacetylated by HDAC6.

ACKNOWLEDGMENTS

This work was supported in part by NIH Grants R01CA169210 (E.S.), R01CA187040 (E.S.), and R21CA184612 (A.V.).

REFERENCES

Carmen, A. A., Rundlett, S. E., & Grunstein, M. (1996). HDA1 and HDA3 are components of a yeast histone deacetylase (HDA) complex. *The Journal of Biological Chemistry*, *271*(26), 15837–15844.

Cheng, F., Lienlaf, M., Perez-Villarroel, P., et al. (2014). Divergent roles of histone deacetylase 6 (HDAC6) and histone deacetylase 11 (HDAC11) on the transcriptional regulation of IL10 in antigen presenting cells. *Molecular Immunology*, *60*(1), 44–53.

Cribbs, A. P., Kennedy, A., Gregory, B., & Brennan, F. M. (2013). Simplified production and concentration of lentiviral vectors to achieve high transduction in primary human T cells. *BMC Biotechnology*, *13*, 98.

Glozak, M. A., Sengupta, N., Zhang, X., & Seto, E. (2005). Acetylation and deacetylation of non-histone proteins. *Gene*, *363*, 15–23.

Hendzel, M. J., Delcuve, G. P., & Davie, J. R. (1991). Histone deacetylase is a component of the internal nuclear matrix. *The Journal of Biological Chemistry*, *266*(32), 21936–21942.

Imai, S., Armstrong, C. M., Kaeberlein, M., & Guarente, L. (2000). Transcriptional silencing and longevity protein Sir2 is an NAD-dependent histone deacetylase. *Nature*, *403*(6771), 795–800.

Inoue, A., & Fujimoto, D. (1969). Enzymatic deacetylation of histone. *Biochemical and Biophysical Research Communications*, *36*(1), 146–150.
Joshi, P., Greco, T. M., Guise, A. J., et al. (2013). The functional interactome landscape of the human histone deacetylase family. *Molecular Systems Biology*, *9*, 672.
Lee, T. I., Johnstone, S. E., & Young, R. A. (2006). Chromatin immunoprecipitation and microarray-based analysis of protein location. *Nature Protocols*, *1*(2), 729–748.
Lee, K. K., & Workman, J. L. (2007). Histone acetyltransferase complexes: One size doesn't fit all. *Nature Reviews Molecular Cell Biology*, *8*(4), 284–295.
Minucci, S., & Pelicci, P. G. (2006). Histone deacetylase inhibitors and the promise of epigenetic (and more) treatments for cancer. *Nature Reviews Cancer*, *6*, 38–51.
Nelson, J. D., Denisenko, O., & Bomsztyk, K. (2006). Protocol for the fast chromatin immunoprecipitation (ChIP) method. *Nature Protocols*, *1*(1), 179–185.
Rezai-Zadeh, N., Tsai, S. C., Wen, Y. D., Yao, Y. L., Yang, W. M., & Seto, E. (2004). Histone deacetylases: Purification of the enzymes, substrates, and assay conditions. *Methods in Enzymology*, *377*, 167–179.
Sun, J. M., Spencer, V. A., Chen, H. Y., Li, L., & Davie, J. R. (2003). Measurement of histone acetyltransferase and histone deacetylase activities and kinetics of histone acetylation. *Methods*, *31*(1), 12–23.
Taunton, J., Hassig, C. A., & Schreiber, S. L. (1996). A mammalian histone deacetylase related to the yeast transcriptional regulator Rpd3p. *Science*, *272*, 408–411.
Villagra, A., Cheng, F., Wang, H. W., et al. (2009). The histone deacetylase HDAC11 regulates the expression of interleukin 10 and immune tolerance. *Nature Immunology*, *10*(1), 92–100.
Wade, P. A., Jones, P. L., Vermaak, D., & Wolffe, A. P. (1999). Purification of a histone deacetylase complex from Xenopus laevis: Preparation of substrates and assay procedures. *Methods in Enzymology*, *304*, 715–725.
Wang, X., & McManus, M. (2009). Lentivirus production. *Journal of Visualized Experiments*, *32*.
Wegener, D., Wirsching, F., Riester, D., & Schwienhorst, A. (2003). A fluorogenic histone deacetylase assay well suited for high-throughput activity screening. *Chemistry & Biology*, *10*, 61.

CHAPTER EIGHT

Recombinant Preparation, Biochemical Analysis, and Structure Determination of Sirtuin Family Histone/Protein Deacylases

B. Suenkel, C. Steegborn[1]
Department of Biochemistry, University of Bayreuth, Bayreuth, Germany
[1]Corresponding author: e-mail address: clemens.steegborn@uni-bayreuth.de

Contents

1. Sirtuins—Evolutionary Conserved Epigenetic Regulators 184
2. Recombinant Sirtuin Preparation—Constructs and Expression Systems 188
3. Purification of a Mammalian Sirtuin Recombinantly Expressed in *E. coli* 189
 3.1 Instruments and Buffers 193
 3.2 Protein Expression 193
 3.3 Preparation of Cleared Lysate 194
 3.4 Affinity Chromatography 194
 3.5 Affinity Tag Removal 195
 3.6 Ion Exchange Chromatography 196
 3.7 Size-Exclusion Chromatography 196
 3.8 Modifications to the Standard Protocol 197
4. Biochemical Characterizations Using Purified, Recombinant Sirtuins 198
 4.1 Sirtuin Activity 198
 4.2 Sirtuin Binding Studies 201
 4.3 Sirtuin Crystallization 202
 4.4 Further Applications 205
Acknowledgments 205
References 205

Abstract

Lysine acetylation is long known as a regulatory posttranslational modification of histone proteins and is emerging as a ubiquitous intracellular protein modification. Additional lysine acylations such as succinylation and glutarylation have also been found on histones and other proteins. Acylations are reversibly attached through nonenzymatic acylation mechanisms and the action of protein acyl transferases and protein deacylases (PDACs). Sirtuins are an evolutionary defined class of PDACs and act as metabolic sensors by catalyzing a unique deacylation reaction that requires the cosubstrate NAD^+.

Methods in Enzymology, Volume 573
ISSN 0076-6879
http://dx.doi.org/10.1016/bs.mie.2015.12.004

Sirtuins are found in all domains of life, and the mammalian sirtuin family comprises seven isoforms in different cellular compartments. They regulate a wide range of cellular targets and functions, such as energy metabolism and stress responses, and they have been implicated in aging processes and aging-related diseases. A large body of functional, biochemical, biophysical, and structural work on isolated sirtuins has provided many important insights that complement the many physiological studies on this enzyme family. They enabled the comprehensive structural and biochemical analysis of sirtuin catalysis, substrate selectivity, and regulation. Here, we describe the recombinant production of sirtuin proteins, with an emphasis on the mammalian isoforms. We then describe their application in activity and binding assays and for crystal structure analysis. We provide protocols for these procedures, and we discuss typical pitfalls in studying this enzyme family and how to avoid them. This information will support further molecular studies on sirtuin mechanisms and functions.

1. SIRTUINS—EVOLUTIONARY CONSERVED EPIGENETIC REGULATORS

Acetylation of histone proteins is long known as a posttranslational modification (PTM) regulating chromatin packaging, DNA replication, and gene expression (Shahbazian & Grunstein, 2007). More recently, however, acetylation was found to be wide spread and to affect thousands of proteins in mammals and lower organisms (Norvell & McMahon, 2010; Scott, 2012). Also, additional protein lysine acylations such as succinylation and crotonylation have been identified on histones and other cellular proteins (Lin, Su, & He, 2012; Rousseaux & Khochbin, 2015). Acylations are attached to protein lysine side chains through both nonenzymatic mechanisms and acetyl transferases, and they are removed by protein lysine deacylases (KDACs) (Scott, 2012; Simic, Weiwad, Schierhorn, Steegborn, & Schutkowski, 2015; Wagner & Payne, 2013). Sirtuins are an evolutionary defined KDAC subfamily also referred to as class III histone deacetylases (HDACs) since histones were their first identified substrates (Shahbazian & Grunstein, 2007). They act as metabolic sensors due to their unique dependence on the cosubstrate NAD^+ (Sauve, Wolberger, Schramm, & Boeke, 2006). Sirtuins are found in all domains of life and the mammalian sirtuin family comprises seven isoforms (Sirt1–Sirt7). All isoforms feature the conserved catalytic core, with N- and C-terminal extensions varying in sequence and length and contributing to their specific cellular localization and function (Fig. 1A; Gertz & Steegborn, 2010; Tennen, Berber, & Chua, 2010; Zhao, Chai, & Marmorstein, 2003). Sirt1, 6, and 7 are mainly nuclear isoforms contributing to the regulation of stress responses, DNA homeostasis, and gene expression

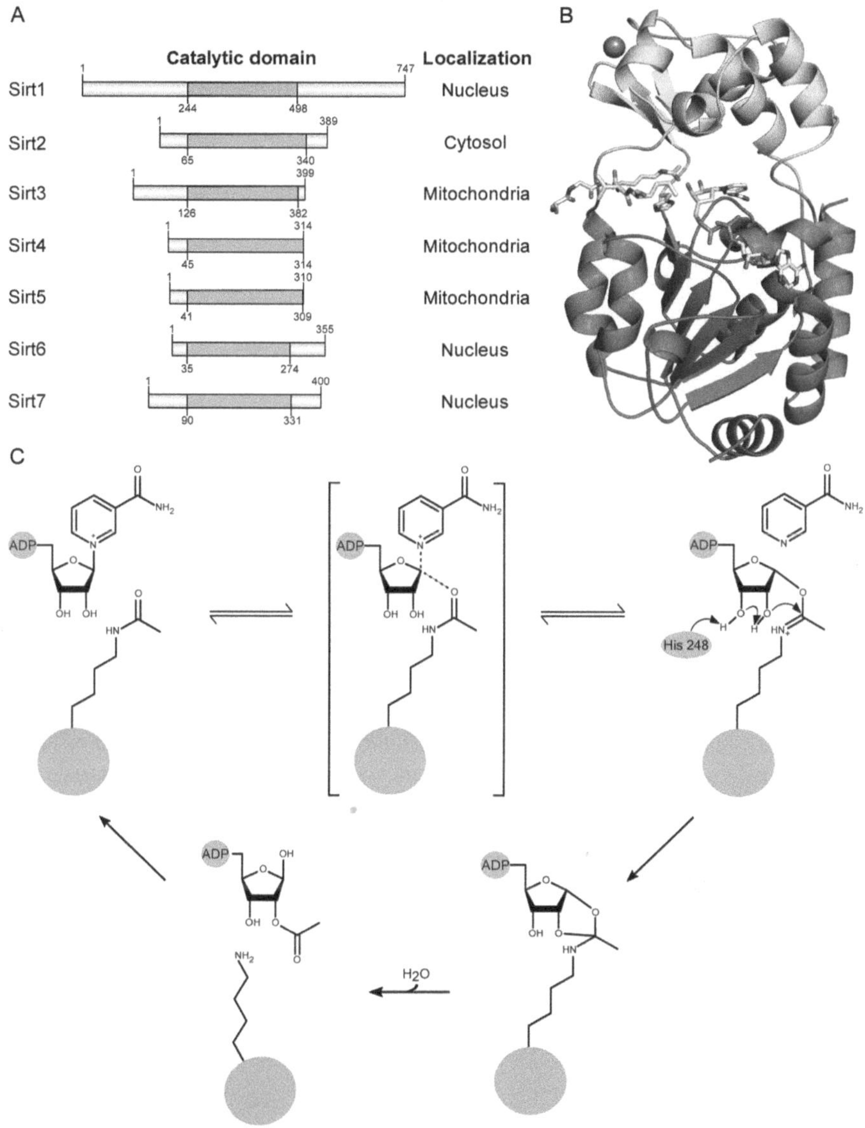

Figure 1 Sirtuin architecture and catalytic mechanism. (A) Domain organization of the seven human sirtuin isoforms. The homolog catalytic domains are flanked by isoform-specific N- and C-terminal extensions. The dominant cellular localization of each isoform is indicated on the right. (B) Overall structure of human Sirt3 in complex with an acetylated peptide and the nonhydrolyzable NAD^+ analog carba-NAD^+ (PDB entry 4FTV; Szczepankiewicz et al., 2012). The Zn^{2+} binding domain is colored cyan (light gray in the print version) and the Rossmann-fold domain magenta (dark gray in the print version). The catalytic His248 as well as the peptide and carba-NAD^+ ligands are shown in stick representation. The Zn^{2+} ion is shown as a sphere. (C) Mechanism for the NAD^+-dependent deacylations catalyzed by sirtuins. His248 refers to the conserved catalytic sirtuin histidine of human Sirt3.

(Cen, Youn, & Sauve, 2011). Sirt2 is the only isoform mainly located to the cytosol and is involved in the regulation of the cell cycle and cellular motility. Sirt3, 4, and 5 are mainly localized to mitochondria and contribute to the regulation, eg, of energy metabolism and redox stress control. Through regulation of this wide range of cellular targets and pathways, sirtuins contribute to many bodily functions and pathophysiologies, and they have been implicated in particular in aging processes and many aging-related diseases (Haigis & Sinclair, 2010). Excellent reviews are available on the targets and specific functions of the mammalian sirtuin isoforms (see, eg, Cen et al., 2011; Haigis & Sinclair, 2010; Michan & Sinclair, 2007).

A large body of functional, biochemical, biophysical, and structural work on isolated sirtuins has provided many important insights that complement the many physiological studies and that contribute to understanding this enzyme family and their biological roles. A set of crystal structures of sirtuin catalytic cores, including those of human isoforms Sirt1, 2, 3, 5, and 6, have revealed the conserved architecture of this domain (Moniot, Weyand, & Steegborn, 2012; Sanders, Jackson, & Marmorstein, 2010). Additional sirtuin structures with bound substrates and intermediates/mimics, together with activity and binding studies, revealed the enzyme family's unique catalytic mechanism (Schutkowski, Fischer, Roessler, & Steegborn, 2014). The conserved sirtuin catalytic core comprises about 270 amino acids, which fold into two subdomains (Fig. 1B). The larger subdomain assumes a Rossmann-fold typical for NAD^+-binding proteins, and the smaller subdomain forms a Zn^{2+}-binding module with four cysteine residues as Zn^{2+} ligands. A groove accommodating the substrate polypeptide chain and the more buried catalytic center are located between these subdomains. The rather shallow polypeptide-binding groove has some isoform-specific features (eg, polarity pattern, shape), leading to very moderate, isoform-specific substrate sequence preferences (Rauh et al., 2013). The Zn^{2+} ion is distant from the catalytic center and appears to have solely a protein-stabilizing function, in contrast to the other three classes of KDAC enzymes (referred to as class I, II, and IV KDACs/HDACs), which feature an active site Zn^{2+} ion that directly contributes to catalyzing hydrolysis of the lysine-ε*N*-acylamide by activating the attacking water molecule. Sirtuins instead exploit a unique mechanism that involves hydrolysis of the cosubstrate NAD^+ (Fig. 1C; Cen et al., 2011; Schutkowski et al., 2014). This NAD^+ hydrolysis takes place in the first reaction step, releasing nicotinamide from the NAD^+ 1′-ribose carbon, which concomitantly forms a new covalent bond to the acyl oxygen of the acylated lysine residue. The resulting 1′-*O*-alkylimidate intermediate appears long lived; its complex with Sirt3 has been structurally characterized, and rebinding of nicotinamide to this complex can lead to reversal of the first reaction step and apparent inhibition of sirtuin-dependent deacylation (Avalos, Bever, & Wolberger, 2005; Gertz et al., 2013; Schutkowski et al., 2014). In the next step, the alkylimidate converts into a second, bicyclic 1′-2′-acetal intermediate. In the final reaction step, the bicyclic intermediate is hydrolyzed to yield the deacylated polypeptide and 2′-*O*-acylated ADP-ribose. During this catalytic cycle, the protein adopts a series of conformational states (Jin et al., 2009; Moniot, Schutkowski, & Steegborn, 2013; Moniot et al., 2012). Small and Rossmann-fold domain reorient relative to each other when the acetyl-lysine containing substrate polypeptide binds to the groove between them; and a so-called cofactor-binding loop participates in NAD^+ binding and changes its conformation upon formation of the two intermediates as well as formation and release of the product acyl-ADP-ribose.

The complexity of sirtuin substrates and catalysis renders routine analyses, such as activity assays or determination of inhibition mechanisms, more challenging than for some other enzyme families. The enzyme's activity can vary due to the substrate peptide sequence and length, for example, and is further sensitive to the concentrations of the cosubstrate NAD^+ and the product inhibitor nicotinamide. To complicate the issue further, nicotinamide inhibition can differ for the same sirtuin depending on the acyl modification of the substrate (Fischer et al., 2012). Several of the emerging protein lysine modifications were shown to be substrates for sirtuins; for some isoforms, they in fact appear to be the main substrates (Schutkowski et al., 2014). In particular for Sirt5, preferential hydrolysis of the physiologically occurring acylations with the dicarboxylates malonylate, succinate, and glutarate was observed and appears to constitute its physiological activity (Du et al., 2011; Fischer et al., 2012; Roessler et al., 2014; Tan et al., 2014). Therefore, succinylated peptides are now routinely used for assaying Sirt5 activity. The prominent desuccinylase activity of Sirt5 is more sensitive to nicotinamide inhibition as compared to its weaker deacetylase activity (Fischer et al., 2012), however, which can influence the observed activities in vitro and possibly also in vivo. Along these same lines, nicotinamide has to be kept low to measure Sirt3 activity, which mainly acts as a deacetylase but shows a pronounced sensitivity to nicotinamide inhibition (Fischer et al., 2012). Even more care has to be taken when studying the influence of pharmacological compounds on sirtuin activity. Artifacts have been observed based on the use of nonphysiological substrates or on compound properties such as absorption and fluorescence interfering with the assay used (Schutkowski et al., 2014). Furthermore, the complex catalytic cycle limits the applicability of concepts from simpler systems, such as the widely used simplified equations for Michaelis–Menten analysis. However, careful in vitro analysis of sirtuins in activity, binding, and structural studies has yielded important insights into their catalytic properties, substrate sequence and acyl specificity, and modulation by pharmacological inhibitors and activators (see, eg, Gertz et al., 2012, 2013; Moniot et al., 2012; Schutkowski et al., 2014).

In this chapter, we provide detailed information for the preparation of highly active, recombinant sirtuins in amounts sufficient for biochemical and biophysical studies. We further describe the application of a widely used activity assay and a binding test system, and we discuss several important issues to consider when planning and conducting experiments, and analyzing data from such experiments. This information should help avoid potential pitfalls when studying the function and/or structure of sirtuins and support further studies to understand the molecular mechanisms regulating sirtuins through PTMs and interacting proteins, and in developing sirtuin-targeting small molecules.

2. RECOMBINANT SIRTUIN PREPARATION—CONSTRUCTS AND EXPRESSION SYSTEMS

The catalytic cores of different sirtuins feature specific N- and C-terminal extensions (Gertz & Steegborn, 2010; Michan & Sinclair, 2007; Fig. 1A). They can contribute to the stability and specific activity of the protein. In particular for mammalian Sirt1, such a role in catalytic core stabilization and activation has been described for a region within its C-terminal extension (Lakshminarasimhan, Curth, et al., 2013; Pan, Yuan, Brent, Ding, & Marmorstein, 2012), and this C-terminal motif should thus generally either be included in the expression construct for Sirt1 production or provided in *trans* as a peptide. However, most sirtuin extensions contribute to cellular localization (eg, Sirt3–Sirt5 N-termini) or interaction with target complexes such as chromatin for Sirt6 (C- and N-terminus). These extensions only need to be included in the sirtuin expression construct if this function or interaction is important for the subsequent application. Otherwise, catalytic core constructs are often preferred because they are easily prepared in large amounts via recombinant expression in *Escherichia coli*, tend to be more stable and soluble, and allow many questions to be fully analyzed. For example, deacylation specificity and modulation by pharmacological inhibitors targeting the catalytic core can generally be more conveniently analyzed using catalytic core constructs. There are also a variety of constructs described for bacterial expression of larger sirtuin fragments (see Table 1 for a selection of frequently used constructs). Several bacterial and yeast sirtuins, mostly with smaller extensions, have been expressed as full-length proteins, and also human Sirt1, 2, 6, and 7. Also varying combinations of catalytic core with complete or partial extensions have been described in particular for Sirt1 and Sirt7 (see Table 1 for a selection).

Since production of a variety of constructs for all mammalian sirtuin isoforms as well as several other sirtuin enzymes in the efficient *E. coli* expression system is established, there are few reports on recombinant sirtuin preparations using yeast and insect cells or other lower eukaryotic expression systems. However, full-length constructs for expression of all human sirtuin isoforms in mammalian cells have been described (see, eg, North, Marshall, Borra, Denu, & Verdin, 2003) and widely used. They are instrumental in cellular studies and allow for the isolation of smaller amounts of sirtuin protein (lower microgram range), mostly via tag-based immunoprecipitation. Such preparations are sufficient for basic, nonquantitative deacylation tests but not for more rigorous biochemical characterizations. We will focus here on the preparation of milligram amounts of mammalian sirtuins using bacterial expression systems, and describe activity tests, binding assays, and crystal structure analysis as key in vitro applications for such proteins in sirtuin studies.

3. PURIFICATION OF A MAMMALIAN SIRTUIN RECOMBINANTLY EXPRESSED IN *E. coli*

Many constructs for heterologous expression of the human sirtuin isoforms, and of their orthologs from other organisms, in the efficient *E. coli* expression host have been described (see above). The first step for protein production is to select an appropriate expression construct for the sirtuin enzyme of interest, or to generate such a construct through cloning in a suitable vector. Typically, the construct features a fusion with a tag that can be exploited for an affinity chromatography step, and a protease recognition site that facilitates specific removal of the affinity tag. Preparation of sirtuins can then be achieved using standard overexpression protocols and chromatography steps for purification as used for many other proteins (Fig. 2).

Table 1 Examples for Sirtuin Constructs for Recombinant Expression in *E.coli*

Sirtuin	Organism	Construct	Tag	Vector	References
Sirt1	*H. sapiens*	Full length	His (N)[a]	pET15b	Lakshminarasimhan, Curth, et al. (2013)
		225–664	His (N)	pET15b	
		186–516 +641–665 (Mini-Sirt1)	His (N)	pET21b	Dai et al. (2015)
Sirt2	*H. sapiens*	34–356	GST (N)	pGEX-4T3	Moniot et al. (2013)
		56–356	His (N)	pET15b (modified)[b]	Rumpf et al. (2015)
Sirt3	*H. sapiens*	102–399	His (N)	pET151/ D-TOPO	Schlicker et al. (2008)
		114–380	His (N)	pET151/ D-TOPO	Gertz et al. (2012)
		118–399	His-Trx (N)	pVFT3S	Gertz et al. (2012)
Sirt4	*H. sapiens*	25–314	His (N)	pET151/ D-TOPO	Rauh et al. (2013)
Sirt5	*H. sapiens*	34–302	His (N)	pET151/ D-TOPO	Schlicker et al. (2008)
	D. rerio	30–298	His (N)	pET151/ D-TOPO	Roessler et al. (2014)
Sirt6	*H. sapiens*	Full length	His (N)	pQE80-L	Pan et al. (2011)
		3–318	His (N)	pET28a-LIC	Pan et al. (2011)
		13–308	His (N)	pET151/ D-TOPO	Schlicker, Boanca, Lakshminarasimhan, and Steegborn (2011)
Sirt7	*H. sapiens*	Full length	His-SUMO (N)	pOPIN S	Rauh et al. (2013)
		14–367	His-SUMO (N)	pOPIN S	Rauh et al. (2013)
Sir2Tm	*T. maritima*	Full length	None	pET11a	Smith et al. (2002)
Hst2	*S. cerevisiae*	Full length	His (N)	pRSET-A	Zhao et al. (2003)

[a]Letters in brackets denote the tag position. N, N-terminus; C, C-terminus.
[b]Modifications: His_{10} and a TEV cleavage site instead of His_6 and a thrombin cleavage site, respectively.

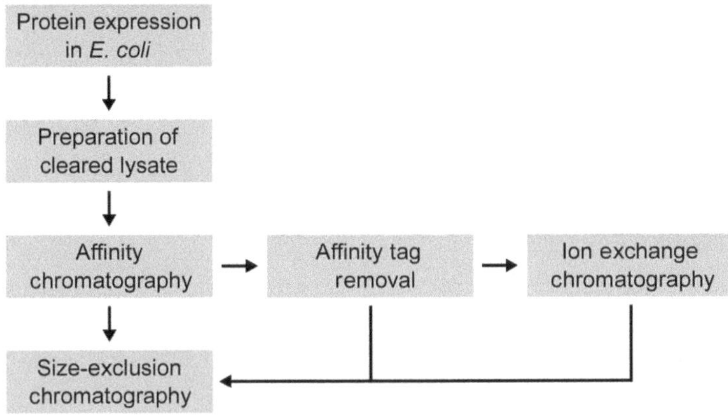

Figure 2 General scheme of sirtuin purification steps. After heterologous overexpression in *E. coli*, cells are disrupted and a cleared lysate is prepared. The standard purification strategy includes affinity chromatography and size-exclusion chromatography. Depending on the desired purity and the planned application, the affinity tag can be removed and an ion exchange chromatography can be included.

Many sirtuins, and especially their catalytic cores, can easily be purified as His-tagged constructs after bacterial overexpression. Examples are human mitochondrial Sirt3 and Sirt5, Sirt6, and several yeast and bacterial sirtuins. The following sections describe the purification of His-tagged human Sirt3 catalytic core as a representative example, many other sirtuin expression constructs can be prepared similarly. We provide details on instruments, buffers, and techniques that allow for the production of milligram amounts of soluble and active protein (Tables 2 and 3). We will then discuss potential pitfalls and adaptations for the successful purification of other isoforms, orthologs, or constructs, and the use of other affinity tags.

The protocol starts with overexpression of His-tagged human Sirt3 catalytic core in *E. coli* under control of the inducible T7 promoter. After cell harvest and lysis, the His-tagged sirtuin is isolated via immobilized metal ion affinity chromatography (IMAC) followed by size-exclusion chromatography (SEC). For higher purity demands, the affinity tag can be removed by proteolytic cleavage and a second, reverse IMAC. Ion exchange chromatography (IEX) can be included prior to SEC for increased purity. In total, this protocol (Fig. 3) requires 2–3 days of work (depending on purity requirements and whether the tag is removed) starting from inoculating 4–6 L of expression culture to yield milligram amounts of pure sirtuin.

Table 2 Instruments and Reagents Needed for Sirtuin Purification

Purification Step	Instrument/Reagent
Throughout whole protocol	Pipettes and tips, thermomixer, magnetic stirrer, centrifuges, centrifugal concentrators, SDS-sample buffer, SDS-PAGE running chamber and generator, UV/vis spectrophotometer
Cell culturing and protein expression	*E. coli* BL21 CodonPlus (DE3)-RIL, LB medium, antibiotics, IPTG, cell incubation shaker
Cell harvest and lysis	Lysozyme, DNase I, PMSF, NaDesoxycholate, $MgCl_2$
Affinity chromatography	Gravity flow column, Ni-NTA or Co-Talon™, imidazole
Dialysis[a]	Dialysis tubes or chambers, protease (eg, TEV)[b]
Ion exchange chromatography[a]	FPLC instrument
Size-exclusion chromatography	FPLC instrument, DTT (or any other reducing agent)

[a]Depending on the needed purity these steps can be omitted.
[b]Depending on the linker sequence other proteases might be needed.

Table 3 Buffers for the Purification of His-Tagged Sirtuins

Purification Step	Buffer	Buffer Composition[a]
Cell homogenization Cell disruption Binding to affinity resin	Lysis buffer	50 m*M* Tris/HCl, pH 7.8–8.5 200 m*M* NaCl 0.1% (v/v) Triton™ X-100
Affinity chromatography	Washing buffer 1	50 m*M* Tris/HCl, pH 7.8–8.5 500 m*M* NaCl
	Washing buffer 2	50 m*M* Tris/HCl, pH 7.8–8.5 200 m*M* NaCl 10 m*M* Imidazole
	Elution buffer	50 m*M* Tris/HCl, pH 7.8–8.5 200 m*M* NaCl 250 m*M* Imidazole
Ion exchange chromatography	IEX buffer A	50 m*M* Tris/HCl, pH 7.8–8.5 50 m*M* NaCl
	IEX buffer B	50 m*M* Tris/HCl, pH 7.8–8.5 1 *M* NaCl
Size-exclusion chromatography	SEC buffer	20 m*M* Tris/HCl, pH 7.8–8.5 150 m*M* NaCl

[a]Buffer pH is adjusted to 0.8–1 pH units below the target sirtuin's pI.

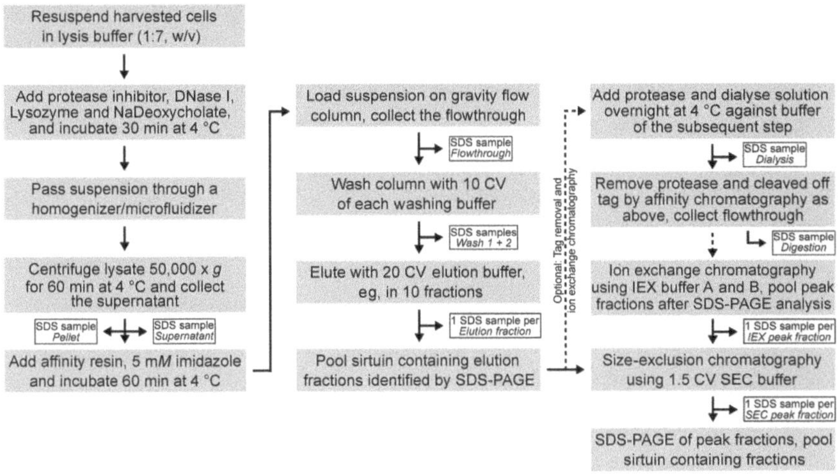

Figure 3 Detailed purification scheme for His-tagged sirtuins. *Dashed arrows* denote optional steps.

3.1 Instruments and Buffers

See Tables 2 and 3

3.2 Protein Expression

Typically, a His-tagged sirtuin construct is expressed from an isopropyl β-D-1-thiogalactopyranoside (IPTG) inducible expression system (eg, pET vector), which yields high amounts of protein in *E. coli*. The protocol described here is based on human Sirt3 residues 114–380 in pET151D/Topo, resulting in a N-terminally His-tagged protein with a tobacco etch virus (TEV) protease cleavage site. The sirtuin coding vector is transformed into an *E. coli* expression strain (eg, BL21 CodonPlus (DE3)-RIL) and plated on Luria broth (LB) agar supplemented with 100 μg/mL ampicillin and 34 μg/mL chloramphenicol. After overnight (ON) incubation at 37°C a single colony is used for inoculating a 50–100 mL preculture (LB medium supplemented with ampicillin and chloramphenicol). The preculture is incubated ON at 37°C in an incubation shaker at 120 rpm. Alternatively, the transformation mix can be used directly for inoculating the preculture, making the procedure one day shorter. For repeated expression, preculture samples can also be stored as glycerol stocks at −80°C and used instead of fresh transformations for inoculating the new preculture.

The following day, the expression culture (4–6 L LB supplemented with ampicillin and chloramphenicol) is inoculated by addition of preculture

(1:200, v/v). Cells are grown at 120 rpm and 37°C in 1-L baffled flasks in an incubation shaker until they reach an optical density at 600 nm of 0.6–0.8. The expression culture is then exposed to a cold shock by placing the flasks on ice for 20 min, and sirtuin expression is subsequently induced by adding 0.5 m*M* IPGT (final concentration). Following ON expression at 18–20°C in the incubation shaker, cells are harvested by centrifugation (15 min, 7000 x *g*, room temperature (RT)). After the mass of the harvested cells is determined, the cells can be flash frozen and stored at –80°C if not immediately processed further.

3.3 Preparation of Cleared Lysate

In order to release the overexpressed His-tagged sirtuin, the cell pellet is resuspended in chilled Lysis buffer (1:7, w/v). To enhance cell disruption, 0.25 mg lysozyme and 5 mg sodium deoxycholate per 1 g cells are added. In addition, the cell suspension is supplemented with a final concentration of 5 m*M* phenylmethylsulfonyl fluoride (PMSF) to inhibit bacterial proteases. To reduce DNA contamination and viscosity 5 μg DNase I per 1 g cell pellet as well as 2 m*M* magnesium chloride (final concentration) are added. After stirring the suspension 30 min on ice, cells are disrupted using a homogenizer/microfluidizer. Alternatively, sonication can be used for cell disruption, but it has to be applied with care to avoid larger increases in the solution's temperature, which would decrease protein yield due to denaturation. The lysate is then cleared from cell debris by centrifugation (60 min, 4°C, 50,000 x *g*). After centrifugation a SDS-PAGE sample of the resulting pellet and the supernatant is prepared.

3.4 Affinity Chromatography

With the lysate being cleared from aggregates and membranes, the overexpressed His-tagged sirtuin is further purified by IMAC. For batch purification, the cleared lysate is supplemented with 5 m*M* imidazole before addition of the affinity resin (Ni-NTA or Co-Talon™, for example), which has been equilibrated in Lysis buffer. The protein is bound to the resin through incubation at 4°C for 60 min, depending on the sample volume in an end-over-end rotator at slow speed or with gentle stirring in a beaker. Afterwards, the lysate/resin slurry is used for gravity flow purification by applying it into a column body. The flowthrough is released and collected for SDS-PAGE analysis. In order to remove contaminating proteins binding non-specifically to the resin, the column is washed with 10 column volumes

(CV) of Washing buffer 1 and Washing buffer 2, respectively. Then, the His-tagged sirtuin is eluted with 20 CV Elution buffer in 10 fractions. SDS-PAGE samples of flowthrough, both wash fractions and of all elution fractions are prepared to determine sample purity and sirtuin amount. Afterwards, all elution fractions containing mostly sirtuin protein are combined.

Depending on the purity and application demands, the pooled elution fractions are then either purified further by removing the affinity tag and/or subjecting the sample to IEX, or the pooled sirtuin fraction is directly subjected to SEC to yield the final product.

3.5 Affinity Tag Removal

Depending on the requirements of the intended application, the affinity tag is cleaved off the sirtuin after affinity chromatography. Preferably, one of the highly specific proteases such as TEV or Prescission© protease is used, in combination with an expression construct harboring the corresponding recognition sequence between tag and target. These proteases are available in His-tagged form, simplifying their removal after cleavage (see below). In our hands the best strategy for tag cleavage is proteolysis during a dialysis step. This approach allows one to combine specific and typically almost complete cleavage of the tag with the removal of imidazole, which can promote aggregation of many proteins if present for extended times at high concentrations. The pooled eluate from IMAC is thus supplemented with an appropriate amount of protease, transferred into a dialyzer/dialysis tube and dialyzed preferably ON at 4°C, or for 4 h at RT.

After dialysis, the sample is recovered and subjected to another affinity chromatography step for removal of released His-tag molecules, uncleaved protein, and the His-tagged protease (reverse IMAC). For this second IMAC, affinity resin is added to the dialysis sample and incubated 60 min at 4°C. Afterwards the sample is loaded into a gravity flow column body and the cleaved sirtuin is obtained by collecting the flowthrough.

In order to verify complete removal of the affinity tag, protease, and uncleaved protein, the samples before and after the second IMAC should be analyzed by SDS-PAGE. If no sirtuin is observed in the flowthrough of the reverse IMAC, the resin is washed with buffer containing a low concentration of imidazole (eg, Washing buffer 1 supplemented with max. 25 m*M* imidazole), which prevents weak nonspecific binding observed for many proteins. A sample of the resin can be included in the SDS-PAGE analysis to check for strong nonspecific attachment to the material, which normally indicates that the protein is unstable under the chosen conditions and aggregates. The buffer conditions then need to be adjusted for stabilization of the protein.

3.6 Ion Exchange Chromatography

The affinity purified sirtuin protein should next be subjected to an IEX step if highly pure protein is required. Otherwise, affinity purified sirtuin is directly subjected to SEC (see Section 3.7). For IEX, the sample is applied to a FPLC-coupled anion exchange column (eg, Mono Q© or HiTrap Q©). The sample can normally be concentrated in spin concentrators without major losses up to 5–10 mg/mL to shorten the loading time. Concentrator membranes should be rinsed thoroughly with IEX buffer A prior to their use and once a suitable volume is reached, concentrator filters should be rinsed thoroughly with the remaining solution in order to wash off any adhering sirtuin protein. The sample is then cleared of possible aggregates by centrifugation (5 min, 4°C, at maximum speed of a table top centrifuge) and loaded on an IEX column equilibrated with 5 CV IEX buffer A. Bound proteins are eluted by a salt gradient of 10 CV from 50 m*M* to 1 *M* sodium chloride (0–100% IEX buffer B) and collected in 0.3–0.5 mL fractions. Sirtuins tend to elute at the beginning of the gradient. SDS-PAGE analysis of peak fractions is used for determining sirtuin amount and purity. Mostly pure sirtuin fractions are pooled and subjected to SEC.

3.7 Size-Exclusion Chromatography

As a final purification step, the sirtuin sample is subjected to SEC. This removes aggregates and higher order oligomers from the final product and is used to transfer the sirtuin into a buffer suitable for storage or a planned application. The pooled eluate of the previous step (IMAC or IEX) is concentrated to a volume that can be loaded onto a FPLC-coupled SEC column (eg, Sephadex 200© or Superose 75©); we use up to 1 mL on columns with 24-mL bed volume and up to 5 mL for columns with 120-mL bed volume to avoid excessive band widths. The sample is applied to a SEC column preequilibrated with 1.5 CV of SEC buffer, and the sirtuin is eluted in 0.3–0.5 mL fractions using 1.5–2 CV SEC buffer. Typically, sirtuins elute in a single peak at a position corresponding to the monomer size, with the exception of larger Sirt1 constructs (Lakshminarasimhan, Curth, et al., 2013). Peaks of higher oligomeric states are observed rarely but can occur as smaller peaks or shoulders, for example with a Sirt5 dimer, and should not be included into the final purification product as in our experience sirtuin oligomers tend to be less active.

Sirtuin peak fractions are analyzed by SDS-PAGE to confirm the protein's purity and are pooled. If necessary for subsequent applications, sirtuins can be concentrated to higher protein concentrations without major losses for most constructs. The final product can be stored for several days at 4°C or shock-frozen in liquid nitrogen and stored for years at –80°C. For storage at –80°C, we recommend small aliquot sizes to avoid repeated freeze–thawing cycles, which can lead to losses in protein and activity.

3.8 Modifications to the Standard Protocol

For most sirtuin catalytic core constructs the above mentioned protocol yields medium to high milligram amounts of soluble and active protein. However, in case of yield problems, different *E. coli* expression strains and/or different cultivation media can be used, for example, *E. coli* BL21 (DE3) pLys and M15 pREP4 or terrific broth and autoinduction medium, respectively. We routinely use *E. coli* BL21 CodonPlus (DE3)-RIL and LB or autoinduction medium for Sirt2 34–356. Details for a suitable expression approach are given in the respective publication for the corresponding construct (see above). Usually, sirtuin constructs are found in the soluble fraction after cell lysis. If the protein is present in the insoluble fraction, solubility enhancing tags have to be employed (eg, GST-, MBP-, SUMO-, or Trx-fusion). We usually use a His-Trx-tag for Sirt3 and a His-SUMO-tag, for Sirt2. Denaturing purification and refolding has rarely been used with great success for sirtuins, but for the bacterial sirtuin Sir2Tm from *Thermotoga maritima* a refolding protocol yields large amounts of properly folded and enzymatically active enzyme (Smith et al., 2002).

In most cases, PMSF is sufficient to avoid sirtuin proteolysis after cell lysis. Nevertheless, in rare cases degradation products can be observed. Supplementing the Lysis buffer with protease inhibitor cocktails normally overcomes this problem, with the exception of full-length Sirt1 were some proteolysis cannot be fully avoided in our experience. To ensure binding of the structurally essential Zn^{2+} ion to sirtuins, Zn^{2+} salts can be added at micromolar concentrations, but this supplement is normally not required to ensure Zn^{2+} saturation due to the high affinity for the ion. Since Zn^{2+} ions can promote aggregation of many proteins, we recommend not including it in absence of experimental evidence for incomplete Zn^{2+} site occupancy. For several sirtuin constructs, however, reducing agents (eg, 5–25 m*M* DTT or 1 m*M* TCEP) are routinely added to the SEC buffer to prevent cysteine oxidation and thereby enhance protein stability.

The catalytic cores of sirtuins have masses around 30 kDa, similar to the widely used proteases for tag cleavage. Thus SEC is not sufficient for removing the protease after affinity tag cleavage and performing a reverse IMAC becomes crucial. However, if larger constructs are purified, the reverse IMAC can be omitted since cleaved sirtuin and protease can be separated by SEC.

Most buffer components will not interfere significantly with sirtuin stability or activity. Thus, Tris can be replaced with HEPES or phosphate in all buffers if the application demands it. However, for isoform- and construct-specific sensitivities or optima with respect to pH ranges or buffer components, the available literature (see above) should be consulted when preparing a particular sirtuin.

4. BIOCHEMICAL CHARACTERIZATIONS USING PURIFIED, RECOMBINANT SIRTUINS

4.1 Sirtuin Activity

Because sirtuins are central regulators in aging processes and age-associated diseases, purified sirtuin enzymes are often used in pharmacological research, in particular in activity studies in the presence of small-molecule compounds to identify sirtuin-targeting drugs. For this purpose, His-tagged sirtuins can normally be used in (semi-)high-throughput assays using fluorogenic substrates or by using coupled reactions to monitor sirtuin activity (Schutkowski et al., 2014). To validate findings, more robust mass spectrometric approaches enable label-free and unbiased sirtuin activity analysis (Fischer et al., 2012; Schutkowski et al., 2014). Larger tags should be removed prior to activity analysis to avoid artifacts commonly observed with, for example, GST-tags, or the fusion protein must be rigorously controlled for in order to avoid artifacts.

A widely used activity assay links the sirtuin reaction to two additional enzyme activities, thereby enabling a continuous optical read-out (Fig. 4; Table 4). In this coupled continuous activity assay first described by Smith et al. (2009) the sirtuin product nicotinamide is consumed by nicotinamidase (NCA) to form ammonia, which in turn is coupled to α-ketoglutarate by glutamate dehydrogenase (GDH) to form glutamate. The latter enzymatic reaction depends on the oxidation of NAD(P)H to $NAD(P)^+$, which can be monitored by an absorption decrease at 340 nm.

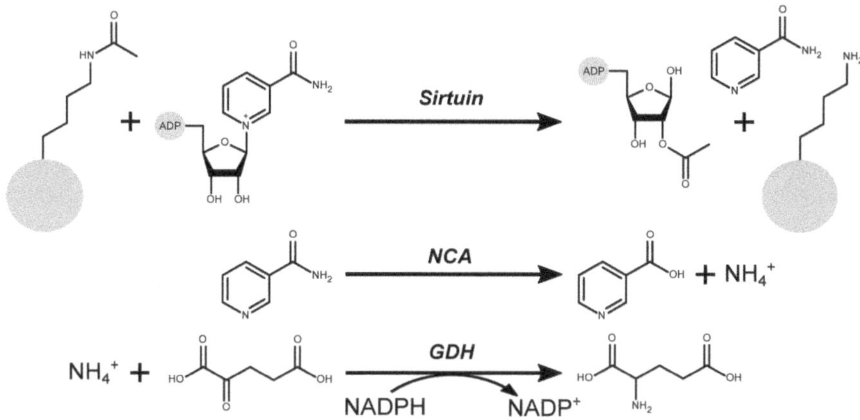

Figure 4 Continuous coupled enzymatic sirtuin activity assay. The sirtuin reaction transfers an acyl moiety to ADP-ribose, which is generated by the release of nicotinamide from NAD^+. Subsequently, nicotinamidase (NCA) degrades nicotinamide to nicotinic acid and ammonia, which is then attached to α-ketoglutarate to form glutamate by glutamate dehydrogenase (GDH). This last reaction is driven by NADPH oxidation to $NADP^+$ and can be monitored by a decrease in the absorption at 340 nm.

Table 4 Buffers and Reagents Needed for the Coupled Continuous Assay (Smith, Hallows, & Denu, 2009)

Buffer and Reagents	**Final Concentration**
Na-phosphate buffer, pH 7.5–7.8 (reaction buffer)	50 m*M*
Sirtuin	1–5 μ*M*
Substrate peptide/protein	100–200 μ*M*
NAD^+	0.5–1 m*M*
Glutamate dehydrogenase	2 U
Nicotinamidase	2 μ*M*
α-Ketoglutarate	3.3 m*M*
NADPH	0.5 m*M*
DTT	2 m*M*

Although sirtuins differ in their substrate and deacylation specificity, all NAD^+-dependent deacylations catalyzed by sirtuins can be monitored by this assay due to formation of the common coproduct nicotinamide. Sirt3 deacetylation of an acetylated peptide is used here as an example for assay preparation.

A coupled continuous assay for Sirt3 activity consists of 1–5 μ*M* Sirt3, 100–200 μ*M* substrate peptide or protein, 0.5–1 m*M* NAD^+, and saturating concentrations of GDH (2U), NCA (2 μ*M*), α-ketoglutarate (3.3 m*M*), and NADPH (500 μ*M*) in 80–100 μl 50 m*M* phosphate buffer pH 7.5–7.8. In order to ensure stability against oxidation throughout the 30–60 min reaction, the sample can be supplemented with 2 m*M* DTT. For better reproducibility, an enzyme master mix and a substrate master mix can be prepared, which are used in a ratio of 6:4 to 8:2. The enzyme master mix contains Sirt3, either NAD^+ or the substrate peptide/protein, DTT, NCA, GDH, NADPH, and α-ketoglutarate. The substrate master mix contains the sirtuin substrate which is missing in the enzyme master mix. Depending on the instrument setup (single chamber or microplate-reading spectrophotometer) the substrate master mix is dispensed in a cuvette or in a 96-well UV plate. The reaction is then started by adding the enzyme master mix and its progress is monitored by the decreasing absorption at a wavelength of 340 nm at room temperature or 37°C for 30–60 min.

Due to the limited sensitivity of this and other sirtuin assays (see below) and due to the relatively low catalytic rates for most sirtuin-dependent reactions, assays often contain considerable enzyme concentrations. Careful analysis of the obtained data is therefore required. It is not uncommon, for example, that the substrate concentration provided decreases significantly during the assay, so that an initial substrate saturation condition changes toward one where substrate availability limits the reaction rate. Also, in tests with potent inhibitors, compound concentrations can go down to—or even below—the enzyme concentrations. Appropriate fitting equations for such situations, such as tight-binding models, need to be chosen for data analysis, since regular models are based on the assumption that inhibitor concentrations significantly exceed enzyme concentration.

Other frequently used activity assays rely on fluorescently labeled substrate peptides or lysine derivatives. Depending on the setup, sirtuin activity is either quantitated by an additional developing reaction step (eg, Fluor-de-Lys, *ENZO Life Sciences*) or it is directly monitored via an increased fluorescence signal (Roessler, Tuting, Meleshin, Steegborn, & Schutkowski, 2015). In addition, P^{32}-labelled NAD^+ can be used to analyze sirtuin activity by quantitating the NAD^+/*O*-acyl-ADP-ribose ratio by autoradiography. Many more assays with specific advantages and disadvantages have been described and are summarized in Schutkowski et al. (2014). Sirtuin-modulating compounds can interfere with many of the above mentioned assay due to their optical properties or by modulating coupled enzymatic reactions. To avoid these artifacts, either the radioactive assay or mass spectrometry based protocols should be used to confirm compound activities, since they enable analysis with nonmodified substrates and via read-outs not sensitive to the presence of small molecules (Schutkowski et al., 2014).

4.2 Sirtuin Binding Studies

Binding affinity measurements with purified sirtuins and their interaction partners are often essential additions to activity analyses to understand the details of the interplay between the complex sirtuin catalytic cycle (and the many sirtuin conformations associated with it) and ligand binding. They further can yield information helpful for the structure determination of the respective sirtuin/ligand complexes (see below). Affinity measurements with purified sirtuins and their interaction partners can often be performed using isothermic titration calorimetry or microscale thermophoresis (MST) (Wienken, Baaske, Rothbauer, Braun, & Duhr, 2010). MST is a relatively new technique, which determines dissociation constants by analyzing the mobility of protein/ligand complexes within a thermal gradient via a fluorescent read-out (either intrinsic protein tryptophan fluorescence or fluorescence-labeled protein, depending on the instrument used). Some sirtuin studies already employed MST, for example, to analyze the interaction between Sirt3, its substrates, and the inhibitor Ex527 to provide insights in the mechanism of inhibition (Gertz et al., 2013). In this case, Sirt3 114–380, still carrying the His-tag, was purified with the final SEC buffer composed of HEPES, allowing its direct use for fluorescence labeling of lysines with fluorescein thioisocyanate (FITC). Since Tris would interfere with this and many other labeling reactions, its replacement with HEPES or phosphate in the SEC buffer avoids a buffer exchange step. For labeling, the sirtuin is incubated with a twofold molar excess of the dye at 4°C ON. Afterwards, free FITC is removed using a NAP25 desalting column (*GE Healthcare*). FITC-Sirt3 samples (15 μL each) are then incubated with varying concentrations of EX-527 on ice for 15 min, and subsequently transferred in glass capillaries and analyzed with a Monolith NT115 (*Nanotemper Technologies*, excitation wavelength 470 nm, emission wavelength 520 nm, LED-power 10–20%, and laser-power 10%). An estimated K_D in the high millimolar range is to be expected for Sirt3 and Ex527 alone, and also in Ex527 titration experiments in the presence of saturating peptide concentrations (Fig. 5A). However, incubating FITC-Sirt3 first with NAD^+ and then with varying EX-527 concentrations enhances compound binding

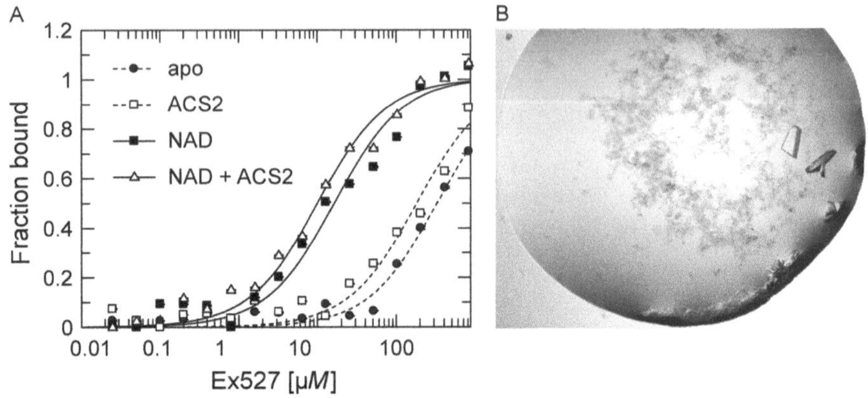

Figure 5 Exemplary experimental results obtained with purified recombinant sirtuins. (A) Binding analyses using MST and fluorophore-labeled Sirt3 to obtain insights into the mechanism of sirtuin inhibition by Ex527 (Gertz et al., 2013). The inhibitor binds weakly to Sirt3 in absence of substrates or in presence of saturating peptide concentrations. The presence of NAD^+ enhances Ex527 binding, and binding appears further enhanced when both substrates are present. (B) Crystals of Sirt3 in complex with acetylated substrate peptide. Crystals were grown in hanging drops from a 11 mg/mL Sirt3 solution in SEC buffer supplemented with 5 m*M* acetyl-ACS2 peptide and mixed 1:1 with reservoir solution (0.1 *M* Bis–Tris, pH 5.5, 21% (w/v) PEG3350, 0.2 *M* $(NH_4)_2SO_4$) as described in Gertz et al. (2013). *Panel (A): The figure was reproduced, in adapted form, from Gertz et al. (2013).*

to a low micromolar K_D, which is enhanced even more in the presence of saturating concentrations of both substrates (Fig. 5A). These results suggest that the inhibitor binds strongly to the sirtuin only if the cosubstrate NAD^+ and/or a product is also bound, which was later confirmed by activity analysis in combination with structure determinations (Gertz et al., 2013).

In many cases, a sirtuin/ligand interaction can also be analyzed with nonlabeled protein. The interaction of Sirt3 with the inhibitor 4′-bromo-resveratrol was studied using this technique (Nguyen, Gertz, & Steegborn, 2013). It avoids protein losses during labeling and potential artifacts caused by the attached labels. We routinely use this approach for sirtuin studies because it combines the low sample requirement of MST in general with a simple yet label-free detection. In cases where the low intrinsic fluorescence of the protein or interfering fluorescence of some ligands prevents its use, we rely on label-based MST analysis.

4.3 Sirtuin Crystallization

Structural biology is another major application for purified sirtuins. Both NMR and X-ray crystallographic measurements demand highly pure and homogeneous protein samples. Hence, the sirtuin preparation protocol should include all purification steps summarized in Fig. 3, since purity and affinity tag removal tend to improve sirtuin crystal growth and its reproducibility.

A series of crystal structures of sirtuin catalytic domains helped in elucidating mechanism of sirtuin catalysis and regulation (see Moniot et al., 2012; Sanders et al., 2010 and above). We use crystallization of Sirt3 in complex with the inhibitor Ex527 as an example (Gertz et al., 2013). For crystallization, an 11 mg/mL Sirt3 solution in SEC buffer is supplemented with 1–1.5 m*M* Ex527 compound (100-m*M* stock solution in DMSO) and 5–10 m*M* NAD^+. This protein ligand mix is mixed in a 1:1 ratio with reservoir solution (Table 5) and placed over 400 μL reservoir solution in standard hanging or sitting drop setups (see, eg, Rupp, 2009 or other standard textbooks on protein crystallography) to yield crystals by vapor diffusion at 20°C. The protein solution can also be mixed with 5 m*M* acetylated or 1 m*M* thioacetylated peptide prior to crystallization to obtain Sirt3/peptide complex crystals (Table 5). Soaking of the Sirt3/acetyl-peptide crystals with NAD^+ and Ex527 yields a mixture of enzyme complexes, in particular a complex with substrates, a complex with reaction intermediate, and a complex with products and inhibitor (Gertz et al., 2013).

Table 5 Examples for Sirt3 Crystallization Conditions from Gertz et al. (2013)

Complex Structure	Protein Solution	Reservoir
Sirt3/NAD^+/EX527	11 mg/mL Sirt3 1 m*M* Peptide 10 m*M* NAD^+ 1 m*M* EX527 10% (v/v) DMSO	20% (w/v) PEG3350 0.2 *M* NH_4NO_3
Sirt3/thio-acetyl peptide	11 mg/mL Sirt3 1 m*M* Thio-acetyl peptide	25% (w/v) PEG3350 0.2 *M* $LiSO_4$ 0.1 *M* Na citrate, pH 5.6
Sirt3/acetyl peptide	11 mg/mL Sirt3 5 m*M* Acetyl peptide	0.2 *M* $(NH_4)_2SO_4$ 0.1 *M* Bis–Tris, pH 5.5 21% (w/v) PEG3350
Sirt3/2′-*O*-acetyl-ADP-ribose/EX-527	11 mg/mL Sirt3 10 m*M* *O*-acetylADP-ribose 1 m*M* EX527 10% (v/v) DMSO	20% (w/v) PEG6000 0.1 *M* HEPES, pH 7.0

In general, most well-behaving catalytic core constructs of sirtuins can be crystallized using 5–10 mg/mL sirtuin protein stock solutions in sitting and hanging drop vapor diffusion experiments at 4°C or 20°C. However, the tendency of sirtuins to crystallize differs. Some enzymes tend to form crystals in apo state or in the presence of substrates (or substrate analogs) or modulators, and conditions are easily identified by using factorial screens (eg, several conformations or ligand states of yeast Hst2, human Sirt2 and Sirt3 (Fig. 5B; Moniot et al., 2013; Nguyen, Gertz et al., 2013; Zhao et al., 2003)). Other sirtuins or sirtuin complexes tend to be more challenging in crystallization experiments, and in fact only recently a larger number of sirtuin/inhibitor complex structures could be determined, sometimes using specific constructs and/or techniques (see Dai et al., 2015; Gertz et al., 2013; Nguyen, Schaefer, Gertz, Weyand, & Steegborn, 2013 for examples). Human Sirt5 has not yet been crystallized in its full-length form, and the catalytic core of the human enzyme often yields twinned crystals (Gertz et al., 2012) and does not produce high-quality crystals reproducibly, even in the presence of seeds. Therefore, its highly similar ortholog from *Danio rerio* is sometimes used for structural analysis of this sirtuin isoform (Roessler et al., 2014). For crystallization of Sirt1 in complex with small molecule activators, a sophisticated construct covering catalytic domain, an N-terminal domain, and a short region within the C-terminus—fused to the catalytic domain via a short, artificial linker—is required (Dai et al., 2015). Many additional crystallization reports for sirtuins and their ligand complexes, including constructs, specific techniques advice, and successful crystallization conditions are available in the literature (see, eg, Du et al., 2011; Gertz et al., 2012; Nguyen, Gertz et al., 2013; Pan et al., 2011; Rumpf et al., 2015). It is strongly advised to review this literature at the start of a sirtuin crystallography project for advice on construct selection, protein purification, and crystallization procedures and conditions.

Once sirtuin crystals are obtained, they can normally be used for structure solution via standard protein crystallography methods (see, eg, Rupp, 2009). Crystals are transferred from the mother liquor to cryoprotection solution, which can be supplemented with small molecules (substrates, inhibitors) for ligand soaking experiments. In most cases, reservoir solution supplemented with 25% (v/v) glycerol is a suitable cryoprotection solution, but in some cases the glycerol content needs to be adjusted or another cryoprotectant has to be used (eg, sucrose, oils) to stabilize the crystals for freezing. Crystals are then flash frozen in liquid nitrogen and used for diffraction experiments either on local X-ray sources or at synchrotron beamlines. Data collection and processing can normally be done using standard software and approaches (McCoy et al., 2007; Rupp, 2009; Winn et al., 2011). With many sirtuin structures now being available, the missing phase information of diffraction experiments can typically be obtained by using pattern search techniques with these structures as search models (molecular replacement). In some cases, however, extensive searches or additional approaches are necessary due to conformational differences, even if the same sirtuin isoenzyme is available as a search model. For human Sirt2, for example, its complex with ADP-ribose was determined with separate Rossmann-fold and Zn^{2+}-binding subdomains as search models, because the ligand induced a relative rotation of the two subdomains (Moniot et al., 2013). Once the phase problem is solved, the model can be fitted to the obtained electron density and refined against the experimental diffraction data using standard methods and software (McCoy et al., 2007; Rupp, 2009; Winn et al., 2011).

4.4 Further Applications

The purified, recombinant sirtuin proteins that can now be obtained for many isoforms and constructs via the procedures described above and in the literature can be used for many other in vitro applications. They are routinely used for melting temperature (T_m) analysis using fluorescent dyes to optimize buffer conditions or to identify or compare the binding properties of ligands (Nguyen, Schaefer, et al., 2013; Schuetz et al., 2007). They can be used for protein/protein interaction studies to understand sirtuin biology and regulation (Lakshminarasimhan, Curth, et al., 2013), or for analyzing the influence of the substrate sequence on compound effects in microarray experiments (Lakshminarasimhan, Rauth, Schutkowski, & Steegborn, 2013). The present chapter should be a good starting point for any such studies employing recombinant sirtuin proteins for studying function and regulation of this unique family of epigenetic regulators.

ACKNOWLEDGMENTS

Work on sirtuins in the authors' lab was supported by Oberfrankenstiftung, Alzheimer Forschung Initiative, and Deutsche Forschungsgemeinschaft.

REFERENCES

Avalos, J. L., Bever, K. M., & Wolberger, C. (2005). Mechanism of sirtuin inhibition by nicotinamide: Altering the NAD(+) cosubstrate specificity of a Sir2 enzyme. *Molecular Cell*, *17*, 855–868.

Cen, Y., Youn, D. Y., & Sauve, A. A. (2011). Advances in characterization of human sirtuin isoforms: Chemistries, targets and therapeutic applications. *Current Medicinal Chemistry*, *18*, 1919–1935.

Dai, H., Case, A. W., Riera, T. V., Considine, T., Lee, J. E., Hamuro, Y., et al. (2015). Crystallographic structure of a small molecule SIRT1 activator-enzyme complex. *Nature Communications*, *6*, 7645.

Du, J., Zhou, Y., Su, X., Yu, J. J., Khan, S., Jiang, H., et al. (2011). Sirt5 is a NAD-dependent protein lysine demalonylase and desuccinylase. *Science*, *334*, 806–809.

Fischer, F., Gertz, M., Suenkel, B., Lakshminarasimhan, M., Schutkowski, M., & Steegborn, C. (2012). Sirt5 deacylation activities show differential sensitivities to nicotinamide inhibition. *PLoS One*, 7, e45098.

Gertz, M., Fischer, F., Nguyen, G. T., Lakshminarasimhan, M., Schutkowski, M., Weyand, M., et al. (2013). Ex-527 inhibits Sirtuins by exploiting their unique NAD+-dependent deacetylation mechanism. *Proceedings of the National Academy of Sciences of the United States of America*, *110*, E2772–E2781.

Gertz, M., Nguyen, G. T., Fischer, F., Suenkel, B., Schlicker, C., Franzel, B., et al. (2012). A molecular mechanism for direct sirtuin activation by resveratrol. *PLoS One*, 7. e49761.

Gertz, M., & Steegborn, C. (2010). Function and regulation of the mitochondrial Sirtuin isoform Sirt5 in Mammalia. *Biochimica et Biophysica Acta*, *1804*, 1658–1665.

Haigis, M. C., & Sinclair, D. A. (2010). Mammalian sirtuins: Biological insights and disease relevance. *Annual Review of Pathology*, *5*, 253–295.

Jin, L., Wei, W., Jiang, Y., Peng, H., Cai, J., Mao, C., et al. (2009). Crystal structures of human SIRT3 displaying substrate-induced conformational changes. *Journal of Biological Chemistry*, *284*, 24394–24405.

Lakshminarasimhan, M., Curth, U., Moniot, S., Mosalaganti, S., Raunser, S., & Steegborn, C. (2013). Molecular architecture of the human protein deacetylase Sirt1 and its regulation by AROS and resveratrol. *Bioscience Reports*, *33*, e00037.

Lakshminarasimhan, M., Rauth, D., Schutkowski, M., & Steegborn, C. (2013). Sirt1 activation by resveratrol is substrate sequence-selective. *Aging (Albany NY)*, *5*, 151–154.

Lin, H., Su, X., & He, B. (2012). Protein lysine acylation and cysteine succination by intermediates of energy metabolism. *ACS Chemical Biology*, 7, 947–960.

McCoy, A. J., Grosse-Kunstleve, R. W., Adams, P. D., Winn, M. D., Storoni, L. C., & Read, R. J. (2007). Phaser crystallographic software. *Journal of Applied Crystallography*, *40*, 658–674.

Michan, S., & Sinclair, D. (2007). Sirtuins in mammals: Insights into their biological function. *Biochemical Journal*, *404*, 1–13.

Moniot, S., Schutkowski, M., & Steegborn, C. (2013). Crystal structure analysis of human Sirt2 and its ADP-ribose complex. *Journal of Structural Biology*, *182*, 136–143.

Moniot, S., Weyand, M., & Steegborn, C. (2012). Structures, substrates, and regulators of mammalian sirtuins—Opportunities and challenges for drug development. *Frontiers in Pharmacology*, *3*, 16.

Nguyen, G. T., Gertz, M., & Steegborn, C. (2013). Crystal structures of sirt3 complexes with 4'-bromo-resveratrol reveal binding sites and inhibition mechanism. *Chemistry and Biology*, *20*, 1375–1385.

Nguyen, G. T., Schaefer, S., Gertz, M., Weyand, M., & Steegborn, C. (2013). Structures of human sirtuin 3 complexes with ADP-ribose and with carba-NAD(+) and SRT1720: Binding details and inhibition mechanism. *Acta Crystallographica. Section D, Biological Crystallography*, *69*, 1423–1432.

North, B. J., Marshall, B. L., Borra, M. T., Denu, J. M., & Verdin, E. (2003). The human Sir2 ortholog, SIRT2, is an NAD+-dependent tubulin deacetylase. *Molecular Cell*, *11*, 437–444.

Norvell, A., & McMahon, S. B. (2010). Cell biology. Rise of the rival. *Science*, *327*, 964–965.

Pan, P. W., Feldman, J. L., Devries, M. K., Dong, A., Edwards, A. M., & Denu, J. M. (2011). Structure and biochemical functions of SIRT6. *Journal of Biological Chemistry*, *286*, 14575–14587.

Pan, M., Yuan, H., Brent, M., Ding, E. C., & Marmorstein, R. (2012). SIRT1 contains N- and C-terminal regions that potentiate deacetylase activity. *Journal of Biological Chemistry*, *287*, 2468–2476.

Rauh, D., Fischer, F., Gertz, M., Lakshminarasimhan, M., Bergbrede, T., Aladini, F., et al. (2013). An acetylome peptide microarray reveals specificities and deacetylation substrates for all human sirtuin isoforms. *Nature Communications*, *4*, 2327.

Roessler, C., Nowak, T., Pannek, M., Gertz, M., Nguyen, G. T., Scharfe, M., et al. (2014). Chemical probing of the human sirtuin 5 active site reveals its substrate acyl specificity and peptide-based inhibitors. *Angewandte Chemie International Edition in English*, *53*, 10728–10732.

Roessler, C., Tuting, C., Meleshin, M., Steegborn, C., & Schutkowski, M. (2015). A novel continuous assay for the deacylase Sirtuin 5 and other deacetylases. *Journal of Medicinal Chemistry*, *58*, 7217–7223.

Rousseaux, S., & Khochbin, S. (2015). Histone acylation beyond acetylation: Terra incognita in chromatin biology. *Cell Journal*, *17*, 1–6.

Rumpf, T., Schiedel, M., Karaman, B., Roessler, C., North, B. J., Lehotzky, A., et al. (2015). Selective Sirt2 inhibition by ligand-induced rearrangement of the active site. *Nature Communications*, *6*, 6263.

Rupp, B. (2009). *Biomolecular crystallography: Principles, practice, and application to structural biology*. New York, NY: Garland Science.

Sanders, B. D., Jackson, B., & Marmorstein, R. (2010). Structural basis for sirtuin function: What we know and what we don't. *Biochimica et Biophysica Acta*, *1804*, 1604–1616.

Sauve, A. A., Wolberger, C., Schramm, V. L., & Boeke, J. D. (2006). The biochemistry of sirtuins. *Annual Review of Biochemistry*, *75*, 435–465.

Schlicker, C., Boanca, G., Lakshminarasimhan, M., & Steegborn, C. (2011). Structure-based development of novel Sirtuin inhibitors. *Aging (Albany NY)*, *3*, 852–857.

Schlicker, C., Gertz, M., Papatheodorou, P., Kachholz, B., Becker, C. F., & Steegborn, C. (2008). Substrates and regulation mechanisms for the human mitochondrial sirtuins Sirt3 and Sirt5. *Journal of Molecular Biology*, *382*, 790–801.

Schuetz, A., Min, J., Antoshenko, T., Wang, C. L., Allali-Hassani, A., Dong, A., et al. (2007). Structural basis of inhibition of the human NAD+-dependent deacetylase SIRT5 by suramin. *Structure*, *15*, 377–389.

Schutkowski, M., Fischer, F., Roessler, C., & Steegborn, C. (2014). New assays and approaches for discovery and design of Sirtuin modulators. *Expert Opinion on Drug Discovery*, *9*, 183–199.

Scott, I. (2012). Regulation of cellular homoeostasis by reversible lysine acetylation. *Essays in Biochemistry*, *52*, 13–22.

Shahbazian, M. D., & Grunstein, M. (2007). Functions of site-specific histone acetylation and deacetylation. *Annual Review of Biochemistry*, *76*, 75–100.

Simic, Z., Weiwad, M., Schierhorn, A., Steegborn, C., & Schutkowski, M. (2015). The epsilon-amino group of protein lysine residues is highly susceptible to nonenzymatic acylation by several physiological acyl-CoA thioesters. *Chembiochem*, *16*, 2337–2347.

Smith, J. S., Avalos, J., Celic, I., Muhammad, S., Wolberger, C., & Boeke, J. D. (2002). SIR2 family of NAD(+)-dependent protein deacetylases. *Methods in Enzymology*, *353*, 282–300.

Smith, B. C., Hallows, W. C., & Denu, J. M. (2009). A continuous microplate assay for sirtuins and nicotinamide-producing enzymes. *Analytical Biochemistry*, *394*, 101–109.

Szczepankiewicz, B. G., Dai, H., Koppetsch, K. J., Qian, D., Jiang, F., Mao, C., et al. (2012). Synthesis of carba-NAD and the structures of its ternary complexes with SIRT3 and SIRT5. *Journal of Organic Chemistry*, *77*, 7319–7329.

Tan, M., Peng, C., Anderson, K. A., Chhoy, P., Xie, Z., Dai, L., et al. (2014). Lysine glutarylation is a protein posttranslational modification regulated by SIRT5. *Cell Metabolism*, *19*, 605–617.

Tennen, R. I., Berber, E., & Chua, K. F. (2010). Functional dissection of SIRT6: Identification of domains that regulate histone deacetylase activity and chromatin localization. *Mechanisms of Ageing and Development*, *131*, 185–192.
Wagner, G. R., & Payne, R. M. (2013). Widespread and enzyme-independent nepsilon-acetylation and nepsilon-succinylation of proteins in the chemical conditions of the mitochondrial matrix. *Journal of Biological Chemistry*, *288*, 29036–29045.
Wienken, C. J., Baaske, P., Rothbauer, U., Braun, D., & Duhr, S. (2010). Protein-binding assays in biological liquids using microscale thermophoresis. *Nature Communications*, *1*, 100.
Winn, M. D., Ballard, C. C., Cowtan, K. D., Dodson, E. J., Emsley, P., Evans, P. R., et al. (2011). Overview of the CCP4 suite and current developments. *Acta Crystallographica. Section D, Biological Crystallography*, *67*, 235–242.
Zhao, K., Chai, X., & Marmorstein, R. (2003). Structure of the yeast Hst2 protein deacetylase in ternary complex with 2'-O-acetyl ADP ribose and histone peptide. *Structure*, *11*, 1403–1411.

CHAPTER NINE

Preparation, Biochemical Analysis, and Structure Determination of SET Domain Histone Methyltransferases

E. Bergamin, J.F. Couture[1]
Ottawa Institute of Systems Biology, University of Ottawa, Ottawa, ON, Canada
[1]Corresponding author: e-mail address: jean-francois.couture@uottawa.ca

Contents

1. Introduction 210
2. Experimental Procedures 218
   2.1 Construct Design Strategies 219
   2.2 Protein Sample Preparation 221
   2.3 Preparation of Selenomethionyl-Labeled RcATXR5 224
   2.4 Protein Crystallization 225
   2.5 Data Collection and Structural Determination 227
   2.6 *In Vitro* MTAs 228
3. Conclusions 231
Acknowledgments 231
References 232

Abstract

In eukaryotes, several lysine residues on histone proteins are methylated. This posttranslational modification is linked to a myriad of nuclear-based transactions such as epigenetic inheritance of heterochromatin, regulation of gene expression, DNA damage repair, and DNA replication. The majority of the enzymes responsible for writing these marks onto chromatin belong to the SET domain family of histone lysine methyltransferases. Although they often share important structural features, including a conserved catalytic domain, SET domain enzymes use different mechanisms to achieve substrate recognition, mono-, di-, or trimethylate lysine residues and some require other proteins to achieve maximal methyltransferase activity. In this chapter, we summarize our efforts to purify, crystallize, and enzymatically characterize SET domain enzymes with a specific focus on the histone H3K27 monomethyltransferase ATXR5.

Methods in Enzymology, Volume 573
ISSN 0076-6879
http://dx.doi.org/10.1016/bs.mie.2016.01.004

1. INTRODUCTION

In eukaryotic cells, the genetic information is encoded in about 1.8 m of DNA packaged into the nucleus in the form of chromatin. The fundamental unit of chromatin is the nucleosome, which consists of 147 base pairs of DNA wrapped around a histone octamer consisting of two copies each of the core histones H2A, H2B, H3, and H4 (Luger, Mader, Richmond, Sargent, & Richmond, 1997). Nucleosomes further assemble in more compact structures to ultimately form chromosomes. Chromatin is a dynamic structure, and its degree of compaction has a great influence on important biological processes such as signal transduction, development, cellular proliferation, differentiation, and gene expression. For instance, a closed chromatin conformation is usually refractory to transcription and instead an open conformation facilitates access of the transcriptional machinery. Elaborate mechanisms have evolved to introduce variations into the structure of the chromatin fiber. Chromatin remodeling by ATP-dependent protein complexes, the deposition of histone variants, and the introduction of covalent histone modifications are three major mechanisms by which chromatin can be altered. Diverse posttranslational modifications (PTMs), including acetylation, phosphorylation, ubiquitination, ADP-ribosylation, and methylation (Fischle, Wang, & Allis, 2003), take place on histone proteins in regions either exposed to the nucleoplasm (ex. tail of histone H3) or located within the nucleosome core particle (NCP) (Peterson & Laniel, 2004). The type of modification and the exact position at which the modifications occur can influence several DNA-based transactions including, but not limited to, the regulation of gene expression, DNA replication, and DNA damage repair.

The enzymes responsible for depositing, reading, and removing these modifications are taking center stage in the epigenomic era. In this chapter, we will focus on lysine methylation, and the major family of enzymes that are capable of writing this histone mark: the SET domain histone lysine methyltransferases (HKMTs). Lysine methylation is particularly interesting for two important reasons: (1) the location of the modification is important for the recruitment of unique protein complexes harboring effector domains able to read the methyllysine mark and (2) the ε-amine group of this residue can be mono-, di-, or trimethylated (me1/me2/me3), and the different degree of methylation has divergent impact on chromatin signaling. Histones have been known to be substrates for methylation for a long time

(Murray, 1964). On histone tails, residues K4 (using the single-letter amino acid code for lysine), K9, K27, K36 of histone H3 and K20 of histone H4 are the most abundant sites of methylation (Strahl, Ohba, Cook, & Allis, 1999). While these studies provided insights into the position of the methylated lysine on these two histone proteins, it is the identification of the family of methyltransferases that marked the beginning of a prolific era during which many research efforts converged toward the elucidation of the roles played by histone lysine methylation. This family of enzymes shares a conserved catalytic domain, the SET domain. SET domain proteins transfer the *S*-adenosyl-L-methionine (AdoMet) methyl group to the ε-amine of a lysine residue on histone or nonhistone proteins, leaving a methylated lysine residue and the cofactor by-product *S*-adenosyl-L-homocysteine (AdoHcy). The SET domain was initially characterized in the *Drosophila* proteins Suppressor of variegation 3-9 [Su(var)3-9], Enhancer of zeste [E(z)], and Trithorax (Trx) (Jenuwein, Laible, Dorn, & Reuter, 1998; Jones & Gelbart, 1993; Stassen, Bailey, Nelson, Chinwalla, & Harte, 1995; Tschiersch et al., 1994), hence the name. As of today, the SMART database (Letunic, Doerks, & Bork, 2015) accounts for more than 14,500 proteins containing a SET domain, distributed among archea, bacteria, eukaryotes, and viruses.

At first, it was shown that the mammalian SUV39H1 specifically methylates K9 on histone H3 (Rea et al., 2000). Then, several other SET domain proteins have been shown to have methyltransferase activity toward specific lysine residues of histone tails but also to nonhistone proteins, such as the plant enzyme Rubisco (Trievel, Beach, Dirk, Houtz, & Hurley, 2002; Trievel, Flynn, Houtz, & Hurley, 2003), mammalian transcription factors (Chuikov et al., 2004; Couture, Collazo, Hauk, & Trievel, 2006; Kouskouti, Scheer, Staub, Tora, & Talianidis, 2004), tumor suppressors (Ferguson et al., 2011), and the kinetochore protein Dam1 (Zhang et al., 2005). Generally, proteins harbouring a SET domain also contain modulatory domains controlling the methyltransferase activity of the enzyme. Further, some SET domain-containing proteins are part of complexes or function with other proteins that regulate their activity (Herz, Garruss, & Shilatifard, 2013). This represents a challenge for the study of these enzymes, as the essential components of the complex or interacting partners all need to be present for the experimenter to be able to perform meaningful enzymatic assays and structural studies.

Histone lysine methylation is a complex process, as it can be involved in a myriad of nuclear processes, depending on the specific lysine methylation site on histones and the methylation state of the ε-amino group of the target lysine. HKMTs can be subdivided based on the nature of their substrate and degree of methylation. H3K4 methyltransferases are well conserved in different species; H3K4 methylation is generally associated with actively transcribed genes and often controls developmental genes. In yeast, Set1, as part of the COMPASS complex (Complex of Proteins Associated with Set1) catalyzes the mono-, di-, and trimethylation of H3K4 (Roguev et al., 2001; Shilatifard, 2012). In *Drosophila*, three proteins are found in the COMPASS family of methyltransferases and include Set1, Trx, and Trithorax-related (Trr). In mammals, yeast COMPASS is represented by six proteins including SET1A, SET1B (homolog of *Drosophila* Set1), MLL1, MLL2 (homolog of *Drosophila* Trx), MLL3, and MLL4 (homolog of *Drosophila* Trr). SET1A/1B and MLL1–4 are present in multiprotein complexes and share some common subunits, such as RbBP5 (retinoblastoma-binding protein 5), Ash2L [(Absent, Small, or Homeotic)-Like (*Drosophila*) 2], and WDR5 (WD repeat-containing protein 5) (Shilatifard, 2012). SET1A and SET1B are di- and trimethyltransferases while MLL3/4 predominantly monomethylates histone H3K4 (Ardehali et al., 2011; Hallson et al., 2012; Mohan et al., 2011). Additionally, SET7/9 methylates the same residue on histone H3 (Xiao, Jing, et al., 2003; Zhang et al., 2003) and several nonhistone targets (Del Rizzo & Trievel, 2011). Methylation of histone H3 on Lys 9 (H3K9) localizes in region that are predominantly transcriptionally silenced. For instance, SUV39H1 (Hwang, Eissenberg, & Worman, 2001; Peters et al., 2003; Rea et al., 2000; Shinkai & Tachibana, 2011), DIM-5 (Zhang et al., 2003), and SETDB1 (Falandry et al., 2010) mediate H3K9 trimethylation at pericentric heterochromatin. *Drosophila* E(z) and the mammalian homologs EZH1 and EZH2 are methyltransferases that, as part of the polycomb repressive complex 2 (PRC2), di- and trimethylate H3K27 and are involved in the negative regulation of the homeotic genes and generally associated with constitutive heterochromatin (Hwang et al., 2001; Margueron & Reinberg, 2011; Simon, Chiang, & Bender, 1992). However the PRC2 complex can also mono- and trimethylate K27 on histone H3 located in actively transcribed genes in mouse ES cells (Ferrari et al., 2014; Young et al., 2011). In plants, ATXR5/6 monomethylate H3K27 on the histone variant H3.1, a mark that is closely associated with gene silencing (Jacob et al., 2009, 2014). In mammalian cells, H3K36 methyltransferases include SET2, NSD1–3, SETMAR, and ASH1L. In yeast, Set2 mono-, di-, and trimethylates H3K36 (Strahl et al., 2002), while in mammals, SET2 is a

H3K36 trimethyltransferase (Edmunds, Mahadevan, & Clayton, 2008) and NSD1–3 mono- and dimethylates H3K36 (Kuo et al., 2011; Li et al., 2009). ASH1L is associated with active genes (Schuettengruber, Martinez, Iovino, & Cavalli, 2011; Tanaka et al., 2011), while NSD1–3 are involved with transcription repression and initiation and the elongation process (Berdasco et al., 2009; Lucio-Eterovic et al., 2010; Yang et al., 2012). In yeast, Set2 prevents spurious transcription, keeps coding regions hypoacetylated, and is involved in splicing (de Almeida et al., 2011; Kim, Kim, Fong, Erickson, & Bentley, 2011; Landry et al., 2003; Wagner & Carpenter, 2012).

Among H4K20 methyltransferases, SET8 is the major monomethyltransferase (Couture, Collazo, Brunzelle, & Trievel, 2005; Xiao et al., 2005) and plays a role in transcriptional repression and chromatin compaction (Houston et al., 2008; Sakaguchi & Steward, 2007). Suv420h1/2 are responsible for di- and trimethylation of H4K20 (Schotta et al., 2004, 2008) and have a repressive role in transcription regulation (Kapoor-Vazirani, Kagey, & Vertino, 2011).

Several crystal structures of SET domain enzymes in complex with peptides representing the cognate histone target are available and include the fungus *Neurospora crassa* DIM-5 (Zhang et al., 2003), the human SET7/9 (Xiao, Jing, et al., 2003), the human SET8 (also named PR-SET7) (Couture et al., 2005; Xiao et al., 2005), the protozoa *Paramecium bursaria* chlorella vSET (Qian et al., 2006), the human MLL1 (Southall, Wong, Odho, Roe, & Wilson, 2009), the plant *Ricinus communis* ATXR5 (Jacob et al., 2014), and the human and fungus *Chaetomium thermophilum* EZH2 (Antonysamy et al., 2013; Jiao & Liu, 2015; Wu et al., 2013).

SET domains share common structural features. They are typically characterized by a unique fold of a core of 10–12 β strands, a region that bisects the SET domain, referred to as the iSET, which contributes to substrate recognition, and extended loops (Fig. 1). The SET domain is flanked by two heterologous regions at its N-and C-terminus, referred to as nSET and cSET, respectively (Couture & Trievel, 2006; Xiao, Wilson, & Gamblin, 2003). The cSET contains conserved residues that are important for substrate binding and for catalysis. Notably, this region folds back onto the SET domain threading through a loop formed by a preceding stretch of the sequence to create a "pseudoknot" (Dillon, Zhang, Trievel, & Cheng, 2005; Taylor, Xiao, Gamblin, & Lin, 2003). This pseudoknot structure packs against the SET domain and contributes to the formation of the cofactor and substrate-binding pockets. The C-terminus in DIM-5 (Zhang et al., 2002, 2003) and MLL1 (Southall et al., 2009) is characterized by a

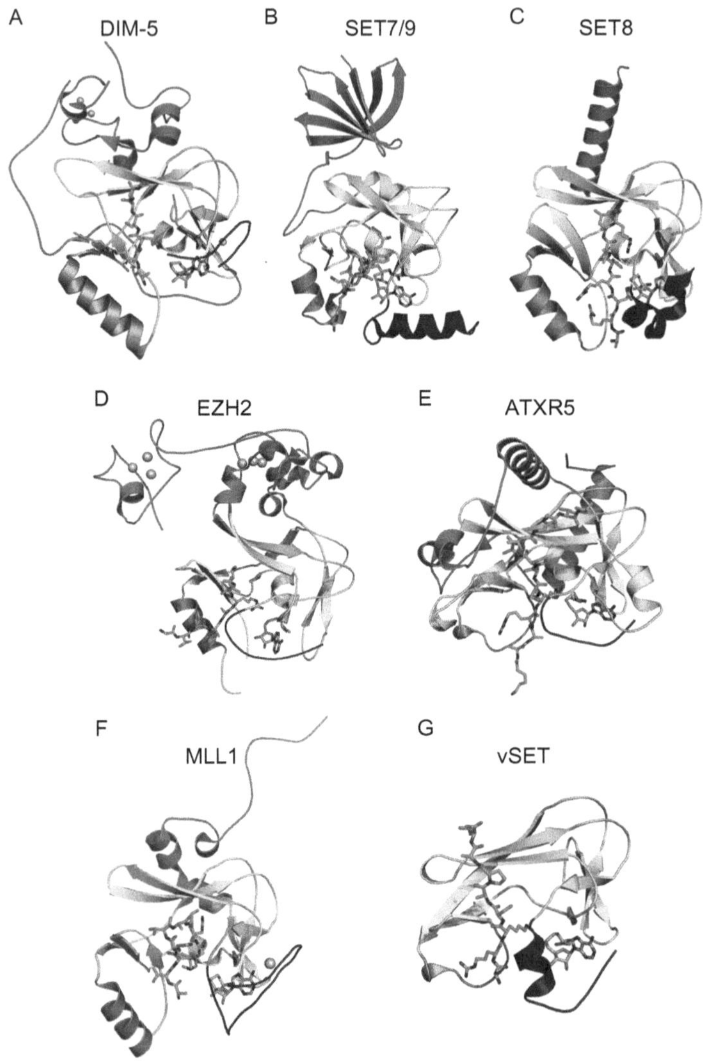

Fig. 1 Crystal structure of SET domain HKMTs in ternary complexes. Ribbon diagrams of (A) DIM-5 (PDB code 1PEG), (B) SET7/9 (PDB code 1O9S), (C) SET8 (PDB code 1ZKK), (D) EZH2 (PDB code 5CH1), (E) ATXR5 (PDB code 4O30), (F) MLL1 (PDB code 2W5Z), (G) vSET (PDB code 3KMT) are represented with the nSET (*orange*), SET (*green*), iSET (*light blue*), and cSET (*dark blue*) regions highlighted. The product cofactor AdoHcy and the substrate histone peptide are shown in stick representation and colored *magenta* and *yellow*, respectively. The Zn^{2+} ions in the nSET and cSET regions are illustrated as gray spheres. (See the color plate.)

Cys_4Zn cluster. In SET7/9 (Xiao, Jing, et al., 2003) and SET8 (Couture et al., 2005; Xiao et al., 2005), the same region folds into an α helix, whereas a loop is found in this region in ATXR5 (Jacob et al., 2014) and EZH2 (Jiao & Liu, 2015). The nSET is a more variable region. In DIM-5, this region contains a Cys_9Zn_3 cluster, which is referred to as pre-SET domain (Zhang et al., 2002, 2003), while in EZH2, the nSET folds as $Cys_8H_1Zn_3$ and Cys_9Zn_3 clusters (Antonysamy et al., 2013; Jiao & Liu, 2015). The analogous region in SET7/9 (Xiao, Jing, et al., 2003) folds as a β sheet (Southall et al., 2009) and in MLL1, SET8 (Couture et al., 2005; Xiao et al., 2005), and ATXR5 (Jacob et al., 2014), the nSET is composed of α helical elements.

A striking feature of SET domains proteins is the narrow hydrophobic channel that connects the cofactor-binding site on one side with the substrate-binding site on the opposite side of the domain. As mentioned earlier, the cSET is essential for the formation of this channel, and it appears to be disordered in the absence of bound cofactor and substrate. For some SET domain enzymes, ordered cSET domains have only been observed for cofactor or ternary complexes of SET proteins. Evidence that the C-flanking domain of SET proteins becomes ordered upon addition of the AdoMet cofactor comes from NMR studies on SET8 (Xiao et al., 2005). The apoenzyme shows a high degree of flexibility in the cSET region, and the flexibility is lost after stoichiometric addition of the cofactor AdoMet, which promotes ordering of the cSET and the proper formation of the substrate lysine channel. This is a very important observation for the crystallization of SET domains and demonstrates that the addition of the cofactor could have a positive impact on the crystallization of SET domain HKMTs.

Typically, as observed in DIM-5 (Zhang et al., 2002, 2003), vSET (Qian et al., 2006), and SET8 (Couture et al., 2005; Xiao et al., 2005), the histone substrate binds in an extended conformation across the domain and forms a short parallel β sheet with the pairing β strand of the SET domain. These enzymes achieve binding and selectivity of the lysine substrate employing the iSET and cSET regions. Specificity for methylation is achieved by the recognition of the residues flanking the N- and C-terminus of the target lysine residue. In contrast to the β-sheet confirmation of the substrate peptide bound to vSET, DIM-5, and SET8, the histone peptide adopts a unique L-shaped conformation in the ATXR5 peptide-binding cleft. This cleft is formed by three loops (L1–L3) (Jacob et al., 2014) located in the iSET, cSET, and nSET regions.

As mentioned earlier, specific HKMTs do not work alone but function in complexes. For instance, MLL1 is part of the mammalian COMPASS-like complex and its activity is regulated by RbBP5, Ash2L, WDR5, and DPY-30 (Avdic et al., 2011; Cao et al., 2010; Odho, Southall, & Wilson, 2010; Patel, Vought, Dharmarajan, & Cosgrove, 2008; Sarvan et al., 2011; Song & Kingston, 2008; Southall et al., 2009; Takahashi et al., 2011; Zhang, Lee, Brunzelle, & Couture, 2012). In the SET domain of MLL1, the iSET and the cSET regions form two separate lobes and as a result create an open active site wherein the histone peptide binds in a conformation that leaves the substrate lysine relatively solvent exposed (Southall et al., 2009). The formation of the proper active site is probably achieved after reorientation of the iSET following the interaction with other components of the MLL complex.

Another example is EZH2 (Enhancer of zeste homolog 2), which requires the cooperation of at least two other members of the PRC2 complex, namely, the embryonic ectoderm development (EED) and the suppressor of zeste 12 (SUZ12) to achieve methyltransferase activity (Cao & Zhang, 2004; Pasini, Bracken, Jensen, Lazzerini Denchi, & Helin, 2004; Yamamoto, Sonoda, Inokuchi, Shirasawa, & Sasazuki, 2004). EZH2, EED, and SUZ12 closely associate with the EZH2 catalytic domain and cooperate in carrying out catalytic activity to produce H3K27me3. The comparison of the crystal structures of EZH2 SET domain in the inactive and active forms shows that the SET domain undergoes a conformational change upon activation. In the absence of regulatory subunits, the iSET and the cSET regions of the enzyme occupy atypical positions (Antonysamy et al., 2013; Jiao & Liu, 2015; Wu et al., 2013) creating an incomplete cofactor-binding site and an occluded peptide-binding cleft. Instead, when cocrystallized in complex with SUZ12, EED, and an histone H3 peptide, EZH2 iSET region rotates of 20 degree from the peptide-binding cleft toward the cofactor-binding pocket, while its cSET region adopts the typical conformation for AdoMet binding (Jiao & Liu, 2015).

The ability of a SET domain KMT to transfer one, two, or three methyl groups to the ε-amine of a lysine side chain is referred to as product specificity. Biochemical and structural studies have shown that an active site tyrosine residue plays a key role in coordinating the binding of a water molecule in proximity of the methyltransfer pore to restrict SET domain PKMT to monomethyltransferase. Observations that upon replacement of the tyrosine with a phenylalanine, a residue found in di-/tri-methyltransferases, switched SET domain enzymes such as SET7/9 and SET8 (Couture, Dirk, Brunzelle, Houtz, & Trievel, 2008) from strict monomethyltransferases to dimethyltransferases gave rise to the Phe/Tyr switch model whereby the loss of a single hydrogen bond to a water molecule controls the product specificity of a SET domain KMT.

SET domain-containing proteins play various roles in methylating histone and nonhistone substrates (Lanouette, Mongeon, Figeys, & Couture, 2014). It is therefore of no surprise that mutations and the malfunctioning of SET domain enzymes are associated with various forms of cancers and other diseases (Albert & Helin, 2010; Greer & Shi, 2012). A remarkable example is the MLL1 gene, for which translocations are correlated to acute myeloid and lymphoid leukemia. Chromosomal translocations affecting the *MLL1* gene result in MLL fusion proteins, where the N-terminal part of MLL1 is juxtaposed with different fusion partners destroying its normal histone methyltransferase activity. The resulting chimeric proteins act as transcriptional regulators that aberrantly control expression of genes like the *HOX* homeobox genes (Thiel et al., 2010). Another example is EZH2, which is involved in prostate, breast, and hematological cancers (Ernst et al., 2010; Kleer et al., 2003; Morin et al., 2010; Varambally et al., 2002). In myeloid neoplasms, nonsense and stop codon mutations of the EZH2 SET protein inactivate its methyltransferase activity (Chase & Cross, 2011). Furthermore, EZH2 overexpression relates to inhibition of tumor suppressor genes and activation of genes involved in metastasis (Wang et al., 2012; Yu et al., 2007). Along the same lines, deregulation of EZH2 has been associated to alterations of micro RNA levels (Varambally et al., 2008), and altered activity of EZH2 has been discovered in B-cell non-Hodgkin's lymphoma (Beguelin et al., 2013; McCabe et al., 2012; Morin et al., 2010). In lymphoma, a heterozygous missense mutation occurs at tyrosine 641 (Y641) within EZH2 SET domain. The Y641 mutant has a higher catalytic activity toward di- and trimethylation of H3K27, compared to wild-type EZH2 that instead has the highest catalytic activity for monomethylation of the substrate. In B-cell lymphoma, the heterozygous mutant Y641 allele works in combination with wild type to increase the overall level of H3K27 trimethylation (Chase & Cross, 2011). As such, EZH2 inactivation can be therapeutically effective in many cancers. Small molecules have been developed that specifically inhibit the catalytic activity of EZH2 (Knutson et al., 2012; McCabe et al., 2012). Further, structural studies on EZH2 lead to the idea of impairing the activity of EZH2 by disrupting the interaction between EZH2 and the partner EED by developing a stabilized EZH2 α-helix peptide that mimics the α-helical EED-binding domain of EZH2. This α-helical peptide binds EED and disrupts the EED–EZH2 enzymatic complex resulting in reduced H3K27 metyltransferase activity of the PRC2 complex (Kim et al., 2013).

The implication of chromatin modifying enzymes is just beginning to be appreciated in clinical oncology. In the future, identification and targeted inhibition of SET domain proteins involved in cancer will be useful to develop selective pharmacological treatments for patients. Gaining structural insight into SET domain proteins is an essential step toward the understanding of the molecular mechanism controlling the action of these enzymes and the development of specific therapeutic molecules. In this chapter, we will provide a general view of the approaches to purify, characterize, and crystallize SET domain proteins in complex with peptide substrates.

2. EXPERIMENTAL PROCEDURES

Most SET domains that have been biochemically characterized were overexpressed in *E. coli* and purified as fusion proteins with an affinity tag (Uhlen, Forsberg, Moks, Hartmanis, & Nilsson, 1992). For high-level protein production, BL21(DE3) is an appropriate *E. coli* strain; it is deficient of *lon* and *omp*T proteases for increased target protein stability, and it is compatible with the T7 *lac*O promoter system (Studier, Rosenberg, Dunn, & Dubendorff, 1990). Sometimes it is useful to use derivatives of BL21 (DE3) such as the Rosetta(DE3), that carry additional tRNAs to overcome the effects of codon bias. The choice of the best tag for each different protein relies on empirical testing (Hammarstrom, Hellgren, van Den Berg, Berglund, & Hard, 2002). Small-scale test expression is widely used as a predictive tool to determine which derivative clone produces soluble protein and to establish the correct conditions for large-scale growth. Many SET domain proteins have been successfully expressed as glutathione *S*-transferase (GST) (Smith, 2000; Smith & Johnson, 1988) or as hexahistidine (His6) (Hochuli, 1988, 1990; Hochuli, Dobeli, & Schacher, 1987) fusion proteins. The fusion tag is subsequently removed following the affinity purification step using a site-specific protease such as the tobacco etch virus (TEV) protease (Carrington & Dougherty, 1988). For instance SET domains of *N. crassa* DIM-5 (Zhang et al., 2002), human SET7/9 (Wilson et al., 2002; Xiao, Jing, et al., 2003), human MLL1 (Southall et al., 2009), *R. communis* ATXR5 (RcATXR5) (Jacob et al., 2014) have been expressed and purified to near homogeneity as GST fusion proteins, while SET7/9 (Couture & Trievel, 2006) and SET8 were overexpressed as N-terminal His-tagged proteins (Couture et al., 2005). Alternative expression systems have to be explored if the fast and inexpensive *E. coli* expression system is unsuccessful. For instance, common eukaryotic alternatives are the baculovirus expression system in insect cells (Smith, Summers, & Fraser, 1983) and the yeast *Saccharomyces cerevisiae* (Holz, Hesse, Bolotina, Stahl, & Lang, 2002). The SET domain of human EZH2 was expressed as a C-terminal His tag fusion protein in insect *Spodoptera frugiperda* (Sf9) cells (Wu et al., 2013). Further, a construct corresponding to residues 191–950 of *C. thermophilum* EZH2 was cooverexpressed in *S. cerevisiae* with EED and SUZ12 as a N-terminal TEV cleavable double Protein A and His-tag protein (Jiao & Liu, 2015). Coexpression of proteins that form a complex is a useful tool to improve solubility when the single components cannot be expressed in soluble form. The hydrophobic patches that often promote insolubility are involved in and protected by the interaction with the cognate interacting partners promoting higher expression level of a correctly folded and stable protein complex (Bullock, Debreczeni, Edwards, Sundstrom, & Knapp, 2006; Romier et al., 2006; Tolia & Joshua-Tor, 2006). The following protocol has been used to produce pure ATXR5 with the purpose of performing enzymatic assays and crystallizing its SET domain in complex with histone H3.1.

2.1 Construct Design Strategies

Sequence alignment among different ATXR5 homologs was performed to provide insights into how to choose the boundaries of a construct. The sequence alignment of ATXR5 and other SET domains is illustrated in Fig. 2. It is a good practice to look at where the homology of the protein of interest to others decreases at the N- and C-terminus of the predicted domain. Additionally, a tool that can help into designing the best constructs suitable for structural studies is the computation of secondary structure prediction using servers such as PROMALS3D (Pei, Kim, & Grishin, 2008) and PSIPRED (Buchan, Minneci, Nugent, Bryson, & Jones, 2013; Jones, 1999; Pei et al., 2008). Further, the most stable domain of the target protein can be identified by their pattern of resistance to limited proteolysis. In the case of SET8, we used limited proteolysis on a purified preparation of SET8 bound to histone H4 to identify a construct of SET8 suitable for crystallization. Traditionally, the protein is incubated with a relatively low concentration of different relatively nonspecific proteases (eg, trypsin or protease V8), which cleave at solvent-exposed recognition sites throughout the protein such as loops and other flexible regions. Following digestion with a protease,

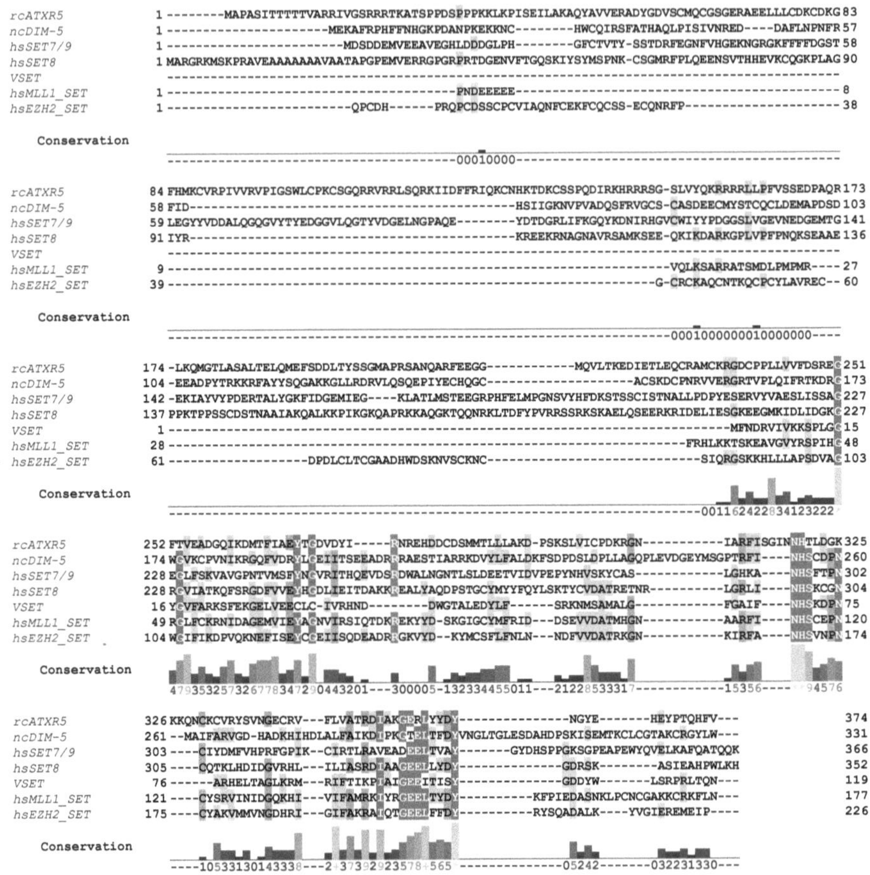

Fig. 2 Sequence alignment of selected protein methyltransferases. All the sequences are from human with the exception for *Neurospora crassa* DIM-5, *Ricinus communis* ATXR5, and *Paramecium bursaria* Chlorella virus 1 vSET. Conserved residues are highlighted in *blue* with darker shades associated with the higher degree of conservation. The program used was JalView (Waterhouse, Procter, Martin, Clamp, & Barton, 2009), and the alignment was produced using Muscle (Edgar, 2004) with default parameters. (See the color plate.)

samples are analyzed by SDS-PAGE to identify cleavage products. The appearance of lower molecular weight bands denotes digestion of the full-length protein, and intensity and appearance/disappearance of bands can be closely monitored and quantified by mass spectrometry while the N-terminus is determined by Edman sequencing. The appearance of similarly sized bands in the presence of different proteases is indicative of inherently disordered or flexible regions of the protein. Another strategy that can be considered initially is the expression of homologous proteins from different species. Generally these proteins differ by only a few amino acid residues, which are typically found on the surface of the protein. Thus, this is a logical way to screen for a collection of point mutations that do not affect protein function. Moreover, at times some species lack nonconserved unstructured loops providing new opportunities to screen protein constructs with lower intrinsic entropy and that are more likely to crystallize. The effect of testing different species on crystallization can be dramatic. It has also been our experience that some proteins crystallize rather easily and under various conditions such as SET7/9, whereas other SET domain proteins, which include SET8 and ATXR5, proved to be very difficult to crystallize. As an example, several species were screened for the ATXR5 SET domain such us *Arabidopsis thaliana* ATXR5, *Vitis vinifera* ATXR5, *Glycine max* ATXR5, *Selaginella moellendorffii* ATXR5, and *Physcomitrella patens* ATXR5 (Fig. 3); however, only the RcATXR5 homolog readily produced preliminary crystals of the apoenzyme and therefore was chosen for further cocrystallization trials with peptides. For this reason, although several species of ATXR5 were expressed, purified, and screened for crystallization, later we are providing the details of the cloning, expression, and purification of RcATXR5 only.

2.2 Protein Sample Preparation

A codon-optimized cDNA for overexpression in *E. coli* encoding residues 158–374 of RcATXR5 was subcloned into the parallel vector pGST2 (Sheffield, Garrard, & Derewenda, 1999). ATXR5 was overexpressed in fusion with a TEV cleavable GST tag in Rosetta(DE3) cells (EMD Millipore). Rosetta(DE3) cells transformed with the ATXR5 construct were grown at 37°C in Luria broth media supplemented with chloramphenicol (50 μg/mL), ampicillin (100 μg/mL), and $ZnSO_4$ (10 μ*M*) to an OD_{600} of 0.6. Protein expression was induced by the addition of isopropyl-thiogalactopyranoside (IPTG, 0.1 m*M*) during 16 h at 18°C. Reducing the growth temperature results in slower protein production but also in slower proteolytic degradation. The overall result is an increased solubility of the target protein (Vera, Gonzalez-Montalban, Aris, & Villaverde, 2007).

Cells were harvested, resuspended in cold PBS buffer (pH 7.3), and lysed by sonication. The lysate was centrifuged at 37,000 × *g* for 1 h at 4°C, and the supernatant was collected. The soluble fraction was incubated with gluthatione sepharose beads for 1 h at 4°C and then washed extensively

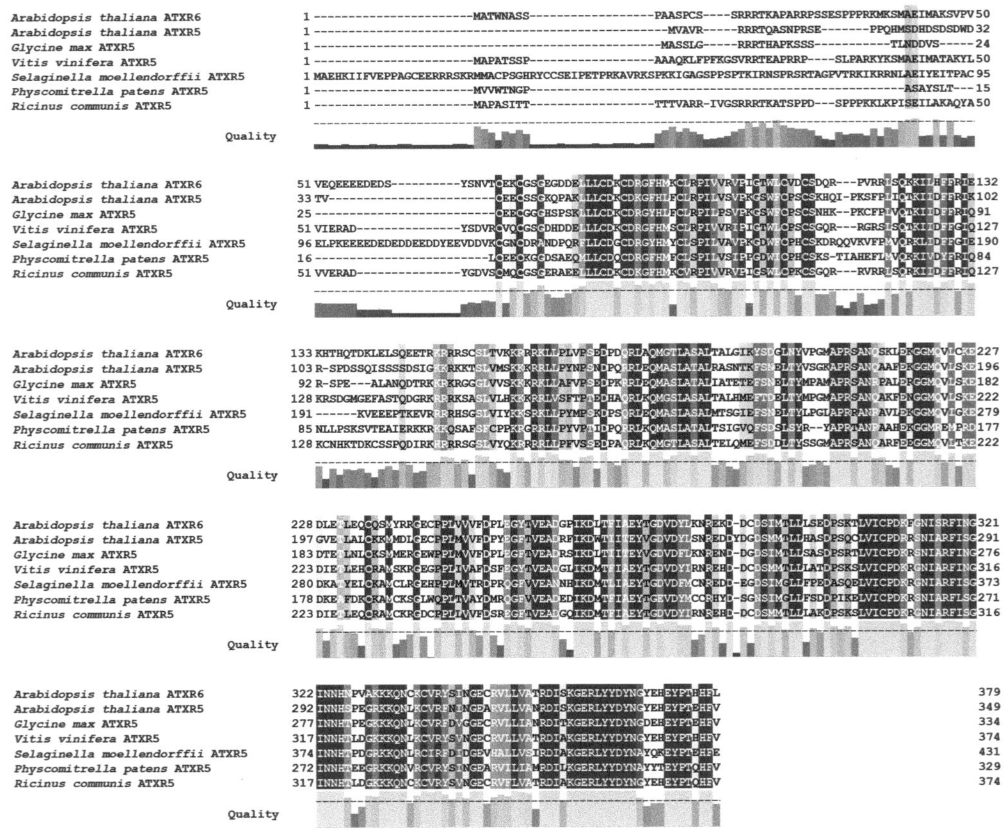

Fig. 3 Sequence alignment of ATXR5 homologs. Conserved residues of aligned sequences of ATXR5/6 homologs are highlighted in *blue* with *darker shades* associated with the higher degree of conservation. The programs used are the same as for Fig. 2. (See the color plate.)

in PBS buffer. GST-ATXR5 was TEV cleaved on beads during 16 h at 4°C in PBS buffer, then concentrated to a final volume of 2 mL, and further purified by gel filtration chromatography (Superdex 200, GE Healthcare) preequilibrated in 20 m*M* Tris, pH 7.5, 250 m*M* NaCl, and 5 m*M* β-mercaptoethanol (BME). Symmetric elution profiles are characteristic of homogeneous protein and can provide information about the quaternary structure of the protein (monomer vs oligomer). As shown in Fig. 4, ATXR5 eluted as a main peak corresponding to a predicted molecular weight of 26 kDa indicating that the protein elutes as a monomer. The purity of the fractions was further confirmed by electrophoresis on a denaturing SDS-PAGE and by staining with Coomassie brilliant blue (Fig. 4). Monodispersity and purity are among the most important prerequisite for the protein's ability to crystallize. Fractions corresponding to pure and monomeric form of ATXR5 were pooled and concentrated to 20 mg/mL

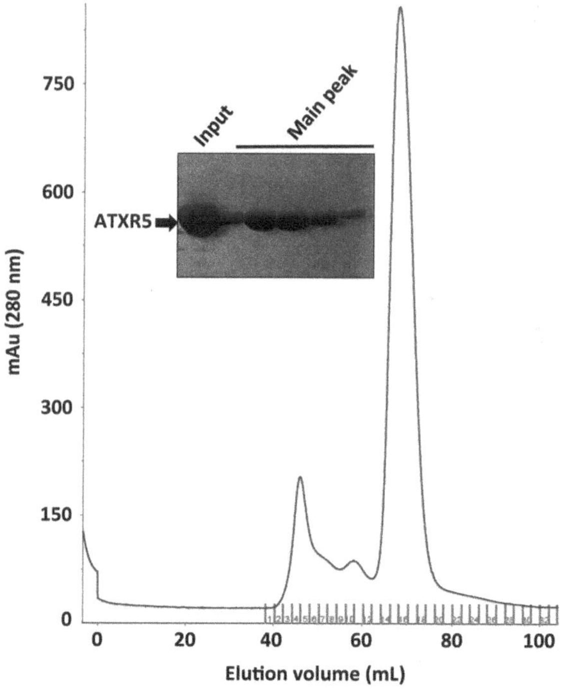

Fig. 4 Protein sample purification. Size exclusion chromatography (Superdex 200, GE Healthcare) profile of ATXR5. The inlet shows the protein composition of the major peak analyzed by SDS-PAGE. The input and the main peak correspond to the protein sample prior to size exclusion chromatography and the fractions eluting at a predicted molecular weight of approximately 26 kDa, respectively.

using centrifugal filters of 10K molecular weight size cutoffs (EMD Millipore). UV absorption was measured at A_{280}, and the concentration of the protein was calculated by using the predicted molar extinction coefficient at A_{280} of $\varepsilon = 0.56$ (http://web.expasy.org/protparam/). To confirm that the selection of a construct does not impair the methyltransferase activity of the SET domain, it is good practice to compare the activity of the shorter constructs with the one corresponding to the full-length protein. For SET domain proteins, the activity can be monitored using a methyltransferase assay (MTA), which will be discussed in detail later in this chapter.

2.3 Preparation of Selenomethionyl-Labeled RcATXR5

Substitution of methionine with selenomethionine is a general method for introducing anomalously scattering atoms into a cloned protein. The use of selenomethionyl-substituted proteins is useful for single or multiple wavelength anomalous dispersion (MAD or SAD) experiments to determine the phase problem in protein crystallography. The protocol for expression of selenomethionyl-substituted RcATXR5 in 1 L of bacterial culture is based on premixed reagents available from AthenaES (Baltimore, MD). The protocol exploits gradual adaptation of bacterial cells to the minimal medium and utilizes the methionine auxotroph *E. coli* strain B834(DE3) to afford expression of ATXR5 with selenium content sufficient for SAD or MAD.

First, the media was prepared according to manufacturer instruction. Briefly 25.6 g of the SelenoMet medium base (AthenaES # 0501) were dissolved into 1 L of DD water and sterilized at 121°C for 15 min. Five gram of SelenoMet Nutrient Mix (AthenaES # 0502) were dissolved into 50 mL of sterile DD water and the mixture was filter sterilized. A freshly prepared solution of L-selenomethionine (Sigma-Aldrich, S3132) (10 mg/mL) was sterilized by filtration and incorporated to the premixed SelenoMet medium base and SelenoMet Nutrient mix solutions.

RcATXR5 was transformed into B834(DE3) cells. A single colony was inoculated into 5 mL of Luria Broth media supplemented with 100 μg/mL of ampicillin and was grown overnight at 37°C. The 5 mL overnight culture was then transferred to 1 L of the prewarmed SelenoMet medium. Cells were grown to mid-log phase before addition of 50 mL of SelenoMet Nutrient Mix and 4 mL of L-selenomethionine solution. Protein expression was induced with the addition of 150 μ*M* IPTG at 19°C for 15 min after addition of the nutrient mix. Purification of the selenomethionyl-labeled

ATXR5 was performed as for native ATXR5. The incorporation of selenium was confirmed by MALDI mass spectrometry, which confirmed incorporation to more than 99%.

2.4 Protein Crystallization

Purified recombinant SET domain proteins can degrade over time producing heterogeneous mixtures of polypeptides. It is therefore recommended to set up crystallization trials the same day as the last step of the purification. This is particularly applicable for SET8 (Couture et al., 2005), which could not be crystallized following a single freeze/thaw cycle or leaving the protein on ice for more than 6 h (data not shown). However, this rule cannot be applied to all the SET domain proteins as SET7/9 can be crystallized in many different conditions and also following several freeze/thaw cycles (data not shown).

Obtaining protein crystals suitable for X-ray diffraction studies always represents a difficult challenge in crystallography. There are several ways to improve the chances of obtaining crystals. For instance many proteins that have not been crystallized alone can be readily crystallized as complexes with their binding partners. Complex formation and cocrystallization screens can be performed with cofactors, peptides representing binding partners, inhibitors as well as antibody fragments. The conformational change induced upon ligand binding may help the crystallization process by stabilizing the protein or by inducing new crystal contacts.

In our initial attempts to crystallize ATXR5, a concentrated solution of ATXR5 SET domain (10 mg/mL) mixed with an equimolar concentration of the product cofactor *S*-adenosylhomocysteine (AdoHcy) was used to carry out crystallization trials. All the screens were set up in 96-well plates using commercially available kits, such as the Crystal, Index, PEG/Ion, SaltRX screens (Hampton Research, Aliso Viejo, CA), the MIDAS, and ProPlex screens (Molecular Dimensions, Altamonte Springs, FL) or the JCSG + Suite and Protein Complex Suite (Qiagen, Valencia, CA). Equal volumes of protein and reservoir solution (1 μL) were mixed in the 96-well plates. Plates were incubated at 19°C or at room temperature and checked regularly. Initial crystals were obtained in less than 24 h in condition A6 of the Index screen (Hampton research), which consists of 0.1 *M* Tris pH 8.5, 2.0 *M* ammonium sulfate. Larger crystals were obtained by varying the concentration of the protein (Fig. 5). Crystals were harvested and cryoprotected in the mother liquor supplemented with 20%

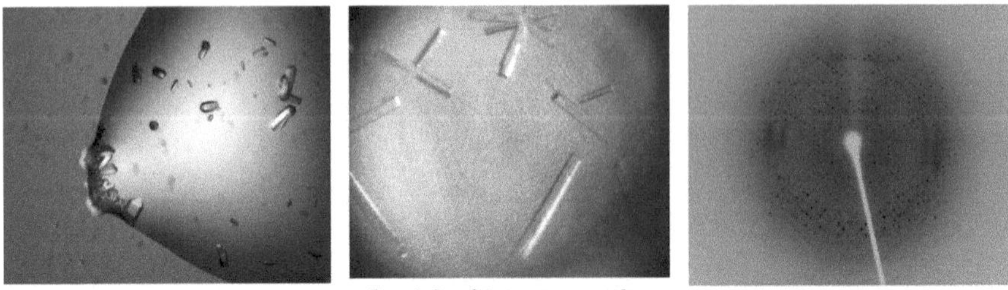

Fig. 5 Crystallization of ATXR5 complexes. The *left panel* shows the hexagonal rod crystals obtained for ATXR5/AdoHcy binary complex; the *center panel* shows the crystals obtained for the ATXR5/H3.1/AdoHcy ternary complex; the *right panel* shows the X-ray diffraction pattern of a single crystal of the ATXR5/H3.1/AdoHcy ternary complex.

1,2-propanediol. These crystals diffracted to 2 Å. Selenomethionyl protein crystals were grown in the same condition and diffracted to 2.7 Å.

In an attempt to obtain crystals of the ternary complex of ATXR5 with the cofactor and the cognate histone H3.1 substrate (ATXR5/H3.1/AdoHcy), three histone peptides of different length corresponding to the region surrounding H3.1 K27 were purchased from Genscript and solubilized in 100 m*M* Tris (pH 8.0) and 150 m*M* NaCl to a final concentration of 100 m*M*:

1. KAARK$_{27}$SAPATG (a.a. 23–33)
2. ATKAARK$_{27}$SAPATGG (a.a. 21–34)
3. KQLATKAARK$_{27}$SAPATGGVK (a.a. 18–36)

In this study, we used a radioactivity-based MTA followed by immobilization on P81, as highlighted in Section 2.6, to determine the differences in activity of ATXR5 for these peptides. These changes in activity were subsequently used as a proxy to determine the best candidate to perform cocrystallization. The results of this MTA are shown in Fig. 6. The activity of ATXR5 is stronger in the presence of the longest H3.1 peptide (residues 18–36) and was therefore subsequently used to carry out crystallization trials to obtain the ATXR5/H3.1/AdoHcy ternary complex. In the crystallization mixture the selected histone peptide and the product cofactor were added in molar excess compared to the SET domain to ensure saturation of the binding sites. Briefly, a concentrated solution of RcATXR5 (20 mg/mL) was incubated with a fivefold molar excess of H3.1 peptide and a twofold molar excess of *S*-Adenosylhomocysteine (AdoHcy). Crystals of RcATXR5/H3.1/AdoHcy ternary complex were grown at 19°C in 50% polypropylene glycol 400, 100 m*M* Na-Hepes pH 6.0, and 5% dimethyl

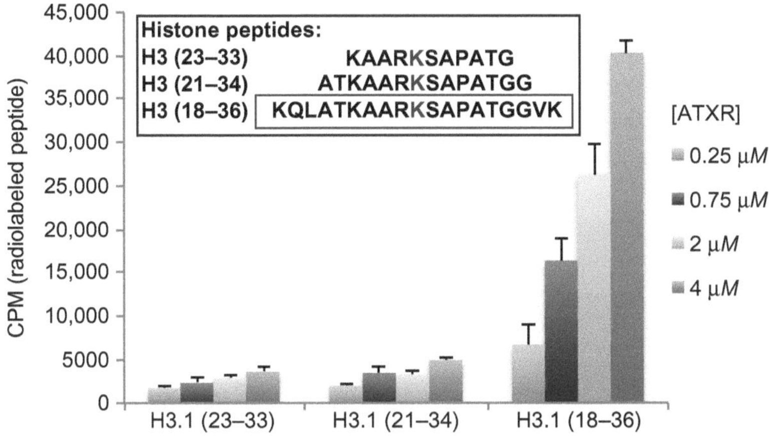

Fig. 6 H3.1 peptide substrate selection for ATXR5. Methyltransferase assay of four different concentrations of ATXR5 (0.25, 0.75, 2, 4 μ*M*, colored in *blue*, *red*, *green*, and *purple*, respectively) incubated with three different constructs of histone H3.1 peptide (their sequence is illustrated in the inset panel). The peptide chosen for crystallization trials is boxed in *red*. (See the color plate.)

sulfoxide (DMSO) (Fig. 5). To obtain the best diffracting crystals, the crystals had to be quickly harvested at 4°C, soaked in mother liquor for 30 s and flash frozen in liquid nitrogen.

2.5 Data Collection and Structural Determination

Owing to important structural differences between the cSET, iSET, and nSET regions of different SET domain enzymes, all new crystal structures of SET HKMT must be phased with heavy/anomalous atoms. Native crystals of the RcATXR/AdoHcy complex were tested on a Rigaku MicroMAX 007-HF X-ray generator. Data for a single SeMet-labeled crystal were collected at the 17-ID-D beamline of LS-CAT (Advanced Photon Source; Argonne National Laboratories) at 0.978 Å, the wavelength corresponding to the peak of K-absorption edge for selenium. Diffraction data were integrated, scaled, and merged using HKL2000 (Otwinowski & Minor, 1997). The crystals of apo-ATXR5 belonged to the space group $P322_1$, containing one molecule per asymmetric unit with unit cell parameters corresponding to $a=b=69.01$ Å and $c=82.20$ Å. A total of eight ordered Se atoms were found and used to build a partial model (>70%) using SOLVE/RESOLVE from Phenix (Zwart et al., 2008). The full dataset of the ternary complex RcATXR5/H3.1/AdoHcy was collected using a

Rigaku MicroMAX 007-HF in-house X-ray source. The reflections were index using d*Trek (Pflugrath, 1999) in a monoclinic centered space group (C2) containing two molecules per asymmetric unit and with unit cell dimensions of $a=101.2$ Å, $b=87.3$ Å, $c=74.5$ Å, and $\beta=127.79$ degree. A solution was found by molecular replacement using Phaser (McCoy et al., 2007) and RcATXR5/AdoHcy partial structure as a search model. The model was completed using interactive rounds of refinement and model building using Buster and Coot, respectively (Blanc et al., 2004; Emsley & Cowtan, 2004; Emsley, Lohkamp, Scott, & Cowtan, 2010). Quality of the model was assessed using Molprobity (Davis et al., 2007).

2.6 *In Vitro* MTAs

Histone SET domain methyltransferases transfer a methyl group from AdoMet to the ε-amino group of a lysine residue on histone proteins, leaving a methylated lysine residue on the substrate and the cofactor by-product AdoHcy. The activity of SET domain enzymes is measured through a MTA. There are multiple parameters to consider when testing the enzymatic activity of a SET domain HKMT. The first one is the choice of substrate. One must have prior knowledge of a suitable substrate of the enzyme of interest. In the absence of this information, one must first identify such a substrate. Several techniques can be used for this purpose. For example, incubation of the enzyme with peptide microarrays and a radioactive AdoMet can yield radioactive microarray spots corresponding to suitable substrates (Rathert, Zhang, Freund, Cheng, & Jeltsch, 2008). Such peptide microarrays can be purchased from commercial sources. In studying enzymes that methylate histones, another consideration is whether the enzyme will be more active on (1) free histones, (2) short peptides representing these histones centered on the lysine residue of interest, or (3) histone found in the form of the octamer or NCP. Several SET domain HKMTs display striking preferences for one type of substrate over the others. We found that ATXR5/6 are much more active on nucleosomal substrates, compared to free histones and peptides (E. Bergamin et al., unpublished). Similar behavior has been observed for instance for SET8, which is also more active in the presence of nucleosomal substrates (Xiao et al., 2005), while EZH2, as part of PRC2 complex, preferentially methylates oligonucleosomes (Cao & Zhang, 2004; Martin, Cao, & Zhang, 2006).

Peptide substrates have several advantages. They are commercially available and can be obtained in large quantity and high purity. These peptides can be easily mutated or chemically modified on residues neighboring the target lysine or can be attached to chromatography resins through terminal biotinylation (Yap et al., 2011). On the other hand, although they are more laborious or difficult to produce, full-length histone proteins, octamers, or nucleosome substrates represent the physiological substrates. Therefore, their use may yield more insights into substrates recognition by SET domain enzymes. Recombinant *Xenopus laevis* histone octamers and nucleosomes can be prepared as described previously (Luger, Rechsteiner, & Richmond, 1999) or recombinant histone octamers can be obtained in native condition by overexpressing all four histones from a single polycistronic vector followed by standard chromatography purification (Shim, Duan, Chen, Smerdon, & Min, 2012). Alternatively, native nucleosome can be prepared from mammalian cells (Schnitzler, 2001); however, owing to the presence of histone-modifying enzymes in these cells, this approach can be prohibitive for structural studies which require homogenous preparation of both, the enzyme and the substrate. To bypass this hurdle, recent methods were developed to introduce PTMs on recombinant histones and nucleosomes (Shimko, Howard, Poirier, & Ottesen, 2013; Simon & Shokat, 2012; Yanagisawa et al., 2014). Finally, NCP's can be purchased from a number of commercial sources (EpiCypher, Active Motif, New England Bio labs, Diagenode, EMD Millipore). The second important parameter to take into consideration is whether the SET domain enzyme of interest requires the presence of partner proteins for its activity. For instance, members of the MLL and EZH1/EHZ2 families function in multiprotein complexes (COMPASS and PRC2, respectively), and display little to no enzymatic activity when assayed on peptide or nucleosome substrate in the absence of their respective regulatory subunits (Herz et al., 2013).

The following is a protocol to test the activity of ATXR5 (purified as shown earlier) on recombinant NCP. The histone octamer is produced by coexpression of the core histones in *E. coli* and purification by binding on Talon resin (Clontech) followed by size exclusion chromatography (Superdex 200, GE Healthcare) (Shim et al., 2012), while the nucleosome-forming DNA is the "Widom 601" sequence of 147 base pairs (Lowary & Widom, 1998). NCPs are assembled using the salt dilution method (Dyer et al., 2004). The octamer and the nucleosome positioning 601 DNA were mixed together at a molar ratio of 1:1 in 2 *M* NaCl, 10 m*M* Tris (pH 8.0), and nucleosomal assembly is achieved by serial salt dilutions down to 250 m*M* NaCl, producing NCPs at a concentration of 0.3 μ*M*. Nucleosomes assembled using this method are stable for more than 1 week if stored at 4°C.

The protocol below details the steps for the characterization of ATXR5 methyltransferase activity:

- Use freshly purified ATXR5 (20 μ*M*) and keep the enzyme on ice.
- Take an aliquot of NCPs and retain on ice.
- Set up a 1.5-mL microcentrifuge tube on ice for each assay.
- To each tube, add the following: 20 μL of NCPs (0.3 μ*M*), 5 μL of reaction buffer [50 m*M* Tris (pH 8.5), 20 m*M* KCl, 10 m*M* $MgCl_2$, 10 m*M* BME, and 10% glycerol], 1 μL of radiolabeled [^{3}H-*methyl*]-AdoMet (17.2 Ci/mmol) (PerkinElmer Life Sciences; Cat. no. NET155V250UC).
- Add increasing amounts of enzyme to each different tube so that the final concentration of ATXR5 ranges from 0.2 to 4 μ*M*.
- Complete the volume to 30 μL using reaction buffer.
- Immediately start the MTA by incubating the reactions at 30°C for 1 h.
- After 1 h, the reactions are stopped by adding Laemmli sample buffer and by boiling at 95°C during 5 min.
- Load samples on a 15% acrylamide SDS-PAGE gel and run until the migration front is nearing the bottom of the gel.
- Stain the gel with Coomassie brilliant blue, destain until the desired contrast is achieved, and take a picture of the stained gel.
- Soak the gel twice for 30 min in 100% DMSO at room temperature.
- Soak the gel for 1 h at room temperature in 22% (wt/vol) 2,5-diphenyloxazole dissolved in 100% DMSO.
- Soak the gel in distilled water for a period of time that ranges from 1 h to overnight.
- Put the gel on Whatman filter paper and vacuum-dry.
- Expose the gel to X-ray film using intensifying screens, for a period ranging from 8 to 48 h, at −80°C.
- Develop the film.

Alternatively, after the HKMTase reactions are completed, the methylated substrate can be separated from the unreacted AdoMet by binding to P81 filter paper. This allows a generally more accurate quantitation of the signal.

- Spot each reaction mixture on an individual filter paper and allow them to dry.
- Dilute a 1 *M* $NaCO_3$ (pH 9.0) stock solution to make 1 L of 50 m*M* $NaCO_3$ (pH 9.0) (referred to as wash buffer). Prepare this solution fresh just before use.

- Transfer the filters to a 1-L beaker filled with 250 mL of wash buffer.
- Incubate 10 min with stirring.
- Carefully discard the wash buffer into radioactive waste and repeat the wash procedure three more times.
- Allow the membranes to dry on paper towels and transfer them individually to scintillation vials filled with 5 mL of scintillation cocktail.
- Measure the radioactive signal on a beta-counter. Ensure that a vial with only scintillation fluid is also included as blank.
- Plot the signal intensity (CPM values, blank value subtracted) as a function of time and/or enzyme amount.

The MTAs described earlier test the activity of ATXR5 using different concentrations of enzyme. Alternatively, the amount of enzyme can be kept constant, and the time of the reaction can be varied. In this case identical reactions need to be set up, and then the reaction can be stopped at established time points either by spotting the reaction on P81 paper or by boiling the sample reaction following the addition of SDS-PAGE sample buffer as described previously.

3. CONCLUSIONS

In this chapter, we have provided insights into the approaches that are used to design suitable constructs for the biochemical and structural studies of SET domain proteins with a specific focus on ATXR5. We also provided details on how to perform MTAs to study the activity of this family of enzymes and how the information gained from these assays help in the identification of a suitable peptide substrate for cocrystallization studies. Considering that the list of SET domain KMTs linked to cancers is growing exponentially, future efforts toward the crystal structure determination of cancer-related SET domain enzymes will provide critical insights for the design of novel inhibitors for the treatment of cancers and further our understanding of epigenetic processes linked to lysine methylation.

ACKNOWLEDGMENTS

We thank Alexandre Blais for providing comments on the manuscript. This work was supported by a Natural Sciences and Engineering Research Council (NSERC) discovery grant awarded to J.F.C. J.F.C. holds a Canada Research Chair in Structural Biology and Epigenetics. Finally, we would like to thank the Canada Foundation for Innovation.

REFERENCES

Albert, M., & Helin, K. (2010). Histone methyltransferases in cancer. *Seminars in Cell and Developmental Biology, 21*(2), 209–220. http://dx.doi.org/10.1016/j.semcdb.2009.10.007.

Antonysamy, S., Condon, B., Druzina, Z., Bonanno, J. B., Gheyi, T., Zhang, F., … Gately Luz, J. (2013). Structural context of disease-associated mutations and putative mechanism of autoinhibition revealed by X-ray crystallographic analysis of the EZH2-SET domain. *PLoS One, 8*(12), e84147. http://dx.doi.org/10.1371/journal.pone.0084147.

Ardehali, M. B., Mei, A., Zobeck, K. L., Caron, M., Lis, J. T., & Kusch, T. (2011). Drosophila Set1 is the major histone H3 lysine 4 trimethyltransferase with role in transcription. *The EMBO Journal, 30*(14), 2817–2828. http://dx.doi.org/10.1038/emboj.2011.194.

Avdic, V., Zhang, P., Lanouette, S., Groulx, A., Tremblay, V., Brunzelle, J., & Couture, J. F. (2011). Structural and biochemical insights into MLL1 core complex assembly. *Structure, 19*(1), 101–108. http://dx.doi.org/10.1016/j.str.2010.09.022.

Beguelin, W., Popovic, R., Teater, M., Jiang, Y., Bunting, K. L., Rosen, M., … Melnick, A. M. (2013). EZH2 is required for germinal center formation and somatic EZH2 mutations promote lymphoid transformation. *Cancer Cell, 23*(5), 677–692. http://dx.doi.org/10.1016/j.ccr.2013.04.011.

Berdasco, M., Ropero, S., Setien, F., Fraga, M. F., Lapunzina, P., Losson, R., … Esteller, M. (2009). Epigenetic inactivation of the Sotos overgrowth syndrome gene histone methyltransferase NSD1 in human neuroblastoma and glioma. *Proceedings of the National Academy of Sciences of the United States of America, 106*(51), 21830–21835. http://dx.doi.org/10.1073/pnas.0906831106.

Blanc, E., Roversi, P., Vonrhein, C., Flensburg, C., Lea, S. M., & Bricogne, G. (2004). Refinement of severely incomplete structures with maximum likelihood in BUSTER-TNT. *Acta Crystallographica Section D: Biological Crystallography, 60*(Pt. 12 Pt. 1), 2210–2221. http://dx.doi.org/10.1107/S0907444904016427.

Buchan, D. W., Minneci, F., Nugent, T. C., Bryson, K., & Jones, D. T. (2013). Scalable web services for the PSIPRED protein analysis workbench. *Nucleic Acids Research, 41*(Web Server issue), W349–W357. http://dx.doi.org/10.1093/nar/gkt381.

Bullock, A. N., Debreczeni, J. E., Edwards, A. M., Sundstrom, M., & Knapp, S. (2006). Crystal structure of the SOCS2-elongin C-elongin B complex defines a prototypical SOCS box ubiquitin ligase. *Proceedings of the National Academy of Sciences of the United States of America, 103*(20), 7637–7642. http://dx.doi.org/10.1073/pnas.0601638103.

Cao, F., Chen, Y., Cierpicki, T., Liu, Y., Basrur, V., Lei, M., & Dou, Y. (2010). An Ash2L/RbBP5 heterodimer stimulates the MLL1 methyltransferase activity through coordinated substrate interactions with the MLL1 SET domain. *PLoS One, 5*(11), e14102. http://dx.doi.org/10.1371/journal.pone.0014102.

Cao, R., & Zhang, Y. (2004). SUZ12 is required for both the histone methyltransferase activity and the silencing function of the EED-EZH2 complex. *Molecular Cell, 15*(1), 57–67. http://dx.doi.org/10.1016/j.molcel.2004.06.020.

Carrington, J. C., & Dougherty, W. G. (1988). A viral cleavage site cassette: Identification of amino acid sequences required for tobacco etch virus polyprotein processing. *Proceedings of the National Academy of Sciences of the United States of America, 85*(10), 3391–3395.

Chase, A., & Cross, N. C. (2011). Aberrations of EZH2 in cancer. *Clinical Cancer Research, 17*(9), 2613–2618. http://dx.doi.org/10.1158/1078-0432.CCR-10-2156.

Chuikov, S., Kurash, J. K., Wilson, J. R., Xiao, B., Justin, N., Ivanov, G. S., … Reinberg, D. (2004). Regulation of p53 activity through lysine methylation. *Nature, 432*(7015), 353–360. http://dx.doi.org/10.1038/nature03117. nature03117 [pii].

Couture, J. F., Collazo, E., Brunzelle, J. S., & Trievel, R. C. (2005). Structural and functional analysis of SET8, a histone H4 Lys-20 methyltransferase. *Genes and Development, 19*(12), 1455–1465. http://dx.doi.org/10.1101/gad.1318405.

Couture, J. F., Collazo, E., Hauk, G., & Trievel, R. C. (2006). Structural basis for the methylation site specificity of SET7/9. *Nature Structural & Molecular Biology*, *13*(2), 140–146. http://dx.doi.org/10.1038/nsmb1045.

Couture, J. F., Dirk, L. M., Brunzelle, J. S., Houtz, R. L., & Trievel, R. C. (2008). Structural origins for the product specificity of SET domain protein methyltransferases. *Proceedings of the National Academy of Sciences of the United States of America*, *105*(52), 20659–20664. http://dx.doi.org/10.1073/pnas.0806712105.

Couture, J. F., & Trievel, R. C. (2006). Histone-modifying enzymes: Encrypting an enigmatic epigenetic code. *Current Opinion in Structural Biology*, *16*(6), 753–760. http://dx.doi.org/10.1016/j.sbi.2006.10.002.

Davis, I. W., Leaver-Fay, A., Chen, V. B., Block, J. N., Kapral, G. J., Wang, X., … Richardson, D. C. (2007). MolProbity: All-atom contacts and structure validation for proteins and nucleic acids. *Nucleic Acids Research*, *35*(Web Server issue), W375–W383. http://dx.doi.org/10.1093/nar/gkm216. gkm216 [pii].

de Almeida, S. F., Grosso, A. R., Koch, F., Fenouil, R., Carvalho, S., Andrade, J., … Carmo-Fonseca, M. (2011). Splicing enhances recruitment of methyltransferase HYPB/Setd2 and methylation of histone H3 Lys36. *Nature Structural & Molecular Biology*, *18*(9), 977–983. http://dx.doi.org/10.1038/nsmb.2123.

Del Rizzo, P. A., & Trievel, R. C. (2011). Substrate and product specificities of SET domain methyltransferases. *Epigenetics*, *6*(9), 1059–1067. http://dx.doi.org/10.4161/epi.6.9.16069.

Dillon, S. C., Zhang, X., Trievel, R. C., & Cheng, X. (2005). The SET-domain protein superfamily: Protein lysine methyltransferases. *Genome Biology*, *6*(8), 227. http://dx.doi.org/10.1186/gb-2005-6-8-227.

Dyer, P. N., Edayathumangalam, R. S., White, C. L., Bao, Y., Chakravarthy, S., Muthurajan, U. M., & Luger, K. (2004). Reconstitution of nucleosome core particles from recombinant histones and DNA. *Methods in Enzymology*, *375*, 23–44.

Edgar, R. C. (2004). MUSCLE: Multiple sequence alignment with high accuracy and high throughput. *Nucleic Acids Research*, *32*(5), 1792–1797. http://dx.doi.org/10.1093/nar/gkh340.

Edmunds, J. W., Mahadevan, L. C., & Clayton, A. L. (2008). Dynamic histone H3 methylation during gene induction: HYPB/Setd2 mediates all H3K36 trimethylation. *The EMBO Journal*, *27*(2), 406–420. http://dx.doi.org/10.1038/sj.emboj.7601967.

Emsley, P., & Cowtan, K. (2004). Coot: Model-building tools for molecular graphics. *Acta Crystallographica Section D: Biological Crystallography*, *60*(Pt. 12 Pt. 1), 2126–2132. http://dx.doi.org/10.1107/S0907444904019158.

Emsley, P., Lohkamp, B., Scott, W. G., & Cowtan, K. (2010). Features and development of Coot. *Acta Crystallographica Section D: Biological Crystallography*, *66*(Pt. 4), 486–501. http://dx.doi.org/10.1107/S0907444910007493. S0907444910007493 [pii].

Ernst, T., Chase, A. J., Score, J., Hidalgo-Curtis, C. E., Bryant, C., Jones, A. V., … Cross, N. C. (2010). Inactivating mutations of the histone methyltransferase gene EZH2 in myeloid disorders. *Nature Genetics*, *42*(8), 722–726. http://dx.doi.org/10.1038/ng.621.

Falandry, C., Fourel, G., Galy, V., Ristriani, T., Horard, B., Bensimon, E., … Magdinier, F. (2010). CLLD8/KMT1F is a lysine methyltransferase that is important for chromosome segregation. *The Journal of Biological Chemistry*, *285*(26), 20234–20241. http://dx.doi.org/10.1074/jbc.M109.052399.

Ferguson, A. D., Larsen, N. A., Howard, T., Pollard, H., Green, I., Grande, C., … Keen, N. (2011). Structural basis of substrate methylation and inhibition of SMYD2. *Structure*, *19*(9), 1262–1273. http://dx.doi.org/10.1016/j.str.2011.06.011.

Ferrari, K. J., Scelfo, A., Jammula, S., Cuomo, A., Barozzi, I., Stutzer, A., … Pasini, D. (2014). Polycomb-dependent H3K27me1 and H3K27me2 regulate active transcription

and enhancer fidelity. *Molecular Cell*, *53*(1), 49–62. http://dx.doi.org/10.1016/j.molcel.2013.10.030.
Fischle, W., Wang, Y., & Allis, C. D. (2003). Histone and chromatin cross-talk. *Current Opinion in Cell Biology*, *15*(2), 172–183.
Greer, E. L., & Shi, Y. (2012). Histone methylation: A dynamic mark in health, disease and inheritance. *Nature Reviews. Genetics*, *13*(5), 343–357. http://dx.doi.org/10.1038/nrg3173.
Hallson, G., Hollebakken, R. E., Li, T., Syrzycka, M., Kim, I., Cotsworth, S., … Honda, B. M. (2012). dSet1 is the main H3K4 di- and tri-methyltransferase throughout Drosophila development. *Genetics*, *190*(1), 91–100. http://dx.doi.org/10.1534/genetics.111.135863.
Hammarstrom, M., Hellgren, N., van Den Berg, S., Berglund, H., & Hard, T. (2002). Rapid screening for improved solubility of small human proteins produced as fusion proteins in Escherichia coli. *Protein Science*, *11*(2), 313–321. http://dx.doi.org/10.1110/ps.22102.
Herz, H. M., Garruss, A., & Shilatifard, A. (2013). SET for life: Biochemical activities and biological functions of SET domain-containing proteins. *Trends in Biochemical Sciences*, *38*(12), 621–639. http://dx.doi.org/10.1016/j.tibs.2013.09.004.
Hochuli, E. (1988). Large-scale chromatography of recombinant proteins. *Journal of Chromatography*, *444*, 293–302.
Hochuli, E. (1990). Purification of recombinant proteins with metal chelate adsorbent. *Genet Eng (N Y)*, *12*, 87–98.
Hochuli, E., Dobeli, H., & Schacher, A. (1987). New metal chelate adsorbent selective for proteins and peptides containing neighbouring histidine residues. *Journal of Chromatography*, *411*, 177–184.
Holz, C., Hesse, O., Bolotina, N., Stahl, U., & Lang, C. (2002). A micro-scale process for high-throughput expression of cDNAs in the yeast Saccharomyces cerevisiae. *Protein Expression and Purification*, *25*(3), 372–378.
Houston, S. I., McManus, K. J., Adams, M. M., Sims, J. K., Carpenter, P. B., Hendzel, M. J., & Rice, J. C. (2008). Catalytic function of the PR-Set7 histone H4 lysine 20 monomethyltransferase is essential for mitotic entry and genomic stability. *The Journal of Biological Chemistry*, *283*(28), 19478–19488. http://dx.doi.org/10.1074/jbc.M710579200.
Hwang, K. K., Eissenberg, J. C., & Worman, H. J. (2001). Transcriptional repression of euchromatic genes by Drosophila heterochromatin protein 1 and histone modifiers. *Proceedings of the National Academy of Sciences of the United States of America*, *98*(20), 11423–11427. http://dx.doi.org/10.1073/pnas.211303598.
Jacob, Y., Bergamin, E., Donoghue, M. T., Mongeon, V., LeBlanc, C., Voigt, P., … Martienssen, R. A. (2014). Selective methylation of histone H3 variant H3.1 regulates heterochromatin replication. *Science*, *343*(6176), 1249–1253. http://dx.doi.org/10.1126/science.1248357.
Jacob, Y., Feng, S., LeBlanc, C. A., Bernatavichute, Y. V., Stroud, H., Cokus, S., … Michaels, S. D. (2009). ATXR5 and ATXR6 are H3K27 monomethyltransferases required for chromatin structure and gene silencing. *Nature Structural & Molecular Biology*, *16*(7), 763–768. http://dx.doi.org/10.1038/nsmb.1611nsmb.1611 [pii].
Jenuwein, T., Laible, G., Dorn, R., & Reuter, G. (1998). SET domain proteins modulate chromatin domains in eu- and heterochromatin. *Cellular and Molecular Life Sciences*, *54*(1), 80–93.
Jiao, L., & Liu, X. (2015). Structural basis of histone H3K27 trimethylation by an active polycomb repressive complex 2. *Science*, *350*(6258), aac4383. http://dx.doi.org/10.1126/science.aac4383.
Jones, D. T. (1999). Protein secondary structure prediction based on position-specific scoring matrices. *Journal of Molecular Biology*, *292*(2), 195–202. http://dx.doi.org/10.1006/jmbi.1999.3091.

Jones, R. S., & Gelbart, W. M. (1993). The Drosophila Polycomb-group gene enhancer of zeste contains a region with sequence similarity to trithorax. *Molecular and Cellular Biology, 13*(10), 6357–6366.

Kapoor-Vazirani, P., Kagey, J. D., & Vertino, P. M. (2011). SUV420H2-mediated H4K20 trimethylation enforces RNA polymerase II promoter-proximal pausing by blocking hMOF-dependent H4K16 acetylation. *Molecular and Cellular Biology, 31*(8), 1594–1609. http://dx.doi.org/10.1128/MCB.00524-10.

Kim, W., Bird, G. H., Neff, T., Guo, G., Kerenyi, M. A., Walensky, L. D., & Orkin, S. H. (2013). Targeted disruption of the EZH2-EED complex inhibits EZH2-dependent cancer. *Nature Chemical Biology, 9*(10), 643–650. http://dx.doi.org/10.1038/nchembio.1331.

Kim, S., Kim, H., Fong, N., Erickson, B., & Bentley, D. L. (2011). Pre-mRNA splicing is a determinant of histone H3K36 methylation. *Proceedings of the National Academy of Sciences of the United States of America, 108*(33), 13564–13569. http://dx.doi.org/10.1073/pnas.1109475108.

Kleer, C. G., Cao, Q., Varambally, S., Shen, R., Ota, I., Tomlins, S. A., ... Chinnaiyan, A. M. (2003). EZH2 is a marker of aggressive breast cancer and promotes neoplastic transformation of breast epithelial cells. *Proceedings of the National Academy of Sciences of the United States of America, 100*(20), 11606–11611. http://dx.doi.org/10.1073/pnas.1933744100.

Knutson, S. K., Wigle, T. J., Warholic, N. M., Sneeringer, C. J., Allain, C. J., Klaus, C. R., ... Kuntz, K. W. (2012). A selective inhibitor of EZH2 blocks H3K27 methylation and kills mutant lymphoma cells. *Nature Chemical Biology, 8*(11), 890–896. http://dx.doi.org/10.1038/nchembio.1084.

Kouskouti, A., Scheer, E., Staub, A., Tora, L., & Talianidis, I. (2004). Gene-specific modulation of TAF10 function by SET9-mediated methylation. *Molecular Cell, 14*(2), 175–182.

Kuo, A. J., Cheung, P., Chen, K., Zee, B. M., Kioi, M., Lauring, J., ... Gozani, O. (2011). NSD2 links dimethylation of histone H3 at lysine 36 to oncogenic programming. *Molecular Cell, 44*(4), 609–620. http://dx.doi.org/10.1016/j.molcel.2011.08.042. S1097-2765(11)00813-6 [pii].

Landry, J., Sutton, A., Hesman, T., Min, J., Xu, R. M., Johnston, M., & Sternglanz, R. (2003). Set2-catalyzed methylation of histone H3 represses basal expression of GAL4 in Saccharomyces cerevisiae. *Molecular and Cellular Biology, 23*(17), 5972–5978.

Lanouette, S., Mongeon, V., Figeys, D., & Couture, J. F. (2014). The functional diversity of protein lysine methylation. *Molecular Systems Biology, 10*, 724. http://dx.doi.org/10.1002/msb.134974.

Letunic, I., Doerks, T., & Bork, P. (2015). SMART: Recent updates, new developments and status in 2015. *Nucleic Acids Research, 43*(Database issue), D257–D260. http://dx.doi.org/10.1093/nar/gku949.

Li, Y., Trojer, P., Xu, C. F., Cheung, P., Kuo, A., Drury, W. J., 3rd., ... Reinberg, D. (2009). The target of the NSD family of histone lysine methyltransferases depends on the nature of the substrate. *The Journal of Biological Chemistry, 284*(49), 34283–34295. http://dx.doi.org/10.1074/jbc.M109.034462.

Lowary, P. T., & Widom, J. (1998). New DNA sequence rules for high affinity binding to histone octamer and sequence-directed nucleosome positioning. *Journal of Molecular Biology, 276*(1), 19–42. http://dx.doi.org/10.1006/jmbi.1997.1494.

Lucio-Eterovic, A. K., Singh, M. M., Gardner, J. E., Veerappan, C. S., Rice, J. C., & Carpenter, P. B. (2010). Role for the nuclear receptor-binding SET domain protein 1 (NSD1) methyltransferase in coordinating lysine 36 methylation at histone 3 with RNA polymerase II function. *Proceedings of the National Academy of Sciences of the United States of America, 107*(39), 16952–16957. http://dx.doi.org/10.1073/pnas.1002653107. 1002653107 [pii].

Luger, K., Mader, A. W., Richmond, R. K., Sargent, D. F., & Richmond, T. J. (1997). Crystal structure of the nucleosome core particle at 2.8 A resolution. *Nature, 389*(6648), 251–260. http://dx.doi.org/10.1038/38444.

Luger, K., Rechsteiner, T. J., & Richmond, T. J. (1999). Preparation of nucleosome core particle from recombinant histones. *Methods in Enzymology, 304*, 3–19.

Margueron, R., & Reinberg, D. (2011). The Polycomb complex PRC2 and its mark in life. *Nature, 469*(7330), 343–349. http://dx.doi.org/10.1038/nature09784.

Martin, C., Cao, R., & Zhang, Y. (2006). Substrate preferences of the EZH2 histone methyltransferase complex. *The Journal of Biological Chemistry, 281*(13), 8365–8370. http://dx.doi.org/10.1074/jbc.M513425200.

McCabe, M. T., Ott, H. M., Ganji, G., Korenchuk, S., Thompson, C., Van Aller, G. S., ... Creasy, C. L. (2012). EZH2 inhibition as a therapeutic strategy for lymphoma with EZH2-activating mutations. *Nature, 492*(7427), 108–112. http://dx.doi.org/10.1038/nature11606.

McCoy, A. J., Grosse-Kunstleve, R. W., Adams, P. D., Winn, M. D., Storoni, L. C., & Read, R. J. (2007). Phaser crystallographic software. *Journal of Applied Crystallography, 40*(Pt. 4), 658–674. http://dx.doi.org/10.1107/S0021889807021206.

Mohan, M., Herz, H. M., Smith, E. R., Zhang, Y., Jackson, J., Washburn, M. P., ... Shilatifard, A. (2011). The COMPASS family of H3K4 methylases in Drosophila. *Molecular and Cellular Biology, 31*(21), 4310–4318. http://dx.doi.org/10.1128/MCB.06092-11.

Morin, R. D., Johnson, N. A., Severson, T. M., Mungall, A. J., An, J., Goya, R., ... Marra, M. A. (2010). Somatic mutations altering EZH2 (Tyr641) in follicular and diffuse large B-cell lymphomas of germinal-center origin. *Nature Genetics, 42*(2), 181–185. http://dx.doi.org/10.1038/ng.518.

Murray, K. (1964). The occurrence of epsilon-N-methyl lysine in histones. *Biochemistry, 3*, 10–15.

Odho, Z., Southall, S. M., & Wilson, J. R. (2010). Characterization of a novel WDR5-binding site that recruits RbBP5 through a conserved motif to enhance methylation of histone H3 lysine 4 by mixed lineage leukemia protein-1. *The Journal of Biological Chemistry, 285*(43), 32967–32976. http://dx.doi.org/10.1074/jbc.M110.159921.

Otwinowski, Z., & Minor, W. (1997). Processing of X-ray diffraction data collected in oscillation mode. *Macromolecular Crystallography, Part A, 276*, 307–326. http://dx.doi.org/10.1016/S0076-6879(97)76066-X.

Pasini, D., Bracken, A. P., Jensen, M. R., Lazzerini Denchi, E., & Helin, K. (2004). Suz12 is essential for mouse development and for EZH2 histone methyltransferase activity. *The EMBO Journal, 23*(20), 4061–4071. http://dx.doi.org/10.1038/sj.emboj.7600402.

Patel, A., Vought, V. E., Dharmarajan, V., & Cosgrove, M. S. (2008). A conserved arginine-containing motif crucial for the assembly and enzymatic activity of the mixed lineage leukemia protein-1 core complex. *The Journal of Biological Chemistry, 283*(47), 32162–32175. http://dx.doi.org/10.1074/jbc.M806317200.

Pei, J., Kim, B. H., & Grishin, N. V. (2008). PROMALS3D: A tool for multiple protein sequence and structure alignments. *Nucleic Acids Research, 36*(7), 2295–2300. http://dx.doi.org/10.1093/nar/gkn072.

Peters, A. H., Kubicek, S., Mechtler, K., O'Sullivan, R. J., Derijck, A. A., Perez-Burgos, L., ... Jenuwein, T. (2003). Partitioning and plasticity of repressive histone methylation states in mammalian chromatin. *Molecular Cell, 12*(6), 1577–1589.

Peterson, C. L., & Laniel, M. A. (2004). Histones and histone modifications. *Current Biology, 14*(14), R546–R551. http://dx.doi.org/10.1016/j.cub.2004.07.007.

Pflugrath, J. W. (1999). The finer things in X-ray diffraction data collection. *Acta Crystallographica Section D: Biological Crystallography, 55*(Pt. 10), 1718–1725.

Qian, C., Wang, X., Manzur, K., Sachchidanand, Farooq, A., Zeng, L., ... Zhou, M. M. (2006). Structural insights of the specificity and catalysis of a viral histone H3 lysine 27 methyltransferase. *Journal of Molecular Biology, 359*(1), 86–96. http://dx.doi.org/10.1016/j.jmb.2006.03.006. S0022-2836(06)00313-5 [pii].

Rathert, P., Zhang, X., Freund, C., Cheng, X., & Jeltsch, A. (2008). Analysis of the substrate specificity of the Dim-5 histone lysine methyltransferase using peptide arrays. *Chemistry and Biology, 15*(1), 5–11. http://dx.doi.org/10.1016/j.chembiol.2007.11.013.

Rea, S., Eisenhaber, F., O'Carroll, D., Strahl, B. D., Sun, Z. W., Schmid, M., ... Jenuwein, T. (2000). Regulation of chromatin structure by site-specific histone H3 methyltransferases. *Nature, 406*(6796), 593–599. http://dx.doi.org/10.1038/35020506.

Roguev, A., Schaft, D., Shevchenko, A., Pijnappel, W. W., Wilm, M., Aasland, R., & Stewart, A. F. (2001). The Saccharomyces cerevisiae Set1 complex includes an Ash2 homologue and methylates histone 3 lysine 4. *The EMBO Journal, 20*(24), 7137–7148. http://dx.doi.org/10.1093/emboj/20.24.7137.

Romier, C., Ben Jelloul, M., Albeck, S., Buchwald, G., Busso, D., Celie, P. H., ... Perrakis, A. (2006). Co-expression of protein complexes in prokaryotic and eukaryotic hosts: Experimental procedures, database tracking and case studies. *Acta Crystallographica Section D: Biological Crystallography, 62*(Pt. 10), 1232–1242. http://dx.doi.org/10.1107/S0907444906031003.

Sakaguchi, A., & Steward, R. (2007). Aberrant monomethylation of histone H4 lysine 20 activates the DNA damage checkpoint in Drosophila melanogaster. *Journal of Cell Biology, 176*(2), 155–162. http://dx.doi.org/10.1083/jcb.200607178.

Sarvan, S., Avdic, V., Tremblay, V., Chaturvedi, C. P., Zhang, P., Lanouette, S., ... Couture, J. F. (2011). Crystal structure of the trithorax group protein ASH2L reveals a forkhead-like DNA binding domain. *Nature Structural & Molecular Biology, 18*(7), 857–859. http://dx.doi.org/10.1038/nsmb.2093.

Schnitzler, G. R. (2001). Isolation of histones and nucleosome cores from mammalian cells. *Current Protocols in Molecular Biology*. http://dx.doi.org/10.1002/0471142727.mb2105s50. chapter 21, Unit 21.5.1–21.5.12.

Schotta, G., Lachner, M., Sarma, K., Ebert, A., Sengupta, R., Reuter, G., ... Jenuwein, T. (2004). A silencing pathway to induce H3-K9 and H4-K20 trimethylation at constitutive heterochromatin. *Genes and Development, 18*(11), 1251–1262. http://dx.doi.org/10.1101/gad.300704.

Schotta, G., Sengupta, R., Kubicek, S., Malin, S., Kauer, M., Callen, E., & Jenuwein, T. (2008). A chromatin-wide transition to H4K20 monomethylation impairs genome integrity and programmed DNA rearrangements in the mouse. *Genes and Development, 22*(15), 2048–2061. http://dx.doi.org/10.1101/gad.476008. 22/15/2048 [pii].

Schuettengruber, B., Martinez, A. M., Iovino, N., & Cavalli, G. (2011). Trithorax group proteins: Switching genes on and keeping them active. *Nature Reviews. Molecular Cell Biology, 12*(12), 799–814. http://dx.doi.org/10.1038/nrm3230.

Sheffield, P., Garrard, S., & Derewenda, Z. (1999). Overcoming expression and purification problems of RhoGDI using a family of "parallel" expression vectors. *Protein Expression and Purification, 15*(1), 34–39. http://dx.doi.org/10.1006/prep.1998.1003. S1046-5928(98)91003-8 [pii].

Shilatifard, A. (2012). The COMPASS family of histone H3K4 methylases: Mechanisms of regulation in development and disease pathogenesis. *Annual Review of Biochemistry, 81*, 65–95. http://dx.doi.org/10.1146/annurev-biochem-051710-134100.

Shim, Y., Duan, M. R., Chen, X., Smerdon, M. J., & Min, J. H. (2012). Polycistronic coexpression and nondenaturing purification of histone octamers. *Analytical Biochemistry, 427*(2), 190–192. http://dx.doi.org/10.1016/j.ab.2012.05.006.

Shimko, J. C., Howard, C. J., Poirier, M. G., & Ottesen, J. J. (2013). Preparing semisynthetic and fully synthetic histones h3 and h4 to modify the nucleosome core. *Methods in Molecular Biology*, *981*, 177–192. http://dx.doi.org/10.1007/978-1-62703-305-3_14.

Shinkai, Y., & Tachibana, M. (2011). H3K9 methyltransferase G9a and the related molecule GLP. *Genes and Development*, *25*(8), 781–788. http://dx.doi.org/10.1101/gad.2027411.

Simon, J., Chiang, A., & Bender, W. (1992). Ten different Polycomb group genes are required for spatial control of the abdA and AbdB homeotic products. *Development*, *114*(2), 493–505.

Simon, M. D., & Shokat, K. M. (2012). A method to site-specifically incorporate methyl-lysine analogues into recombinant proteins. *Methods in Enzymology*, *512*, 57–69. http://dx.doi.org/10.1016/B978-0-12-391940-3.00003-2.

Smith, D. B. (2000). Generating fusions to glutathione S-transferase for protein studies. *Methods in Enzymology*, *326*, 254–270.

Smith, D. B., & Johnson, K. S. (1988). Single-step purification of polypeptides expressed in Escherichia coli as fusions with glutathione S-transferase. *Gene*, *67*(1), 31–40.

Smith, G. E., Summers, M. D., & Fraser, M. J. (1983). Production of human beta interferon in insect cells infected with a baculovirus expression vector. *Molecular and Cellular Biology*, *3*(12), 2156–2165.

Song, J. J., & Kingston, R. E. (2008). WDR5 interacts with mixed lineage leukemia (MLL) protein via the histone H3-binding pocket. *The Journal of Biological Chemistry*, *283*(50), 35258–35264. http://dx.doi.org/10.1074/jbc.M806900200.

Southall, S. M., Wong, P. S., Odho, Z., Roe, S. M., & Wilson, J. R. (2009). Structural basis for the requirement of additional factors for MLL1 SET domain activity and recognition of epigenetic marks. *Molecular Cell*, *33*(2), 181–191. http://dx.doi.org/10.1016/j.molcel.2008.12.029.

Stassen, M. J., Bailey, D., Nelson, S., Chinwalla, V., & Harte, P. J. (1995). The Drosophila trithorax proteins contain a novel variant of the nuclear receptor type DNA binding domain and an ancient conserved motif found in other chromosomal proteins. *Mechanisms of Development*, *52*(2-3), 209–223.

Strahl, B. D., Grant, P. A., Briggs, S. D., Sun, Z. W., Bone, J. R., Caldwell, J. A., … Allis, C. D. (2002). Set2 is a nucleosomal histone H3-selective methyltransferase that mediates transcriptional repression. *Molecular and Cellular Biology*, *22*(5), 1298–1306.

Strahl, B. D., Ohba, R., Cook, R. G., & Allis, C. D. (1999). Methylation of histone H3 at lysine 4 is highly conserved and correlates with transcriptionally active nuclei in Tetrahymena. *Proceedings of the National Academy of Sciences of the United States of America*, *96*(26), 14967–14972.

Studier, F. W., Rosenberg, A. H., Dunn, J. J., & Dubendorff, J. W. (1990). Use of T7 RNA polymerase to direct expression of cloned genes. *Methods in Enzymology*, *185*, 60–89.

Takahashi, Y. H., Westfield, G. H., Oleskie, A. N., Trievel, R. C., Shilatifard, A., & Skiniotis, G. (2011). Structural analysis of the core COMPASS family of histone H3K4 methylases from yeast to human. *Proceedings of the National Academy of Sciences of the United States of America*, *108*(51), 20526–20531. http://dx.doi.org/10.1073/pnas.1109360108.

Tanaka, Y., Kawahashi, K., Katagiri, Z., Nakayama, Y., Mahajan, M., & Kioussis, D. (2011). Dual function of histone H3 lysine 36 methyltransferase ASH1 in regulation of Hox gene expression. *PLoS One*. *6*(11). e28171http://dx.doi.org/10.1371/journal.pone.0028171.

Taylor, W. R., Xiao, B., Gamblin, S. J., & Lin, K. (2003). A knot or not a knot? SETting the record 'straight' on proteins. *Computational Biology and Chemistry*, *27*(1), 11–15.

Thiel, A. T., Blessington, P., Zou, T., Feather, D., Wu, X., Yan, J., … Hua, X. (2010). MLL-AF9-induced leukemogenesis requires coexpression of the wild-type Mll allele. *Cancer Cell*, *17*(2), 148–159. http://dx.doi.org/10.1016/j.ccr.2009.12.034.

Tolia, N. H., & Joshua-Tor, L. (2006). Strategies for protein coexpression in Escherichia coli. *Nature Methods*, *3*(1), 55–64. http://dx.doi.org/10.1038/nmeth0106-55.

Trievel, R. C., Beach, B. M., Dirk, L. M., Houtz, R. L., & Hurley, J. H. (2002). Structure and catalytic mechanism of a SET domain protein methyltransferase. *Cell*, *111*(1), 91–103.

Trievel, R. C., Flynn, E. M., Houtz, R. L., & Hurley, J. H. (2003). Mechanism of multiple lysine methylation by the SET domain enzyme Rubisco LSMT. *Nature Structural Biology*, *10*(7), 545–552. http://dx.doi.org/10.1038/nsb946. nsb946 [pii].

Tschiersch, B., Hofmann, A., Krauss, V., Dorn, R., Korge, G., & Reuter, G. (1994). The protein encoded by the Drosophila position-effect variegation suppressor gene Su(var) 3-9 combines domains of antagonistic regulators of homeotic gene complexes. *The EMBO Journal*, *13*(16), 3822–3831.

Uhlen, M., Forsberg, G., Moks, T., Hartmanis, M., & Nilsson, B. (1992). Fusion proteins in biotechnology. *Current Opinion in Biotechnology*, *3*(4), 363–369.

Varambally, S., Cao, Q., Mani, R. S., Shankar, S., Wang, X., Ateeq, B., … Chinnaiyan, A. M. (2008). Genomic loss of microRNA-101 leads to overexpression of histone methyltransferase EZH2 in cancer. *Science*, *322*(5908), 1695–1699. http://dx.doi.org/10.1126/science.1165395.

Varambally, S., Dhanasekaran, S. M., Zhou, M., Barrette, T. R., Kumar-Sinha, C., Sanda, M. G., … Chinnaiyan, A. M. (2002). The polycomb group protein EZH2 is involved in progression of prostate cancer. *Nature*, *419*(6907), 624–629. http://dx.doi.org/10.1038/nature01075.

Vera, A., Gonzalez-Montalban, N., Aris, A., & Villaverde, A. (2007). The conformational quality of insoluble recombinant proteins is enhanced at low growth temperatures. *Biotechnology and Bioengineering*, *96*(6), 1101–1106. http://dx.doi.org/10.1002/bit.21218.

Wagner, E. J., & Carpenter, P. B. (2012). Understanding the language of Lys36 methylation at histone H3. *Nature Reviews. Molecular Cell Biology*, *13*(2), 115–126. http://dx.doi.org/10.1038/nrm3274.

Wang, C., Liu, Z., Woo, C. W., Li, Z., Wang, L., Wei, J. S., … Thiele, C. J. (2012). EZH2 Mediates epigenetic silencing of neuroblastoma suppressor genes CASZ1, CLU, RUNX3, and NGFR. *Cancer Research*, *72*(1), 315–324. http://dx.doi.org/10.1158/0008-5472.CAN-11-0961.

Waterhouse, A. M., Procter, J. B., Martin, D. M., Clamp, M., & Barton, G. J. (2009). Jalview Version 2—A multiple sequence alignment editor and analysis workbench. *Bioinformatics*, *25*(9), 1189–1191. http://dx.doi.org/10.1093/bioinformatics/btp033.

Wilson, J. R., Jing, C., Walker, P. A., Martin, S. R., Howell, S. A., Blackburn, G. M., … Xiao, B. (2002). Crystal structure and functional analysis of the histone methyltransferase SET7/9. *Cell*, *111*(1), 105–115.

Wu, H., Zeng, H., Dong, A., Li, F., He, H., Senisterra, G., … Schapira, M. (2013). Structure of the catalytic domain of EZH2 reveals conformational plasticity in cofactor and substrate binding sites and explains oncogenic mutations. *PLoS One*. *8*(12). e83737http://dx.doi.org/10.1371/journal.pone.0083737.

Xiao, B., Jing, C., Kelly, G., Walker, P. A., Muskett, F. W., Frenkiel, T. A., … Wilson, J. R. (2005). Specificity and mechanism of the histone methyltransferase Pr-Set7. *Genes and Development*, *19*(12), 1444–1454. http://dx.doi.org/10.1101/gad.1315905. gad.1315905 [pii].

Xiao, B., Jing, C., Wilson, J. R., Walker, P. A., Vasisht, N., Kelly, G., … Gamblin, S. J. (2003). Structure and catalytic mechanism of the human histone methyltransferase SET7/9. *Nature*, *421*(6923), 652–656. http://dx.doi.org/10.1038/nature01378.

Xiao, B., Wilson, J. R., & Gamblin, S. J. (2003). SET domains and histone methylation. *Current Opinion in Structural Biology*, *13*(6), 699–705.

Yamamoto, K., Sonoda, M., Inokuchi, J., Shirasawa, S., & Sasazuki, T. (2004). Polycomb group suppressor of zeste 12 links heterochromatin protein 1alpha and enhancer of zeste 2. *The Journal of Biological Chemistry*, *279*(1), 401–406. http://dx.doi.org/10.1074/jbc.M307344200.

Yanagisawa, T., Takahashi, M., Mukai, T., Sato, S., Wakamori, M., Shirouzu, M., ... Yokoyama, S. (2014). Multiple site-specific installations of Nepsilon-monomethyl-L-lysine into histone proteins by cell-based and cell-free protein synthesis. *Chembiochem*, *15*(12), 1830–1838. http://dx.doi.org/10.1002/cbic.201402291.

Yang, P., Guo, L., Duan, Z. J., Tepper, C. G., Xue, L., Chen, X., ... Chen, H. W. (2012). Histone methyltransferase NSD2/MMSET mediates constitutive NF-kappaB signaling for cancer cell proliferation, survival, and tumor growth via a feed-forward loop. *Molecular and Cellular Biology*, *32*(15), 3121–3131. http://dx.doi.org/10.1128/MCB.00204-12.

Yap, D. B., Chu, J., Berg, T., Schapira, M., Cheng, S. W., Moradian, A., ... Aparicio, S. A. (2011). Somatic mutations at EZH2 Y641 act dominantly through a mechanism of selectively altered PRC2 catalytic activity, to increase H3K27 trimethylation. *Blood*, *117*(8), 2451–2459. http://dx.doi.org/10.1182/blood-2010-11-321208.

Young, M. D., Willson, T. A., Wakefield, M. J., Trounson, E., Hilton, D. J., Blewitt, M. E., ... Majewski, I. J. (2011). ChIP-seq analysis reveals distinct H3K27me3 profiles that correlate with transcriptional activity. *Nucleic Acids Research*, *39*(17), 7415–7427. http://dx.doi.org/10.1093/nar/gkr416.

Yu, J., Cao, Q., Mehra, R., Laxman, B., Yu, J., Tomlins, S. A., ... Chinnaiyan, A. M. (2007). Integrative genomics analysis reveals silencing of beta-adrenergic signaling by polycomb in prostate cancer. *Cancer Cell*, *12*(5), 419–431. http://dx.doi.org/10.1016/j.ccr.2007.10.016.

Zhang, P., Lee, H., Brunzelle, J. S., & Couture, J. F. (2012). The plasticity of WDR5 peptide-binding cleft enables the binding of the SET1 family of histone methyltransferases. *Nucleic Acids Research*, *40*(9), 4237–4246. http://dx.doi.org/10.1093/nar/gkr1235.

Zhang, K., Lin, W., Latham, J. A., Riefler, G. M., Schumacher, J. M., Chan, C., ... Dent, S. Y. (2005). The Set1 methyltransferase opposes Ipl1 aurora kinase functions in chromosome segregation. *Cell*, *122*(5), 723–734. http://dx.doi.org/10.1016/j.cell.2005.06.021.

Zhang, X., Tamaru, H., Khan, S. I., Horton, J. R., Keefe, L. J., Selker, E. U., & Cheng, X. (2002). Structure of the Neurospora SET domain protein DIM-5, a histone H3 lysine methyltransferase. *Cell*, *111*(1), 117–127.

Zhang, X., Yang, Z., Khan, S. I., Horton, J. R., Tamaru, H., Selker, E. U., & Cheng, X. (2003). Structural basis for the product specificity of histone lysine methyltransferases. *Molecular Cell*, *12*(1), 177–185.

Zwart, P. H., Afonine, P. V., Grosse-Kunstleve, R. W., Hung, L. W., Ioerger, T. R., McCoy, A. J., ... Adams, P. D. (2008). Automated structure solution with the PHENIX suite. *Methods in Molecular Biology*, *426*, 419–435. http://dx.doi.org/10.1007/978-1-60327-058-8_28.

CHAPTER TEN

Expression, Purification, and Biochemical Analysis of the LSD1/KDM1A Histone Demethylase

B. Laurent[*,†], **Y. Shi**[*,†,1]
[*]Boston Children's Hospital, Boston, MA, United States
[†]Harvard Medical School, Boston, MA, United States
[1]Corresponding author: e-mail addresses: yang_shi@hms.harvard.edu; yang.shi@childrens.harvard.edu

Contents

1. Introduction 242
2. Expression and Purification of LSD1 Proteins 245
   2.1 Expression and Purification of LSD1 in Bacteria 245
   2.2 Expression and Purification of LSD1 from Sf9 Insect Cells 246
   2.3 LSD1 Purification for Structure Determination 248
   2.4 Purification of LSD1 Complex from Mammalian Cells 250
3. In Vitro Enzymatic Assays 251
   3.1 Enzymatic Assay on Peptides and Histones 251
   3.2 Enzymatic Assay on Nucleosomes 252
   3.3 Enzymatic Assays with LSD1 Complex 253
4. Analysis of LSD1 Demethylase Activity 253
   4.1 Analysis by Mass Spectrometry 253
   4.2 Analysis by Western Blot 254
   4.3 Additional Methods for Detecting LSD1 Activity 255
   4.4 In vivo Analysis 256
Acknowledgments 257
References 258

Abstract

Posttranslational modifications (PTMs) of histones play important roles in the regulation of chromatin architecture and gene transcription. A decade ago, it was still believed that methyl groups could not be removed from histones, until the first histone demethylase LSD1 (lysine-specific demethylase 1; also known as KDM1A) was identified. This discovery initiated an era in the understanding of chromatin dynamic regulation by active histone demethylation. Since then, the repertoire of histone demethylases has expanded, and our understanding of the molecular mechanisms, structures, and macromolecular complexes of the demethylases has grown significantly. Histone demethylases have emerged as important players in developmental processes and have been linked to human diseases and cancer. Studies highlighting the functions of LSD1 have

Methods in Enzymology, Volume 573
ISSN 0076-6879
http://dx.doi.org/10.1016/bs.mie.2016.02.001

significantly increased our understanding of chromatin biology and have revealed that new facets of histone demethylases remain to be discovered. In vitro methods have been developed to assess the biochemistry, structure, and enzymology of lysine demethylases. Here, we describe the methods of expression, purification, and biochemical analysis that we have successfully used in characterizing the functions of LSD1.

1. INTRODUCTION

The histone tails are extensively posttranslationally modified, including by phosphorylation, ubiquitination, acetylation, and methylation, and these modifications collectively serve as regulators for chromatin template-based events including gene transcription and changes in chromatin architecture (Jenuwein & Allis, 2001; Strahl & Allis, 2000). Histone methylation was considered an irreversible PTM since its discovery in the 1960s, but this dogma was overturned by the discovery of the first histone demethylase, LSD1 (Lysine-Specific Demethylase 1; also known as KDM1A), in 2004 (Shi et al., 2004). Since then, the repertoire of histone demethylases has been greatly expanded, and today, it consists of 33 histone demethylases (Allis et al., 2007; Pedersen & Helin, 2010).

Two classes of histones demethylases have thus far been identified: the Jumonji C (JmjC) domain-containing proteins and the amine oxidase domain (AOD)-containing proteins. They have distinct enzymatic mechanisms and substrate specificity. The JmjC domain is conserved through evolution and belongs to the dioxygenase superfamily. In the presence of oxygen, and Fe^{2+} and α-ketoglutarate as cofactors, the JmjC domain-containing histone demethylases (JHDMs) convert the methyl group in the methyl-lysine to an unstable hydroxymethyl ammonium intermediate, which is then released as formaldehyde (Fig. 1A). This reaction mechanism allows JHDMs to remove the mono-, di-, and trimethyl groups from their substrates. Whereas many members of the JHDM family have been reported, the class of the AOD-containing histone demethylases is solely represented by LSD1/KDM1A and LSD2/KDM1B. In contrast to the JHDMs, LSD1 and LSD2 use the flavin adenine dinucleotide (FAD) as a cofactor to oxidize the methyl-lysine to form an imine intermediate, which is then hydrolyzed to release a formaldehyde molecule (Fig. 1B). The formation of the imine intermediate requires a protonated nitrogen as a hydrogen donor. Since the nitrogen of a trimethylated lysine cannot be

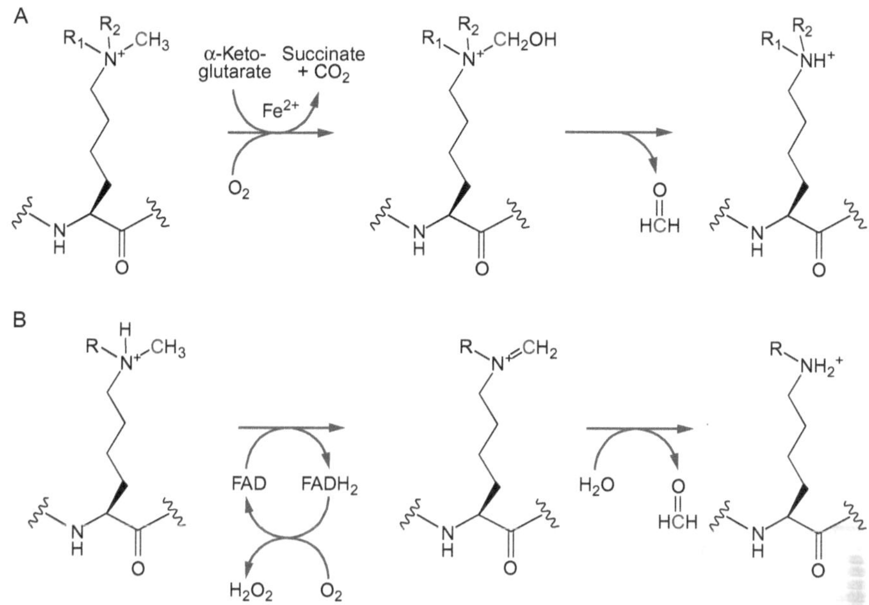

Fig. 1 (A) Chemical mechanism of lysine demethylation catalyzed by Jumonji C (JmjC) domain-containing proteins. (B) Chemical mechanism of LSD1/KDM1A-catalyzed lysine demethylation. The carbons in *red* are subjected to demethylation in each reaction. *Adapted from Mosammaparast, N., & Shi, Y. (2010). Reversal of histone methylation: Biochemical and molecular mechanisms of histone demethylases.* Annual Review of Biochemistry, *79, 155–179. doi: 10.1146/annurev.biochem.78.070907.103946.* (See the color plate.)

protonated, the substrate specificity of this family is limited to mono- and dimethylated lysine residues.

The biochemical activity of LSD1 plays an important role in the regulation of gene expression and has been shown to be highly complex as LSD1 exhibits diverse substrate specificities. LSD1 was initially identified as a component of a transcriptional repressor complex (Ballas et al., 2001; Shi et al., 2005). Consistent with its repressive role, LSD1 was found to remove both mono- and dimethylation of histone H3 at lysine 4 (H3K4me1/2) (Shi et al., 2004), modifications that are associated with active promoters and enhancers (Shi et al., 2004; Whyte et al., 2012). LSD1 also interacts with nuclear hormone receptors, eg, androgen receptor and estrogen receptor (Garcia-Bassets et al., 2007; Metzger et al., 2005; Nair et al., 2010). Interestingly, in association with nuclear receptor complexes, LSD1 appears to function as a coactivator and mediates demethylation of the repressive H3K9me1/2 marks (Garcia-Bassets et al., 2007;

Metzger et al., 2005; Nair et al., 2010; Perillo et al., 2008). At the atomic level, the LSD1-histone peptide cocrystal structures cannot explain how LSD1 can mediate both H3K4 and H3K9 demethylation (Forneris, Binda, Adamo, Battaglioli, & Mattevi, 2007; Forneris et al., 2006; Hou & Yu, 2010; Yang et al., 2006). Our group and others recently described that alternative splicing can provide a mechanism by which LSD1 acquires differential substrate specificities (Laurent et al., 2015; Wang et al., 2015).

The LSD1 gene contains 19 exons that are highly conserved among vertebrates. Through RNA alternative splicing, two additional exons, exon E2a and exon E8a, can be included in the mature mRNA, generating four possible LSD1 isoforms named LSD1, LSD1+2a, LSD1+8a, and LSD1+2a+8a (Fig. 2). While the inclusion of exon E2a can occur in all tissues, LSD1 transcripts containing the microexon E8a are mainly found in brain and testis (Zibetti et al., 2010). The insertion of the 60-nucleotide-long exon E2a in the LSD1 transcript occurs upstream of a SWIRM domain, which is a domain often found in chromatin-interacting proteins (Fig. 2). These additional 20 amino-acids within the LSD1 protein are not predicted to modify its enzymatic activity. In contrast, the insertion of the 12-nucleotide-long exon E8a occurs within the catalytic amino-oxidase domain in the final LSD1 transcript (Fig. 2), raising the possibility that LSD1+8a isoforms could exhibit distinct enzymatic activity, compared to the canonical form of LSD1.

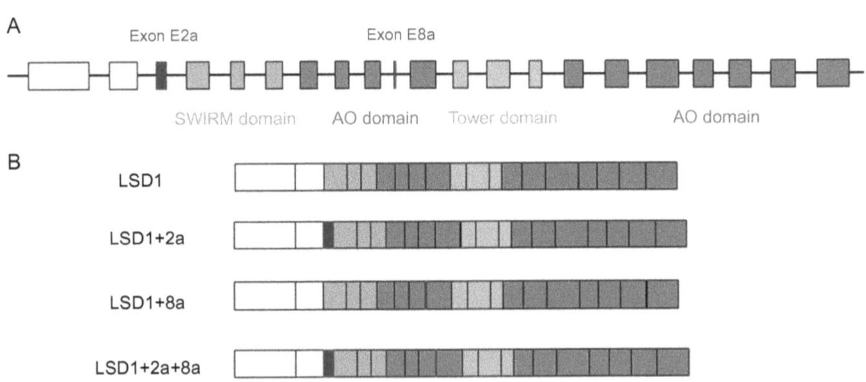

Fig. 2 (A) Schematic representation of the human LSD1 gene with its 19 constitutive exons; alternative exons E2a and E8a are represented in *red*. Different colors indicate LSD1 functional domains: the SWIRM domain and the SWIRM-oxidase connector (*green*) coded by exons 3–5, the amine oxidase (AO) domain (*blue*) coded by exons 6–9 and exons 13–19, the tower domain (*orange*) coded by exons 10–12. (B) Exon structure of the four LSD1 transcript variants. (See the color plate.)

Indeed, our group recently showed that the specific LSD1+8a isoform plays a role in mediating H3K9 demethylation when associated with its SVIL and other yet to be identified cofactors (Laurent et al., 2015). Moreover, the LSD1+8a isoform has also been reported to function as a histone H4K20 demethylase when associated with the cofactor CoREST (Wang et al., 2015). LSD1+8a demethylase activity toward H4K20 is linked to the regulation of transcriptional elongation, while its H3K9 demethylase function is associated with gene activation (Laurent et al., 2015; Wang et al., 2015).

Therefore, by regulating the dynamics of different histone methylation marks, the LSD1 demethylase plays important functions in a number of chromatin-based cellular processes. Further research on LSD1 may require the in vitro characterization of its biochemical activity. This chapter describes the methods that we have used to study LSD1 catalytic activity in vitro. Even though this methodology is focused on LSD1, most of the biochemical advice outlined here is also applicable to the histone demethylase LSD2/KDM1B.

2. EXPRESSION AND PURIFICATION OF LSD1 PROTEINS

2.1 Expression and Purification of LSD1 in Bacteria

For the production of LSD1 recombinant proteins in bacteria, the full-length human LSD1 cDNA [2559 base pairs (bp); GenBank accession number NM_ 015013] was amplified by polymerase chain reaction (PCR) from human cDNA and sequentially cloned into the N-terminal 6× Histidine (HIS)-tag bacterial expression vector pET15b. As the LSD1+2a and LSD1+8a isoforms do not show high expression in mammalian cells, we separately generated PCR fragments containing either the exon E2a (60 bp) or the exon E8a (12 bp), using a LSD1+2a+8a template (GenBank accession number NM_ 001009999) purchased at Open Biosystems (Dharmacon/GE Healthcare). Each PCR fragment was subcloned into the LSD1 pET15b expression vector. The LSD1 catalytic mutants (LSD1 K661A for the LSD1 canonical form) were generated using the QuickChange Site-directed Mutagenesis Kit (Stratagene).

To express LSD1 proteins, the pET15b plasmid was transformed into Rosetta (DE3) competent cells (Invitrogen). 2 mL of Lysogeny Broth media was inoculated with a single colony from a freshly transformed plate as the starting culture, cultured at 37°C for 8 h, and then expanded in a larger culture volume at the same temperature. When the culture reached an OD_{600} of ~0.6, the expression of the recombinant proteins was induced by addition of IPTG (isopropyl β-D-1-thiogalactopyranoside) to a final concentration of 0.2 m*M* and incubated at 16°C overnight. The next day, the cell suspension was collected and spun down for 15 min at 5000 × *g* and 4°C. The cell pellet was resuspended in a cold pH 8 buffer containing 50 m*M* NaH_2PO_4, 300 m*M* NaCl, 10 m*M* imidazole (lysis buffer), supplemented with complete EDTA-free protease inhibitors (Roche). The cell suspension was lysed using a high-pressure homogenizer (Avestin EmulsiFlex-C3) for two to three rounds and then centrifuged for at least 30 min at 21,000 × *g* and 4°C. The supernatant was applied to a column of HisPur Cobalt resin (Thermo Scientific) previously equilibrated with the lysis buffer. The column was extensively washed with the lysis buffer until the Bradford test (Protein Assay Dye Reagent Concentrate; Bio-Rad) with the wash fraction gave no blue color. HIS-tagged LSD1 proteins were then eluted from the resin in several fractions using a buffer containing 50 m*M* NaH_2PO_4, 300 m*M* NaCl, and 250 m*M* imidazole (elution buffer). The fractions were next run on SDS-PAGE gel, and fractions containing the recombinant proteins were identified by Coomassie blue staining. The concentration of the LSD1 proteins was estimated using bovine serum albumin (BSA) as standard. This method routinely provided recombinant LSD1 with a purity of greater than 90% with an average yield of 1–2 mg proteins per liter of culture. Glycerol was added to a final concentration of 10% to enhance the solubility and stability of LSD1 proteins. Aliquots of the recombinant LSD1 proteins were frozen in liquid nitrogen and stored at −80°C. Concentrated proteins (>1 mg/mL) tend to be more stable to freeze–thaw cycles; the method to concentrate LSD1 proteins can be found in Section 2.3.

2.2 Expression and Purification of LSD1 from Sf9 Insect Cells

Bacteria are one of the most widely used hosts for protein expression; however, the use of insect cell culture for protein expression has increased over the last decades to become a common expression system for both basic research and large-scale commercial applications. The key factor to the popularity of this system is the ability of insect cells to produce relatively large quantities of posttranslationally modified proteins in a relatively short period of time.

For production of LSD1 recombinant proteins in insect cells, cDNAs encoding LSD1 and its isoforms were amplified and cloned into a modified pFastBac-HTa (Invitrogen) insect cell expression vector in frame with a HIS-tag at the N-terminus. Constructs were transformed into DH10Bac bacteria (Invitrogen) to generate the bacmid DNAs. The recombinant bacmid DNAs are too large (>135 kbp) to be purified on silica-gel-membrane-based or anion-exchange-resin-based columns from traditional plasmid miniprep kits. The alternative method is to purify the bacmid DNAs using a conventional alkaline lysis followed by isopropanol precipitation. To verify the successful transposition of LSD1 cDNAs into the bacmid DNA, a PCR analysis is performed with pUC/M13 primers. A PCR band of 5000 bp (2400 bp + the size of the LSD1 cDNA) indicates a successful insertion, while a 300 bp band will be indicative of an empty bacmid DNA. Once a correct bacmid DNA is generated, baculovirus can be produced in Sf9 insect cells. The bacmid DNAs should be stored at −20°C, and frequent freeze/thaw cycles should be avoided as it decreases the transfection efficiency.

Sf9 insect cells were maintained at 25°C in Insect-Xpress medium (Lonza) supplemented with 10% heat-inactivated fetal bovine serum, 2 m*M* L-glutamine, and 1% penicillin–streptomycin. Before transfection, 8×10^5 cells/well were seeded in a six-well plate and allowed to attach for 15 min at room temperature. 1 μg of LSD1 bacmid DNA and 8 μL of Cellfectin II (Invitrogen) were separately diluted in 100 μL of un-supplemented Insect-Xpress medium. The diluted bacmid and the diluted Cellfectin II reagent were then mixed, vortexed, and incubated at room temperature for 20 min. The mixture was gently added onto the cells. Cells were incubated for 4 h at 25°C, before replacing the transfection mixture with 2 mL of complete growth medium. After 72 h, the medium was collected and centrifuged at 500 × *g* for 5 min to remove floating cells. The P1 viral stock is a small-scale and low-titer stock. To amplify the P1 viral stock, we seeded 2×10^6 cells/well in a six-well plate, added the totality of the P1 viral stock onto the cells, and supplemented with the medium to have a final volume of 3 mL/well. Cells were incubated for 72 h at 25°C, and the P2 viral stock was collected as per the P1 stock. Once a higher-titer P2 stock is generated, the amplification procedure can be scaled to any volume of choice. Usually, we further amplified the P2 viral stock to generate P3 viral stock of higher titer by seeding 9×10^6 cells/dish in a 10-cm dish, adding the entire P2 baculoviral stock onto the cells, supplementing with the medium to a final volume of 10-mL/dish, and incubating for 72 h at 25°C. Virally infected insect cells display characteristics that can be observed using an inverted phase microscope: increase of nuclei size, cessation of cell growth and granular appearance. Healthy, low passage (5–20) cells with >95% viability are necessary to ensure efficient amplification of the viral stock.

We advise against attempting to amplify the LSD1 baculoviruses indefinitely, as they may acquire deleterious mutations with each passage. Usually, the P5 viral stock is the highest usable passage for LSD1. Viral stocks can be stored at 4°C protected from light for up to 6 months; for long-term storage, viral stocks can be stored at −80°C.

For the purification of LSD1 recombinant proteins, Sf9 insect cells were collected and washed once with ice-cold phosphate buffered saline (PBS). Cell pellet was then resuspended for 30 min at 4°C in a lysis buffer containing 10 m*M* HEPES, pH 7.6, 3 m*M* $MgCl_2$, 300 m*M* KCl, 5% glycerol, 0.5% NP-40, 10 m*M* imidazole, 1 m*M* Na_2VO_4, 20 m*M* NaF, 1 m*M* sodium pyrophosphate, 25 m*M* β-glycerophosphate, and complete EDTA-free protease inhibitors (Roche). The cell lysate was cleared by centrifugation for at least 20 min at 21,000 × *g* and 4°C. While centrifuging the supernatant a second time in the same condition, a column of HisPur Cobalt resin (Thermo Scientific) was equilibrated in the lysis buffer. The column was extensively washed with the lysis buffer until the Bradford test (Protein Assay Dye Reagent Concentrate; Bio-Rad) with the wash fraction gave no blue color. HIS-tagged LSD1 proteins were then eluted from the resin in several fractions using a buffer containing 10 m*M* HEPES, pH 7.6, 3 m*M* $MgCl_2$, 300 m*M* KCl, 10% glycerol, and 250 m*M* imidazole. The elution fractions were next assayed and stored as previously described in Section 2.1. This method routinely provided recombinant LSD1 with a purity of greater than 90% with an average yield of 2 mg proteins per 100×10^6 cells.

2.3 LSD1 Purification for Structure Determination

The crystal structures of LSD1, alone or in complex with its cofactor CoREST, have been determined (Chen et al., 2006; Forneris et al., 2007; Mimasu, Sengoku, Fukuzawa, Umehara, & Yokoyama, 2008; Stavropoulos, Blobel, & Hoelz, 2006; Yang et al., 2006). LSD1 is a molecule with three distinct structural domains (Fig. 2). Structural studies revealed that two of the domains, the N-terminal SWIRM domain and the AOD, pack with each other to form a globular structural core from which the tower domain forms a long stalk of two antiparallel helices (Forneris et al., 2007). The tower domain provides the binding platform for the cofactor CoREST; this interaction is important for LSD1 chromatin accessibility (Shi et al., 2005). The structures of the free and CoREST-bound LSD1 are identical, with only a small difference in the orientation of helices relative to the tower domain. In this section, we describe a method of purification to determine the crystal structure of free LSD1; details related to the purification of the LSD1–CoREST complex can be found in Section 3.2.

Recombinant LSD1 proteins purified from bacteria or Sf9 insect cells can be used to determine the structure of LSD1 by crystallography. The HIS-tag was removed either by adding TEV protease (1 OD_{280} TEV protease per 100 OD_{280} recombinant LSD1 protein) overnight at 4°C for LSD1 proteins purified from insect cells or by adding thrombin (GE Healthcare; 1 unit/100 μg of recombinant LSD1 protein) overnight at room temperature for LSD1 proteins purified from bacteria, into the protein fraction. The cleavage reaction was monitored the next day on SDS-PAGE by Coomassie blue staining.

The cleavage reaction was diluted with size exclusion chromatography (SEC) buffer (20 m*M* Hepes-NaOH, pH 7.5, 150 m*M* NaCl, and 1 m*M* TCEP-HCl buffer) to reduce the imidazole concentration to 25 m*M*. Diluted samples were then injected into a Superdex 200 (10/300 GL) gel filtration column (GE Healthcare) equilibrated with SEC buffer. The volume of the sample loaded should be no more than 4% of the column volume and contain no more than 10 mg protein. If this volume is larger than recommended, the sample should be concentrated before injection, using centrifuge-driven filters (Amicon Ultra centrifugal filters; Millipore) with adequate molecular weight size cutoffs. The peak fractions of pure LSD1 proteins from the gel filtration column were pooled together, and as shown in Fig. 3, the color of the solution is *yellow* (*gray* in the print version). This color is due to the FAD cofactor, and it is informative about the quantity of LSD1 recombinant proteins, as the intensity of the color is proportional to

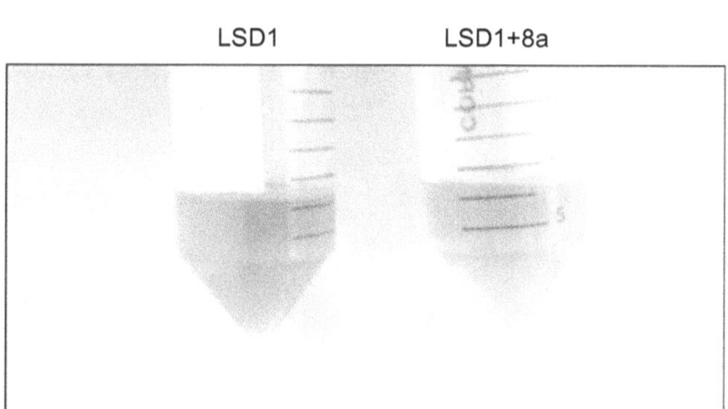

Fig. 3 Elution fractions of LSD1 (*left tube*) and LSD1+8a (*right tube*) proteins after the gel filtration column. The *yellow* (*gray* in the print version) color is due to the flavin adenine dinucleotide (*FAD*) cofactor.

the concentration of LSD1 (Fig. 3). LSD1 proteins can be concentrated to 1–5 mg/mL, using an Amicon Ultra centrifugal filter (Millipore). Care should be taken during centrifugation (at 500 × *g* and 4°C) to avoid local overconcentration and irreversible precipitation or aggregation of the LSD1 proteins on the filtration membrane. Particular care should be taken with LSD1+8a isoforms which are more susceptible to precipitation and aggregation than the conventional LSD1. Aliquots of the recombinant LSD1 proteins should be flash-frozen in liquid nitrogen and stored at −80°C.

2.4 Purification of LSD1 Complex from Mammalian Cells

Biochemical analysis of LSD1 demethylase activity on specific histone marks requires the use of purified LSD1 complexes (see Section 3.3). For the stable expression of tagged LSD1 proteins in mammalian cells, cDNAs encoding LSD1 and its isoforms were amplified and cloned into a pENTR-D-TOPO vector (Invitrogen). The LSD1 pENTR vectors were then recombined into the pHAGE-CMV-Flag-HA Gateway destination vector for lentiviral expression, which introduces Flag-HA tag at the N-terminus of LSD1. LSD1 lentiviruses were produced by cotransfection of the pHAGE LSD1 plasmids with helper vectors (pHDM-VSV-G, pHDM-tat1b, pHDM-HgPM2, and pRC-CMV-RaII) into HEK293T cells. The viral supernatants were collected 60–72 h after transfection, filtered with a 0.45-μm filter, and incubated with cells of our interest. Cells infected with LSD1 lentiviruses were selected 72 h after transduction with 1 μg/mL puromycin.

For the purification of LSD1 complex, 400×10^6 cells were washed once with ice-cold PBS. The cell pellet was then lysed for 30 min at 4°C in a buffer (buffer A) containing 10 m*M* HEPES, pH 7.6, 3 m*M* $MgCl_2$, 300 m*M* KCl, 5% glycerol, 0.5% NP-40, 1 m*M* Na_2VO_4, 20 m*M* NaF, 1 m*M* sodium pyrophosphate, 25 m*M* β-glycerophosphate, and complete EDTA-free protease inhibitors (Roche). The cell lysate was centrifuged for at least 20 min at 21,000 × *g* and 4°C; the supernatant was collected and diluted by half with a buffer (buffer B) containing 10 m*M* HEPES, pH 7.6, 3 m*M* $MgCl_2$, 10 m*M* KCl, 5% glycerol, 0.5% NP-40, 1 m*M* Na_2VO_4, 20 m*M* NaF, 1 m*M* sodium pyrophosphate, 25 m*M* β-glycerophosphate, and complete EDTA-free protease inhibitors (Roche). Flag M2 beads (Sigma) were equilibrated in the buffer B and incubated with the cell lysate for at least 4 h under rotation at 4°C. Beads were then washed two or three times with buffer B before being used for an in vitro demethylase assay (see Section 3.3).

3. IN VITRO ENZYMATIC ASSAYS

3.1 Enzymatic Assay on Peptides and Histones

LSD1 demethylase assays can be performed on various substrates: peptides, histones, or nucleosomes (see Section 3.2 for nucleosomes). For a reaction with peptides, we recommend the use of nonbiotinylated peptides corresponding to residues 1–21 of the histone H3 tail (Anaspec). The enzymatic reaction works on biotinylated peptides, but the biotin group impedes an optimal purification of the peptides for the mass spectrometry analysis (see Section 4.1). For a reaction with histones, we recommend the use of histones from calf thymus (Sigma Aldrich or Santa Cruz Biotechnology). Isolated histones or nucleosomes from any cell culture or tissue sample are probably the best substrates to use in demethylase assays as they display a multitude of PTMs. Recombinant histones with site-specific methylation can also be used as substrates in LSD1 functional assays; however, LSD1 catalytic activity might not be optimal as some PTMs on histone tails can enhance LSD1 substrate recognition, eg, acetylation of H3K9 increases LSD1-binding affinity to H3K4me1/2 (Forneris, Binda, Vanoni, Battaglioli, & Mattevi, 2005).

LSD1 demethylase assays were carried out at 37°C for 4 h with the following demethylation buffer (DM buffer): 50 m*M* Tris, pH 8.5, 50 m*M* KCl, 5 m*M* $MgCl_2$, 0.5% BSA, 5% glycerol, and complete EDTA-free protease inhibitors (Roche), using either histone peptides or calf histones as substrates. For an assay with peptides, the volume of the reaction was 25 μL (up to 50 μL if the volume of LSD1 protein is large), in which 3 μg of modified histone peptides were used as substrates. 5–10 μg of LSD1 recombinant proteins were used in each assay. The reaction was stopped by adding sodium dodecyl sulfate (SDS) to a final concentration of 0.1% and boiling the mixture at 95°C for 5 min. The reaction mixture can be frozen at this step at −20°C before the purification of peptides. For a demethylase assay with histones, the volume of the reaction was 50 μL (up to 100 μL if the volume of LSD1 proteins is large), in which 30 μg of calf histones were used as substrates. Different amounts of LSD1 recombinant proteins ranging from 10 to 20 μg were used in the reaction. The reaction was stopped by adding Laemmli buffer (63 m*M* Tris–HCl, pH 6.8, 10% glycerol, 2% SDS, 0.05% bromophenol blue, 50 m*M* dithiothreitol) and boiling the mixture at 95°C for 5 min. The reaction mixture can be frozen at this step at −20°C before loading on the SDS-PAGE system.

3.2 Enzymatic Assay on Nucleosomes

LSD1 alone demethylates H3K4me1/2 in peptides or bulk histones, but not in nucleosomes (Lee, Wynder, Cooch, & Shiekhattar, 2005; Shi et al., 2005). Only the LSD1–CoREST complex efficiently demethylates H3K4me1/2 in nucleosomal substrates. CoREST stimulates the binding and activity of LSD1 toward nucleosomes by binding to the nucleosomal DNA (Yang et al., 2006). The tower domain of LSD1 provides the binding platform for CoREST. The LSD1-binding domain of CoREST mainly consists of the linker between the two SANT domains; to investigate the enzymatic activity of LSD1 toward nucleosomes, it is necessary to have this CoREST domain in the reaction mixture.

To do so, the LSD1-binding domain of CoREST (residues 305–482) was amplified by PCR from human cDNA and sequentially cloned into the pET26b bacterial expression vector. A stop codon was included at the C-terminus of the CoREST coding sequence to ensure that the protein was not tagged by HIS at the C-terminus. The stoichiometry of the LSD1–CoREST complex is 1:1; the complex can be purified by coexpressing LSD1 and the LSD1-binding domain of CoREST. To do so, the LSD1 pET15b and CoREST pET26b plasmids were transformed into Rosetta (DE3) competent cells (Invitrogen), and colonies were obtained by double ampicillin–kanamycin selection. The LSD1–CoREST complex was purified on HisPur Cobalt resin following the method previously described in Section 2.1. An alternative method is to independently express the LSD1-binding domain of CoREST and to prepare a bacterial lysate to which HIS-tagged LSD1 recombinant proteins are added. The LSD1–CoREST complex is then purified on HisPur Cobalt resin as previously described in Section 2.1.

For both methods, the fractions containing the LSD1–CoREST complex were next identified on SDS-PAGE by Coomassie blue staining, and the concentration of the LSD1 proteins was estimated using BSA as standard. LSD1–CoREST demethylase assays on nucleosomes were then carried out at 37°C for 4 h within the DM buffer (see Section 3.1). The volume of the reaction was 50 μL (up to 100 μL if the volume of LSD1–CoREST proteins is large), in which 10 μg of mono- or polynucleosomes (EpiCypher) were used as substrates. Different amounts of LSD1–CoREST recombinant proteins ranging from 10 to 20 μg were used in the reaction (estimation based on the quantification of the LSD1 proteins). The reaction was stopped by adding Laemmli buffer (see Section 3.1) and boiling the mixture at 95°C for 5 min. The reaction can be frozen at this step at −20°C before loading on the SDS-PAGE system.

3.3 Enzymatic Assays with LSD1 Complex

LSD1+8a isoforms do not have the intrinsic ability to demethylate H3K4me1/2, but the LSD1+8a protein complex purified from neuronal cells exhibits robust H3K9 demethylase activity (Laurent et al., 2015). Therefore, enzymatic assays of LSD1 activity toward specific histone marks, such as H3K9me1/2, require purification of the native complex rather than just the LSD1 recombinant protein. In Section 2.4, we described the isolation of LSD1 complexes on Flag M2 beads (Sigma). These beads were then resuspended in the DM buffer (see Section 3.1) containing 500 μg/mL Flag peptide and incubated in a thermocycler at 37°C for 4 h with a constant shaking speed at 800–900 rpm. Demethylase assays with LSD1 complexes were carried out on calf histones (Sigma Aldrich or Santa Cruz Biotechnology) or nucleosomes (EpiCypher). The volume of the reaction correlated with the volume of Flag M2 beads, following the respective ratio of 3:1. For example, 50 μL of Flag M2 beads was incubated in a final reaction volume of 150 μL. The reaction was stopped by adding Laemmli buffer (see Section 3.1) and boiling the mixture at 95°C for 5 min. The reaction can be frozen at this step at −20°C before loading on SDS-PAGE.

The LSD1+8a isoform is not only a H3K9 histone demethylase, but it also functions as a histone H4K20 demethylase (Wang et al., 2015). This H4K20 demethylase activity requires LSD1+8a association with its cofactor CoREST, meaning that any demethylase assay toward H4K20me1/2 should be performed following the method described in Section 3.2.

4. ANALYSIS OF LSD1 DEMETHYLASE ACTIVITY

4.1 Analysis by Mass Spectrometry

The demethylase activity of LSD1 on peptides was analyzed by mass spectrometry, and it required an additional step of peptide purification. The enzymatic reaction was previously stopped by adding SDS to a final concentration of 0.1% (see Section 3.1). In order to purify the peptides, the reaction mixture was first diluted with three volumes of 0.1% trifluoroacetic acid (TFA) to obtain a final SDS concentration of 0.025%. The peptides were then purified by passing the diluted reaction mixture through a C18 ZipTip (Millipore) that had been preactivated and equilibrated using 20 μL of 70% acetonitrile/0.1% TFA (three times), followed by 20 μL of 50% acetonitrile/0.1% TFA (three times) and finally by 20 μL of 0.1% TFA (10 times). The diluted reaction mixture was then loaded onto the activated ZipTips and passed through between 20 and 30 times. The ZipTips were next washed with 20 μL of 0.1% TFA (three times), and the bound material was eluted from the ZipTip using 2 μL of 70% acetonitrile/0.1% TFA containing 1 mg/mL α-cyano-4-hydroxycinnamic acid (matrix). The eluates were spotted onto a circle of open target areas to allow solvent evaporation and peptide/matrix cocrystallization. The samples were finally analyzed by matrix-assisted laser desorption/ionization (MALDI) mass spectroscopy, using a MALDI-TOF/TOF mass spectrometer.

The analysis of LSD1 enzymatic activity by mass spectrometry presents advantages compared to an analysis by Western blot (see Section 4.2). Mass spectrometry enables a rapid sample analysis, with high sensitivity and very accurate mass measurements. This technology is excellent for peptides and proteins, and it is particularly well suited for the quantification of enzymatic reactions. As an example for this methodology, we describe the analysis of LSD1 enzymatic activity toward H3K4me2 peptides. By mass spectrometry, demethylation of H3K4me2 to H3K4me1 will lead to a shift of 14 daltons (Da) in the spectrum, while a demethylation to H3K4me0 will correspond to a loss of 28 Da for the peptide. We performed a demethylation reaction on H3K4me2 peptides using LSD1 recombinant proteins purified from bacteria or Sf9 insect cells, to compare their respective demethylase efficiencies. As shown in Fig. 4, the mass spectrometry analysis demonstrated that LSD1 proteins purified from bacteria led to mono- and di-demethylation of the H3K4me2 peptides, with only 35% of the peptides showing a total demethylation to H3K4me0 (Fig. 4). When a similar assay was performed using the same amount of LSD1 proteins purified from insect cells, the totality of the H3K4me2 peptides were demethylated to H3K4me0. This observation highlights the high sensitivity of the mass spectrometry and the advantage of using this technology to quantify the enzymatic activity of LSD1 proteins. It further suggests that LSD1 proteins purified from insect cells are catalytically more active than those purified from bacteria, probably due to the presence of PTMs on LSD1 proteins in Sf9 cells.

4.2 Analysis by Western Blot

Although the enzymatic activity of LSD1 proteins on histones and nucleosomes can be analyzed by mass spectrometry, it is more commonly detected by SDS-PAGE/Western blotting using methyl-specific antibodies. Demethylase assays are typically performed on 30 μg of calf histones or 10 μg of nucleosomes (see Sections 3.1 and 3.2).

In order to easily detect LSD1 demethylase activity by Western blot, 10% of the mixture reaction should be loaded on a 15% polyacrylamide running

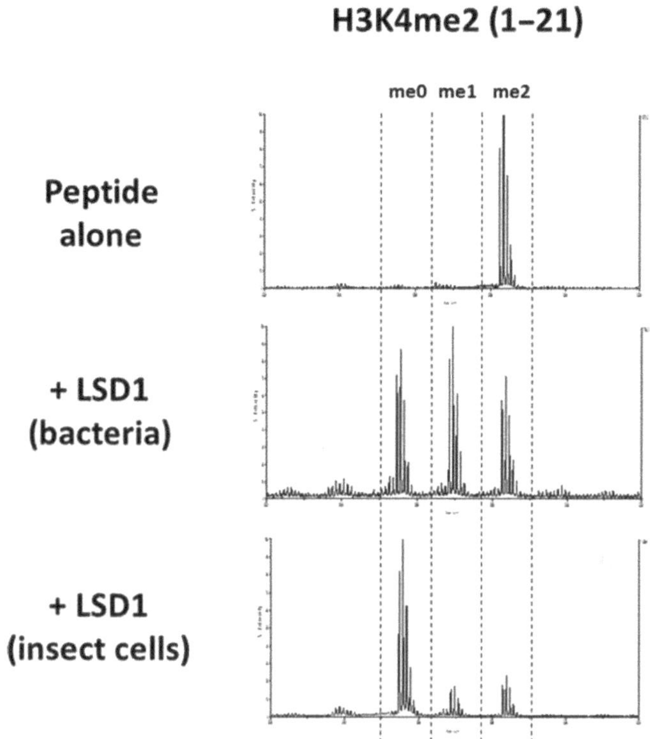

Fig. 4 Histone demethylase activity of LSD1 proteins purified from bacteria and Sf9 insect cells, on H3K4me2 peptides (residues 1–21 of histone H3 tail). Activity was examined by MALDI-TOF mass spectrometry. *Adapted from Laurent, B., Ruitu, L., Murn, J., Hempel, K., Ferrao, R., Xiang, Y., … Shi, Y. (2015). A specific LSD1/KDM1A isoform regulates neuronal differentiation through H3K9 demethylation.* Molecular Cell, *57(6), 957–970. doi: 10.1016/j.molcel.2015.01.010.*

gel. We recommend the use of these methyl-specific antibodies to further perform the analysis: H3K4me2 (07-030; Millipore), H3K4me1 (ab8895; Abcam), H3K9me2 (ab1220; Abcam), H3K9me1 (ab8896; Abcam), H3 (ab1791; Abcam), H4K20 me2 (ab9052; Abcam), H4K20me1 (ab9051; Abcam), H4 (39269; Active Motif).

4.3 Additional Methods for Detecting LSD1 Activity

Additional methods have been developed to indirectly assay the LSD1 enzymatic activity. These assays are based on the events related to the demethylation reaction, ie, the release of formaldehyde and hydrogen peroxide (Fig. 1).

Our lab originally used a formaldehyde dehydrogenase (FDH)-coupled assay to investigate LSD1 enzymatic activity (Shi et al., 2004). The formaldehyde is a by-product of LSD1-mediated demethylation that can be detected by coupling a FDH assay to the enzymatic reaction. The formaldehyde produced during the demethylation-FDH-coupled reaction is oxidized to formic acid by the FDH, along with the reduction of nicotinamide adenine dinucleotide (NAD); NADH is monitored using a spectrophotometer at 340 nm. Importantly, the detection of formaldehyde as a readout for demethylase activity has successfully been used in an unbiased biochemical screen to identify additional histone demethylases (Tsukada et al., 2006). An alternative method developed in our lab used the conversion of the formaldehyde to formaldemethone (FDM) by the addition of dimedone in the assay (Shi et al., 2004). The FDM produced during the demethylation-dimedone-coupled reaction is detected by a high-pressure liquid chromatography system directly coupled to a mass spectrometer.

An alternative assay developed to quantify LSD1 enzymatic activity detects the production of hydrogen peroxide (H_2O_2) by a horseradish peroxidase-coupled demethylation reaction. In the presence of peroxidase and H_2O_2, 4-aminoantipyrine and 3,5-dichloro-2-hydroxybenzenesulfonic acid are oxidized to produce a red chromophore whose concentration is measured at 515 nm (Forneris, Binda, Vanoni, Mattevi, & Battaglioli, 2005).

4.4 In vivo Analysis

The best manner to address LSD1 enzymatic functions in vivo is to abrogate its expression by knockdown (KD) with shRNA or by knockout (KO) with the CRISPR/Cas9 technology. However, many investigations on LSD1 also made use of its chemical inhibition. As the catalytic domain of LSD1 shares high similarities with other members of the amine oxidase family, monoamine oxidases (MAO) inhibitors, such as pargyline and tranylcypromine (TCP, also known as 2-phenylcyclopropylamine or 2-PCPA), can be used to inhibit LSD1. However, these inhibitors induce toxicity in vivo (eg, TCP is an irreversible inhibitor forming a covalent adduct with FAD), confounding the interpretation of the in vivo functions of LSD1. As the selectivity of MAO inhibitors for LSD1 are also very low, new TCP-analogues were recently developed, and they show more specific LSD1 inhibitory activity, eg, LSD1 Inhibitor II, S2101 (Calbiochem) and LSD1 Inhibitor IV, RN-1 (Calbiochem). Another class of LSD1 inhibitors was developed to mimic the peptide structure of LSD1 substrates, such as histone H3 tail. These inhibitors are potent and reversible substrate competitive inhibitors for LSD1, eg, LSD1 Inhibitor III, CBB1007 (Calbiochem).

LSD1 is overexpressed in several cancer types and interestingly, all these inhibitors proved successful in suppressing tumor growth, hinting that LSD1 inhibition may have therapeutic potential for treating cancer.

Inhibiting LSD1 expression (by KD or KO) or activity (by the use of pharmacological inhibitors) in mammalian cells does not lead to a detectable change in H3K4 or H3K9 methylation levels by Western blotting or immunofluorescence, as LSD1 loss is apparently compensated by other histone demethylases (data not shown). Only a slight increase of the global H4K20 methylation level can be observed in LSD1+8a KO neurons (Wang et al., 2015). However, KD/KO or chemical inhibition of LSD1 leads to localized epigenetic changes at specific target loci, which can be investigated by performing chromatin immunoprecipitation followed by DNA sequencing (ChIP-seq) or quantitative PCR (ChIP-qPCR), on specific histone marks. LSD1 and its isoforms regulate local changes of H3K4, H3K9, and H4K20 methylation at target promoters or enhancers (Laurent et al., 2015; Wang et al., 2015; Whyte et al., 2012), and broader changes of H4K20 methylation at transcribed coding regions (Wang et al., 2015).

LSD1 enzymatic activity can also be assessed in vitro and in vivo on substrates other than histones, as LSD1 was also described to remove methyl group on nonhistones proteins, such as p53, STAT3, ERα, or DNMT1 (Huang et al., 2007; Nicholson & Chen, 2009).

It is a major challenge to understand how LSD1 fulfills its multiple roles in many chromatin-based cellular processes. Tackling this challenging problem requires the characterization of the biochemical functions of LSD1. We hope that the protocols provided here, on the purification and functional analysis of LSD1 proteins from various sources, will help further analyses to understand the biology of this enzyme, and whether LSD1 enzymatic activity can be modulated for therapeutic benefit in cancer and other diseases.

ACKNOWLEDGMENTS

We thank Dr Jamie Anastas, Dr Aimee Badeaux, Dr Ryan Ferrao, and Dr Ashwini Jambhekar for helpful discussions and critical reading of the manuscript. In particular, we wish to thank James Lee (Dana Farber Cancer Institute, Boston, MA) for help with the mass spectrometer and Dr Ryan Ferrao (Boston Children's Hospital, Boston, MA) for help with the gel filtration column. This work was supported by a Grant from the NIH to Y.S. (R01 CA118487). Y.S. is an American Cancer Society Research Professor.

Conflict of Interest Statement: Y.S. is a cofounder of Constellation Pharmaceuticals, Inc. and is a member of its scientific advisory board. Y.S. is also a consultant for Active Motif, Inc.

REFERENCES

Allis, C. D., Berger, S. L., Cote, J., Dent, S., Jenuwien, T., Kouzarides, T., … Zhang, Y. (2007). New nomenclature for chromatin-modifying enzymes. *Cell*, *131*(4), 633–636. http://dx.doi.org/10.1016/j.cell.2007.10.039.

Ballas, N., Battaglioli, E., Atouf, F., Andres, M. E., Chenoweth, J., Anderson, M. E., … Mandel, G. (2001). Regulation of neuronal traits by a novel transcriptional complex. *Neuron*, *31*(3), 353–365.

Chen, Y., Yang, Y., Wang, F., Wan, K., Yamane, K., Zhang, Y., & Lei, M. (2006). Crystal structure of human histone lysine-specific demethylase 1 (LSD1). *Proceedings of the National Academy of Sciences of the United States of America*, *103*(38), 13956–13961. http://dx.doi.org/10.1073/pnas.0606381103.

Forneris, F., Binda, C., Adamo, A., Battaglioli, E., & Mattevi, A. (2007). Structural basis of LSD1-CoREST selectivity in histone H3 recognition. *The Journal of Biological Chemistry*, *282*(28), 20070–20074. http://dx.doi.org/10.1074/jbc.C700100200.

Forneris, F., Binda, C., Dall'Aglio, A., Fraaije, M. W., Battaglioli, E., & Mattevi, A. (2006). A highly specific mechanism of histone H3-K4 recognition by histone demethylase LSD1. *The Journal of Biological Chemistry*, *281*(46), 35289–35295. http://dx.doi.org/10.1074/jbc.M607411200.

Forneris, F., Binda, C., Vanoni, M. A., Battaglioli, E., & Mattevi, A. (2005a). Human histone demethylase LSD1 reads the histone code. *The Journal of Biological Chemistry*, *280*(50), 41360–41365. http://dx.doi.org/10.1074/jbc.M509549200.

Forneris, F., Binda, C., Vanoni, M. A., Mattevi, A., & Battaglioli, E. (2005b). Histone demethylation catalysed by LSD1 is a flavin-dependent oxidative process. *FEBS Letters*, *579*(10), 2203–2207. http://dx.doi.org/10.1016/j.febslet.2005.03.015.

Garcia-Bassets, I., Kwon, Y. S., Telese, F., Prefontaine, G. G., Hutt, K. R., Cheng, C. S., … Rosenfeld, M. G. (2007). Histone methylation-dependent mechanisms impose ligand dependency for gene activation by nuclear receptors. *Cell*, *128*(3), 505–518. http://dx.doi.org/10.1016/j.cell.2006.12.038.

Hou, H., & Yu, H. (2010). Structural insights into histone lysine demethylation. *Current Opinion in Structural Biology*, *20*(6), 739–748. http://dx.doi.org/10.1016/j.sbi.2010.09.006.

Huang, J., Sengupta, R., Espejo, A. B., Lee, M. G., Dorsey, J. A., Richter, M., … Berger, S. L. (2007). p53 is regulated by the lysine demethylase LSD1. *Nature*, *449*(7158), 105–108. http://dx.doi.org/10.1038/nature06092.

Jenuwein, T., & Allis, C. D. (2001). Translating the histone code. *Science*, *293*(5532), 1074–1080. http://dx.doi.org/10.1126/science.1063127.

Laurent, B., Ruitu, L., Murn, J., Hempel, K., Ferrao, R., Xiang, Y., … Shi, Y. (2015). A specific LSD1/KDM1A isoform regulates neuronal differentiation through H3K9 demethylation. *Molecular Cell*, *57*(6), 957–970. http://dx.doi.org/10.1016/j.molcel.2015.01.010.

Lee, M. G., Wynder, C., Cooch, N., & Shiekhattar, R. (2005). An essential role for CoREST in nucleosomal histone 3 lysine 4 demethylation. *Nature*, *437*(7057), 432–435. http://dx.doi.org/10.1038/nature04021.

Metzger, E., Wissmann, M., Yin, N., Müller, J. M., Schneider, R., Peters, A. H., … Schüle, R. (2005). LSD1 demethylates repressive histone marks to promote androgen-receptor-dependent transcription. *Nature*, *437*(7057), 436–439. http://dx.doi.org/10.1038/nature04020.

Mimasu, S., Sengoku, T., Fukuzawa, S., Umehara, T., & Yokoyama, S. (2008). Crystal structure of histone demethylase LSD1 and tranylcypromine at 2.25 A. *Biochemical and Biophysical Research Communications*, *366*(1), 15–22. http://dx.doi.org/10.1016/j.bbrc.2007.11.066.

Nair, S. S., Nair, B. C., Cortez, V., Chakravarty, D., Metzger, E., Schüle, R., ... Vadlamudi, R. K. (2010). PELP1 is a reader of histone H3 methylation that facilitates oestrogen receptor-alpha target gene activation by regulating lysine demethylase 1 specificity. *EMBO Reports*, *11*(6), 438–444. http://dx.doi.org/10.1038/embor.2010.62.

Nicholson, T. B., & Chen, T. (2009). LSD1 demethylates histone and non-histone proteins. *Epigenetics*, *4*(3), 129–132.

Pedersen, M. T., & Helin, K. (2010). Histone demethylases in development and disease. *Trends in Cell Biology*, *20*(11), 662–671. http://dx.doi.org/10.1016/j.tcb.2010.08.011.

Perillo, B., Ombra, M. N., Bertoni, A., Cuozzo, C., Sacchetti, S., Sasso, A., ... Avvedimento, E. V. (2008). DNA oxidation as triggered by H3K9me2 demethylation drives estrogen-induced gene expression. *Science*, *319*(5860), 202–206. http://dx.doi.org/10.1126/science.1147674.

Shi, Y., Lan, F., Matson, C., Mulligan, P., Whetstine, J. R., Cole, P. A., & Casero, R. A. (2004). Histone demethylation mediated by the nuclear amine oxidase homolog LSD1. *Cell*, *119*(7), 941–953. http://dx.doi.org/10.1016/j.cell.2004.12.012.

Shi, Y. J., Matson, C., Lan, F., Iwase, S., Baba, T., & Shi, Y. (2005). Regulation of LSD1 histone demethylase activity by its associated factors. *Molecular Cell*, *19*(6), 857–864. http://dx.doi.org/10.1016/j.molcel.2005.08.027.

Stavropoulos, P., Blobel, G., & Hoelz, A. (2006). Crystal structure and mechanism of human lysine-specific demethylase-1. *Nature Structural and Molecular Biology*, *13*(7), 626–632. http://dx.doi.org/10.1038/nsmb1113.

Strahl, B. D., & Allis, C. D. (2000). The language of covalent histone modifications. *Nature*, *403*(6765), 41–45. http://dx.doi.org/10.1038/47412.

Tsukada, Y., Fang, J., Erdjument-Bromage, H., Warren, M. E., Borchers, C. H., Tempst, P., & Zhang, Y. (2006). Histone demethylation by a family of JmjC domain-containing proteins. *Nature*, *439*(7078), 811–816. http://dx.doi.org/10.1038/nature04433.

Wang, J., Telese, F., Tan, Y., Li, W., Jin, C., He, X., ... Rosenfeld, M. G. (2015). LSD1n is an H4K20 demethylase regulating memory formation via transcriptional elongation control. *Nature Neuroscience*, *18*(9), 1256–1264. http://dx.doi.org/10.1038/nn.4069.

Whyte, W. A., Bilodeau, S., Orlando, D. A., Hoke, H. A., Frampton, G. M., Foster, C. T., ... Young, R. A. (2012). Enhancer decommissioning by LSD1 during embryonic stem cell differentiation. *Nature*, *482*(7384), 221–225. http://dx.doi.org/10.1038/nature10805.

Yang, M., Gocke, C. B., Luo, X., Borek, D., Tomchick, D. R., Machius, M., ... Yu, H. (2006). Structural basis for CoREST-dependent demethylation of nucleosomes by the human LSD1 histone demethylase. *Molecular Cell*, *23*(3), 377–387. http://dx.doi.org/10.1016/j.molcel.2006.07.012.

Zibetti, C., Adamo, A., Binda, C., Forneris, F., Toffolo, E., Verpelli, C., ... Battaglioli, E. (2010). Alternative splicing of the histone demethylase LSD1/KDM1 contributes to the modulation of neurite morphogenesis in the mammalian nervous system. *Journal of Neuroscience*, *30*(7), 2521–2532. http://dx.doi.org/10.1523/JNEUROSCI.5500-09.2010.

CHAPTER ELEVEN

LSD1 Histone Demethylase Assays and Inhibition

D. Hayward, P.A. Cole[1]
Johns Hopkins School of Medicine, Baltimore, MD, United States
[1]Corresponding author: e-mail address: pcole@jhmi.edu

Contents

1. Introduction and Background 262
2. LSD1 Assays 264
   2.1 Coupled Enzyme Assays 265
   2.2 Radiolabeled Assays 266
   2.3 Antibody-Based Assays 266
   2.4 Mass Spectrometry-Based Assays 267
3. Inhibitors of LSD1 268
4. Applications of LSD1 Inhibitors 272
   4.1 Bizine 273
   4.2 NCL1 273
   4.3 GSK2879552 274
5. Summary and Future Directions 274
Acknowledgments 275
References 275

Abstract

The lysine-specific demethylase (LSD1) is a flavin-dependent amine oxidase that selectively removes one or two methyl groups from histone H3 at the Lys4 position. Along with histone deacetylases 1 and 2, LSD1 is involved in epigenetically silencing gene expression. LSD1 has been implicated as a potential therapeutic target in cancer and other diseases. In this chapter, we discuss several approaches to measure LSD1 demethylase activity and their relative strengths and limitations for inhibitor discovery and mechanistic characterization. In addition, we review the principal established chemical functional groups derived from monoamine oxidase inhibitors that have been investigated in the context of LSD1 as demethylase inhibitors. Finally, we highlight a few examples of recently developed LSD1 mechanism-based inactivators and their biomedical applications.

Methods in Enzymology, Volume 573
ISSN 0076-6879
http://dx.doi.org/10.1016/bs.mie.2016.01.020

1. INTRODUCTION AND BACKGROUND

The histone code features reversible Lys modifications as a major mechanism for the regulation of chromatin accessibility, gene expression, and cellular growth. Lys side chain acetylation and methylation are considered the dominant and best-studied PTMs in histones. Lys acetylation is regulated by histone acetyltransferases (HATs or KATs) and histone deacetylases (HDACs or KDACs), whereas Lys methylation is controlled by histone Lys methyltransferases (HMTs or KMTs) and histone demethylases (KDMs) (Cole, 2008). Whereas acetylation of the Lys side chain only occurs once per Lys residue, Lys methylation can occur as mono-, di-, and trimethylation forms. Until the report of LSD1 (lysine-specific demethylase 1) in 2004, there was some uncertainty as to whether protein Lys methylation was reversible (Shi et al., 2004). It is now generally accepted that there are at least 18 Lys demethylases, including two flavoenzymes LSD1 (KDM1A) and LSD2 (KDM1B) and the rest being nonheme iron, α-ketoglutarate-dependent JMJ oxygenases (Culhane & Cole, 2007; Thinnes et al., 2014). Common features among the histone demethylases are that they utilize molecular oxygen, catalyze oxidative demethylation, and produce formaldehyde as a by-product (Culhane & Cole, 2007).

LSD1, and its less well-studied paralog LSD2, is members of the amine oxidase enzyme family that depend on a flavin cofactor (Hou & Yu, 2010). This family includes monoamine oxidases that act to metabolize norepinephrine and related neurotransmitters and polyamine oxidases that metabolize spermidine, spermine, and other alkylamines (Edmondson, Mattevi, Binda, Li, & Hubalek, 2004). Although the precise chemical details of oxidation by amine oxidases are still being debated, functionally the reactions can be viewed as involving hydride transfer between the substrate nitrogen and the flavin cofactor (Culhane & Cole, 2007). Consequently, LSD1 and LSD2, which catalyze demethylation reactions on mono- and dimethyl Lys substrates, are incapable of demethylating trimethyl-Lys substrates because of their lack of an available electron lone pair. This contrasts the JMJ demethylase enzymes that typically process trimethyl-Lys substrates since they directly oxidize methyl groups (Hou & Yu, 2010). Upon LSD1-mediated hydride transfer, the corresponding unstable imine intermediate likely spontaneously hydrolyzes to formaldehyde and the demethylated amine (Fig. 1). In order for there to be multiple catalytic turnovers, the

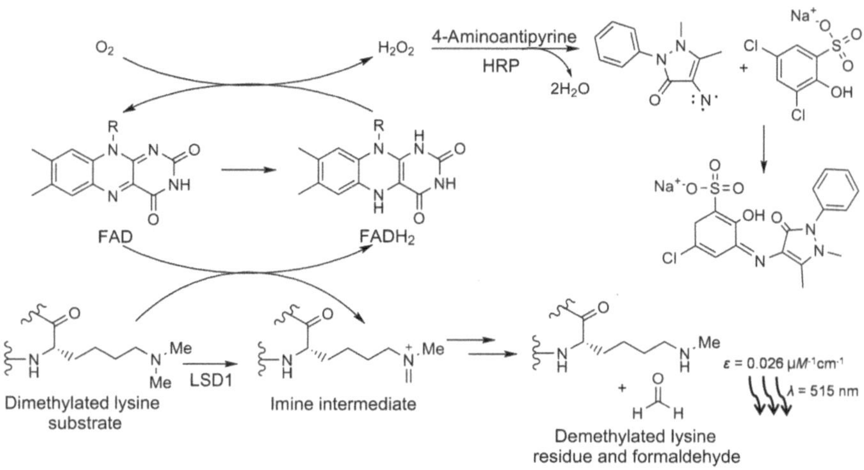

Fig. 1 Hydrogen peroxide (HOOH) detection assay for LSD1. When a dimethylated lysine substrate (*bottom left*) is demethylated by LSD1, the FAD cofactor is reduced to $FADH_2$ (*center*). The cofactor is reoxidized by molecular oxygen to form HOOH (*top center*). Horseradish peroxidase (HRP) uses this to couple 4-aminoantipyrine and dichlorohydroxybenzenesulfonate (*top right*) to form the chromophore (*center right*) that is detected at 515 nm (*bottom right*).

reduced flavin must be reoxidized, and this involves reaction with molecular oxygen, extracted out of the aerobic environment, leading to stoichiometric hydrogen peroxide as a by-product.

In contrast to many histone modifying enzymes that are often pleiotropic with regard to site specificity, LSD1 is highly selective for demethylation of the Lys4 site in histone H3, avoiding thousands of other methylated lysines in the cell (Forneris et al., 2006). Since H3K4 methylation is a mark of gene activation, LSD1 is most often associated with effecting gene silencing (Shi et al., 2004). In this way, LSD1 can act in concert with HDACs that are also commonly considered to serve as turn off switches for gene expression. In fact, LSD1 and HDAC1 (or its highly similar paralog HDAC2) are subunits of the well-studied CoREST repressor complex, with both enzymes interacting with the two different SANT domains of CoREST1 (Lee, Wynder, Cooch, & Shiekhattar, 2005; Yang et al., 2006). LSD1 is a 90 kDa protein that contains a SWIRM domain for targeting nucleosomes and an insert in its amine oxidase domain known as a "tower domain" that engages the SANT2 domain of CoREST (Yang et al., 2006). LSD2 lacks this tower domain, does not complex with CoREST, and has distinct biological roles (Fang et al., 2012). Recent biochemical and structural studies on the

LSD1/CoREST heterodimer suggest the multilayer interactions of the demethylase with its native nucleosome substrate that include an interaction between one section of the SANT2 domain and nucleosomal DNA (Kim, Chatterjee, Jennings, Bartholomew, & Tan, 2015; Pilotto et al., 2015). Using purified LSD1, it is established that a minimum of about 20 residues of the histone H3 tail are necessary for efficient LSD1-catalyzed demethylation (Forneris, Binda, Antonietta Vanoni, Battaglioli, & Mattevi, 2005; Forneris et al., 2006). However, several different crystal structure complexes of LSD1 bound to peptides substrate analogs have offered alternative models on how such substrate recognition occurs (Forneris, Binda, Adamo, Battaglioli, & Mattevi, 2007; Yang et al., 2007).

There have been a smattering of reports that LSD1 can act on sites beyond methyl-H3K4, the most commonly discussed site being Lys9 of histone H3 (Laurent et al., 2015; Nicholson & Chen, 2009). However, these reports on targeting H3K9Me are typically based on cell-based experiments where the role of LSD1 on H3K9 demethylation may be indirect and noncatalytic. Compounding the assignment of site specificity is that antibodies for H3K9Me have been reported as lacking in reliability (Hattori et al., 2013). Although a recent report suggested that a splice variant of LSD1 may be capable of histone H3K9Me demethylation (Laurent et al., 2015), this possibility was previously rendered unlikely by an earlier enzymologic study with the purified LSD1 splice variant protein that showed similar catalytic activity and site specificity to the standard LSD1 form (Toffolo et al., 2014; Zibetti et al., 2010).

A number of lines of evidence point to LSD1 as having important functions in biology and disease. Knockout of LSD1 is embryonically lethal, and LSD1 is particularly associated with neuronal development (Zibetti et al., 2010). LSD1 may participate in mechanisms of neurologic disease and viral pathogenesis (Hill et al., 2014; Liang, Vogel, Narayanan, Peng, & Kristie, 2009; Rusconi et al., 2015; Sun et al., 2010; Wang, Telese, et al., 2015). LSD1 levels have been found to be elevated in many cancers (Wu, Hu, Du, Kong, & Pan, 2015). Consequently, there has been an intense interest in the development of LSD1 inhibitors.

2. LSD1 ASSAYS

Kinetic analysis of the LSD1 demethylase reaction is important both for understanding the mechanistic features of the reaction and for the screening and analysis of LSD1 inhibitors. Several in vitro biochemical LSD1 enzymatic assays have been developed including HOOH detection through a peroxidase-coupled assay, formaldehyde detection through a formaldehyde dehydrogenase assay, radioactive measurements to monitor changes in a radiolabeled peptide substrate, and mass spectrometry analysis to measure changes in the unlabeled substrate (Blair et al., 2011; Forneris et al., 2005; Kokura, Sun, & Fang, 2015). In addition, cellular assays have been adapted to quantify LSD1-mediated effects on histone H3K4 methylation using antibody detection (Schmitt et al., 2014).

2.1 Coupled Enzyme Assays

Mattevi and colleagues described the use of horseradish peroxidase-coupled assays for accurately measuring LSD1 activity with methylated H3 tail peptide substrates (Forneris et al., 2005). This approach uses well-established methodologies that have been applied decades ago to the study of monoamine oxidases. HOOH is an obligate side product of LSD1 catalyzed demethylation and a substrate for horseradish peroxidase. In the presence of 4-aminoantipyrine (4-AP) along with 3,5-dichloro-2-hydroxybenzenesulfonate (DHBS), horseradish peroxidase converts HOOH into a spectrophotometric absorbance change that can be monitored at 515 nm with a sizable extinction coefficient of 26,000 M^{-1} cm^{-1} (Forneris et al., 2005; Fig. 1). This is a relatively sensitive and reliable assay that can be done continuously. A more sensitive version of this assay replaces 4-AP/DHBS with Amplex red, which reacts with HOOH to generate the fluorophore resorufin that has fluorescence emission detected at 585 nm (Schmitt et al., 2014).

The other major side product of the LSD1 histone demethylase reaction is formaldehyde. Formaldehyde release can be monitored via its oxidation to formate using formaldehyde dehydrogenase where cofactor NADH formation is measured spectrophotometrically at 340 nm (Kokura et al., 2015). Both the horseradish peroxidase and formate dehydrogenase assays are relatively low cost and can be done continuously to measure LSD1 activity in real time. However, formaldehyde dehydrogenase is somewhat more complex as a coupling enzyme than horseradish peroxidase and the UV change for NAD reduction is less sensitive than HOOH detection and so the horseradish peroxidase assay is typically preferred. Because these coupled enzyme assays provide real-time kinetic information, they are particularly powerful for analyzing time-dependent inactivation, a common phenomenon of LSD1 inhibitors. Disadvantages of both the horseradish peroxidase and formaldehyde dehydrogenase coupled assays are that they are indirect and changes in reaction conditions or addition of inhibitors can be misinterpreted.

2.2 Radiolabeled Assays

Radioactive LSD1 assays typically involve the use of radiolabeled peptide substrates, usually tritiated methyl groups attached to the H3K4 side chain and scintillation counting is applied to quantify the radioactivity in the substrate and/or formaldehyde product. One example relies on lysine methyltransferases and tritiated *S*-adenosylmethionine (SAM) as the methyl donor to attach a radiolabeled methyl group to a synthetic histone H3 tail peptide (Kokura et al., 2015). Subsequently the radiolabeled peptide is reacted with LSD1 to afford tritiated formaldehyde, which is then converted into diacetyldihydrolutidine, extracted with pentanol and the radioactive material quantified by scintillation counting. A complementary method employs tritiated SAM and the KMT7 methyltransferase to attach a radiolabeled methyl group after LSD1 demethylation of an unlabeled methyl-substrate peptide/product mixture, to "remethylate" the demethylated site (Yu et al., 2014). These assays also allow for kinetic analysis of LSD1 and can measure both substrate and product levels concurrently. However, they are technically demanding, expensive, and discontinuous. They are not readily applicable for high-throughput assays but can be used to validate findings with the spectrophotometric coupled assays described earlier.

2.3 Antibody-Based Assays

Antibody-based assays can be used to analyze LSD1 activity on peptide and histone protein substrates. A recent version of this assay involves a heterogeneous format where biotinylated peptide substrates are incubated with LSD1 and then bound to a streptavidin microplate using primary antibodies to the site of interest and secondary antibodies conjugated to europium in an assay called DELFIA (dissociation-enhanced lanthanide fluorescent immunoassay) (Schmitt et al., 2014). An advantage of this assay is the use of the biotinylated peptide, which allows for washout of all other assay components before treatment with antibody. A related assay which uses the europium antibody is called LANCE Ultra where FRET occurs between the donor europium-labeled primary antibody to the site of interest and the acceptor ULight (Gale & Yan, 2015). Here, the peptide is N-terminally bound by primary antibody and FRET occurs between the europium and ULight at the C-terminal end. An Alpha Screen that is analogous to LANCE Ultra has also been reported (Gale & Yan, 2015). Library screening can be pursued with the DELFIA assay since no fluorescence or quenching of the compounds is likely to occur. However, these methods have somewhat limited quantitative robustness and depend heavily on the reliability of the Ab reagents.

2.4 Mass Spectrometry-Based Assays

Mass spectrometry has been used for many years to discover new posttranslational modifications and is frequently employed to measure demethylation in peptide and histone substrates (Huang, Lin, Garcia, & Zhao, 2015). Quantifying the four states of lysine modification: unmodified, mono-, di-, and trimethylation can be a challenge. Several methods have been developed to address the methyl-Lys population states. One of these approaches involves the reductive methylation with heavy (dideuterio) formaldehyde to convert all states to the same chemical species such that there are no intrinsic ionization differences but the mass differences are detectable (Blair et al., 2011; Blair, Avaritt, & Tackett, 2012). This method, called MassSQUIRM, which uses MALDI-TOF (matrix-assisted laser desorption ionization-time-of-flight) is a highly quantitative technique and has been used to assess LSD1 inhibition (Blair et al., 2011).

A second method for analyzing multiple methylation states uses a flow-injection-based time-of-flight mass spectrometry involving electrospray ionization. The multiply charged species of each type of modification can be separated to accurately determine abundance of each peptide form (Sakane, Ohta, & Shidoji, 2015). A third mass spectrometry method for analyzing demethylase activity applies SAMDI (self-assembled monolayer desorption/ionization) that involves biotinylated peptide absorption onto a neutravidin monolayer surface followed by release using MALDI as the ionization method and TOF–TOF as the separation method (Wigle et al., 2015). SAMDI has been used in a high-throughput screen to identify LSD1 inhibitors. Another mass spectrometry-based assay to analyze demethylation is RapidFire that uses tagged peptides, which are then pooled (Gale & Yan, 2015). Mass spectrometry-based methods are sensitive, can be made label free, and require very little amounts of enzyme and substrate. However, they are intrinsically technically demanding and expensive in requiring specialized instrumentation.

Beyond peptide substrates, some of these LSD1 approaches are applicable to more complex substrates and conditions including nucleosome demethylation and cell-based assays. For example, the formaldehyde release method using radioactive SAM incorporation can be performed on nucleosomes or histone octamers (Kokura et al., 2015). Antibody-based assays are now validated to assess LSD1 activity in cells, with the monitoring of demethylation of histone H3 being particularly diagnostic.

3. INHIBITORS OF LSD1

The first reports of LSD1 inhibitors appeared within about 18 months of the enzyme's functional identification as a histone demethylase (Culhane et al., 2006; Lee, Wynder, Schmidt, McCafferty, & Shiekhattar, 2006; Schmidt & McCafferty, 2007; Szewczuk et al., 2007). Experiments with peptide substrate analogs and established monoamine oxidase inhibitor antidepressant drugs led to the identification of several LSD1 suicide inactivating compounds (Culhane et al., 2006; Lee et al., 2006; Paech, Salach, & Singer, 1980). Propargyl-Lys-containing peptide substrate analogs, inspired by a class of MAO inhibitors such as pargyline, are potent LSD1 inhibitors and display characteristic time-dependent mechanism-based inactivation (Culhane et al., 2006).

Mechanism-based enzyme inactivators, also known as suicide inhibitors, are latently reactive molecules that exploit the target enzymes normal chemistry to unmask a warhead that can covalently modify the active sites of these protein catalysts, leading to irreversible inhibition (Silverman, 1995; Walsh, 1984). These inhibitors are characterized using the Kitz–Wilson equation and potencies most precisely expressed as a ratio of k_{inact} over $K_{i(inact)}$ inhibitory constants (Kitz & Wilson, 1962; Walsh, 1984). The k_{inact} is the first-order rate constant of enzyme inactivation at saturating inhibitor concentration and is analogous to the k_{cat} for a standard enzyme reaction catalyzing conversion of substrate to product. The $K_{i(inact)}$ corresponds to the apparent affinity of the inhibitor for the enzyme in the initial encounter complex, prior to enzyme processing, and is akin to the K_m for a substrate in a standard enzyme reaction. The ratio $k_{inact}/K_{i(inact)}$ is a measure of overall efficiency of inactivation by a given compound (Silverman, 1995) and can be used to rank order a series of suicide inhibitors. In practice, given the complexities of these measurements, many labs report suicide inactivator potencies as IC_{50}s, making it difficult to compare inhibitors from one study to the next (Wang, Huang, Suzuki, Liu, & Zhan, 2015). Regardless, the pharmacologic effects in the cellular environments depend on additional factors among which are cell permeability, rate of resynthesis of LSD1 protein, and partition ratio of the enzyme-generated reactive species toward modification of the target vs diffusion off of the enzyme (Silverman, 1995).

Propargyl-Lys-containing peptides are potent suicide inactivators of LSD1 that generate a covalent adduct with the active site flavin cofactor but are likely too large to enter cells, limiting their in vivo pharmacologic potential (Culhane et al., 2006; Szewczuk et al., 2007). Nevertheless, these propargyl-Lys peptides have shown utility in structural biology studies. An H3 tail peptide containing a propargyl-Lys was used to generate a crystallographic complex with LSD1 as discussed earlier (Yang et al., 2007). In more recent studies, a propargyl-Lys analog was incorporated into histone H3 and ultimately nucleosomes and was elegantly applied to capture a low-resolution structure of an LSD1/CoREST heterodimer stably bound to nucleosomes (Pilotto et al., 2015).

Although the propargyl-containing antidepressant MAO inhibitor pargyline itself is not an LSD1 inhibitor (Culhane, Wang, Yen, & Cole, 2010; Lee et al., 2006), despite cell-based pharmacologic reports that have suggested otherwise (Metzger et al., 2005), tranylcypromine and phenelzine MAO inhibitors do act as suicide inhibitors of LSD1 via their cyclopropylamine and hydrazine functional groups (Culhane et al., 2010; Lee et al., 2006; Schmidt & McCafferty, 2007). Tranylcypromine and phenelzine themselves are relatively weak LSD1 inhibitors compared to their actions vs MAOs, but in the context of elaboration into peptides or more complex small molecule analogs, these cyclopropylamine and hydrazine functionalities can confer high potency and specificity for LSD1 (Prusevich et al., 2014; Vianello et al., 2014). For tranylcypromine analogs, it has been shown that attachment of an aryl group via flexible linkers in the para position of the phenyl ring can greatly enhance the mechanism-based inactivation of LSD1 relative to other amine oxidases (Vianello et al., 2014). Such aryl-alkyl attachment also has been used to increase the potency and specificity of phenelzine compounds (Prusevich et al., 2014). N-substitution of the cyclopropylamine of tranylcypromine is another approach to improve LSD1 targeting. X-ray crystal structures have been somewhat informative in establishing the mechanistic basis for the improved LSD1 potencies of these synthetic analogs of tranylcypromine (Vianello et al., 2014).

In general, cyclopropylamine analogs undergo ring opening, likely generating a benzyl radical that is stabilized by aromatic conjugation. Several different fates have been reported for the covalent bond formation of the radical intermediate and the flavin cofactor as determined by X-ray crystal structures (Binda et al., 2010; Mohammad et al., 2015; Yang et al., 2007; Fig. 2).

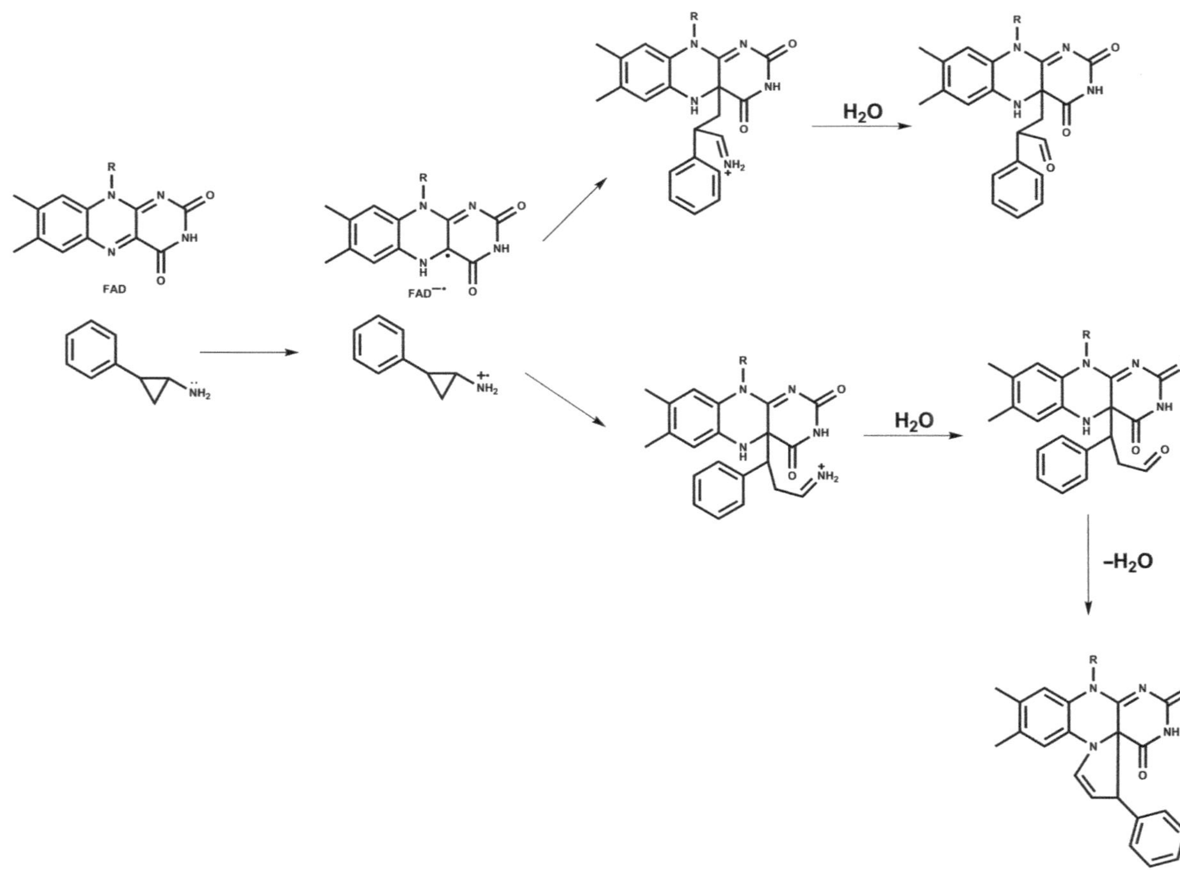

Fig. 2 Mechanism of LSD1 inactivation by cyclopropylamine analogs. Cyclopropylamine analogs form covalent bonds with the flavin cofactor of LSD1 through ring opening. The C4a and N5 atoms (*top right* and *bottom right*) serve as proposed points of attachment that occur after cyclopropyl ring opening (*center*).

Both the N5 and C4a atoms of the flavin have been invoked as attachment points for the benzyl scaffold, driving ring formation as well as acyclic structures (Binda et al., 2010; Yang et al., 2007). It seems plausible that more than one of these covalent adducts can be generated from a given inhibitor and X-ray crystallography of the LSD1 complexes is limited in its ability to distinguish such mixed molecular populations. For inactivation by phenelzine analogs, the precise mechanism of inhibition is also unclear. One proposed pathway for inhibition involves oxidation of the hydrazine functionality to the diazonium species, with subsequent attack by the flavin or perhaps an LSD1 enzyme residue (Culhane et al., 2010; Fig. 3). Such oxidation to the diazonium, however, requires a net four-electron transfer, which may be kinetically challenging. It is therefore possible that the two-electron oxidation, generating the aza analog or its functional equivalent, is the key species for enzyme inactivation. There have not yet been X-ray crystal structure of LSD1 inactivated by phenelzine analogs, although mass spectroscopy has suggested that a flavin adduct can be formed, at least as a minor product (Culhane et al., 2010).

Other compounds beyond tranylcypromine and phenelzine analogs have been reported as LSD1 inhibitors including polyamines (Nowotarski et al., 2015) and hydrazone HCI-2509 but whose specificity and mechanisms of inhibition remain less well characterized (Wang, Huang, et al., 2015). Given that many of the in vitro LSD1 demethylase assays employ peroxidase as an indirect measure of LSD1 enzymatic activity, and the peroxidase activity can be interfered with by particular compounds, it is critical to use secondary

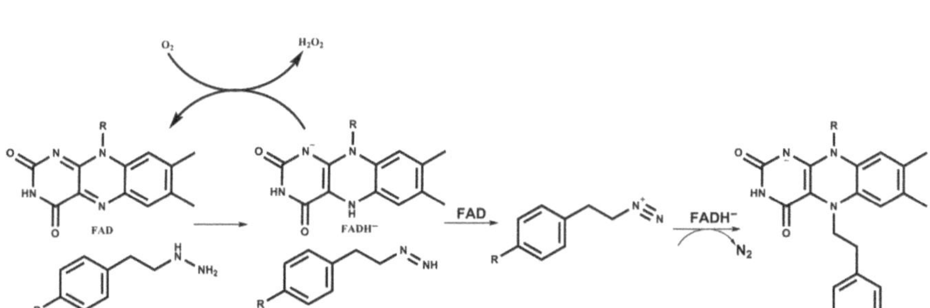

Fig. 3 Potential mechanism of LSD1 inactivation by hydrazine analogs. A possible mechanism of hydrazine-mediated inactivation of LSD1 involves formation of a covalent bond with the flavin cofactor. When the hydrazine moiety initially encounters the FAD cofactor (*left*), it may undergo a four-electron oxidation to form the diazonium species (*center*) which can be attacked by the cofactor or another nucleophile in the vicinity. When the flavin attacks (as shown), a covalent bond forms which inactivates the enzyme.

assays such as mass spectrometry analysis that directly monitors peptide methylation status to ensure the reliability of a particular LSD1 inhibitor finding.

4. APPLICATIONS OF LSD1 INHIBITORS

Applications of LSD1 inhibitors can be considered in the context of stem cell differentiation (Eliazer et al., 2014), neurobiology (Neelamegam et al., 2012), oxidative stress (Prusevich et al., 2014), viral infectivity (Hill et al., 2014; Sakane et al., 2011), and many forms of cancer. There are now numerous reports of synthetic LSD1 inhibitors of varying mechanisms of inhibition, potencies, and selectivities being applied to biomedical discovery. Basic features including effects on histone marks and gene expression as well as functional effects on cell growth and physiologic processes have been assessed with these compounds. Here we highlight a select group of recent findings involving well-characterized LSD1 inhibitors with an emphasis on cancer (Fig. 4).

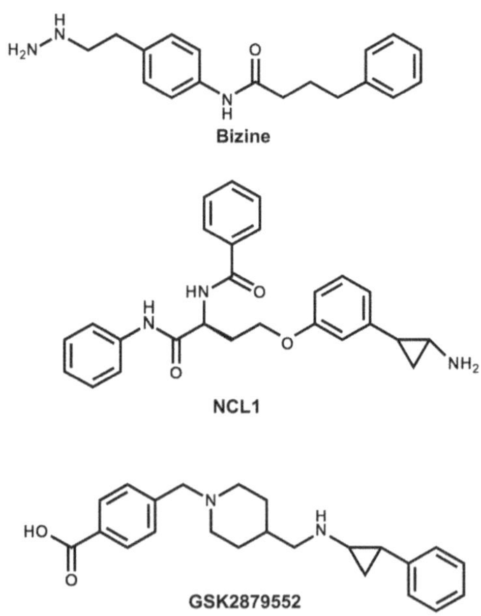

Fig. 4 Structures of representative LSD1 inhibitors. Bizine, a selective phenelzine analog (*top*); NCL1, a tranylcypromine analog (*middle*); and GSK2879552, a recent tranylcypromine analog (*bottom*), have been shown to be potent and selective LSD1 inhibitors.

4.1 Bizine

Bizine is a selective and potent LSD1 inhibitor based on the monoamine oxidase inhibitor phenelzine. This inhibitor showed specific effects on H3K4Me2 in a variety of cancer cell lines including LnCAP, androgen-dependent prostate cancer, and H460 lung cancer (Prusevich et al., 2014). Interestingly, the effects on H3K4Me2 showed an oscillating kinetic behavior with increases in H3K4Me2 detected by 6 h, then absent at 12 h, and reappearing at 24 h and persisting to 72 h (Prusevich et al., 2014). The relatively rapid appearance of H3K4Me2 at 6 h suggests that basal cellular activity of LSD1 is substantial and is in equilibrium with one or more methyltransferases. The basis for cellular oscillation in H3K4Me2 after bizine treatment is not known but the rebound to baseline at 12 h suggests that compensatory effects after LSD1 inhibition may occur involving changes in methyltransferase levels or localization. Interestingly, levels of mono- and trimethylation of H3K4 were not affected by bizine treatment (Prusevich et al., 2014). These findings suggest that globally stimulating dimethylation of H3K4 does not drive H3K4 trimethylation per se. While HeK4Me3 is generally considered most critical for gene activation, perhaps a global increase in H3K4Me2 in chromatin may serve biologically key functions that are less appreciated.

Whole genome chromatin immunoprecipitation (ChIP-seq) analyzing H3K4Me2 positioning after bizine treatment suggested widespread effects on numerous gene promoters, including many tumor suppressor genes (Prusevich et al., 2014). Despite these findings, the antitumor proliferation effects of bizine were of modest potency (Prusevich et al., 2014), considerably weaker than its effects on histone marks. Combination studies of bizine with some histone deacetylase inhibitors did suggest the potential of additivity or mild synergy of antitumor effects with particular epigenetic agents (Prusevich et al., 2014). Earlier studies in glioblastoma with another LSD1 inhibitor also pointed to the potential of such combination therapies (Singh et al., 2011). Of note, bizine also resulted in neuronal protection against oxidative stress (Prusevich et al., 2014), although the mechanism for this remains unclear.

4.2 NCL1

The tranylcypromine derivative NCL1 is a potent and selective LSD1 mechanism-based inhibitor developed by Suzuki and colleagues (Ueda et al., 2009). This compound was shown to exhibit with relatively low toxicity but significant antiproliferative effects against LnCAP cells in vitro and in mouse xenograft studies (Etani et al., 2014). While the cellular potency of NCL1 on LnCAP growth was modest, the in vivo antitumor effect was relatively strong. The precise mechanisms of NCL1's antitumor effects were unclear but appear to involve induction of apoptosis and autophagy.

4.3 GSK2879552

The GSK pharmaceutical company effort led by Kruger reported on a tranylcypromine analog—GSK2879552 (Mohammad et al., 2015) as a specific and fairly potent LSD1 inhibitor. These investigators showed GSK2879552 is a suicide inactivator and obtained a crystal structure with LSD1. GSK2879552 is most noteworthy for its very high specificity vs MAOs, likely residing from the extra substitution on the cyclopropylamine nitrogen. GSK2879552 was tested for its antitumor effects across a wide range of tumor cell lines and was shown to be potent against most acute myeloid leukemia lines investigated and a subset of small cell lung cancer lines (Mohammad et al., 2015). Like those of bizine, effects of GSK2879552 on H3K4Me2 were detected globally, and these effects as well as influence on gene expression occurred relatively rapidly compared to antiproliferative effects. It was especially noteworthy that for many cell lines, antiproliferative effects did not appear until 6 days after treatment with GSK2879552 (Mohammad et al., 2015). In some cases, cancer cell growth inhibition was incomplete and plateaued at high GSK2879552 concentration, suggesting that cells were still dividing but at a slower rate (Mohammad et al., 2015). Taken together with GSK2879552-induced gene expression changes, these data were consistent with a model of GSK2879552-induced cellular differentiation but not cell death in such cases. GSK investigators were unable to identify a gene expression signature from one cancer cell line to the next conferred by GSK2879552. After careful bioinformatics analysis, however, the GSK team was able to identify a DNA hypomethylation signature that was predictive of response to GSK2879552. The molecular mechanism for such a response was unclear, however.

5. SUMMARY AND FUTURE DIRECTIONS

Since LSD1's discovery as a histone demethylase about a dozen years ago, our understanding of LSD1's structure and enzymatic mechanism has advanced relatively rapidly, but elucidating the biological consequences of its catalytic activity have proved challenging. Major strides in the development of enzymatic assays and inhibitor discovery have begun to allow for the development of powerful chemical probes being brought to bear on LSD1's apparent myriad of cellular functions. Still, how LSD1 contributes to cancer and other diseases, how LSD1 inhibitors might synergize with other epigenetic agents, and how subunits of the CoREST complex modulate LSD1 function are just some of the unresolved topics under active investigation. Other poorly understood issues relate to LSD1's regulation by alternative splicing and posttranslational modifications. We anticipate that the vibrant research efforts devoted to LSD1 investigation will drive important discoveries in the coming years with translational and clinical impact.

ACKNOWLEDGMENTS

We thank Dr. Jay Kalin for helpful advice and the NIH, FAMRI Foundation, and V Foundation for financial support.

REFERENCES

Binda, C., Valente, S., Romanenghi, M., Pilotto, S., Cirilli, R., Karytinos, A., et al. (2010). Biochemical, structural, and biological evaluation of tranylcypromine derivatives as inhibitors of histone demethylases LSD1 and LSD2. *Journal of the American Chemical Society*, *132*, 6827–6833.

Blair, L. P., Avaritt, N. L., Huang, R., Cole, P. A., Taverna, S. D., & Tackett, A. J. (2011). An assay for quantitative measurement of lysine demethylase activity. *Epigenetics*, *6*, 490–499.

Blair, L. P., Avaritt, N. L., & Tackett, A. J. (2012). Application of MassSQUIRM for quantitative measurement of lysine demethylase activity. *Journal of Visualized Experiment*, *61*, e3604.

Cole, P. A. (2008). Chemical probes for histone-modifying enzymes. *Nature Chemical Biology*, *4*(10), 590–597.

Culhane, J. C., & Cole, P. A. (2007). LSD1 and the chemistry of histone demethylation. *Current Opinion in Chemical Biology*, *11*, 561–568.

Culhane, J. C., Szewczuk, L. M., Liu, X., Da, G., Marmorstein, R., & Cole, P. A. (2006). A mechanism-based inactivator for histone demethylase LSD1. *Journal of the American Chemical Society*, *128*, 4536–4537.

Culhane, J. C., Wang, D., Yen, P. M., & Cole, P. A. (2010). Comparative analysis of small molecules and histone substrate analogues as LSD1 lysine demethylase inhibitors. *Journal of the American Chemical Society*, *132*, 3164–3176.

Edmondson, D. E., Mattevi, A., Binda, C., Li, M., & Hubalek, F. (2004). Structure and mechanism of monoamine oxidase. *Current Medicinal Chemistry*, *11*, 1983–1993.

Eliazer, S., Palacios, V., Wang, Z., Kollipara, R. K., Kittler, R., & Buszczak, M. (2014). Lsd1 restricts the number of germline stem cells by regulating multiple targets in escort cells. *PLoS Genetics*, *10*, e1004200.

Etani, T., Suzuki, T., Naiki, T., Naiki-Ito, A., Ando, R., Iida, K., et al. (2014). NCL1, a highly selective lysine-specific demethylase 1 inhibitor, suppresses prostate cancer without adverse effect. *Oncotarget*, *6*, 2865–2878.

Fang, R., Chen, F., Dong, Z., Hu, D., Barbera, A. J., Clark, E. A., et al. (2012). LSD2/KDM1B and its cofactor NPAC/GLYR1 endow a structural and molecular model for regulation of H3K4 demethylation. *Molecular Cell*, *49*, 558–570.

Forneris, F., Binda, C., Adamo, A., Battaglioli, E., & Mattevi, A. (2007). Structural basis of LSD1-CoREST selectivity in histone H3 recognition. *The Journal of Biological Chemistry*, *282*, 20070–20074.

Forneris, F., Binda, C., Antonietta Vanoni, M., Battaglioli, E., & Mattevi, A. (2005). Human histone demethylase LSD1 reads the histone code. *The Journal of Biological Chemistry*, *280*, 41360–41365.

Forneris, F., Binda, C., Dall'Aglio, A., Fraajie, M. W., Battaglioli, E., & Mattevi, A. (2006). A highly specific mechanism of histone H3-K4 recognition by histone demethylase LSD1. *The Journal of Biological Chemistry*, *281*(46), 35289–35295.

Gale, M., & Yan, Q. (2015). High-throughput screening to identify inhibitors of lysine demethylases. *Epigenomics*, *7*, 57–65.

Hattori, T., Taft, J. M., Swist, K. M., Luo, H., Witt, H., Slattery, M., et al. (2013). Recombinant antibodies to histone post-translational modifications. *Nature Methods*, *10*, 992–998.

Hill, J. M., Quenelle, D. C., Cardin, R. D., Vogel, J. L., Clement, C., Bravo, F. J., et al. (2014). Inhibition of LSD1 reduces herpesvirus infection, shedding, and recurrence by promoting epigenetic suppression of viral genomes. *Science Translational Medicine*, *6*, 265ra169.

Hou, H., & Yu, H. (2010). Structural Insights into histone lysine demethylation. *Current Opinion in Structural Biology*, *20*, 739–748.

Huang, H., Lin, S., Garcia, B. A., & Zhao, Y. (2015). Quantitative proteomic analysis of histone modifications. *Chemical Reviews*, *115*, 2376–2418.

Kim, S.-A., Chatterjee, N., Jennings, M. J., Bartholomew, B., & Tan, S. (2015). Extranucleosomal DNA enhances the activity of the LSD1/CoREST histone demethylase complex. *Nucleic Acids Research*, *43*(10), 4868–4880.

Kitz, R., & Wilson, I. B. (1962). Esters of methanesulfonic acid as irreversible inhibitors of acetylcholinesterase. *Journal of Biological Chemistry*, *237*, 3245–3249.

Kokura, K., Sun, L., & Fang, J. (2015). In vitro histone demethylase assays. In S. P. Chellappan (Ed.), *Chromatin protocols, methods in molecular biology: Vol. 1288*. (pp. 109–122). New York: Springer.

Laurent, B., Ruitu, L., Mum, J., Hempei, K., Ferrao, R., Xiang, Y., et al. (2015). A specific LSD1/KDM1A isoform regulates neuronal differentiation through H3K9 demethylation. *Molecular Cell*, *57*, 1–14.

Lee, M. G., Wynder, C., Cooch, N., & Shiekhattar, R. (2005). An essential role for CoREST in nucleosomal histone 3 lysine 4 demethylation. *Nature*, *437*, 432–435.

Lee, M. G., Wynder, C., Schmidt, D. M., McCafferty, D. G., & Shiekhattar, R. (2006). Histone H3 lysine 4 demethylation is a target of nonselective antidepressive medications. *Chemistry and Biology*, *13*, 565–567.

Liang, Y., Vogel, J. L., Narayanan, A., Peng, H., & Kristie, T. M. (2009). Inhibition of the histone demethylase LSD1 blocks α-herpesvirus lytic replication and reactivation from latency. *Nature Medicine*, *15*, 1312–1318.

Metzger, E., Wissmann, M., Yin, N., Muller, J. M., Schneider, R., Peters, A. H. F. M., et al. (2005). LSD1 demethylates repressive histone marks to promoter androgen-receptor-dependent transcription. *Nature*, *437*, 436–439.

Mohammad, H., Smitherman, K. N., Kamat, C. D., Soong, D., Federowicz, K. E., Van Aller, G. S., et al. (2015). A DNA hypomethylation signature predicts antitumor activity of LSD1 inhibitors in SCLC. *Cancer Cell*, *28*, 57–69.

Neelamegam, R., Ricq, E. L., Malvaez, M., Patnaik, D., Norton, S., Carlin, S. M., et al. (2012). Brain-penetrant LSD1 inhibitors can block memory consolidation. *ACS Chemical Neuroscience*, *3*, 120–128.

Nicholson, T. B., & Chen, T. (2009). LSD1 demethylates histone and non-histone proteins. *Epigenetics*, *4*(3), 129–132.

Nowotarski, S. L., Pachaiyappan, B., Holshouser, S. L., Kutz, C. J., Li, Y., Huang, Y., et al. (2015). Structure-activity study for (bis)ureidopropyl- and (bis)thioureidopropyldiamine LSD1 inhibitors with 3-5-3 and 3-6-3 carbon backbone architectures. *Bioorganic & Medicinal Chemistry*, *23*, 1601–1612.

Paech, C., Salach, J. I., & Singer, T. P. (1980). Suicide inactivation of monoamine oxidase by trans-phenylcyclopropylamine. *Journal of Biological Chemistry*, *10*, 2700–2704.

Pilotto, S., Speranzini, V., Tortorici, M., Durand, D., Fish, A., Valente, S., et al. (2015). Interplay among nucleosomal tails, histone tails, and corepressor CoREST underlies LSD1-mediated H3 demethylation. *Proceedings of the National Academy of Sciences of the United States of America*, *112*(9), 2752–2757.

Prusevich, P., Kalin, J. H., Ming, S. A., Basso, M., Givens, J., Li, X., et al. (2014). A selective phenelzine analogue inhibitor of histone demethylase LSD1. *ACS Chemical Biology*, *9*, 1284–1293.

Rusconi, F., Paganini, L., Braida, D., Ponzoni, L., Toffolo, E., Maroli, A., et al. (2015). LSD1 neurospecific alternative splicing controls neuronal excitability in mouse models of epilepsy. *Cerebral Cortex*, *25*, 2729–2740.

Sakane, N., Kwon, H.-S., Pagans, S., Kaehlcke, K., Mizusawa, Y., Kamada, M., et al. (2011). Activation of HIV transcription by the viral Tat protein requires a demethylation step mediated by lysine-specific demethylase 1 (LSD1/KDM1). *PLoS Pathogens*, *7*, e1002184.

Sakane, C., Ohta, H., & Shidoji, Y. (2015). Measurement of lysine-specific demethylase-1 activity in the nuclear extracts by flow-injection based time-of-flight mass spectrometry. *Journal of Clinical Biochemical Nutrition*, *56*, 123–131.

Schmidt, D. M. Z., & McCafferty, D. G. (2007). *Trans*-2-phenylcyclopropylamine is a mechanism based inactivator of demethylase LSD1. *Biochemistry*, *46*, 4408–4416.

Schmitt, M. L., Ladwein, K. I., Carlino, L., Schulz-Fincke, J., Willmann, D., Metzger, E., et al. (2014). Heterogeneous antibody-based activity assay for lysine specific demethylase I (LSD1) on a histone peptide substrate. *Journal of Biomolecular Screening*, *19*, 973–978.

Shi, Y., Lan, F., Matson, C., Mulligan, P., Whetstine, J. R., Cole, P. A., et al. (2004). Histone demethylation mediated by the nuclear amine oxidase homolog LSD1. *Cell*, *119*, 941–953.

Silverman, R. B. (1995). Mechanism-based enzyme inactivators. In J. N. Abelson, M. I. Simon, & D. L. Purich (Eds.), *Methods in enzymology: Vol. 249* (pp. 240–283). Academic Press: San Diego.

Singh, M. M., Manton, C. A., Bhat, K. P., Tsai, W.-W., Aldape, K., Barton, M. C., et al. (2011). Inhibition of LSD1 sensitizes glioblastoma cells to histone deacetylase inhibitors. *Neuro-oncology*, *13*, 894–903.

Sun, G., Alzayady, K., Stewart, R., Ye, P., Yang, S., Wendong, L., et al. (2010). Histone demethylase LSD1 regulates neural stem cell proliferation. *Molecular and Cellular Biology*, *30*, 1997–2005.

Szewczuk, L. M., Culhane, J. C., Yang, M., Majumdar, A., Yu, H., & Cole, P. A. (2007). Mechanistic analysis of a suicide inactivator of histone demethylase LSD1. *Biochemistry*, *46*, 6892–6902.

Thinnes, C. C., England, K. S., Kawamura, A., Chowdhury, R., Schofield, C., & Hopkinson, R. J. (2014). Targeting histone lysine demethylases—Progress, challenges, and the future. *Biochimica et Biophysica Acta*, *1839*, 1416–1432.

Toffolo, E., Rusconi, F., Paganini, L., Tortorici, M., Pilotto, S., Heisi, C., et al. (2014). Phosphorylation of neuronal lysine-specific demethylase 1 LSD1/KDM1A impairs transcriptional repression by regulating interaction with CoREST and histone deacetylases HDAC1/2. *Journal of Neurochemistry*, *128*, 603–616.

Ueda, R., Suzuki, T., Mino, K., Tsumoto, H., Nakagawa, H., Hasegawa, M., et al. (2009). Identification of cell-active lysine specific demethylase 1-selective inhibitors. *Journal of the American Chemical Society*, *131*, 17536–17537.

Vianello, P., Botrugno, O. A., Cappa, A., Ciossani, G., Dessanti, P., Mai, A., et al. (2014). Synthesis, biological activity and mechanistic insights of 1-substituted cyclopropylamine derivatives: A novel class of irreversible inhibitors of histone demethylase KDM1A. *European Journal of Medicinal Chemistry*, *86*, 352–363.

Walsh, C. T. (1984). Suicide substrates, mechanism-based enzyme inactivators: Recent developments. *Annual Review of Biochemistry*, *53*, 493–535.

Wang, X., Huang, B., Suzuki, T., Liu, X., & Zhan, P. (2015). Medicinal chemistry insights in the discovery of novel LSD1 inhibitors. *Epigenomics*, *7*, 1379–1396.

Wang, J., Telese, F., Tan, Y., Li, W., Jin, C., He, X., et al. (2015). LSD1n is an H4K20 demethylase regulating memory formation via transcriptional elongation control. *Nature Neuroscience*, *18*, 1256–1268.

Wigle, T. J., Swinger, K. K., Campbell, J. E., Scholle, M. D., Sherrill, J., Admirand, E. A., et al. (2015). A high-throughput mass spectrometry assay coupled with redox activity testing reduces artifacts and false positives in lysine demethylase screening. *Journal of Biomolecular Screening*, *20*, 810–820.

Wu, J., Hu, L., Du, Y., Kong, F., & Pan, Y. (2015). Prognostic role of LSD1 in various cancers: Evidence from a meta-analysis. *Oncotargets and Therapy*, *8*, 2565–2570.

Yang, M., Culhane, J. C., Szewczuk, L. M., Gocke, C. B., Brautigam, C. A., Tomchick, D. R., et al. (2007). Structural basis of histone demethylation by LSD1 revealed by suicide inactivation. *Nature Structural and Molecular Biology*, *14*, 535–539.

Yang, M., Gocke, C. B., Luo, X., Borek, D., Tomchick, D. R., Machius, M., et al. (2006). Structural basis for CoREST-dependent demethylation of nucleosome by the human LSD1 histone demethylase. *Molecular Cell*, *23*, 377–387.

Yu, W., Eram, M. S., Hajian, T., Szykowska, A., Burgess-Brown, N., Vedadi, M., et al. (2014). A scintillation proximity assay for histone demethylases. *Analytical Biochemistry*, *463*, 54–60.

Zibetti, C., Adamo, A., Binda, C., Forneris, F., Toffolo, E., Verpelli, C., et al. (2010). Alternative splicing of the histone demethylase LSD1/KDM1 contributes to the modulation of neurite morphogenesis in the mammalian nervous system. *The Journal of Neuroscience*, *30*(7), 2521–2532.

CHAPTER TWELVE

Purification, Biochemical Analysis, and Structure Determination of JmjC Lysine Demethylases

S. Krishnan[1], R.C. Trievel[2]
University of Michigan, Ann Arbor, MI, United States
[2]Corresponding author: e-mail address: rtrievel@umich.edu

Contents

1. Introduction 280
2. Purification 282
   2.1 Expression and Purification of Hexahistidine-Tagged JMJD2 KDMs 283
   2.2 Expression and Purification of Strep(II)-Tagged JMJD2 KDMs 284
   2.3 Analysis of the Metal Content of JMJD2 KDMs 285
3. KDM Assay Methodology 287
   3.1 Design of Histone Peptide Substrates 289
   3.2 Purification of FDH 289
   3.3 FDH-Coupled Fluorescent Assay 290
4. Structure Determination of JmjC KDMs 293
   4.1 Crystallization of JMJD2 KDMs with Methylated Histone Peptides 293
   4.2 Structure Determination of JMJD2 KDM Peptide Complexes 295
5. Summary and Future Perspectives 295
Acknowledgment 296
References 296

Abstract

Jumonji C (JmjC) lysine demethylases (KDMs) catalyze the site- and state-specific demethylation of lysine residues in histone and nonhistone protein substrates. These enzymes have been implicated in diverse genomic processes, including epigenetic gene regulation, DNA damage response, DNA replication, and regulation of heterochromatin structure. In addition, a number of JmjC KDMs contribute to the incidence of numerous cancers, rendering them targets for the development of novel chemotherapeutic drugs. Using the JMJD2 KDM subfamily as representative examples, this chapter outlines strategies for purifying highly active, recombinant JmjC KDMs lacking inhibitory transition metal ions, characterizing kinetic parameters of these enzymes using a

[1] Current address: Howard Hughes Medical Institute, Department of Biochemistry and Molecular Pharmacology, NYU Langone School of Medicine, New York, NY, 10016, USA.

Methods in Enzymology, Volume 573
ISSN 0076-6879
http://dx.doi.org/10.1016/bs.mie.2016.01.023

coupled fluorescent assay, and determining crystal structures of the enzymes in complex with methylated histone peptides. Together, these approaches provide a foundation for structural and biochemical characterization of the JmjC KDMs and facilitate efforts to identify small molecule inhibitors through high-throughput screening and structure-guided design.

1. INTRODUCTION

Protein lysine methylation is an important posttranslational modification that has fundamental roles in regulating epigenetic gene expression, DNA transactions, cellular signaling, and development. Within the context of chromatin, the major sites of lysine methylation occur at Lys4, Lys9, Lys27, Lys36, and Lys79 in histone H3 (H3K4, H3K9, H3K27, H3K36, and H3K79) and Lys20 in histone H4 (H4K20) (Fick & Trievel, 2013; Lanouette, Mongeon, Figeys, & Couture, 2014). In addition, numerous nonhistone nuclear proteins are subject to lysine methylation, including transcription factors, histone-modifying enzymes, and subunits of the basal transcriptional machinery (Moore & Gozani, 2014; Zhang, Huang, & Shi, 2015). Further, lysine methylation has emerged as a key regulatory modification in JAK-STAT and MAP signaling pathways, illuminating its contributions in cytoplasmic signal transduction (Biggar & Li, 2015; Mazur et al., 2014). In contrast to most posttranslational modifications wherein a single modification state can exist on a given amino acid, the lysine ε-amine group can undergo mono-, di-, and trimethylation, imparting an additional level of regulation in signal transduction. Lysine methylation frequently mediates protein–protein interactions through effector proteins that possess domains which specifically recognize both the site and state of lysine methylation within their binding partner, thereby transducing the signal associated with the modification (Musselman, Khorasanizadeh, & Kutateladze, 2014).

Lysine methylation status is dynamically governed through the concerted activities of protein lysine methyltransferases (KMTs) and lysine demethylases (KDMs) (Black, Van Rechem, & Whetstine, 2012). KMTs methylate specific lysines within their substrates and also catalyze a distinct degree of methylation of the lysine ε-amine group, a property referred to as product specificity (Zhang et al., 2003). In contrast, KDMs exhibit site and state specificity for methyllysine residues within their protein substrates (Mosammaparast & Shi, 2010). Currently, there are two known families of KDMs: (1) the LSD1 (KDM1) family of flavin-dependent amino oxidases and (2) the larger and more diverse Jumonji C (JmjC) family (Mosammaparast & Shi, 2010; Shi et al., 2004; Tsukada et al., 2006).

The JmjC enzymes belong to the cupin superfamily and are classified as Fe(II)- and 2-oxoglutarate (2-OG)-dependent oxygenases (Aik, McDonough, Thalhammer, Chowdhury, & Schofield, 2012). In humans, the JmjC family comprises six subfamilies of KDMs, as well as protein and RNA hydroxylases (Markolovic, Wilkins, & Schofield, 2015). The mammalian JmjC KDM subfamilies are categorized by sequence and structural homology and exhibit distinct methylation site and state specificities for lysines in histone H3: FBXL11/FBXL10 (mono- and dimethyl H3K36 (H3K36me1/2)), JMJD1 (H3K9me1/2), JMJD2 (H3K9me2/3 with certain homologs also exhibiting activity toward H3K36me2/3), UTX/UTY/JMJD3 (H3K27me2/3), JARID1 (H3K4me2/3), and PHF8/KIAA1718 (H3K9me1/2 or H3K27me1/2) (Horton et al., 2010; Mosammaparast & Shi, 2010; Walport et al., 2014). The JmjC KDMs share a common catalytic mechanism in which the methyllysine methyl group is hydroxylated via a radical-based reaction using the cosubstrates 2-OG and oxygen, producing succinate, carbon dioxide, and ε-hydroxymethyllysine (Fig. 1; Schneider & Shilatifard, 2006; Tsukada et al., 2006). The latter product represents an unstable hemiaminal intermediate that spontaneously decomposes to yield formaldehyde and the demethylated lysine. JmjC KDMs have been implicated in numerous genomic processes, including transcriptional activation, gene silencing, DNA replication, DNA damage response, cell cycle control, and the epigenetic regulation of cellular differentiation and developmental programs (Alam, Gu, & Lee, 2015; Cheng & Trievel, 2015; Dimitrova, Turberfield, & Klose,

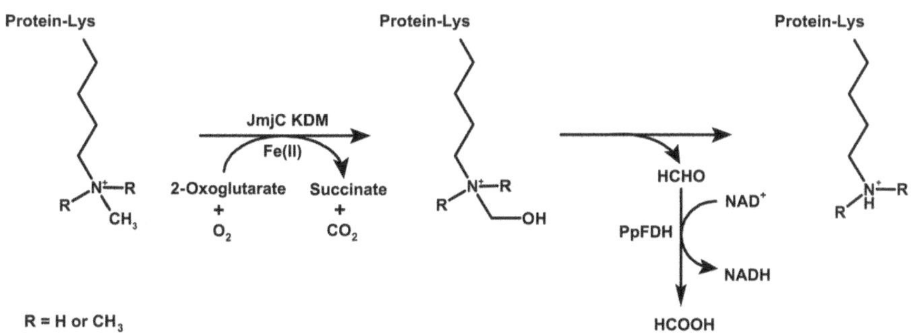

Fig. 1 Scheme for the formaldehyde dehydrogenase (*FDH*)-coupled KDM assay. Formaldehyde produced from the demethylation reaction is oxidized to formic acid by *Pseudomonas putida* FDH (*PpFDH*) with the concomitant reduction of NAD^+ to NADH. Demethylase activity is measured as a function of NADH fluorescence.

2015). Moreover, the aberrant expression of many JmjC KDMs has been reported to contribute to the onset or progression of multiple cancers, and thus have emerged as clinical targets for chemotherapeutic drug design (Black & Whetstine, 2013; Maes, Carceller, Salas, Ortega, & Buesa, 2015; McGrath & Trojer, 2015; Thinnes et al., 2014).

In this chapter, we outline methods for purifying highly active JmjC KDMs for biochemical characterization, assaying KDM kinetics using a coupled fluorescent assay (Fig. 1), and determining the crystal structures of these enzymes in complex with methylated histone peptides. These methods will be described using the members of the JMJD2 (KDM4) subfamily as representative examples. These enzymes were the first JmjC KDMs for which crystal structures bound to methylated peptide substrates were reported, yielding seminal insights into the molecular basis of substrate recognition (Chen et al., 2007; Couture, Collazo, Ortiz-Tello, Brunzelle, & Trievel, 2007; Ng et al., 2007). In addition, initial efforts to identify small molecule inhibitors of JmjC KDMs were undertaken utilizing these enzymes (Rose et al., 2008). Thus, the JMJD2 KDMs represent archetypes for characterizing the substrate specificities, catalytic mechanism, and inhibition of the JmjC KDM family.

2. PURIFICATION

Initial kinetic studies of recombinant JMJD2 KDMs reported turnover numbers (k_{cat} values) that were at least two orders of magnitude lower than that measured for the flavoprotein demethylase LSD1 (Couture et al., 2007; Forneris, Binda, Vanoni, Battaglioli, & Mattevi, 2005; Gaweska, Henderson Pozzi, Schmidt, McCafferty, & Fitzpatrick, 2009; Schmidt & McCafferty, 2007; Szewczuk et al., 2007). The low apparent catalytic activity of JMJD2 enzymes led to speculation as to whether JmjC KDMs require binding partners to stimulate their enzymatic activity or possess alternative substrates. Subsequently, several JmjC KDMs were shown to be potently inhibited by transition metal ions, such as Co(II), Ni(II), Cu(II), and Zn(II), in vitro and in vivo (Chen & Costa, 2009; Chen, Giri, et al., 2010; Chen, Kluz, Zhang, & Costa, 2010; Li, Ke, & Costa, 2009; Sekirnik, Rose, Mecinovic, & Schofield, 2010). These metal ions compete with Fe(II) for coordination to the His-X-Glu/Asp...His catalytic triad within the active site, inhibiting the Fe(II)-dependent hydroxylation of the methyllysine substrate (Couture et al., 2007; Ng et al., 2007). Given that biochemical studies of many JmjC KDMs are frequently conducted using hexahistidine-tagged enzymes, we postulated that the purification of the enzymes using Ni or Co immobilized metal affinity chromatography (IMAC) could inadvertently introduce inhibitory transition metal ions into their active sites, analogous to the findings reported for a divalent metal ion-dependent protein deacetylase (Gantt, Gattis, & Fierke, 2006). To test this hypothesis, we devised an alternative purification strategy using Strep(II) affinity-tagged JMJD2 KDMs and Strep-Tactin chromatography that eliminates the need for IMAC, thus minimizing transition metal ion contamination during protein purification (Krishnan, Collazo, Ortiz-Tello, & Trievel, 2012). The following section (Section 2.1) describes the expression and purification of hexahistidine-tagged and Strep(II)-tagged JMJD2A (KDM4A) and JMJD2D (KDM4D) and quantitative analysis of the transition metal content of the purified recombinant enzymes using inductively coupled plasma mass spectrometry (ICP-MS). The ICP-MS data demonstrate that purification of hexahistidine-tagged JmjC KDMs introduces variable quantities of contaminating transition metals derived from the IMAC column and that Strep-Tactin purification of Strep(II)-tagged enzymes yields highly pure protein that is essentially devoid of these inhibitory metals.

2.1 Expression and Purification of Hexahistidine-Tagged JMJD2 KDMs

The DNA sequences encoding the catalytic domains of human JMJD2A (residues 1–350) and human JMJD2D (residues 12–342) were cloned into an engineered variant of the pET15b plasmid (Novagen) in which the thrombin cleavage site was substituted with a tobacco etch virus (TEV) protease cleavage site, permitting the removal of the N-terminal hexahistidine tag (Couture et al., 2007; Krishnan & Trievel, 2013). The JMJD2A and JMJD2D expression plasmids were then transformed into Rosetta2 DE3 competent cells (Novagen). The transformed cells were grown in 2xYT media at 37°C until their absorbance at 600 nm (A_{600}) reached 0.4–0.6 units, at which time the cultures were induced with 1.0 m*M* isopropyl β-D-1-thiogalactopyranoside (IPTG). The induced cultures were then grown overnight at 18°C for JMJD2A and at 16°C for JMJD2D. In the morning, cells were harvested by centrifugation at 4°C, and cell pellets were stored at −20°C.

Hexahistidine-tagged JMJD2A and JMJD2D were purified at 4°C as follows (Krishnan et al., 2012). All buffers were prepared from water purified using a Milli-Q Biocel system (Millipore) to minimize contamination by transition metal ions. A cell pellet corresponding to 1 L of culture was resuspended in Buffer A (100 m*M* TRIS pH 7.5 and 500 m*M* NaCl) supplemented with ethylenediaminetetraacetic acid (EDTA)-free HALT Protease Inhibitor Cocktail (Thermo Scientific), and the cells were lysed by sonication. The cell lysate was centrifuged at high speed (15,000 rpm for 30 min in a Sorvall SS-34 rotor) to separate the insoluble matter, and the soluble fraction of the lysate was applied to a column containing 4 mL Ni Sepharose High Performance resin (GE Healthcare). After washing the column with at least four column volumes of Buffer A, the hexahistidine-tagged protein was eluted with a linear gradient from Buffer A to Buffer B (Buffer A supplemented with 500 m*M* imidazole). His-tagged TEV protease S219V mutant (Kapust et al., 2001) was added to the pooled peak fractions containing the hexahistidine-tagged JMJD2 protein, and the sample was dialyzed overnight in TEV protease cleavage buffer (50 m*M* TRIS pH 7.0 and 150 m*M* NaCl). The following morning, the sample was applied to 5 mL of Ni Sepharose High Performance resin preequilibrated with Buffer A to bind the His-tagged TEV protease and any uncleaved hexahistidine-tagged JMJD2 protein. After batch binding with the Ni Sepharose resin, the sample was centrifuged at low speed, and the supernatant containing the hexahistidine-tag cleaved JMJD2 protein was isolated from the resin. The supernatant was then concentrated to 2 mL using an Amicon Ultra-15 concentrator with a 30 kDa cutoff membrane (Millipore) and loaded onto a Superdex 200 gel filtration column (GE Healthcare) preequilibrated with Buffer C (20 m*M* TRIS pH 7.5 and 150 m*M* NaCl). The peak fractions from the gel filtration chromatography were judged to be at least 95% pure by SDS-PAGE and were pooled. The protein was then concentrated to 10–20 mg/mL as determined by its absorbance at 280 nm (A_{280}) in 6 *M* guanidinium chloride and 100 m*M* HEPES pH 7.5 buffer. The concentrated protein was dispensed into small aliquots, flash-frozen in liquid nitrogen, and stored at −80°C.

2.2 Expression and Purification of Strep(II)-Tagged JMJD2 KDMs

The catalytic domains of human JMJD2A (residues 1–350) and JMJD2D (12–342) were cloned into a variant of the pET15b expression vector described in Section 2.1 in which the N-terminal hexahistidine tag was replaced with the DNA sequence encoding the Strep(II) affinity tag (Trp-Ser-His-Pro-Gln-Phe-Glu-Lys) (Krishnan et al., 2012). The TEV protease cleavage site is maintained in these plasmids, enabling facile removal of the Strep(II) tag during purification. Expression plasmids for the

Strep(II)-tagged catalytic domains of human JMJD2B (KDM4B) and JMJD2C (KDM4C) were also generated, but this chapter will focus on JMJD2A and JMJD2D, which were the focus of our structural and functional studies of JMJD2 substrate specificity (Couture et al., 2007; Krishnan et al., 2012; Krishnan & Trievel, 2013). Expression of the Strep(II)-tagged JMJD2A and JMJD2D was carried out as described for the respective hexahistidine-tagged constructs (see Section 2.1).

The Strep(II)-tagged JMJD2 KDMs were purified at 4°C according to the following protocol (Krishnan et al., 2012). All buffers were prepared from water purified using a Milli-Q Biocel System (Millipore) to minimize contamination by transition metal ions. A cell pellet corresponding to 1 L of culture was resuspended in Buffer A supplemented with EDTA-free HALT Protease Inhibitor cocktail and lysed by sonication. The inclusion of the protease inhibitor cocktail is essential to maintaining the integrity of the Strep-Tactin column, enabling it to be used for multiple purifications of Strep(II)-tagged proteins. The soluble fraction of the cell lysate was separated from the insoluble matter via high-speed centrifugation and then applied to a 4-mL column containing Strep-Tactin Superflow Plus resin (Qiagen). The column was washed with a minimum of four column volumes of Buffer A to remove contaminants prior to the step elution with Buffer D (Buffer A supplemented with 2.5 m*M* desthiobiotin). After elution of the protein, the Strep-Tactin column was regenerated three times with five column volumes of Regeneration Buffer (50 m*M* NaH_2PO_4 pH 8.0, 300 m*M* NaCl, and 1 m*M* 4-hydroxyazobenzene-2-carboxylic acid) and then washed with at least four column volumes of Buffer A. The fractions of the Strep-Tactin column were analyzed by SDS-PAGE, and the peak fractions were pooled, incubated with TEV protease, and dialyzed overnight against TEV protease cleavage buffer to remove the Strep(II) tag. The following morning, the dialyzed protein was concentrated to 2 mL and loaded onto a Superdex 200 gel filtration column preequilibrated with Buffer C. Fractions from the column were analyzed by SDS-PAGE. The peak fractions were subsequently pooled, concentrated to between 10 and 20 mg/mL, flash-frozen in liquid nitrogen, and stored at −80°C. Protein concentration was determined by its A_{280} value in 6 *M* guanidinium chloride and 100 m*M* HEPES pH 7.5 buffer.

2.3 Analysis of the Metal Content of JMJD2 KDMs

After completing the purifications of the hexahistidine-tagged and Strep(II)-tagged JMJD2A and JMJD2D (see Sections 2.1 and 2.2), we investigated their metal content to ascertain whether purification of the hexahistidine-tagged enzymes using Ni Sepharose led to contamination with Ni(II). Analysis of the transition metal content of purified JMJD2 enzymes was performed using a Thermo Scientific Element Inductively Coupled Plasma High-Resolution Mass Spectrometer (W.M. Keck Elemental Geochemistry Laboratory, University of Michigan Department of Geological Sciences). Representative percent transition metal contents for duplicate purifications of hexahistidine- and Strep(II)-tag JMJD2D are presented in Table 1. The data reveal a Zn content of between 82% and 92% for the different purifications, in agreement with the structural Zn motif that stabilizes the catalytic domains of the JMJD2 KDMs (Chen et al., 2006). In contrast, the Fe content of the hexahistidine- and Strep(II)-tag JMJD2D was below 10%. This finding is consistent with the oxygen lability of Fe(II), implying that the Fe(II) oxidized to Fe(III) and was lost from the enzyme during aerobic purification. In examining the other transition metals analyzed, there is a notable difference in the Ni content between the hexahistidine- and Strep(II)-tag JMJD2D. The Ni content ranges from 40–70% for the duplicate purifications of the hexahistidine-tag JMJD2D, whereas the content of the Strep(II)-tag enzyme is below 3% for both purifications. Notably, the Ni copurified with JMJD2D, after removal of the hexahistidine-tag and subsequent gel filtration chromatography (see Section 2.1), indicating that the Ni was tightly bound within the JmjC active site.

Table 1 ICP-MS Analysis of Transition Metal Content of Two Independent Purifications of Strep(II)-Tag and Hexahistidine-Tag JMJD2D

Percent Metal (%)	Strep(II)-JMJD2D Batch 1[a]	Strep(II)-JMJD2D Batch 2[a]	His-JMJD2D Batch 1	His-JMJD2D Batch 2
Zn	84	90	82	92
Fe	5.8	5.9	6.9	8.6
Ni	2.4	1.8	35	70
Co	0.15	0.19	N.D.	N.D.
Mn	0.5	0.4	1.3	0.6
Cu	0.8	0.5	0.3	0.2

[a]Data previously reported in Krishnan et al. (2012).
Percent metal contents were calculated from the molar ratio of the metal to enzyme concentrations. Metal concentration in the protein sample was determined by subtracting the concentration of metal in the protein sample from the concentration of the corresponding metal in the buffer used for ICP-MS analysis. *N.D.*, not detected.

Consistent with these findings, recent studies of recombinant hexahistidine-tagged JARID1A purified using a Co(II) affinity column reported 24% Co in the purified enzyme, as determined by ICP-MS (Torres et al., 2015). Correlatively, analysis of the X-ray anomalous data for the crystal structure of FBXL11 (KDM2A) purified using a Ni(II)-NTA column revealed that the predominant metal bound in the enzyme's active site was Ni with a relatively small contribution in the anomalous signal for Fe (Cheng et al., 2014). It is important to note that inhibitory transition metal ions cannot be readily eliminated by treatment with EDTA or other metal chelators because many JmjC KDMs contain structural Zn motifs either within their catalytic domains or in neighboring accessory domains that contribute to their enzymatic activity (Chen et al., 2006; Goda, Isagawa, Chikaoka, Kawamura, & Aburatani, 2013; Horton et al., 2010; Krishnan & Trievel, 2013; Kruidenier et al., 2012; Sengoku & Yokoyama, 2011; Torres et al., 2015; Walport et al., 2014). These motifs would be disrupted through chelation of their Zn ions, resulting in the denaturation and inactivation of the enzyme. Taken together, these results demonstrate that purification of hexahistidine-tagged JmjC KDMs using IMAC introduces inhibitory transition metal ions into the enzymes and that Strep-Tactin chromatography affords an effective strategy for minimizing transition metal contamination, yielding highly pure protein.

3. KDM ASSAY METHODOLOGY

Since their discovery, a diverse set of biochemical assays for JmjC KDMs has been developed for characterizing their kinetic properties and for high-throughput screening to identify small molecule inhibitors. These assays utilize an array of different methods for quantifying KDM activity, including radioactivity (Tsukada et al., 2006), fluorescence (Cloos et al., 2006; Shi et al., 2004), time-resolved fluorescence energy transfer (TR-FRET) (Wang et al., 2012; Yu et al., 2012), amplified luminescent proximity homogeneous assay (Gauthier et al., 2012; Kawamura et al., 2010), scintillation proximity (Yu et al., 2014), NMR spectroscopy (Hopkinson, Hamed, Rose, Claridge, & Schofield, 2010), peptide-based microarrays with antibody-based detection (Cornett et al., 2016), and quantitative mass spectrometry (Leveridge et al., 2014; Mackeen et al., 2010). Among the most widely used methods is a nicotinamide adenine dinucleotide (NAD^+)-dependent formaldehyde dehydrogenase (FDH)-coupled assay in which the formaldehyde generated during demethylation is oxidized to formic acid with the concomitant reduction of NAD^+ to NADH (Fig. 1; Cloos et al., 2006; Shi et al., 2004). The NADH produced is detected by its intrinsic fluorescence, permitting KDM activity to be monitored continuously over time. This coupled assay offers several advantages over other KDM assays in that it is nonradiometric, does not require specialized instrumentation or reagents, is relatively inexpensive, and can be performed using a standard spectrofluorometer or fluorescence microplate reader. In addition, the versatility of this assay has enabled it to be adapted for high-throughput screening for KDM inhibitors (King et al., 2010). This section will outline the methodology of the FDH-coupled fluorescent assay and its application in measuring the steady-state kinetic parameters of the JMJD2 KDMs.

In designing a KDM assay, one must consider the type of substrate that will be utilized. Synthetic peptides incorporating mono-, di-, and trimethyllysine are among the most commonly used substrates, owing to the relative ease of their synthesis, their commercial availability, and the ability to generate libraries with mutations or other posttranslational modifications in the peptide sequence. However, recent advances in molecular and chemical biology have afforded methods for site-specifically installing methyllysines into full-length proteins, permitting KDM assays to be performed using methylated histone or nucleosomal substrates. Shokat and colleagues have devised a synthetic strategy for site-specifically installing methyllysine analogs (MLAs) into peptides and recombinant proteins wherein the target lysine is mutated to cysteine and then alkylated with (2-chloroethyl)-methylammonium, (2-chloroethyl)-dimethylammonium, and 2-bromoethyl-trimethylammonium salts to yield the respective mono-, di-, and trimethyllysine analogs ($K_C$me1, $K_C$me2, and $K_C$me3) (Simon et al., 2007). Kinetic studies have demonstrated that JMJD2A is active toward histone H3 peptide and nucleosomal substrates containing the H3K_C9me3 (Krishnan et al., 2012; Shiau et al., 2013). However, the catalytic efficiency (k_{cat}/K_M value) of the enzyme toward the H3K_C9me3 peptide was approximately an order of magnitude lower compared to the cognate H3K9me3 peptide substrate, illustrating that MLAs are imperfect mimics of their methyllysine counterparts. Alternatively, *bona fide* mono- and dimethyllysines can be site-specifically incorporated in recombinant proteins using genetic and chemical approaches employing unnatural lysine analogs, such as ε-*tert*-butyl-oxycarbonyl-lysine and allylcarbamoyl methyllysine, thus circumventing the drawbacks associated with MLAs (Ai, Lee, & Schultz, 2010; Nguyen, Garcia Alai, Kapadnis, Neumann, & Chin, 2009; Nguyen, Garcia Alai, Virdee, & Chin, 2010). In addition, native chemical ligation and enzyme-mediated ligation techniques can be utilized to site-specifically introduce methyllysines and other posttranslational modifications into histones and other proteins (He et al., 2003; Piotukh et al., 2011; Shogren-Knaak, Fry, & Peterson, 2003). In summary, numerous methods are available for preparing methylated peptide or protein substrates for JmjC KDMs that can be tailored to the specific enzyme and assay type.

3.1 Design of Histone Peptide Substrates

Careful consideration should be taken in designing peptide substrates for structural and biochemical characterization of JmjC KDMs. In structure–function studies of the JMJD2 enzymes, histone H3 peptides representing the K9 (residues 1–15) and K36 (residues 28–42) methylation sites were obtained from commercial sources, primarily AnaSpec, Inc. and New England Peptide, Inc. (Couture et al., 2007; Krishnan et al., 2012; Krishnan & Trievel, 2013). Peptides were designed with N-terminal acetyl and C-terminal amide groups to mimic the properties of the peptide backbone, with the exception of histone H3K9me peptides whose N-termini start at residue A1 and were thus left unblocked. Peptides were purified with chloride as the counterion, as opposed to the more commonly used trifluoroacetate ion, and were then lyophilized to dryness prior to shipment. These steps aid in diminishing the overall acidity of the histone peptide stock solutions prepared in either water or buffer. To ensure accurate peptide concentrations in the KDM assays, net peptide content for each sample was determined by amino acid analysis.

3.2 Purification of FDH

The expression vector encoding the His-tagged Zn(II)- and (NAD^+)-dependent FDH from *Pseudomonas putida* (PpFDH) was kindly provided by A. Bhagwat (Wayne State University) (Roy & Bhagwat, 2007). In performing the coupled fluorescent KDM assay, recombinant PpFDH is preferable to the commercially available form of the enzyme that includes bovine serum albumin (BSA) and other stabilizers (Krishnan et al., 2012). BSA is known to chelate divalent metal ions and has been reported to oxidize Fe(II) to Fe(III) (Bar-Or, Curtis, Rao, Bampos, & Lau, 2001; Masuoka, Hegenauer, Van Dyke, & Saltman, 1993; Masuoka & Saltman, 1994; Xu, Zhang, Shen, Wu, & Liu, 2008) and thus could inhibit JmjC KDMs. The PpFDH expression vector was transformed into *E. coli* Rosetta 2 DE3 cells, and the transformed cells were grown in 2xYT media at 37°C. Once the A_{600} of the cells reached 0.4–0.6, protein production was induced by the addition of 0.5 m*M* IPTG to the media, and the cells were allowed to grow overnight at 25°C. The following morning, cells were harvested by centrifugation at 4°C and stored at −20°C as previously described (see Section 2.1).

The His-tagged PpFDH was purified at 4°C according to the following protocol (Krishnan et al., 2012). A cell pellet corresponding to 1 L of culture was resuspended in 50 m*M* HEPES pH 8.0 and 500 m*M* NaCl (Buffer E) supplemented with EDTA-free HALT Protease Inhibitor Cocktail and lysed by sonication. The soluble fraction of the cell lysate was isolated by high-speed centrifugation and loaded onto a 4-mL column of IMAC Sepharose High Performance resin (GE Healthcare) that was charged with Zn(II) and preequilibrated in Buffer E. Purification using the Zn IMAC column assured the presence of the correct metal in the active site of PpFDH and precluded potential contamination with other transition metal ions commonly used in IMAC chromatography, such as Co and Ni. After washing the Zn column with at least four column volumes of Buffer E, the His-tagged protein was eluted using a linear gradient from Buffer E to Buffer F (Buffer E supplemented 500 m*M* imidazole) over 10 column volumes. The peak fractions containing the protein were then concentrated and loaded onto a Superdex 200 gel filtration column preequilibrated in 50 m*M* HEPES pH 8.0 and 150 m*M* NaCl. The fractions from the gel filtration column were analyzed by SDS-PAGE, and the peak fractions containing PpFDH were pooled and concentrated to 13 mg/mL, as determined by the A_{280} value in 6.0 *M* guanidinium chloride and 100 m*M* HEPES pH 7.5. The protein was then divided into small aliquots, flash-frozen in liquid nitrogen, and stored at −80°C.

3.3 FDH-Coupled Fluorescent Assay

The FDH-coupled KDM assay is based on the method initially reported by Tipton and coworkers who developed it to quantify methylamine oxidase (MAO) activity (Lizcano, Unzeta, & Tipton, 2000). Notably, they demonstrated that the FDH-coupled assay yielded kinetic parameters for MAO that were nearly identical to the values measured using a radiochemical assay, validating the FDH-coupled reaction as an accurate and reliable method of quantifying formaldehyde generation. The protocol for performing the FDH-coupled KDM assay with the JMJD2 KDMs and a 15mer H3K9me3 peptide substrate is presented in Table 2. In the assay cocktail, L-ascorbic acid is used as the reducing reagent, and $(NH_4)_2Fe(SO_4)_2$ supplies the Fe(II) cofactor required for the JmjC-catalyzed demethylase reaction (Fig. 1).

Table 2 Protocol for Performing the FDH-Coupled KDM Assay

Reagent	**Final Concentration (in 100 μL)**
Assay Cocktail 1	
HEPES pH 7.5	50 m*M*
NaCl	50 m*M*
Ammonium iron(II) sulfate [$(NH_4)_2Fe(SO_4)_2$]	50 μ*M*
L-Ascorbic acid	1 m*M*
NAD^+	1 m*M*
Formaldehyde dehydrogenase (FDH)	100 n*M*
KDM	Assay dependent
Water	Up to 75 μL
Assay Cocktail 2	
H3K9me3 peptide (or other substrate)	Assay dependent
2-Oxoglutarate	1 m*M*
Water	Up to 25 μL

Protocol

1. Set up microplate reader for continuous measurement of NADH fluorescence
2. Add 75 μL of cocktail 1 in one well and 25 μL of cocktail 2 in another well, of a black 96-well plate
3. Incubate at 37°C for 3–4 min
4. Initiate reaction by pipetting 25 μL of cocktail 2 into 75 μL of cocktail 1, avoiding bubbles
5. Monitor NADH fluorescence ($\lambda_{excitation}$ = 340 nm; $\lambda_{emission}$ = 490 nm) at 30-s intervals, for a total assay time of 5–10 min (until saturation)
6. Plot relative fluorescence units (RFUs) vs time and calculate slope by linear regression

To obtain optimal and reproducible JMJD2 activities, the L-ascorbic acid and $(NH_4)_2Fe(SO_4)_2$ stock solutions were freshly prepared every 2 h to minimize their oxidation (Krishnan et al., 2012; Sakurai et al., 2010). Reactions were performed in 96-well black polypropylene microplates (Corning), and NADH fluorescence was measured continuously using excitation and emission wavelengths of 340 and 490 nm, respectively, with a Safire2 Microplate Reader (Tecan). Initial velocities were measured over a period of 5–10 min, and a four-point calibration curve of NADH concentrations (0, 3.3, 6.6, and

10 μ*M*) was performed in triplicate in each microplate to convert the values of the initial velocities from relative fluorescent units (RFU) per minute to units of μ*M* min^{-1}.

Prior to characterizing the kinetic properties of JMJD2A and JMJD2D, a series of control assays were conducted to ensure optimal activity for the JMJD2 KDMs and that the concentration of FDH was not rate limiting in the coupled reaction (Krishnan et al., 2012). Coupled assays were performed with increasing concentrations of FDH and 2-OG to determine the concentrations that produced the optimal rates using 1.0 μ*M* JMJD2A and 700 μ*M* H3K9me3 peptide substrate. The optimal concentrations for FDH and 2-OG were 100 n*M* and 1.0 m*M*, respectively. Additional control assays were carried out to verify that the initial velocities were linear as a function of JMJD2A and JMJD2D concentration using fixed substrate concentrations (1.0 m*M* 2-OG and 700 μ*M* H3K9me3 peptide), demonstrating that activities of the KDMs were rate limiting in the FDH-coupled reaction.

After completing these controls, we characterized the steady-state kinetic parameter of Strep-Tactin-purified JMJD2A and JMJD2D (see Section 2.2) using the H3K9me3 peptide (Krishnan et al., 2012). Assays were performed at varying concentrations of the H3K9me3 peptide substrate (0–700 μ*M*). Initial velocities were measured and then plotted as a function of H3K9me3 peptide concentration. The Michaelis–Menten equation was fit to the data using Prism (GraphPad) to determine the k_{cat} and K_M values of JMJD2A and JMJD2D for the H3K9me3 peptide, as previously reported (Krishnan et al., 2012).

The k_{cat} values measured for the Strep-tagged JMJD2A and JMJD2D (mean values of 1.8 and 4.0 min^{-1}, respectively) were substantially higher than the values previously reported for the hexahistidine-tagged enzymes assayed using the same H3K9me3 peptide substrate (0.015 and 0.074 min^{-1}, respectively) (Couture et al., 2007; Krishnan et al., 2012). These turnover numbers correlate with the presence of inhibitory Ni(II) in the hexahistidine-tagged JMJD2 KDMs introduced during IMAC purification, whereas the Strep(II)-tagged enzymes purified by Strep-Tactin affinity chromatography lacked contaminating transition metal ions (see Table 1 and Section 2.3). In addition, the hexahistidine-tagged JMJD2A and JMJD2D were assayed using commercially available FDH that contained BSA (Couture et al., 2007), potentially contributing to the inhibition of their demethylase activities (see Section 3.2). The turnover numbers measured for the Strep(II)-tagged JMJD2 KDMs are in general agreement with the k_{cat} values reported for LSD1, which range between 2 and 12 min^{-1}

(Forneris et al., 2005; Gaweska et al., 2009; Schmidt & McCafferty, 2007; Szewczuk et al., 2007). Taken together, these findings illustrate that the optimized purification and assay conditions for the JMJD2 KDMs yield activities that are comparable to LSD1 in vitro, suggesting that these KDMs exhibit similar activities in vivo.

4. STRUCTURE DETERMINATION OF JmjC KDMs

Crystal structures of JmjC KDMs bound to methylated histone substrates have provided fundamental insights into their methylation site and state specificities, as well as their catalytic mechanism. To date, structures of the catalytic domains of FBXL11 (Cheng et al., 2014), JMJD2A (Chen et al., 2007; Couture et al., 2007; Ng et al., 2007), JMJD2D (Krishnan & Trievel, 2013), UTX (KDM6A) (Sengoku & Yokoyama, 2011), JMJD3 (KDM6B) (Kruidenier et al., 2012), and PHF8 (KDM7B) (Horton et al., 2010) have been determined in complex with cognate methylated histone peptides. These studies have taken advantage of the fact that ternary complexes of JmjC KDMs bound to methylated peptides and 2-OG can be obtained by substituting the active site Fe(II), which is required for the radical-based hydroxylation reaction, with inhibitory transition metal ions, such as Ni(II) and Co(II). Alternatively, ternary complexes in the presence of Fe(II) can be obtained by substituting 2-OG with the isosteric analog inhibitor *N*-oxalylglycine (NOG). In this section, we will provide an overview of methods for crystallizing JmjC KDMs with methylated substrates peptides using JMJD2A and JMJD2D as examples.

4.1 Crystallization of JMJD2 KDMs with Methylated Histone Peptides

Crystallization trials for JMJD2A and JMJD2D were conducted using the hexahistidine-tagged catalytic domains purified by Ni IMAC (see Section 2.1), as the inhibitory Ni(II) bound in the Fe(II) coordination site permitted us to capture ternary complexes bound to methylated histone H3 peptides (Couture et al., 2007; Krishnan & Trievel, 2013). All crystallization experiments were conducted at 4°C, as attempts to crystalize the enzymes at 20°C resulted in precipitation in the majority of the conditions screened. Crystallization trials were conducted using the vapor diffusion method with a variety of commercially available screens, including the Index (Hampton), JCSG+ (Qiagen), Wizard (Emerald Biosystems), Structure (Molecular Dimensions), PEGs I and II (Qiagen), Cation (Qiagen), and

Anion (Qiagen) screens. For JMJD2D, trials were conducted with 10–12 mg/mL protein, 1.5 m*M* H3K9me3 peptide (residues 1–15), and 1.5 m*M* 2-OG (Krishnan & Trievel, 2013). JMJD2A was screened at 10–12 mg/mL in the presence of 1.0 m*M* $NiSO_4$ and a twofold molar excess of H3K9me3, H3K36me2, or H3K36me3 (residues 26–41) peptide and cofactor (2-OG, NOG, or succinate) (Couture et al., 2007).

Initial crystals of JMJD2A bound to the methylated histone H3 peptides were obtained in 200 m*M* sodium/potassium phosphate pH 6.5 and 20% (w/v) PEG 3350 (Couture et al., 2007). Microseeding experiments were used to obtain larger, diffraction-quality crystals in 100 m*M* HEPES pH 7.5, 7% (v/v) Tacsimate (Hampton Research), and 12% (w/v) PEG monomethyl ether 5000. Crystals were harvested via serial transfer in the mother liquor supplemented with 5%, 10%, 15%, and 22.5% ethylene glycol as the cryoprotectant. The crystals were then flash-frozen and stored in liquid nitrogen.

Crystals of the JMJD2D · 2-OG · H3K9me3 peptide complex were initially obtained in 300 m*M* KNO_3 and 16% (w/v) PEG 3350 (Krishnan & Trievel, 2013). An Additive Screen (Hampton Research) was used to improve the quality of the initial crystals, and NaSCN was found to enhance their size and morphology. The final condition that yielded diffraction-quality crystals was 7% (w/v) PEG 3350, 100 m*M* NaSCN, and 350 m*M* KNO_3 (Fig. 2A). Crystals were harvested through serial transfer in the

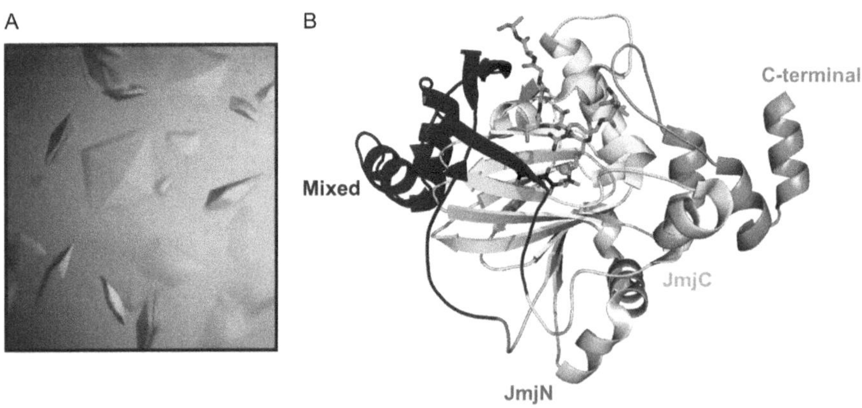

Fig. 2 Structure determination of JMJD2D bound to an H3K9me3 peptide. (A) Crystals of the JMJD2D · 2-OG · H3K9me3 peptide complex grown in 7% PEG 3350, 100 m*M* NaSCN, and 350 m*M* KNO_3. (B) Ribbon diagram rendering of the JMJD2D · 2-OG · H3K9me3 peptide ternary complex. The JmjN, mixed, JmjC, and C-terminal domains of JMJD2D are colored and labeled. The H3K9me3 peptide (*green* carbon atoms) and 2-OG (*black* carbon atoms) are shown in stick representation. The active site Ni(II) ion is depicted as a *cyan* sphere. The Zn(II) ion bound in the Zn-binding motif is omitted for clarity. (See the color plate.)

mother liquor supplemented with 5%, 15%, and 25% 1,3-propanediol as the cryoprotectant and then flash-frozen and stored in liquid nitrogen.

4.2 Structure Determination of JMJD2 KDM Peptide Complexes

X-ray diffraction data for JMJD2A and JMJD2D crystals were collected at the Advanced Photon Synchrotron Source, sectors 21 (LS-CAT) and 23 (GM/CA-CAT). Data were indexed, integrated, and scaled using HKL-2000 (HKL Research) (Otwinowski & Minor, 1997). Crystals of JMJD2A bound to H3K9me and H3K36me peptides belong to space group $P2_12_12$ and diffracted to between 2.05 and 2.3 Å resolution. Crystals of JMJD2D in complex with the H3K9me3 peptide belonged to space group $P2_12_12_1$ and diffracted to 1.8 Å resolution. Structures of the JMJD2A and JMJD2D complexes were determined by molecular replacement using MOLREP with a previously determined structure of JMJD2A used as a search model (Winn et al., 2011). In each structure, two molecules of the enzyme were found in the asymmetric unit. Structures of JMJD2A and JMJD2D were refined using REFMAC with model building in Coot (Emsley & Cowtan, 2004; Emsley, Lohkamp, Scott, & Cowtan, 2010; Murshudov, Vagin, & Dodson, 1997). With respect to the JMJD2A complexes, the electron density of the peptides was improved using translation libration screw refinement in REFMAC. Histone H3 peptide interactions with JMJD2A and JMJD2D were analyzed using PyMOL (Schrödinger) (Fig. 2B).

5. SUMMARY AND FUTURE PERSPECTIVES

The methods presented in this chapter outline strategies for the purification, kinetic characterization, and structure determination of the JMJD2 KDMs and can be readily adapted to other JmjC KDMs and KDM-containing protein complexes. Importantly, the pitfalls and caveats associated with biochemical characterization of hexahistidine-tagged JmjC KDMs purified by IMAC can be ameliorated using Strep(II)-tagged enzymes. Studies of the JMJD2, JARID1, and FBXL11 enzymes have demonstrated that IMAC purification of these enzymes results in contamination with inhibitory transition metal ions (Cheng et al., 2014; Krishnan et al., 2012; Torres et al., 2015), which can result in substantially diminished activity. Purification of JmjC KDMs using a Strep(II) tag or other affinity tags that afford transition metal-free purification, such as maltose-binding protein or glutathione *S*-transferase, should be employed for kinetic characterization of these enzymes and in high-throughput screening for KDM inhibitors.

X-ray crystallography has proven to be an essential tool in elucidating the molecular basis for the catalytic mechanism and substrate specificities of JmjC KDMs. Presently, structures of mammalian KDMs bound to methylated peptide substrates have been reported for the majority of the JmjC KDM subfamilies, with the exception of the JARID1 (KDM5) and JMJD1 (KDM3) enzymes. The crystallization methods described here as well as those reported for other JmjC KDMs should serve as a guide for determining the structures of the JARID1 and JMJD1 KDMs in complex with their cognate substrates. Indeed, the cocrystal structure of the catalytic domain of human JARID1A (KDM5A) bound to the inhibitor GST-J1 was recently reported (Horton et al., 2015), and the structure of the human JMJD1B (KDM3B) catalytic domain in complex with NOG has been deposited into the Protein Data Bank. Beyond characterization of substrate-bound structures, certain JmjC KDMs have also been shown to function in multi-subunit complexes (Mosammaparast & Shi, 2010). Structure determination of these complexes by crystallography, cryo-electron microscopy, or other approaches will provide insights into how accessory proteins can modulate the enzymatic activity, substrate specificity, and localization of JmjC KDMs.

ACKNOWLEDGMENT

Research into biological methylation in the Trievel laboratory is supported by NSF Grant CHE-1508492.

REFERENCES

Ai, H. W., Lee, J. W., & Schultz, P. G. (2010). A method to site-specifically introduce methyllysine into proteins in E. coli. *Chemical Communications (Cambridge)*, *46*(30), 5506–5508. http://dx.doi.org/10.1039/c0cc00108b.

Aik, W., McDonough, M. A., Thalhammer, A., Chowdhury, R., & Schofield, C. J. (2012). Role of the jelly-roll fold in substrate binding by 2-oxoglutarate oxygenases. *Current Opinion in Structural Biology*, *22*(6), 691–700. http://dx.doi.org/10.1016/j.sbi.2012.10.001.

Alam, H., Gu, B., & Lee, M. G. (2015). Histone methylation modifiers in cellular signaling pathways. *Cellular and Molecular Life Sciences*, *72*(23), 4577–4592. http://dx.doi.org/10.1007/s00018-015-2023-y.

Bar-Or, D., Curtis, G., Rao, N., Bampos, N., & Lau, E. (2001). Characterization of the Co(2+) and Ni(2+) binding amino-acid residues of the N-terminus of human albumin. An insight into the mechanism of a new assay for myocardial ischemia. *European Journal of Biochemistry*, *268*(1), 42–47. ejb1846 [pii].

Biggar, K. K., & Li, S. S. (2015). Non-histone protein methylation as a regulator of cellular signalling and function. *Nature Reviews Molecular Cell Biology*, *16*(1), 5–17. http://dx.doi.org/10.1038/nrm3915.

Black, J. C., Van Rechem, C., & Whetstine, J. R. (2012). Histone lysine methylation dynamics: Establishment, regulation, and biological impact. *Molecular Cell*, *48*(4), 491–507. http://dx.doi.org/10.1016/j.molcel.2012.11.006.

Black, J. C., & Whetstine, J. R. (2013). Tipping the lysine methylation balance in disease. *Biopolymers*, *99*(2), 127–135. http://dx.doi.org/10.1002/bip.22136.

Chen, H., & Costa, M. (2009). Iron- and 2-oxoglutarate-dependent dioxygenases: An emerging group of molecular targets for nickel toxicity and carcinogenicity. *Biometals*, *22*(1), 191–196. http://dx.doi.org/10.1007/s10534-008-9190-3.

Chen, H., Giri, N. C., Zhang, R., Yamane, K., Zhang, Y., Maroney, M., & Costa, M. (2010). Nickel ions inhibit histone demethylase JMJD1A and DNA repair enzyme ABH2 by replacing the ferrous iron in the catalytic centers. *The Journal of Biological Chemistry*, *285*(10), 7374–7383. http://dx.doi.org/10.1074/jbc.M109.058503. M109.058503 [pii].

Chen, H., Kluz, T., Zhang, R., & Costa, M. (2010). Hypoxia and nickel inhibit histone demethylase JMJD1A and repress Spry2 expression in human bronchial epithelial BEAS-2B cells. *Carcinogenesis*, *31*(12), 2136–2144. http://dx.doi.org/10.1093/carcin/bgq197. bgq197 [pii].

Chen, Z., Zang, J., Kappler, J., Hong, X., Crawford, F., Wang, Q., … Zhang, G. (2007). Structural basis of the recognition of a methylated histone tail by JMJD2A. *Proceedings of the National Academy of Sciences of the United States of America*, *104*(26), 10818–10823.

Chen, Z., Zang, J., Whetstine, J., Hong, X., Davrazou, F., Kutateladze, T. G., … Zhang, G. (2006). Structural insights into histone demethylation by JMJD2 family members. *Cell*, *125*(4), 691–702.

Cheng, Z., Cheung, P., Kuo, A. J., Yukl, E. T., Wilmot, C. M., Gozani, O., & Patel, D. J. (2014). A molecular threading mechanism underlies Jumonji lysine demethylase KDM2A regulation of methylated H3K36. *Genes & Development*, *28*(16), 1758–1771. http://dx.doi.org/10.1101/gad.246561.114.

Cheng, X., & Trievel, R. C. (2015). JmjC lysine demethylases. In R. P. Hausinger & C. J. Schofield (Eds.), *2-Oxoglutarate dependent hydroxylases* (pp. 210–233). Cambridge, UK: Royal Society of Chemistry.

Cloos, P. A., Christensen, J., Agger, K., Maiolica, A., Rappsilber, J., Antal, T., … Helin, K. (2006). The putative oncogene GASC1 demethylates tri- and dimethylated lysine 9 on histone H3. *Nature*, *442*(7100), 307–311.

Cornett, E. M., Dickson, B. M., Vaughan, R. M., Krishnan, S., Trievel, R. C., Strahl, B. D., & Rothbart, S. B. (2016). Substrate specificity profiling of histone-modifying enzymes by peptide microarray. *Methods in Enzymology*, *574*, 31–52.

Couture, J. F., Collazo, E., Ortiz-Tello, P. A., Brunzelle, J. S., & Trievel, R. C. (2007). Specificity and mechanism of JMJD2A, a trimethyllysine-specific histone demethylase. *Nature Structural & Molecular Biology*, *14*(8), 689–695.

Dimitrova, E., Turberfield, A. H., & Klose, R. J. (2015). Histone demethylases in chromatin biology and beyond. *EMBO Reports*, *16*(12), 1620–1639. http://dx.doi.org/10.15252/embr.201541113.

Emsley, P., & Cowtan, K. (2004). Coot: Model-building tools for molecular graphics. *Acta Crystallographica. Section D, Biological Crystallography*, *60*, 2126–2132. http://dx.doi.org/10.1107/S0907444904019158.

Emsley, P., Lohkamp, B., Scott, W. G., & Cowtan, K. (2010). Features and development of Coot. *Acta Crystallographica. Section D, Biological Crystallography*, *66*, 486–501. http://dx.doi.org/10.1107/S0907444910007493.

Fick, R. J., & Trievel, R. C. (2013). An overview of chromatin modifications. *Biopolymers*, *99*(2), 95–97. http://dx.doi.org/10.1002/bip.22158.

Forneris, F., Binda, C., Vanoni, M. A., Battaglioli, E., & Mattevi, A. (2005). Human histone demethylase LSD1 reads the histone code. *The Journal of Biological Chemistry*, *280*(50), 41360–41365.

Gantt, S. L., Gattis, S. G., & Fierke, C. A. (2006). Catalytic activity and inhibition of human histone deacetylase 8 is dependent on the identity of the active site metal ion. *Biochemistry*, *45*(19), 6170–6178. http://dx.doi.org/10.1021/bi060212u.

Gauthier, N., Caron, M., Pedro, L., Arcand, M., Blouin, J., Labonte, A., … Rodriguez-Suarez, R. (2012). Development of homogeneous nonradioactive methyltransferase and demethylase assays targeting histone H3 lysine 4. *Journal of Biomolecular Screening*, *17*(1), 49–58. http://dx.doi.org/10.1177/1087057111416659.

Gaweska, H., Henderson Pozzi, M., Schmidt, D. M., McCafferty, D. G., & Fitzpatrick, P. F. (2009). Use of pH and kinetic isotope effects to establish chemistry as rate-limiting in oxidation of a peptide substrate by LSD1. *Biochemistry*, *48*(23), 5440–5445. http://dx.doi.org/10.1021/bi900499w.

Goda, S., Isagawa, T., Chikaoka, Y., Kawamura, T., & Aburatani, H. (2013). Control of histone H3 lysine 9 (H3K9) methylation state via cooperative two-step demethylation by Jumonji domain containing 1A (JMJD1A) homodimer. *The Journal of Biological Chemistry*, *288*(52), 36948–36956. http://dx.doi.org/10.1074/jbc.M113.492595.

He, S., Bauman, D., Davis, J. S., Loyola, A., Nishioka, K., Gronlund, J. L., … McCafferty, D. G. (2003). Facile synthesis of site-specifically acetylated and methylated histone proteins: Reagents for evaluation of the histone code hypothesis. *Proceedings of the National Academy of Sciences of the United States of America*, *100*(21), 12033–12038. http://dx.doi.org/10.1073/pnas.2035256100.

Hopkinson, R. J., Hamed, R. B., Rose, N. R., Claridge, T. D., & Schofield, C. J. (2010). Monitoring the activity of 2-oxoglutarate dependent histone demethylases by NMR spectroscopy: Direct observation of formaldehyde. *Chembiochem*, *11*(4), 506–510. http://dx.doi.org/10.1002/cbic.200900713.

Horton, J. R., Engstrom, A., Zoeller, E. L., Liu, X., Shanks, J. R., Zhang, X., … Cheng, X. (2015). Characterization of a linked Jumonji domain of the KDM5/JARID1 family of histone H3 lysine 4 demethylases. *The Journal of Biological Chemistry*, *291*, 2631–2646. http://dx.doi.org/10.1074/jbc.M115.698449.

Horton, J. R., Upadhyay, A. K., Qi, H. H., Zhang, X., Shi, Y., & Cheng, X. (2010). Enzymatic and structural insights for substrate specificity of a family of jumonji histone lysine demethylases. *Nature Structural & Molecular Biology*, *17*(1), 38–43. http://dx.doi.org/10.1038/nsmb.1753. nsmb.1753 [pii].

Kapust, R. B., Tozser, J., Fox, J. D., Anderson, D. E., Cherry, S., Copeland, T. D., & Waugh, D. S. (2001). Tobacco etch virus protease: Mechanism of autolysis and rational design of stable mutants with wild-type catalytic proficiency. *Protein Engineering*, *14*(12), 993–1000. Retrieved from, http://www.ncbi.nlm.nih.gov/pubmed/11809930.

Kawamura, A., Tumber, A., Rose, N. R., King, O. N., Daniel, M., Oppermann, U., … Schofield, C. (2010). Development of homogeneous luminescence assays for histone demethylase catalysis and binding. *Analytical Biochemistry*, *404*(1), 86–93. http://dx.doi.org/10.1016/j.ab.2010.04.030.

King, O. N., Li, X. S., Sakurai, M., Kawamura, A., Rose, N. R., Ng, S. S., … Simeonov, A. (2010). Quantitative high-throughput screening identifies 8-hydroxyquinolines as cell-active histone demethylase inhibitors. *PLoS One*, *5*(11), e15535. http://dx.doi.org/10.1371/journal.pone.0015535.

Krishnan, S., Collazo, E., Ortiz-Tello, P. A., & Trievel, R. C. (2012). Purification and assay protocols for obtaining highly active Jumonji C demethylases. *Analytical Biochemistry*, *420*(1), 48–53.

Krishnan, S., & Trievel, R. C. (2013). Structural and functional analysis of JMJD2D reveals molecular basis for site-specific demethylation among JMJD2 demethylases. *Structure*, *21*(1), 98–108. http://dx.doi.org/10.1016/j.str.2012.10.018.

Kruidenier, L., Chung, C. W., Cheng, Z., Liddle, J., Che, K., Joberty, G., … Wilson, D. M. (2012). A selective jumonji H3K27 demethylase inhibitor modulates the

proinflammatory macrophage response. *Nature*, *488*(7411), 404–408. http://dx.doi.org/10.1038/nature11262. nature11262 [pii].
Lanouette, S., Mongeon, V., Figeys, D., & Couture, J. F. (2014). The functional diversity of protein lysine methylation. *Molecular Systems Biology*, *10*, 724. http://dx.doi.org/10.1002/msb.134974.
Leveridge, M., Buxton, R., Argyrou, A., Francis, P., Leavens, B., West, A., … Chung, C. W. (2014). Demonstrating enhanced throughput of RapidFire mass spectrometry through multiplexing using the JmjD2d demethylase as a model system. *Journal of Biomolecular Screening*, *19*(2), 278–286. http://dx.doi.org/10.1177/1087057113496276.
Li, Q., Ke, Q., & Costa, M. (2009). Alterations of histone modifications by cobalt compounds. *Carcinogenesis*, *30*(7), 1243–1251. http://dx.doi.org/10.1093/carcin/bgp088. bgp088 [pii].
Lizcano, J. M., Unzeta, M., & Tipton, K. F. (2000). A spectrophotometric method for determining the oxidative deamination of methylamine by the amine oxidases. *Analytical Biochemistry*, *286*(1), 75–79. http://dx.doi.org/10.1006/abio.2000.4782. S0003-2697(00)94782-1 [pii].
Mackeen, M. M., Kramer, H. B., Chang, K. H., Coleman, M. L., Hopkinson, R. J., Schofield, C. J., & Kessler, B. M. (2010). Small-molecule-based inhibition of histone demethylation in cells assessed by quantitative mass spectrometry. *Journal of Proteome Research*, *9*(8), 4082–4092. http://dx.doi.org/10.1021/pr100269b.
Maes, T., Carceller, E., Salas, J., Ortega, A., & Buesa, C. (2015). Advances in the development of histone lysine demethylase inhibitors. *Current Opinion in Pharmacology*, *23*, 52–60. http://dx.doi.org/10.1016/j.coph.2015.05.009.
Markolovic, S., Wilkins, S. E., & Schofield, C. J. (2015). Protein hydroxylation catalyzed by 2-oxoglutarate-dependent oxygenases. *The Journal of Biological Chemistry*, *290*(34), 20712–20722. http://dx.doi.org/10.1074/jbc.R115.662627.
Masuoka, J., Hegenauer, J., Van Dyke, B. R., & Saltman, P. (1993). Intrinsic stoichiometric equilibrium constants for the binding of zinc(II) and copper(II) to the high affinity site of serum albumin. *The Journal of Biological Chemistry*, *268*(29), 21533–21537.
Masuoka, J., & Saltman, P. (1994). Zinc(II) and copper(II) binding to serum albumin. A comparative study of dog, bovine, and human albumin. *The Journal of Biological Chemistry*, *269*(41), 25557–25561.
Mazur, P. K., Reynoird, N., Khatri, P., Jansen, P. W., Wilkinson, A. W., Liu, S., … Gozani, O. (2014). SMYD3 links lysine methylation of MAP3K2 to Ras-driven cancer. *Nature*, *510*(7504), 283–287. http://dx.doi.org/10.1038/nature13320.
McGrath, J., & Trojer, P. (2015). Targeting histone lysine methylation in cancer. *Pharmacology & Therapeutics*, *150*, 1–22. http://dx.doi.org/10.1016/j.pharmthera.2015.01.002.
Moore, K. E., & Gozani, O. (2014). An unexpected journey: Lysine methylation across the proteome. *Biochimica et Biophysica Acta*, *1839*(12), 1395–1403. http://dx.doi.org/10.1016/j.bbagrm.2014.02.008.
Mosammaparast, N., & Shi, Y. (2010). Reversal of histone methylation: Biochemical and molecular mechanisms of histone demethylases. *Annual Review of Biochemistry*, *79*, 155–179. http://dx.doi.org/10.1146/annurev.biochem.78.070907.103946.
Murshudov, G. N., Vagin, A. A., & Dodson, E. J. (1997). Refinement of macromolecular structures by the maximum-likelihood method. *Acta Crystallographica. Section D, Biological Crystallography*, *53*(Pt. 3), 240–255.
Musselman, C. A., Khorasanizadeh, S., & Kutateladze, T. G. (2014). Towards understanding methyllysine readout. *Biochimica et Biophysica Acta*, *1839*(8), 686–693. http://dx.doi.org/10.1016/j.bbagrm.2014.04.001.
Ng, S. S., Kavanagh, K. L., McDonough, M. A., Butler, D., Pilka, E. S., Lienard, B. M., … Oppermann, U. (2007). Crystal structures of histone demethylase JMJD2A reveal basis for substrate specificity. *Nature*, *448*(7149), 87–91.

Nguyen, D. P., Garcia Alai, M. M., Kapadnis, P. B., Neumann, H., & Chin, J. W. (2009). Genetically encoding N(epsilon)-methyl-L-lysine in recombinant histones. *Journal of the American Chemical Society*, *131*(40), 14194–14195. http://dx.doi.org/10.1021/ja906603s.

Nguyen, D. P., Garcia Alai, M. M., Virdee, S., & Chin, J. W. (2010). Genetically directing varepsilon-N, N-dimethyl-L-lysine in recombinant histones. *Chemistry & Biology*, *17*(10), 1072–1076. http://dx.doi.org/10.1016/j.chembiol.2010.07.013.

Otwinowski, Z., & Minor, W. (1997). Processing of X-ray diffraction data collected in oscillation mode. *Methods in Enzymology*, *276*, 307–326.

Piotukh, K., Geltinger, B., Heinrich, N., Gerth, F., Beyermann, M., Freund, C., & Schwarzer, D. (2011). Directed evolution of sortase A mutants with altered substrate selectivity profiles. *Journal of the American Chemical Society*, *133*(44), 17536–17539. http://dx.doi.org/10.1021/ja205630g.

Rose, N. R., Ng, S. S., Mecinovic, J., Lienard, B. M., Bello, S. H., Sun, Z., … Schofield, C. J. (2008). Inhibitor scaffolds for 2-oxoglutarate-dependent histone lysine demethylases. *Journal of Medicinal Chemistry*, *51*(22), 7053–7056. http://dx.doi.org/10.1021/jm800936s.

Roy, T. W., & Bhagwat, A. S. (2007). Kinetic studies of Escherichia coli AlkB using a new fluorescence-based assay for DNA demethylation. *Nucleic Acids Research*, *35*(21), e147. http://dx.doi.org/10.1093/nar/gkm1031. gkm1031 [pii].

Sakurai, M., Rose, N. R., Schultz, L., Quinn, A. M., Jadhav, A., Ng, S. S., … Simeonov, A. (2010). A miniaturized screen for inhibitors of Jumonji histone demethylases. *Molecular Biosystems*, *6*(2), 357–364. http://dx.doi.org/10.1039/b912993f.

Schmidt, D. M., & McCafferty, D. G. (2007). trans-2-Phenylcyclopropylamine is a mechanism-based inactivator of the histone demethylase LSD1. *Biochemistry*, *46*(14), 4408–4416. http://dx.doi.org/10.1021/bi0618621.

Schneider, J., & Shilatifard, A. (2006). Histone demethylation by hydroxylation: Chemistry in action. *ACS Chemical Biology*, *1*(2), 75–81.

Sekirnik, R., Rose, N. R., Mecinovic, J., & Schofield, C. J. (2010). 2-Oxoglutarate oxygenases are inhibited by a range of transition metals. *Metallomics*, *2*(6), 397–399. http://dx.doi.org/10.1039/c004952b.

Sengoku, T., & Yokoyama, S. (2011). Structural basis for histone H3 Lys 27 demethylation by UTX/KDM6A. *Genes & Development*, *25*(21), 2266–2277. http://dx.doi.org/10.1101/gad.172296.111.

Shi, Y., Lan, F., Matson, C., Mulligan, P., Whetstine, J. R., Cole, P. A., & Casero, R. A. (2004). Histone demethylation mediated by the nuclear amine oxidase homolog LSD1. *Cell*, *119*(7), 941–953.

Shiau, C., Trnka, M. J., Bozicevic, A., Ortiz Torres, I., Al-Sady, B., Burlingame, A. L., … Fujimori, D. G. (2013). Reconstitution of nucleosome demethylation and catalytic properties of a Jumonji histone demethylase. *Chemistry & Biology*, *20*(4), 494–499. http://dx.doi.org/10.1016/j.chembiol.2013.03.008.

Shogren-Knaak, M. A., Fry, C. J., & Peterson, C. L. (2003). A native peptide ligation strategy for deciphering nucleosomal histone modifications. *The Journal of Biological Chemistry*, *278*(18), 15744–15748. http://dx.doi.org/10.1074/jbc.M301445200.

Simon, M. D., Chu, F., Racki, L. R., de la Cruz, C. C., Burlingame, A. L., Panning, B., Shokat, K. M. … (2007). The site-specific installation of methyl-lysine analogs into recombinant histones. *Cell*, *128*(5), 1003–1012. http://dx.doi.org/10.1016/j.cell.2006.12.041. S0092-8674(07)00115-8 [pii].

Szewczuk, L. M., Culhane, J. C., Yang, M., Majumdar, A., Yu, H., & Cole, P. A. (2007). Mechanistic analysis of a suicide inactivator of histone demethylase LSD1. *Biochemistry*, *46*(23), 6892–6902.

Thinnes, C. C., England, K. S., Kawamura, A., Chowdhury, R., Schofield, C. J., & Hopkinson, R. J. (2014). Targeting histone lysine demethylases—Progress, challenges, and the future. *Biochimica et Biophysica Acta*, *1839*(12), 1416–1432. http://dx.doi.org/10.1016/j.bbagrm.2014.05.009.

Torres, I. O., Kuchenbecker, K. M., Nnadi, C. I., Fletterick, R. J., Kelly, M. J., & Fujimori, D. G. (2015). Histone demethylase KDM5A is regulated by its reader domain through a positive-feedback mechanism. *Nature Communications*, *6*, 6204. http://dx.doi.org/10.1038/ncomms7204.

Tsukada, Y., Fang, J., Erdjument-Bromage, H., Warren, M. E., Borchers, C. H., Tempst, P., & Zhang, Y. (2006). Histone demethylation by a family of JmjC domain-containing proteins. *Nature*, *439*(7078), 811–816.

Walport, L. J., Hopkinson, R. J., Vollmar, M., Madden, S. K., Gileadi, C., Oppermann, U., … Johansson, C. (2014). Human UTY(KDM6C) is a male-specific N{epsilon}-methyl lysyl-demethylase. *The Journal of Biological Chemistry*, *289*(26), 18302–18313. http://dx.doi.org/10.1074/jbc.M114.555052.

Wang, C., Caron, M., Burdick, D., Kang, Z., Auld, D., Hill, W. A., … Zhang, J. H. (2012). A sensitive, homogeneous, and high-throughput assay for lysine-specific histone demethylases at the H3K4 site. *Assay and Drug Development Technologies*, *10*(2), 179–186. http://dx.doi.org/10.1089/adt.2011.0395.

Winn, M. D., Ballard, C. C., Cowtan, K. D., Dodson, E. J., Emsley, P., Evans, P. R., … Wilson, K. S. (2011). Overview of the CCP4 suite and current developments. *Acta Crystallographica. Section D, Biological Crystallography*, *67*(Pt. 4), 235–242. http://dx.doi.org/10.1107/S0907444910045749.

Xu, X., Zhang, L., Shen, D., Wu, H., & Liu, Q. (2008). Oxygen-dependent oxidation of Fe(II) to Fe(III) and interaction of Fe(III) with bovine serum albumin, leading to a hysteretic effect on the fluorescence of bovine serum albumin. *Journal of Fluorescence*, *18*(1), 193–201. http://dx.doi.org/10.1007/s10895-007-0263-4.

Yu, W., Eram, M. S., Hajian, T., Szykowska, A., Burgess-Brown, N., Vedadi, M., & Brown, P. J. (2014). A scintillation proximity assay for histone demethylases. *Analytical Biochemistry*, *463*, 54–60. http://dx.doi.org/10.1016/j.ab.2014.06.023.

Yu, V., Fisch, T., Long, A. M., Tang, J., Lee, J. H., Hierl, M., … Emkey, R. (2012). High-throughput TR-FRET assays for identifying inhibitors of LSD1 and JMJD2C histone lysine demethylases. *Journal of Biomolecular Screening*, *17*(1), 27–38. http://dx.doi.org/10.1177/1087057111418228.

Zhang, X., Huang, Y., & Shi, X. (2015). Emerging roles of lysine methylation on non-histone proteins. *Cellular and Molecular Life Sciences*, *72*(22), 4257–4272. http://dx.doi.org/10.1007/s00018-015-2001-4.

Zhang, X., Yang, Z., Khan, S. I., Horton, J. R., Tamaru, H., Selker, E. U., & Cheng, X. D. (2003). Structural basis for the product specificity of histone lysine methyltransferases. *Molecular Cell*, *12*(1), 177–185. Retrieved from <Go to ISI>://000184528100018.

CHAPTER THIRTEEN

Preparation and Analysis of Native Chromatin-Modifying Complexes

Y. Doyon*[,1], J. Côté*[,†,1]
*Centre Hospitalier Universitaire de Québec Research Center and Faculty of Medicine, Laval University, Quebec City, QC, Canada
†St-Patrick Research Group in Basic Oncology and Laval University Cancer Research Center, Quebec City, QC, Canada
[1]Corresponding authors: e-mail address: yannick.doyon@crchudequebec.ulaval.ca; jacques.cote@crhdq.ulaval.ca

Contents

1. Introduction 304
2. Methods 305
   2.1 Endogenous Tagging of EPC1 C-Terminus Using CRISPR/Cas9 305
   2.2 Rapid Generation of Isogenic Cell Lines Expressing TAP-Tagged EPC1 from the AAVS1 Safe Harbor Locus 308
   2.3 Large-Scale Expansion of K562 Cells 309
   2.4 Preparation of NEs 309
   2.5 TAP Purification of EPC1 310
   2.6 Analysis of EPC1 Complex Subunits by SDS–Polyacrylamide Gel Electrophoresis and Silver Staining 312
   2.7 Sample Preparation for Mass Spectrometry Analysis 313
   2.8 Endogenous Tagging of EZH2 N-Terminus Using TALENs 313
   2.9 Expression of TAP-Tagged EZH2 Variants from the AAVS1 Safe Harbor Locus 313
   2.10 Dox-Inducible Expression of EZH2 from the AAVS1 Safe Harbor Locus 315
   2.11 TAP Purification of EZH2 315
   2.12 HAT/HMT Assays 315
3. Conclusion 317
Acknowledgments 317
References 318

Abstract

Nucleosomes, the basic units of chromatin, are decorated with a myriad of posttranslational modifications (PTMs) by the action of chromatin modifiers. These enzymes function almost exclusively as part of stable protein complexes that assist their recruitment to specific genomic loci, specify their substrate, and provide allosteric control. By altering the interactions within nucleosomes or with neighboring nucleosomes and serving as a platform to engage effector proteins, PTMs deposited by histone-modifying complexes influence virtually every nuclear process and are at the heart of the epigenetic

Methods in Enzymology, Volume 573
ISSN 0076-6879
http://dx.doi.org/10.1016/bs.mie.2016.01.017

mechanisms. Hence, it is critical to identify their components, define their structures, and characterize their biochemical activities. Here we describe protocols for tandem affinity purification (TAP) of native histone acetyltransferase (HAT) and methyltransferase (HMT) complexes from human cells engineered to express bait proteins from a genomic safe harbor or their endogenous chromosomal genes, using zinc-finger nucleases (ZFNs), TAL effector nucleases (TALENs), and clustered regularly interspaced short palindromic repeats (CRISPR)/Cas9 systems. The approaches presented aim to preserve natural transcriptional and posttranscriptional regulation and minimize biochemical artifacts due to ectopic expression. Near homogenous preparations of native complexes are obtained in sufficient amounts to perform biochemical assays and characterize their components.

1. INTRODUCTION

The nucleosome, the fundamental repeating unit of chromatin, wraps nearly two turns of DNA around a core histone octamer formed by one H3/H4 tetramer flanked by two H2A/H2B dimers. It acts as a signal integration platform that regulates all DNA transactions. Posttranslational modifications (PTMs) generated by chromatin-modifying enzymes influence gene expression profiles, DNA repair, and replication. Often composed of modular units these protein machines contain specific PTM readers, transcription factor interacting interfaces, scaffolding subunits, and catalysis-stimulating factors. Determining how these parts are assembled and coordinated is crucial to our understanding of cellular physiology.

The NuA4 complex is a good example of chromatin modifiers as it contains more than 16 subunits, including the KAT5/TIP60 histone acetyltransferase, several histone PTM reader modules in addition to another enzymatic subunit, EP400, responsible for replacing H2A with variant H2A.Z in chromatin. NuA4 is recruited to specific genes to stimulate transcription as well as to DNA damage sites to assist repair (Steunou, Rossetto, & Cote, 2014). The polycomb repressive complex-2 (PRC2) is another example of well-known chromatin modifiers as it contains the EZH2 histone methyltransferase and is important for transcription silencing (Margueron & Reinberg, 2011). NuA4 and PRC2 are essential for embryonic stem cell maintenance and development, and are deregulated in a very large number of cancers.

In this chapter, we describe the isolation and characterization of native NuA4 and PRC2 complexes from human cells using standard biochemical conditions that can be generally applied to other chromatin-modifying complexes. Furthermore, we describe genome-editing-based methods to engineer human cell lines expressing the bait protein of interest from isogenic backgrounds and endogenous genes.

2. METHODS

To facilitate the production of human cell lines expressing tagged subunits of chromatin-modifying complexes we describe methods to tag endogenous genes or express cDNAs ectopically from a genomic safe harbor. Detailed protocols for nuclear extract (NE) preparations and tandem affinity purification (TAP) are also presented.

2.1 Endogenous Tagging of EPC1 C-Terminus Using CRISPR/Cas9

Large-scale proteomic screens in yeast highlighted the importance of maintaining natural regulation of gene expression for bait proteins in order to obtain high-quality datasets. With the advent of customizable nucleases such as zinc-finger nucleases (ZFNs), TAL effector nucleases (TALENs), and clustered regularly interspaced short palindromic repeats (CRISPR)/Cas9, it has become routine to edit genes in human cells. We took advantage of these systems to tag endogenous genes with a combination of 3 × FLAG and 2 × STREP tags to perform TAP under gentle conditions and obtain highly purified native/endogenous protein complexes (Dalvai et al., 2015). This is illustrated by the C-terminal tagging of the Enhancer of Polycomb homolog 1 (EPC1), a 100-kDa subunit of the TIP60/NuA4 complex responsible for interaction with the nucleosome and stimulation of catalysis.

2.1.1 Design and Cloning of sgRNA Targeting EPC1

1. Scan your target sequence using a web-based CRISPR design tool such as http://crispr.mit.edu/ to identify sgRNA that will dictate cleavage in proximity to the STOP codon. The target sequence should span a ~100-bp window centered on your target, eg, STOP codon for C-terminal tagging.
2. Generate and order nonphosphorylated oligonucleotides compatible with expression of sgRNA from the U6 promoter in the MLM3636 vector (Addgene #43860) using ZiFiT Targeter (http://zifit.partners.org/ZiFiT/ChoiceMenu.aspx). For *EPC1* the target site is 5′-GTGACGT AGCTTCCTCCGAG-3′ (Fig. 1A). If the nucleotide at position

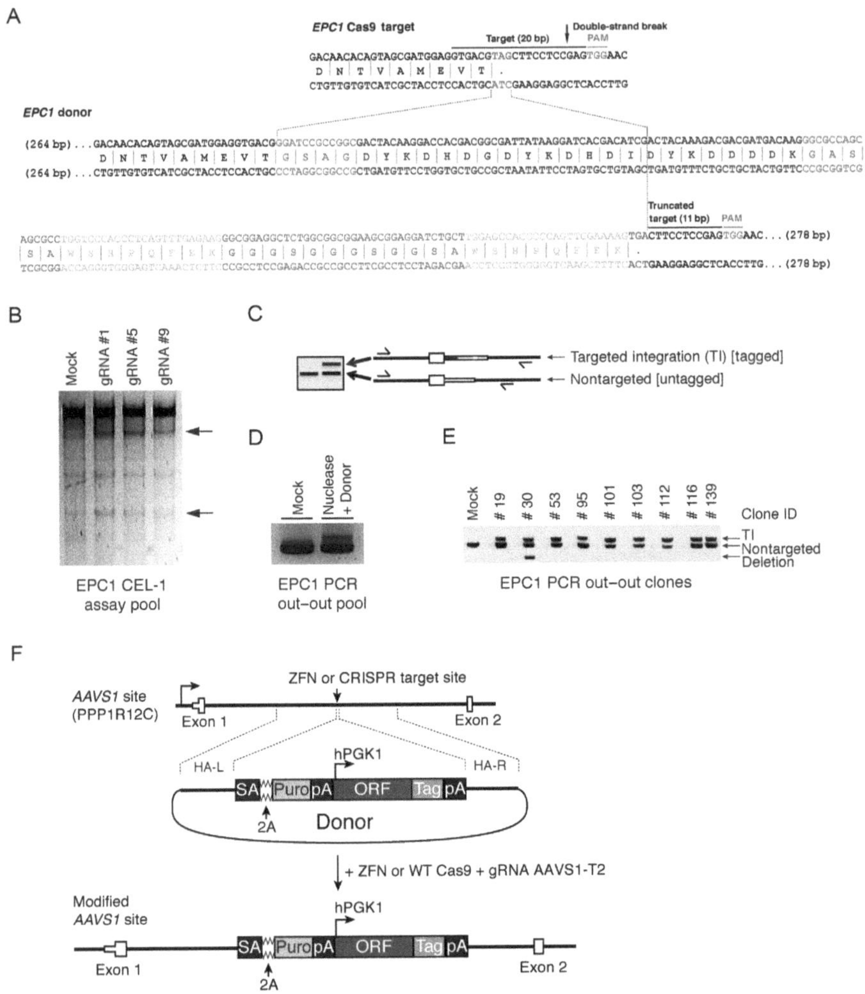

Fig. 1 Strategies to establish cell lines expressing tagged proteins from endogenous genes or a genomic safe harbor locus. (A) Schematic of the EPC1 locus, wild-type Cas9 target site, and donor construct. Annotated are the positions of the stop codon (TAG), the target site, and the protospacer adjacent motif (PAM) that specifies the cleavage site. The 3′ end of the left homology arm (HA-L) starts just before the STOP and the 5′ end of the right homology arm (HA-R) starts after the STOP. Note that in this case, the STOP codon in the donor was changed from TAG to TGA to reduce the possibility of cleavage by the CRISPR/Cas9 enzyme. This way, the donor sequence contains a truncated target sequence of just 11 bp, which should be sufficient to prevent DSB formation in the episomal donor and in the chromosome after targeted integration (TI) of the tag. We avoid repetitive sequences in HA-L and HA-R. The optimal guide RNA is selected based on several criteria: (i) high activity based on CEL1 assay, (ii) induction of DSB in proximity to the STOP codon, (iii) option to prevent donor cleavage (see above), and, if possible, (iv) cleavage in the 3′ UTR of the gene to prevent NHEJ-mediated mutagenesis

20 (20 bp 5′ to the protospacer adjacent motif (PAM) sequence) of your favorite gene is A, C, or T, then modify it to a G to accommodate the transcription initiation requirements of the human U6 promoter.

3. Mix 2 μg of each oligonucleotide in 50 μL annealing buffer (10 m*M* Tris pH 7.4, 1 m*M* EDTA, 50 m*M* NaCl), heat for 2 min at 95°C, and chill at −1°C/s in a thermocycler.
4. Ligate 1 μL of annealed oligos into pMLM3636 digested with BsmBI.
5. Sequence plasmids using OS280 5′-CAGGGTTATTGTCTCATGAG CGG-3′.

Note: The strength and specificity of different sgRNAs can be tested by transient transfection with CAS9 nuclease and CEL-1 assay/Surveyor assay to detect indels (insertions–deletions) created by nonhomologous end-joining of the break (Dalvai et al., 2015; Guschin et al., 2010). Extract genomic DNA from 2.5E5 cells with 250 μL of QuickExtract DNA extraction solution (Epicentre) per manufacturer's recommendations 3 days post-transfection. Amplify genomic DNA by PCR for 30 cycles with the Phusion polymerase (Thermo Scientific) using primers 5′-TGATGGTAATGTAG TTGACTGTGG-3′ and 5′-GAGCATTGCTGTCAAGTCCC-3′ for *EPC1* and perform the Surveyor mutation detection assay (Transgenomics) according to the manufacturer's protocol. Load samples on 10% PAGE gels in TBE buffer (Fig. 1B).

2.1.2 Design and Cloning of Tagging Donor

1. Synthesize a nonphosphorylated gBlock gene fragment (Integrated DNA Technologies) containing the TAP tag (3 × FLAG–2 × STREP tags, Fig. 1A) (Addgene #68375) flanked by genomic DNA sequences

Fig. 1—Cont'd of the coding sequence of the untargeted allele. (B) Result of a Cel-1 assay to detect the activity of different sgRNA targeting EPC1. (C) Schematic of a PCR-based assay (out–out PCR) to detect TI of the tag sequence. Primers are located outside of the homology arms and are designed to yield a longer PCR product if the tag is inserted. (D) Results of a out–out PCR assay conducted on genomic DNA from K562 cells transfected with (i) wild-type Cas9 expression vector and EPC1 donor (Mock) and (ii) wild-type Cas9 expression vector, gRNA #1 and EPC1 donor (nuclease+ donor). Cells were collected 3 days posttransfection. (E) Same as in (D), but on single-cell-derived clones. (F) Schematic of the donor construct and of the *AAVS1* locus following ORF addition. The first two exons of the *PPP1R12C* gene are shown as *open boxes*. Also annotated are the locations of the splice acceptor site (SA), 2A self-cleaving peptide sequence (2A), puromycin resistance gene (Puro), polyadenylation sequence (pA), human phosphoglycerate kinase 1 promoter (h*PGK1*), 3 × FLAG–2 × STREP tandem affinity tag (Tag), and homology arms left and right (HA-L and HA-R are, respectively, 800 and 840 bp). (See the color plate.)

5′ and 3′ of the STOP codon, 291 and 295 bp, respectively, in the case of *EPC1*. To adapt to your favorite gene, the homology arms can be anywhere between 250 bp and 1 kb as long as they do not contain repetitive elements. Make sure that the sgRNA will not dictate cleavage of the donor by splitting the target sequence on each side of the tag or mutating the PAM sequence (Fig. 1A).
2. Assemble using Zero Blunt TOPO cloning kit (Life Technologies).
3. Alternatively, the homology arms can be cloned by PCR directly into the *AAVS1*-TAP tagging plasmid (Addgene #68375).

2.1.3 Gene Targeting in K562 Cells

1. K562 cells (ATCC CCL-243) are maintained at 37°C under 5% CO_2 in RPMI medium supplemented with 10% FBS, penicillin–streptomycin, and GlutaMAX.
2. Electroporate 1 million K562 cells with 2 μg of gRNA plasmid, 2 μg of wild-type Cas9 expression vector (Addgene #41815), and 4 μg of donor using the Amaxa 4D-Nucleofector (Lonza) per manufacturer's recommendations.
3. Clone and expand cells via limiting dilution 3 days posttransfection in 96-well plates.
4. Extract genomic DNA from 2.5E5 cells with 250 μL of QuickExtract DNA extraction solution (Epicentre) per manufacturer's recommendations.
5. Amplify genomic DNA by PCR for 30 cycles with the Phusion polymerase (Thermo Scientific) using primers 5′-AAGCCTGACACAAA TCTCAGT-3′ and 5′-CCAAGGAGTCCACAGCTACC-3′ in the case of *EPC1*, to identify targeted clones. These primers are designed to bind outside of the homology arms and yield a longer PCR product when the tag is integrated. Load products in 1% agarose gels in TAE buffer (Fig. 1C–E).
6. Sequence both alleles to confirm that accurate gene modification occurred. Be careful to avoid selecting clones with indels (insertions/deletions) present on the nontargeted allele.
7. Confirm expression of EPC1-tag by Western blot analysis using an anti-FLAG antibody. In our hands, efficiency of integration and expression of the tagged protein has varied between 1% and 30%.

2.2 Rapid Generation of Isogenic Cell Lines Expressing TAP-Tagged EPC1 from the AAVS1 Safe Harbor Locus

As an alternative to endogenous tagging, it is possible to generate cell lines expressing near physiological levels of tagged proteins via nuclease-driven targeted integration into the human *PPP1R12C* gene, a safe harbor genomic locus known as *AAVS1* (Dalvai et al., 2015). In this system, the use of a gene-trap vector allowing puromycin selection of targeted cells greatly simplifies the isolation of the cell line of interest (Fig. 1F).

1. Clone *EPC1* cDNA (or your favorite cDNA) into AAVS1_Puro_PGK1_3xFLAG_Twin_Strep (Addgene #68375).
2. Electroporate 1 million K562 cells with 1 μg of ZFN expression vector (Hockemeyer et al., 2009) and 4 μg of *AAVS1* donor construct. Alternatively, use wild-type Cas9 expression vector (Addgene #41815) in combination with gRNA_AAVS1-T2 (Addgene #41818).
3. Select and clone cells simultaneously for 10 days in methylcellulose-based semisolid RPMI medium supplemented with 0.25 μg/mL puromycin starting 3 days posttransfection. Alternatively, clone and expand cells via limiting dilution 3 days posttransfection in 96-well plates. Pools of cells can also be selected in suspension cultures.
4. Pick and expand clones in 96 wells for 3 days and then transfer to 12-well plates for another 3–4 days.
5. Harvest cells for Western blot analysis and confirm expression of EPC1-tag by Western blot analysis using an anti-FLAG antibody.

2.3 Large-Scale Expansion of K562 Cells

1. Grow 1–3 L of K562 cells in spinner or Erlenmeyer flasks using gentle agitation in basal media supplemented with 25 m*M* HEPES–NaOH, pH 7.4.
2. Monitor cell growth and viability every day. For optimal growth, maintain cultures between 3E5 and 1E6 cells per mL.
3. Harvest cells when they reach saturation at 1E6 per mL.

2.4 Preparation of NEs

1. Soluble proteins from the nuclei are extracted using a high-salt buffer (300 m*M* KCl) exactly as described (Abmayr, Yao, Parmely, & Workman, 2006). However, at step 11, the NE is diluted to 30 mL using the following buffer (20 m*M* HEPES–KOH, pH 7.9 at 4°C, 25% glycerol v/v, 300 m*M* KCl, 1.5 m*M* $MgCl_2$, 0.2 m*M* EDTA, supplemented with 1× Halt protease and phosphatase inhibitor cocktail without EDTA (Pierce)) for the extraction.
2. Flash-freeze NE in liquid nitrogen and store at −80°C.

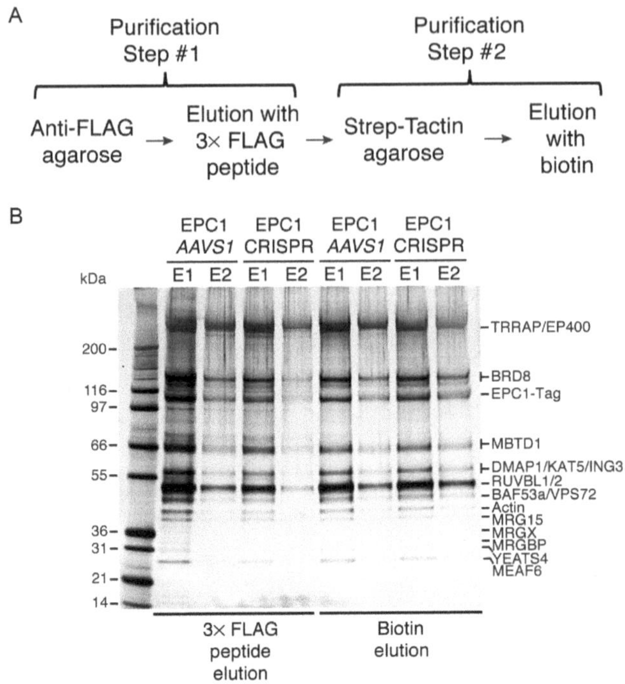

Fig. 2 Tandem affinity purification (TAP) of the EPC1 complex. (A) TAP purification scheme. (B) Silver-stained SDS–PAGE showing the comparison of the EPC1 complex obtained when expressed from the endogenous (EPC1-CRISPR) and the *AAVS1* (EPC1-*AAVS1*) loci. Eluted fractions (E1, E2) from the two purification steps are shown.

2.5 TAP Purification of EPC1

The TAP tag selected contains 59 AA for a predicted molecular weight of 6 kDa and does not necessitate proteolytic cleavage for elution. The first step of purification is performed on anti-FLAG M2 agarose affinity resin due to its high binding capacity, while the second step uses Strep-Tactin resin to increase the purity of the fractions (Fig. 2A).

2.5.1 Anti-FLAG M2 Affinity Purification

1. Prechill rotor, centrifuge tubes, tabletop centrifuge, and ultracentrifuge to 4°C.
2. Rapidly thaw NE in a 37°C water bath and immediately transfer to ice/water bath when it becomes slushy.
3. Determine the exact volume of NE and add 0.1% Tween 20 (10% stock). Mix promptly but gently.

4. Ultracentrifuge NE at 100,000 × *g* for 45 min at 4°C.
5. Immediately transfer NE by decanting to a prechilled 50-mL conical tube. Rapid transfer helps to reduce the amount of contaminating lipids.
6. In the meantime, split 300 μL of Sepharose CL-6B resin (400 μL of 75% slurry) between three polyprep chromatography columns (Bio-Rad). Wash with 2 mL of PBS and then equilibrate with 2 mL of the following buffer (20 m*M* HEPES–KOH pH 7.9 at 4°C, 25% glycerol v/v, 300 m*M* KCl, 1.5 m*M* $MgCl_2$, 0.2 m*M* EDTA, supplemented with 1 × Halt protease and phosphatase inhibitor cocktail without EDTA (Pierce)). Use end-caps to prevent resin from drying and place the three columns into a 50-mL conical tube in an ice bucket.
7. Pour the ultracentrifuged NE over the CL-6B columns to preclear the extract (~8 mL of NE per column). We use three columns instead of one to decrease processing time.
8. In the meantime, prepare 250 μL of anti-FLAG M2 agarose beads (Sigma) (50% slurry). Pipet 500 μL of slurry into a 5-mL low-protein binding "microcentrifuge" tube. Add 2.5 mL PBS and spin at 750 × *g* for 1 min. Remove PBS and wash with 2.5 mL of 100 m*M* glycine–HCl, pH 3.5 (mix for 30 s and spin as above). Wash twice with 2.5 mL PBS and one time with 2.5 mL (20 m*M* HEPES–KOH, pH 7.9 at 4°C, 25% glycerol v/v, 300 m*M* KCl, 1.5 m*M* $MgCl_2$, 0.2 m*M* EDTA, supplemented with 1 × Halt protease and phosphatase inhibitor cocktail without EDTA (Pierce)). Resuspend resin with 250 μL of the same buffer.
9. Transfer anti-FLAG beads to the precleared NE in the 50-mL conical.
10. Rotate on wheel at 4°C for 2 h.
11. Decant into one polyprep chromatography column and harvest flow through. Make sure to collect all resin.
12. Wash with 10 mL (40 column volumes (CV)) of buffer #1 (20 m*M* HEPES–KOH, pH 7.9, 10% glycerol, 300 m*M* KCl, 0.1% Tween 20, 1 m*M* DTT, 1 × Halt protease and phosphatase inhibitor cocktail without EDTA (Pierce)).
13. Wash with 10 mL (40 CV) of buffer #2 (20 m*M* HEPES–KOH, pH 7.9, 10% glycerol, 150 m*M* KCl, 0.1% Tween 20, 1 m*M* DTT, 1 × Halt protease and phosphatase inhibitor cocktail without EDTA (Pierce)).
14. Place end-cap on column and transfer resin to 1.5-mL low-protein binding microcentrifuge tube. Spin for 10 s at 14,000 rpm and remove the supernatant.

15. Elute resin with 625 μL FLAG elution buffer (buffer #2 plus 150 μg/mL 3 × FLAG peptide). Rotate for 30 min at 4°C on wheel. Spin for 10 s at 14,000 rpm and collect the supernatant.
16. Repeat step 15 once.
17. Pool both fractions and transfer eluate over microspin column (Bio-Rad) placed into a 2-mL low-protein binding microcentrifuge. Spin for 10 s at 14,000 rpm to collect FLAG eluate. This step aims to prevent contamination of eluate with beads.
18. Collect a 15-μL aliquot for analysis by silver staining.

2.5.2 Strep-Tactin Affinity Purification

1. Pipet 125 μL of Strep-Tactin agarose beads (IBA) (250 μL of 50% slurry) into a 1.5-mL low-protein binding microcentrifuge tube. Wash with 1 mL PBS and spin for 10 s at 14,000 rpm.
2. Repeat step 1.
3. Wash with 1 mL of buffer #2, spin for 10 s at 14,000 rpm, and resuspend resin with 125 μL of buffer #2.
4. Add resin to pooled FLAG elution fractions; rotate for 1 h at 4°C in a 2-mL low-protein binding microcentrifuge tube.
5. Transfer resin to mini Bio-Spin column (Bio-Rad) and collect FT.
6. Wash resin with 5 mL (40 CV) of buffer #2.
7. Add end-caps and transfer resin to a 1.5-mL low-protein binding microcentrifuge tube.
8. Elute with 200 μL of buffer #2 supplemented with 2.5 m*M* D-biotin. Rotate for 30 min at 4°C on wheel.
9. Spin for 10 s at 14,000 rpm and carefully transfer supernatant over microspin column placed into a 2-mL low-protein binding microcentrifuge tube. Centrifuge for 10 s at 14,000 rpm to collect eluate. This is to prevent contamination of eluate with beads.
10. Repeat steps 8–9 with 300 μL of buffer #2 supplemented with 2.5 m*M* D-biotin.
11. Collect a 15-μL aliquot for analysis by silver staining.
12. Flash-freeze in liquid nitrogen and store at −80°C.

2.6 Analysis of EPC1 Complex Subunits by SDS–Polyacrylamide Gel Electrophoresis and Silver Staining

1. Typically, 15 μL of the first FLAG and biotin elution fractions are loaded on Bolt 4–12% Bis-Tris gels (Life Technologies) and ran for 45 min at 200 V in MOPS buffer.

2. Silver staining is performed using the SilverQuest kit (Life Technologies).
3. A typical profile of purification is shown in Fig. 2B.

2.7 Sample Preparation for Mass Spectrometry Analysis

1. Load 30 μL samples from the first biotin elution fraction into a Bolt 12% Bis-Tris gels and run over 1 cm to stack all proteins into one band.
2. Wash gel briefly with ultrapure water.
3. Incubate for 30 min in 10% MeOH/7% acetic acid.
4. Incubate overnight in Sypro Ruby (Bio-Rad) gel stain.
5. Rince twice with ultrapure water.
6. Cut the protein band under UV light.
7. Rinse band with 70% acetonitrile, repeat once.
8. Store samples at −80°C.

2.8 Endogenous Tagging of EZH2 N-Terminus Using TALENs

1. Design and clone a recombination donor to tag EZH2 at its N-terminus using the rules described in Section 2.1.2 (Fig. 3A).
2. Electroporate 1 million K562 cells with 2.5 μg of each TALEN expression vector (Addgene #36775 and #36776) and 4 μg of donor using the Amaxa 4D-Nucleofector (Lonza) per manufacturer's recommendations.
3. Isolate targeted clones as described in Section 2.1.3. Use the following primers for the out–out PCR assays: 5′-GTGGCACAAGAGGC AAAAAT-3′ and 5′-CGATTGCCATCCTTTCTTTG-3′.

2.9 Expression of TAP-Tagged EZH2 Variants from the AAVS1 Safe Harbor Locus

The *AAVS1* ectopic expression system is useful to rapidly and reliably generate panels of isogenic cell lines expressing protein variants (eg, splice variants) (Dalvai et al., 2015). To illustrate this, we established multiple clonal cell lines expressing wild-type EZH2 or EZH2 Y641F, a gain-of-function mutation in the SET domain, occurring in follicular lymphoma and aggressive diffuse large B-cell lymphoma (Yap et al., 2011). As shown in Fig. 3B, the increase in histone H3 Lys27 trimethylation (H3K27me3) found in these cancer cells can be recapitulated. To generate such cells lines, follow instructions in Section 2.2.

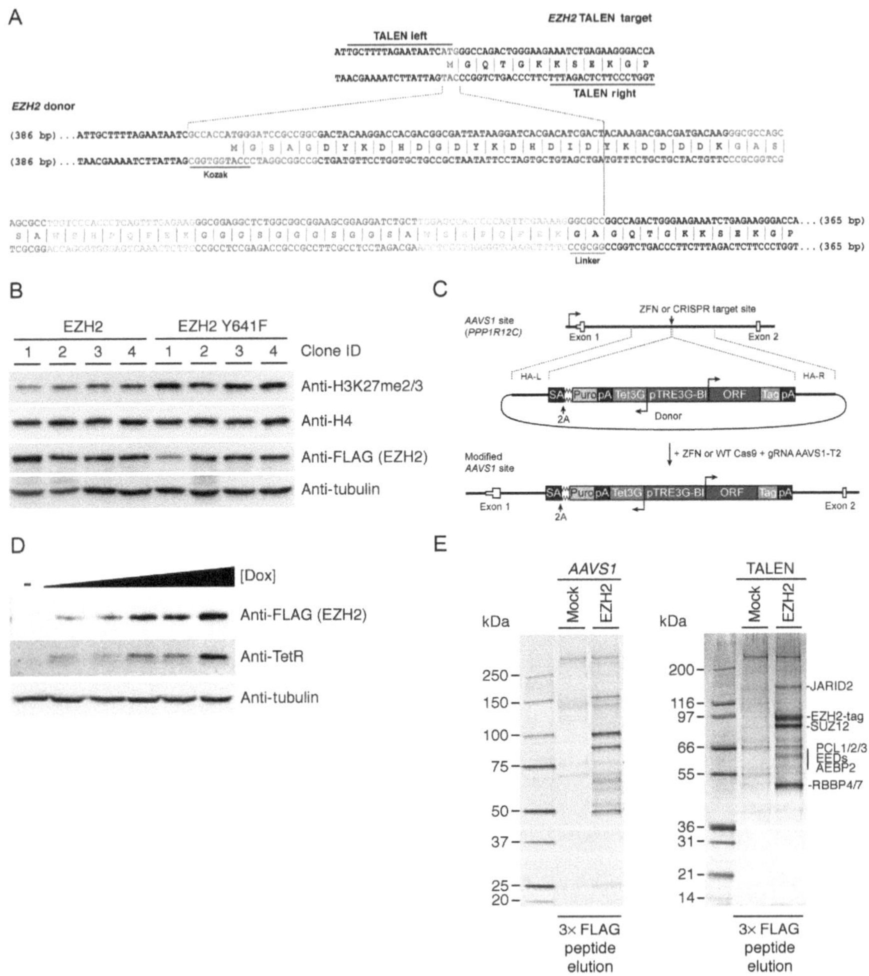

Fig. 3 Production of cell lines for tandem affinity purification (TAP) of the EZH2/PRC2 complex. (A) Strategy for TALEN-driven insertion of the TAP tag to the N-terminus of the EZH2 protein within the endogenous gene. Schematic of the EZH2 locus, TALEN target sites, and donor construct. The position of the start codon (ATG) is shown. The 3′ end of the left homology arm (HA-L) starts just before the ATG and the 5′ end of the right homology arm (HA-R) starts after the ATG. A Kozak consensus sequence was placed upstream of the tag sequence for efficient translation initiation at this site. Note that in this case, the TALEN-binding sites are separated by the tag sequence in the donor, which should be sufficient to prevent DSB formation in the episomal donor and in the chromosome after targeted integration of the tag. We avoid repetitive sequences in HA-L and HA-R. (B) Western blots showing the levels of H3K27me3 on bulk chromatin in K562 clones expressing EZH2 ± Y641F mutation from the *AAVS1* locus. (C) Schematic of donor construct and of the AAVS1 locus following ORF addition for the robust induction of target proteins upon doxycycline treatment. The first two exons of the PPP1R12C

2.10 Dox-Inducible Expression of EZH2 from the AAVS1 Safe Harbor Locus

Ectopic expression of some chromatin modifiers is "toxic" for the cell. In order to bypass this issue, we generated a single-vector autoregulated Tet-On expression system permitting tightly controlled inducible expression of target proteins (Dalvai et al., 2015) (Fig. 3C). The *PGK1* promoter in AAVS1_Puro_PGK1_3xFLAG_Twin_Strep (Addgene #68375) was first replaced by the bidirectional Tet-On 3G inducible expression systems (pTRE3G-BI) (Clonetech). Then, the Tet-On 3G transactivator was cloned on one side of the bidirectional promoter. Under these settings, simultaneous inducible expression of the Tet-On 3G transactivator and the transgene of interest with minimal background expression are achieved. As an example, Fig. 3D shows the dox-dependent expression of EZH2 from this all-in-one vector. To generate such cells lines, follow instructions in Section 2.2.

2.11 TAP Purification of EZH2

Purification of the PRC2 complex was performed exactly as described in Section 2.5. Fig. 3E shows the first step of purification (anti-FLAG IP) of PRC2 from TALEN-derived cell lines and the *AAVS1* locus along with their mock control.

2.12 HAT/HMT Assays

These assays measure the transfer of radiolabeled acetyl and methyl groups from a donor molecule, acetyl-coenzyme A, on the Nε of lysines present on histone N terminal tails, or from *S*-adenosylmethionine (SAM) on the Nε of lysines to produce mono-, di-, or tri-methylated residues or on the terminal (or ω) guanidino nitrogen atoms of arginines to give mono- or di-methylated residues on histones, the latter being possibly symmetric or asymmetric.

Fig. 3—Cont'd gene are shown as *open boxes*. Also annotated are the locations of the splice acceptor site (SA), 2A self-cleaving peptide sequence (2A), puromycin resistance gene (Puro), polyadenylation sequence (pA), bidirectional Tet-responsive promoter (pTRE3G-BI), Tet3G transactivator (Tet3G), 3× FLAG–2× STREP tandem affinity tag (Tag), and homology arms left and right (HA-L, HA-R). (D) Western blots showing tag-EZH2 and Tet-On 3G transactivator protein expression in a K562 clone treated for 24 h with increasing doses of doxycycline. (E) Silver-stained SDS–PAGE showing the comparison of the EZH2 complex obtained when expressed from the endogenous (TALEN) and the *AAVS1* loci after only the first purification step (FLAG elution, before Streptavidin resin). (See the color plate.)

1. Use 0.5 μg of purified native human chromatin or core histones (Côté, Utley, & Workman, 1995) with up to 3 μL of HAT/HMT complex in a 15 μL reaction volume. Larger volumes of HAT/HMT complexes are usually detrimental to the reaction linearity.
2. Mix chromatin or core histones, (a) the HAT fraction, 0.125 μCi of [3H] acetyl-CoA (4.7 Ci/mmol, PerkinElmer), 3 μL of 5 × HAT buffer (250 m*M* Tris–HCl, pH 8, 25% glycerol, 0.5 m*M* EDTA, pH 8, 5 m*M* DTT, 5 m*M* PMSF), and 1 m*M* Na-butyrate; (b) the HMT fraction, 3 μL of 5 × HMT buffer (100 m*M* Tris–HCl, pH 8, 25% glycerol, 0.5 m*M* EDTA, pH 8, 5 m*M* DTT, 1 m*M* PMSF) and 0.5 μCi of SAM (*S*-adenosyl-L-[methyl-3H] methionine; PerkinElmer).
3. KCl/NaCl total concentration should be adjusted to 50 m*M* taking into account all reaction components. Total salt concentration can be higher, but we found that 100 m*M* inhibits most HAT/HMT activities.
4. Incubate the reaction mix at 30°C for 30–60 min.
5. For liquid HAT/HMT assay, at the end of the incubation spot each reaction onto p81 phosphocellulose membranes (Millipore).
6. Air-dry the membranes and wash three times for 5 min in 50 m*M* carbonate ($NaHCO_3$–$NaCO_3$, pH 9.2) buffer. The highly basic histones will stay charged at this pH and remain bound.
7. Rinse the membranes briefly with acetone, air-dry, and place in scintillation vials.
8. Cover with scintillation fluid and measure counts in scintillation counter for 30 min.
9. Histone tail specificity can be analyzed by loading the HAT/HMT reaction on SDS–polyacrylamide gel electrophoresis (SDS–PAGE) followed by fluorography as described below.
10. Stop reactions by adding 5 μL of 4 × Laemmli sample buffer and boil for 5 min.
11. Load samples on 18% polyacrylamide gel.
12. Run for 2 h at 160 V in Tris–glycine buffer.
13. After coomassie staining, destain until histones are readily visible (to verify equivalent histone loading between the reactions).
14. Incubate gel in EN^3HANCE for 30 min and then washed in H_2O for 30 min. Dry gel for 90 min at 60°C and place under an X-ray film for 2 or 3 days.

3. CONCLUSION

In this chapter, we describe efficient methods to purify and characterize two important chromatin-modifying complexes, namely the TIP60 histone acetyltransferase and PRC2 histone methyltransferase. The TAP strategy presented facilitates the isolation of highly pure material from a limited amount of cells, even for low-abundance protein complexes. We have successfully purified the TIP60/NuA4, PRC2/EZH2, HBO1, MOZ/MORF, histone deacetylase, and deubiquitinase complexes from multiple tagged subunits (data not shown), illustrating the robustness and versatility of the approach.

Furthermore, we describe how to generate cell lines, in which the endogenous gene is tagged using the widely available and customizable TALENs and CRISPR/Cas9 systems. This allows purification of truly endogenous complexes that are natively regulated at the transcriptional and posttranscriptional levels (promoters/5′UTR/introns/3′UTR preserved). In addition, tagging of endogenous genes allows standardized chromatin immunoprecipitation experiments, alleviating dependency on antibodies of variable quality (Dalvai et al., 2015). In parallel, we demonstrate the utility of the *AAVS1* safe-harbor-targeting system to rapidly generate isogenic human cell lines expressing near physiological levels of tagged proteins. To fully exploit the utility of this system, it can be coupled to CRISPR/Cas9-mediated knockout of the endogenous alleles of the gene of interest in order to study the function of the ectopically expressed gene variants/splicing isoforms/mutants. We believe that our approach using endogenous activities along with native substrates will help to decipher the exact histone and residue specificity of chromatin-modifying enzymes as well as to understand their structure and function in physiological conditions.

ACKNOWLEDGMENTS

We thank members of our laboratories for implementing these approaches and fruitful discussions. This work was supported by grants from the Natural Sciences and Engineering Research Council of Canada (RGPIN-2014-059680) to Y.D. and the Canadian Institutes of Health Research (MOP-64289 and FDN-143314) to J.C. Salary support for Y.D. was provided by the Fonds de la recherche du Québec-Santé (FRQS). J.C. holds a Canada Research Chair in Chromatin Biology and Molecular Epigenetics.

REFERENCES

Abmayr, S. M., Yao, T., Parmely, T., & Workman, J. L. (2006). Preparation of nuclear and cytoplasmic extracts from mammalian cells. *Current Protocols in Molecular Biology*, *75*, 12.1, 12.1.1–12.1.10. http://dx.doi.org/10.1002/0471142727.mb1201s75.

Côté, J., Utley, R. T., & Workman, J. L. (1995). Basic analysis of transcription factor binding to nucleosomes. *Methods in Molecular Genetics*, *6*, 108–128. http://dx.doi.org/10.1016/S1067-2389(06)80009-9.

Dalvai, M., Loehr, J., Jacquet, K., Huard, C. C., Roques, C., Herst, P., ... Doyon, Y. (2015). A scalable genome-editing-based approach for mapping multiprotein complexes in human cells. *Cell Reports*, *13*(3), 621–633. http://dx.doi.org/10.1016/j.celrep.2015.09.009.

Guschin, D. Y., Waite, A. J., Katibah, G. E., Miller, J. C., Holmes, M. C., & Rebar, E. J. (2010). A rapid and general assay for monitoring endogenous gene modification. *Methods in Molecular Biology*, *649*, 247–256. http://dx.doi.org/10.1007/978-1-60761-753-2_15.

Hockemeyer, D., Soldner, F., Beard, C., Gao, Q., Mitalipova, M., DeKelver, R. C., ... Jaenisch, R. (2009). Efficient targeting of expressed and silent genes in human ESCs and iPSCs using zinc-finger nucleases. *Nature Biotechnology*, *27*(9), 851–857. http://dx.doi.org/10.1038/nbt.1562.

Margueron, R., & Reinberg, D. (2011). The polycomb complex PRC2 and its mark in life. *Nature*, *469*(7330), 343–349. http://dx.doi.org/10.1038/nature09784.

Steunou, A.-L., Rossetto, D., & Cote, J. (2014). Regulating chromatin by histone acetylation. In J. L. Workman & S. M. Abmayr (Eds.), *Fundamentals of chromatin* (pp. 147–212). New York: Springer. http://dx.doi.org/10.1007/978-1-4614-8624-4_4.

Yap, D. B., Chu, J., Berg, T., Schapira, M., Cheng, S. W., Moradian, A., ... Aparicio, S. A. (2011). Somatic mutations at EZH2 Y641 act dominantly through a mechanism of selectively altered PRC2 catalytic activity, to increase H3K27 trimethylation. *Blood*, *117*(8), 2451–2459. http://dx.doi.org/10.1182/blood-2010-11-321208.

PART III

Histone Modification Binders (Readers)

CHAPTER FOURTEEN

Preparation, Biochemical Analysis, and Structure Determination of the Bromodomain, an Acetyl-Lysine Binding Domain

C. Ren, L. Zeng, M.-M. Zhou[1]
Icahn School of Medicine at Mount Sinai, New York, NY, United States
[1]Corresponding author: e-mail address: ming-ming.zhou@mssm.edu

Contents

1. Introduction 322
2. General Equipment 324
3. General Materials 324
   3.1 Bacterial Protein Expression Vector and Host Strain 324
   3.2 Bacteria Growth Medium 324
   3.3 Antibiotics 325
   3.4 Purification Columns and Resins 325
   3.5 Protein Purification Buffers 325
   3.6 Crystallization Reagents and Tools 325
   3.7 Mounting Tools and Shipping Dewar 326
   3.8 NMR Reagents and Tools 326
   3.9 LC/MS Reagents and Tools 326
4. Preparation of BrDs 326
   4.1 Transformation 326
   4.2 Expression 329
   4.3 Purification 329
5. Biochemical Analysis of BrDs 330
   5.1 NanoDrop to Determine Protein Concentration and Check Protein Purity 330
   5.2 SDS-PAGE to Check Protein Purity and MW 331
   5.3 LC/MS to Determine Protein MW 331
   5.4 ^{1}H–^{15}N-HSQC to Detect Ligand Binding 331
   5.5 Fluorescence Polarization to Determine Protein–Ligand Binding 332
6. Structural Determination of BrDs 334
   6.1 Structural Determination Using X-Ray Crystallography 334
   6.2 Structural Determination Using NMR 336
7. Conclusions and Perspectives 338
Acknowledgments 340
References 340

Methods in Enzymology, Volume 573
ISSN 0076-6879
http://dx.doi.org/10.1016/bs.mie.2016.01.018

Abstract

The bromodomain (BrD) represents an evolutionarily conserved protein domain whose function mostly is to recognize acetylated lysine residues in histones and nuclear proteins in regulation of gene transcription in chromatin. The highly conserved BrD structure features an unusual left-handed, antiparallel four-helix bundle and a hydrophobic pocket between the interhelical ZA and BC loops important for acetyl-lysine binding. Many proteins, particularly transcriptional activators, contain BrDs, and mutation or deletion of the BrDs impairs the protein function, implying their critical role in human biology and disease. In this chapter, we provide general protocols of the preparation, biochemical analysis, and structure determination of BrDs, aiming to offer a general guideline for structural and biochemical functional characterization of BrD-containing proteins.

1. INTRODUCTION

Histone lysine acetylation is a posttranslational modification on the ε-NH_2 of lysine residues, which can neutralize the positive charge on histone tails and decrease affinity between histone tails and DNA, leading to a less compact and more accessible chromatin conformation that is typically associated with gene transcriptional activation (Grunstein, 1997; Struhl, 1998). Various histone acetyltransferases (HATs) and deacetylases (HDACs) are known to be responsible for reversibly adding or removing lysine acetylation marks in histones (Strahl & Allis, 2000), but protein domains that can recognize these marks are largely restricted to the bromodomain (BrD) (Sanchez & Zhou, 2009; Zeng & Zhou, 2002), and only recently also attributed to a tandem PHD finger in DPF3b (Zeng et al., 2010) and the YEATS domain of AF9 (Li et al., 2014). BrD originates from the Brahma protein (Kennison & Tamkun, 1988) and bears a structural motif shared by fsh, SNF2, SPT7, and CCG1 proteins, and many other transcriptional coactivators, implying its role in gene transcriptional activation (Jeanmougin, Wurtz, LeDouarin, Chambon, & Losson, 1997; Tamkun et al., 1992). The first BrD structure of the HAT transcriptional coactivator P/CAF revealed that BrD is a histone acetylation reader domain, which has an unusual left-handed, antiparallel four-helix bundle (αZ, αA, αB, αC) and a long ZA loop. A hydrophobic pocket between the ZA and BC loops, opposite to the N- and C-termini, is responsible for acetyl-lysine (Kac) binding (Dhalluin et al., 1999). Studies of BrDs in other BrD-containing proteins, such as TAFII250 (Jacobson, Ladurner, King, & Tjian, 2000),

Gcn5p (Owen et al., 2000), and CBP (Mujtaba et al., 2004), confirmed the same left-handed structural fold, hydrophobic binding pocket between ZA and BC loops, as well as critical residues involved in Kac binding (Sanchez, Meslamani, & Zhou, 2014; Zeng & Zhou, 2002).

The critical roles of BrD in chromatin assembly and transcriptional activation have been validated by mutation and deletion studies in many proteins including the Swi2/Snf2 BrD that anchors the SWI/SNF ATP-dependent remodeling complex to chromatin (Syntichaki, Topalidou, & Thireos, 2000), the Gcn5 BrD that anchors the SAGA HAT to acetylated nucleosomes (Hassan et al., 2002), the BRD4-BrDs that mediate BRD4 chromatin association and localization during mitosis (Dey, Chitsaz, Abbasi, Misteli, & Ozato, 2003), the yeast RSC BrDs that recognize histone H3 acetylated lysine 14 (H3K14ac) (Kasten et al., 2004), and the p300/CBP BrD that binds with the acetylated N-terminal region of Stat3 (Hou, Ray, Lee, & Brasier, 2008). Comprehensive BrD structural studies can be found in the literature (Filippakopoulos et al., 2012), and its classification, substrates, and roles in various diseases are also reviewed in the literature (Dawson, Kouzarides, & Huntly, 2012; Muller, Filippakopoulos, & Knapp, 2011; Sanchez et al., 2014).

Given the critical role of BrD/acetyl-lysine binding in gene transcriptional activation in chromatin, the idea of modulating BrD-mediated protein functions in chromatin association and transcriptional activation through chemical inhibition of BrD/acetyl-lysine binding has been an area of significant interest and development. BrD inhibitors basically fall into two classes as nonacetylated and acetylated lysine mimetics (Filippakopoulos & Knapp, 2014), among which BET (bromo and extra terminal) family BrDs including BRD2, BRD3, BRD4, and BRDT have received particular attention (Muller et al., 2011). Studies using these small molecules provide therapeutic potential in treating various diseases, such as the antitumor effects of JQ1 (Delmore et al., 2011; Filippakopoulos et al., 2010; Mertz et al., 2011; Ott et al., 2012; Zuber et al., 2011) and I-BET151 (Chaidos et al., 2014; Dawson et al., 2011; Gallagher et al., 2014; Pastori et al., 2014), and the antiinflammatory potential of I-BET762 (Nicodeme et al., 2010) and MS417 (Zhang et al., 2012). A comprehensive review of BrD inhibitors, highlighting their classification, application in cancer, inflammation and viral infection, and use in clinical trials, can be found elsewhere (Filippakopoulos & Knapp, 2014).

Here, we use the BrD1 of BRD4 as an example to summarize general protocols for protein preparation, biochemical analysis, and structure determination of BrDs, aiming to provide general guidelines for structural and biochemical characterization of BrD-containing proteins.

2. GENERAL EQUIPMENT

Incubating shaker temperature controlled (New Brunswick I2500 Series)
Microfluidizer (Microfluidics M-110P)
Centrifuge (temperature controlled, ie, Sorvall RC-5C Plus with SS-34 rotor)
ÄKTA purifier (GE Life Sciences)
NanoDrop 2000 (Thermo Scientific)
Liquid chromatography–mass spectrometer (LC–MS) (Agilent 1200 Series Q-TOF LC/MS)
NMR spectrometer (Bruker 500 MHz, 600 MHz)
Microplate reader (Tecan SAFIRE II monochromator-based microplate reader)
Synchrotron (eg, X6A beamline, NSLS, Brookhaven National Laboratory)

3. GENERAL MATERIALS

3.1 Bacterial Protein Expression Vector and Host Strain

pNIC28-Bsa4 (SGC, kanamycin, His-tag, TEV cleavage site)
BL21-CodonPlus (DE3)-RIPL competent cells (Agilent)

3.2 Bacteria Growth Medium

Lysogeny broth (LB) medium (ThermoFisher)
Terrific broth (TB) medium (ThermoFisher)
M9 minimal medium:
0.5 g NaCl
3 g KH_2PO_4
5.3 g Na_2HPO_4
1.2 mL 5% Thiamine
1.2 mL 1 *M* $MgSO_4$
1.2 mL 100 m*M* $CaCl_2$
1 g ^{15}N–NH_4Cl
4 g Regular glucose or 2 g $^{13}C_6$-glucose
Dissolved in 1 L ddH_2O and filtered with 0.22 μ*M* sterile filter

3.3 Antibiotics

Kanamycin stock, 50 mg/mL (1:1000 dilution)
Chloramphenicol stock, 30 mg/mL in ethanol (1:1000 dilution)

3.4 Purification Columns and Resins

Affinity chromatography column
HiTrap IMAC FF (GE 17-0921-04)
Gel filtration/size-exclusion chromatography column
XK 26/70 (GE 28-9889-50)
Resins:
Ni Sepharose 6 Fast Flow (GE 17-5318-01)
Superdex 75 prep grade (GE 17-1044-02) for molecular weight (MW) 3–70K

3.5 Protein Purification Buffers

Protein lysis buffer:
100 mL 10 × PBS (Fisher BP399-1)
100 mL 5 *M* NaCl
100 mL 50% glycerol
Dissolved in 1 L ddH_2O and kept at 4°C
IMAC buffer A:
30 m*M* Tris–HCl (pH 8.0), 500 m*M* NaCl
IMAC buffer B:
30 m*M* Tris–HCl (pH 8.0), 500 m*M* NaCl, 250 m*M* imidazole
Superdex buffer: 30 m*M* HEPES (pH 7.4), 500 NH_4Ac

3.6 Crystallization Reagents and Tools

Crystallization blocks:
Crystal screen (Hampton Research HR2-130)
Index (Hampton Research HR2-134)
Wizard I/II (Rigaku 1009532)
Cryo regents:
Glycerol
Ethylene glycol
Crystallization plate (Intelli-plate 96, Hampton Research HR3-299)
Clear seal film and applicator (Hampton Research HR4-521 and HR4-525)

3.7 Mounting Tools and Shipping Dewar

Micro-loops (depends on synchrotron and beamline)
Cryo-puck (depends on synchrotron and beamline, ie, uni-puck)
Puck holder
Dry shipping dewar (Hampton Research HR1-004)
Shipper container (Hampton Research HR1-006)

3.8 NMR Reagents and Tools

^{1}H–^{15}N-Heteronuclear Single Quantum Coherence (HSQC) buffer:
PBS containing 10% D_2O
Shigemi NMR microtube

3.9 LC/MS Reagents and Tools

LC column
Discovery BioWide Pore C18 HPLC column (Sigma 568202-U)
LC/MS solvent A:
H_2O, 0.1% formic acid
LC/MS solvent B:
Acetonitrile, 0.1% formic acid

4. PREPARATION OF BrDs

BrD can be defined using "identify Conserved Domain" option for a protein sequence on NCBI website: http://www.ncbi.nlm.nih.gov/Structure/cdd/wrpsb.cgi?INPUT_TYPE=live&SEQUENCE=. Construct boundaries are usually extended for 3–15 residues (see Fig. 1). Choose the appropriate expression vector (typically His-tag) and use the corresponding cleavage enzyme to remove the tag if necessary. Make sure the protein sequence does not contain the cleavage site.

4.1 Transformation

1. Add 1 μL pNIC28-BrD plasmid (or another BrD construct) into 50 μL ice-cold RIPL-BL21 (DE3)-CodonPlus competent cells, mix well, and leave on ice for 30 min.
2. Heat shock in 42°C water bath for 90 s, leave on ice for 1 min, add 500 μL LB medium (no antibiotics), and shake at 37°C for 40 min.

A

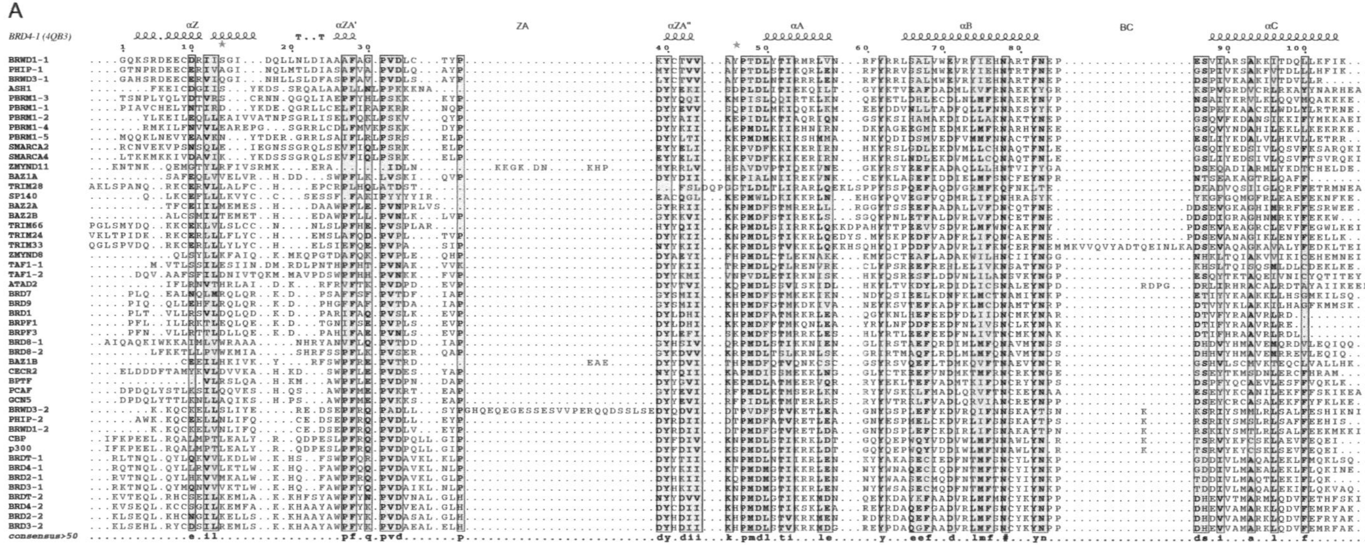

Fig. 1 Protein sequence alignment of human bromodomains. (A) Human BrD sequences were aligned using EMBL Clustal Omega (http://www.ebi.ac.uk/Tools/msa/clustalo/). Secondary structure depiction was visualized based on the BRD4-BD1 structure (PDB: 4QB3) in ESPript 3.0. (B) Phylogenetic tree was subsequently generated using EMBL ClustalW2 Phylogeny (http://www.ebi.ac.uk/Tools/phylogeny/clustalw2_phylogeny/; Li et al., 2015). Sequence alignment and phylogenetic tree (unrooted format) were visualized using ESPript 3.0 (http://espript.ibcp.fr/ESPript/cgi-bin/ESPript.cgi; Robert & Gouet, 2014) and iTOL (http://itol.embl.de/index.shtml; Letunic & Bork, 2011), respectively. (See the color plate.)

(Continued)

B

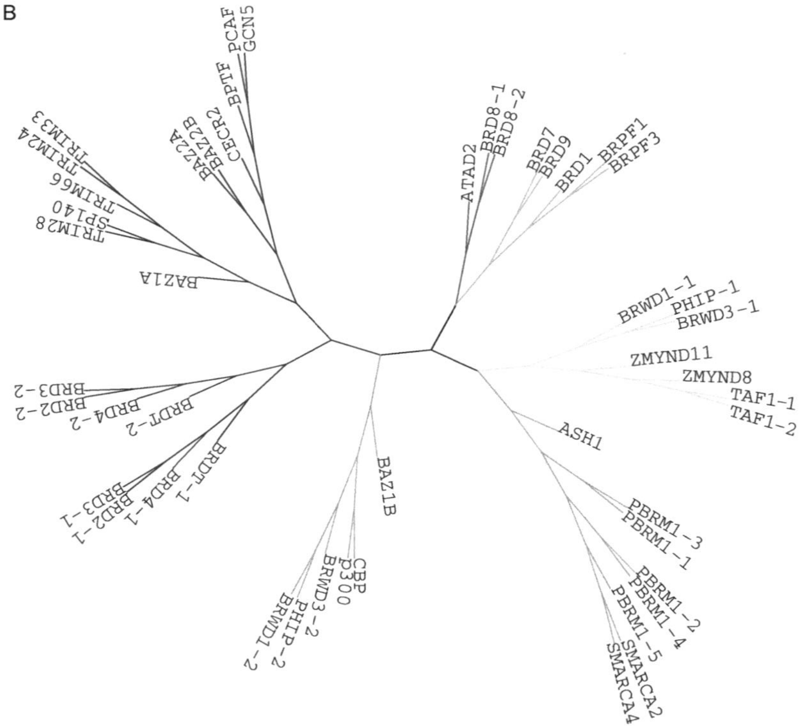
PCAF
GCN5
BPTF
CECR2
BAZ2B
BAZ2A
TRIM33
TRIM24
TRIM66
SP140
TRIM28
BAZ1A
ATAD2
BRD8-1
BRD8-2
BRD7
BRD9
BRD1
BRPF1
BRPF3
BRWD1-1
PHIP-1
BRWD3-1
ZMYND11
ZMYND8
TAF1-1
TAF1-2
ASH1
PBRM1-3
PBRM1-1
PBRM1-2
PBRM1-4
PBRM1-5
SMARCA2
SMARCA4
BAZ1B
CBP
p300
BRWD3-2
PHIP-2
BRWD1-2
BRD3-2
BRD2-2
BRD4-2
BRDT-2
BRDT-1
BRD4-1
BRD2-1
BRD3-1

Fig. 1—Cont'd

3. Spread the transformed bacteria on LB-agar plate containing 50 μg/mL kanamycin and 30 μg/mL chloramphenicol, and place in 37°C incubator overnight till clones appear.
4. Pick one single clone to inoculate 5 mL LB medium containing 50 μg/mL kanamycin and 30 μg/mL chloramphenicol, and shake overnight at 37°C.

4.2 Expression

1. Inoculate the small-scale 5 mL LB culture into 1 L TB medium containing 50 μg/mL kanamycin in 2-L flask and grow at 37°C for around 3 h.
2. Take a small amount (around 2 mL) and measure the absorbance at OD_{600nm} every 10–20 min till OD reaches 1.6–1.8.
3. Move the flask to a 15°C shaker and continue to grow for 1 h till the temperature of bacteria culture becomes fully constant at 15°C.
4. Add 1 mL 1 *M* IPTG (isopropyl β-D-1-thiogalactopyranoside) into each 1 L culture to induce protein expression and continue to grow around 16–20 h.
5. Harvest bacteria by centrifuge at 4°C.

4.3 Purification

Bacteria disruption

1. Vortex or use a homogenizer to fully homogenize the cell paste in ice-cold lysis buffer containing 0.5% Igepal CA-630 and 100 μ*M* ABESF (freshly added).
2. Disrupt bacteria using microfluidizer at 20K Psi. Pass the sample multiple times if necessary till the lysate becomes translucent.
3. Centrifuge the lysate at 20,000 × *g* for at least 40 min and collect the supernatant.

Affinity chromatography

4. Wash the ÄKTA system pump A and B with IMAC buffer A and B, respectively.
5. Equilibrate IMAC column with 2 column volume (CV) buffer A.
6. Load the supernatant to IMAC column.
7. Wash IMAC column with buffer A mixed with 0% (2 CV), 4% (5 CV), 20% (5 CV), and 100% (5 CV) buffer B, respectively, and collect fractions as 5 mL for each tube.

8. Check the fractions by SDS-PAGE. Usually large target protein peaks appear immediately at 100% buffer B elution and early to middle stage of 20% buffer B.
9. Remove the His-tag using TEV overnight at 4°C if necessary.

Gel filtration/size-exclusion chromatography

10. Equilibrate the superdex 75 column with 1.2 CV superdex buffer.
11. Concentrate the elute from IMAC column to less than 7 mL using concentrator with appropriate cutoff (ie, BRD4-BrD1 MW ~15 kDa, then use 10 kDa concentrators).
12. Load the concentrated elute to a superdex 75 column.
13. Prewash the column with 0.3 CV of superdex buffer first and elute the column with 0.8 CV of superdex buffer. Collect the fractions as 5 mL for each tube.
14. Check fractions by SDS-PAGE.
15. Combine the designated fractions and concentrate to appropriate concentration (around 20–30 mg/mL).
16. Aliquot the concentrated protein and flash freeze in liquid nitrogen. Typically the yield is 100 mg protein from 2 L culture.
17. Store the protein at −80°C.

Note: ^{15}N-labeled or ^{15}N–^{13}C double-labeled protein for NMR analysis is prepared by growing bacterial cells in M9 medium containing ^{15}N–NH_4Cl with regular glucose or with $^{13}C_6$-glucose, respectively. Switch the bacteria to a 15°C shaker till OD_{600nm} reaches 0.6–0.8. Follow the same purification procedure. Since the yield is much less in M9 medium than in TB medium, please increase the culture volume when expressing isotope-labeled protein.

5. BIOCHEMICAL ANALYSIS OF BrDs

5.1 NanoDrop to Determine Protein Concentration and Check Protein Purity

1. Calculate the molecular weight (MW) and extinction coefficients (EC) of the protein using the website: http://web.expasy.org/protparam/.

   For example, BRD4-BrD1 (residues 44–168) with Ser and Met from the cleaved tag sequence has a MW of 15,083.4 Da, a molar EC of 27,055–26,930 *M*/cm, and an EC represented as Abs 0.1% (=1 g/L) of 1.794–1.785, which is equal to 1/10 of the percent EC (EC 1%) = 10 × molar EC/MW.

2. When using NanoDrop, enter the protein mode, choose the "other protein E 1%" option, input 10 × Abs 0.1% as the percent EC, and the concentration will be measured directly as mg/mL.

Note: The $OD_{260/280}$ ratio should be around 0.56. Higher ratios most likely indicate the contamination from DNA or RNA. Buffer components, such as imidazole, also increase the ratio.

5.2 SDS-PAGE to Check Protein Purity and MW

1. Mix protein with equal volume of Laemmli sample buffer (Bio-Rad 161-0737) and heat at 95°C for 5 min.
2. Load the protein sample on 10% SDS-PAGE gel and run at constant amplitude of 50 mA for around 30 min till the frontier reaches the end.
3. Wash the gel with ddH_2O first for 15 min.
4. Stain the gel with GelCode blue stain reagent (ThermoFisher 24592) around 1 h till band appears.
5. Wash the gel with ddH_2O again to clear the gel.
6. Pure protein should appear as one band with few or almost invisible residue bands.

Note: A clear target protein band of increased expression should be visible after IPTG induction compared with noninduced bacteria culture. Fractions from chromatography can be mixed directly with Laemmli buffer (ie, 10 μL of each) to check if the fractions contain target protein. Concentrated proteins can be loaded around 1–2 μg to check the purity or cleavage state.

5.3 LC/MS to Determine Protein MW

1. Dilute the protein sample in ddH_2O.
2. Inject 2 μL into LC/MS system.
3. Elute the column with LC/MS solvent 1% B (1 min), gradient 1–99% B (1–4 min), and 99% B (4–8 min) sequentially at a flow rate of 0.4 mL/min.
4. Detect and deconvolute the peaks using the TOF protein analysis software. Protein spectrum is a cluster of peaks as a bell-shaped distribution. Set the appropriate MW range and deconvolute. One peak should show up with the desired MW.

5.4 ^{1}H–^{15}N-HSQC to Detect Ligand Binding

1. ^{15}N-labeled protein samples are prepared in PBS buffer at pH 7.4 containing 10% D_2O.

2. Make sure the additives are constant among different samples. For example, DSMO is added to help dissolve chemical compounds and same amount of DMSO should be added to the free protein sample as a control.
3. Record ^{1}H–^{15}N-HSQC spectra on Bruker NMR spectrometers at 298 K (see Fig. 2A).
4. Process the spectra using NMRPipe (Delaglio et al., 1995).
5. Analyze the spectra using NMRViewJ software (Johnson, 2004). Chemical shift or line broadening caused by ligand can be visualized when overlapping the ligand spectrum with the control one.
6. Binding affinity (K_d) measurement is performed using titration analysis in NMRViewJ. First, prepare a series of spectra in the presence of different ligand concentrations. Open all the spectra and define the corresponding concentrations for protein and ligand in NMRViewJ. Plot well-defined chemical shift paths against ligand concentrations to yield individual K_d for each residue, or an overall K_d for multiple residues.

Note: To distinguish each peak on a ^{1}H–^{15}N-HSQC spectrum, typically two 3D NMR experiments NHCACB (Wittekind & Mueller, 1993) and CBCA(CO)NH (Grzesiek & Bax, 1992) using ^{15}N–^{13}C double-labeled and fractionally deuterated protein sample should be performed first to assign backbone.

5.5 Fluorescence Polarization to Determine Protein–Ligand Binding

5.5.1 *Saturation Binding Assay to Determine Binding Affinity, Measured as Dissociation Constant, K_d*

1. Prepare decreasing concentrations of BrDs in fluorescence polarization (FP) assay buffer (PBS and 0.005% Igepal CA-630) by serial dilution in 384-well plate. Each well contains 20 μL protein solution.
2. Prepare 2× 10 n*M* FITC-labeled probe (MS574, FITC-conjugated MS417 at position) in FP assay buffer.
3. Add 20 μL probe solution into protein solution. The final reaction system contains 10 n*M* FITC-labeled probe and various concentrations of protein in a twofold gradient.
4. Read the FP signal using SafireII microplate reader.
5. Determine the IC_{50} using one-site saturation model in GraphPad Prism software (GraphPad Software, CA).

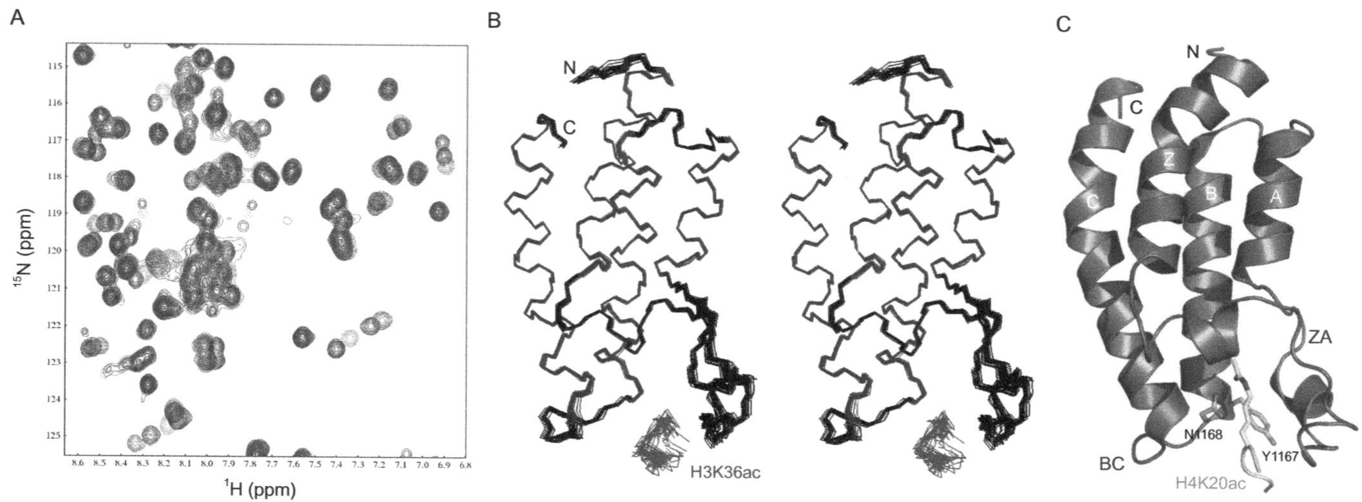

Fig. 2 Biochemical analysis of PCAF bromodomain binding to lysine-acetylated histone H3 peptides using NMR spectroscopy. (A) ^{1}H–^{15}N-HSQC spectra of the PCAF bromodomain showing backbone amide resonances in the free state (*black*) or in the presence of a histone peptide containing H3K9ac (*green*), H3K14ac (*red*), and H3K36ac (*blue*). The extent of chemical shift perturbation and line broadening indicates the strength of protein/peptide binding. (B) Stereoview of the backbone (N, Ca, and C′) atoms of 25 superimposed NMR structures of the PCAF bromodomain in complex with an H3K36ac peptide. (C) Crystal structure of CBP bromodomain bound to an H4K20ac peptide. The image was prepared with the structure coordinates of CBP bromodomain–PHD finger tandem module (PDB: 4N3W) in complex with H4K20ac peptide. The PHD finger is not shown for clarity (Plotnikov et al., 2014). *Panel (B) images were modified from Zeng, L., Zhang, Q., Gerona-Navarro, G., Moshkina, N., & Zhou, M. M. (2008). Structural basis of site-specific histone recognition by the bromodomains of human coactivators PCAF and CBP/p300.* Structure, 16, *643–652.* (See the color plate.)

5.5.2 Competition Binding Assay to Determine IC_{50}

1. Prepare decreasing concentrations of ligand solution in FP assay buffer.
2. Prepare protein–probe mixture using $2\times$ K_d concentration of protein and $2\times$ 10 nM fluorescein-labeled probe.
3. Add 20 μL protein–probe mixture into ligand plate. The final reaction system contains 10 nM FITC-labeled probe, the protein at its K_d concentration, and various concentrations of ligand in a twofold gradient. The DMSO concentration is constant among all wells (typically 1%, v/v).
4. Wait for 1 h and read the FP signal using SafireII microplate reader.
5. Determine the IC_{50} using one-site competition for IC_{50} in GraphPad Prism software (GraphPad Software, CA).
6. Determine the K_i using the website: http://sw16.im.med.umich.edu/software/calc_ki/ by inputting the related parameters.

Note: Check the original readings since fluorescent ligands will produce false-positive effects.

6. STRUCTURAL DETERMINATION OF BrDs

6.1 Structural Determination Using X-Ray Crystallography

6.1.1 Crystallization

1. Mix purified BRD4-BrD1 with ligand at a molar ratio of 1:5.
2. Dispense the screening solutions into crystallization plate and mix equal amount of protein–ligand mixture with reservoir solution (ie, 1 μL each).
3. Seal the plate with film using applicator (sitting-drop method) and leave at 290K.
4. Pick the best crystallization condition where single and big crystals appear.
5. Optimize the crystallization conditions by varying salt concentrations, pH, additives, or other possible reagents as well as other methods such as seeding, if necessary.

Note: Use a UV microscope to check if crystals are protein crystals or crystals from salt.

6.1.2 Crystal Mounting

1. Typical mounting solution can be the original crystallization reservoir solution plus 20% ethylene glycol or glycerol. Optimize it if necessary.
2. Preimmerse the pucks in liquid nitrogen to chill them.

3. Use the appropriate cryo-loops (depends on synchrotron, crystal size) to fish individual crystal under the microscope.
4. Flash freeze the crystal in liquid nitrogen and store the pin in the cryo-pucks.
5. Transfer the pucks in the shipping dewar prechilled with liquid nitrogen. Refill liquid nitrogen every 2 days when storing pucks inside.

Note:

Be patient.

Perform procedures as quickly as possible.

Use appropriate micro-tools (Hampton Research HR4-811) such as prober and needle to puncture any skin surrounding the crystal (no need to remove), spade, and spatula to detach crystals from the surface and to cut single crystals from a crystal cluster.

6.1.3 Data Collection at Synchrotron

1. Follow the instructions of the synchrotron staff and synchrotron guideline. Keep yourself safe and do not damage instruments.
2. Choose the appropriate data collection strategy, and detector distance (depends on resolution), exposure time (not to overflow), oscillation angle (typically 1 degree), and total degrees collected (make sure to achieve enough redundancy depending on the space group).

6.1.4 Structure Determination

Most computer software programs are integrated into the CCP4 suite (Winn et al., 2011).

1. Data reduction

Process diffraction images using HKL2000 (Otwinowski & Minor, 1997) or iMOSFLM (Battye, Kontogiannis, Johnson, Powell, & Leslie, 2011) and Scala in CCP4 to get an mtz file containing information for all the reflections. This progress involves indexing, lattice refinement, integration, and scaling. Make sure that the data show a relatively low mosaicity/2 (around 0.2–0.4) after integration and high I/σ (>2) at the resolution cutoff.

2. Phasing

Molecular replacement is most likely sufficient for solving the phase problem for BrDs since multiple BrD structures can be found in PDB database and the general fold of these domains is very similar. Solve the phases using Phaser (Mccoy et al., 2007) with appropriate search models or using automatic pipelines such as BALBES (Long, Vagin, Young, & Murshudov, 2008).

3. Model building, refinement, and validation
Visualize and build coordinates using Coot (Emsley & Cowtan, 2004). Use Refmac (Murshudov, Vagin, & Dodson, 1997) or Phenix (Adams et al., 2010) for refinement. Use PROCHECK (Laskowski, Macarthur, Moss, & Thornton, 1993) in CCP4 to check the structure geometry. One example of X-ray data collection and refinement statistics using the structure of CBP BrD with H4K20ac (PDB code 4N3W) (Fig. 2B; Plotnikov et al., 2014) is listed in Table 1.

6.2 Structural Determination Using NMR

6.2.1 Backbone Assignment

1. Record NHCACB and HN(CO)CACB spectra using ^{15}N–^{13}C double-labeled and fractionally deuterated protein sample (Yamazaki, Lee, Arrowsmith, Muhandiram, & Kay, 1994).

6.2.2 Side-Chain Assignment

1. Record 3D HCCH-TOCSY (Kay, Xu, Singer, Muhandiram, & Formankay, 1993) and HC(CO)NH-TOCSY (Logan, Olejniczak, Xu, & Fesik, 1993) spectra using ^{15}N–^{13}C double-labeled protein sample.
2. Stereospecific assignments of the methyl groups of Val and Leu residues can be obtained from ^{13}C–^{1}H–COSY and TOCSY-relayed ^{13}C–^{1}H–COSY spectra using a nonrandomly fractionally ^{13}C-labeled sample (Neri, Szyperski, Otting, Senn, & Wuthrich, 1989).

6.2.3 Restraints Collection

1. Record NOE-derived distance restraints using ^{15}N- or ^{13}C-edited 3D NOESY spectra (Clore & Gronenborn, 1994).
2. Record ϕ-angle restraints based on the ^{3}JHN,Hα coupling constants measured in a 3D water-flip-back HNHA spectrum (Kuboniwa, Grzesiek, Delaglio, & Bax, 1994).
3. Detect intermolecular NOEs between protein and ligand using ^{13}C-edited (F_1), $^{13}C/^{15}N$-filtered (F_3) 3D NOESY spectra (Clore & Gronenborn, 1994).
4. Identify slowly exchanging amide protons for hydrogen bonding from a series of 2D ^{1}H–^{15}N-HSQC spectra recorded by changing H_2O buffer to D_2O buffer.

Table 1 Crystallography Data Collection and Refinement Statistics of 4N3W

	4N3W (CBP-H4K20ac)
Data collection	
Space group	C121
Cell dimensions	
a, *b*, *c* (Å)	92.46, 59.31, 53.44
α, β, γ (degree)	90.00, 102.96, 90.00
Resolution (Å) (highest resolution shell)	30.00–1.90 (1.95–1.90)
Unique reflections	22,003
R_{merge} (%)	5.1 (32.2)
I/σ	35.9 (5.6)
Completeness (%)	98.7 (93.8)
Redundancy	7.4 (7.1)
Refinement	
Resolution (Å)	19.58–1.90
No. reflections	20,877
R_{work}/R_{free} (%)	23.3/28.6
No. atoms	
Protein	1522
Ligand	51
Water	127
B-factors (Å^2)	
Protein	44.5
Ligand	59.0
Water	48.9
RMSD	
Bond lengths (Å)	0.017
Bond angles (degree)	1.897
Ramachandran plot % residues	
Favored	97.8
Allowed	1.1
Outlier	1.1

Exemplary table from Plotnikov, A. N., Yang, S., Zhou, T. J., Rusinova, E., Frasca, A., & Zhou, M. M. (2014). Structural insights into acetylated-histone H4 recognition by the bromodomain-PHD finger module of human transcriptional coactivator CBP. *Structure, 22*, 353–360.

6.2.4 Structure Calculations

1. Process the data using NMRPipe (Delaglio et al., 1995) and analyze with NMRView (Johnson, 2004).
2. Apply distance geometry and simulated annealing strategies using the CNS (Brünger et al., 1998) or X-PLOR program for structure calculations (Brünger, 1992).
3. Assign NOE-derived distance restraints obtained from the ^{15}N- or ^{13}C-edited 3D NOESY spectra manually or automatically using ARIA2 (Rieping et al., 2007) and CNS (Brünger et al., 1998) programs for intraresidue, sequential, medium-range, and long-range NOEs.
4. Dihedral angle restraints and hydrogen bond restraints can be predicted from TALOS based on resonance assignments and conserved secondary structure elements determined by TALOS (Cornilescu, Delaglio, & Bax, 1999) or TALOS+ (Shen, Delaglio, Cornilescu, & Bax, 2009).
5. Determine the complex structure using the free-form structure and additional intermolecular and intraligand NOE-derived distance restraints.
6. Evaluate the geometry and restraint violation of the ensemble using PROCHECK-NMR (Laskowski, Rullmannn, MacArthur, Kaptein, & Thornton, 1996). For the ensemble of the final 20 out of 200 structures (see Fig. 2C), the RMSD of the distance restraint and torsional angle restraint violations should be no greater than 0.5 Å and 5 degree, respectively. The Lennard-Jones potential is not used during any refinement stage. One example of NMR restraints and statistics of the final NMR structures using the structure of PCAF-H3K36ac (PDB code 2RNX) (Zeng, Zhang, Gerona-Navarro, Moshkina, & Zhou, 2008) is listed in Table 2.

7. CONCLUSIONS AND PERSPECTIVES

BrDs are present in many proteins associated with gene transcription and chromatin and play important role in regulation of gene transcription in biology. Structural studies of BrDs provide valuable information for mechanistic characterization of the functional roles of BrD proteins in gene regulation, and also guide rational design and development of novel small molecule chemical inhibitors, which can in turn aid the mechanistic studies of BrDs. Given that some BrD proteins, particularly the BET proteins, have been implicated in the development of human diseases, systematical analysis of structure and function of BrDs can further facilitate discovery, validation,

Table 2 Restraints and Statistics of the Final NMR Structures of 2RNX

	2RNX (PCAF-H3K36ac)
NMR distance and dihedral constraints	
Distance constraints	3541
Intramolecular constraints	3441
Total NOE	3394
Intraresidue	1078
Inter-residue	2316
Sequential ($\|i-j\|=1$)	662
Medium range ($\|i-j\|\leq 4$)	774
Long range ($\|i-j\|>4$)	880
Hydrogen bonds	47
Intermolecular constraints	100
Total dihedral angle restraints	
ϕ angle	70
Ψ angle	70
Ramachandran map analysis (%)	
Most favored region	86.1 ± 1.6
Additional allowed region	12.3 ± 1.5
Generously allowed region	1.6 ± 0.6
Disallowed region	0.06 ± 0.24
Structure statistics	
Violations (mean ± s.d.)	
Distance constraints (Å)	0.077 ± 0.0049
Dihedral angle constraints (degree)	1.05 ± 0.11
Max. distance constraint violation (Å)	0.088
Max. dihedral angle violation (degree)	1.33
Deviations from idealized geometry	
Bond lengths (Å)	0.0093 ± 0.00014
Bond angles (degree)	0.92 ± 0.016
Impropers (degree)	0.77 ± 0.017
Average pairwise r.m.s. deviation (Å)	
Heavy	0.64 ± 0.066
Backbone	0.20 ± 0.035

Exemplary table from Zeng, L., Zhang, Q., Gerona-Navarro, G., Moshkina, N., & Zhou, M. M. (2008). Structural basis of site-specific histone recognition by the bromodomains of human coactivators PCAF and CBP/p300. *Structure*, *16*, 643–652.

and development of new therapeutic treatments for numerous human diseases including cancer and inflammatory disorders through selective chemical modulation of BrD/acetyl-lysine binding in gene transcription.

ACKNOWLEDGMENTS

We thank Drs. Lei Zeng and Guangtao Zhang for helpful discussion. The work is supported in part by grants from NIH to M.-M.Z.

REFERENCES

Adams, P. D., Afonine, P. V., Bunkoczi, G., Chen, V. B., Davis, I. W., Echols, N., et al. (2010). PHENIX: A comprehensive Python-based system for macromolecular structure solution. *Acta Crystallographica D*, *66*, 213–221.

Battye, T. G. G., Kontogiannis, L., Johnson, O., Powell, H. R., & Leslie, A. G. W. (2011). iMOSFLM: A new graphical interface for diffraction-image processing with MOSFLM. *Acta Crystallographica D*, *67*, 271–281.

Brünger, A. T. (1992). *X-PLOR, version 3.1: A system for X-ray crystallography and NMR*. New Haven: Yale University Press.

Brünger, A. T., Adams, P. D., Clore, G. M., DeLano, W. L., Gros, P., Grosse-Kunstleve, R. W., et al. (1998). Crystallography & NMR system: A new software suite for macromolecular structure determination. *Acta Crystallographica. Section D, Biological Crystallography*, *54*, 905–921.

Chaidos, A., Caputo, V., Gouvedenou, K., Liu, B., Marigo, I., Chaudhry, M. S., et al. (2014). Potent antimyeloma activity of the novel bromodomain inhibitors I-BET151 and I-BET762. *Blood*, *123*, 697–705.

Clore, G. M., & Gronenborn, A. M. (1994). Multidimensional heteronuclear nuclear-magnetic-resonance of proteins. *Methods in Enzymology*, *239*, 349–363.

Cornilescu, G., Delaglio, F., & Bax, A. (1999). Protein backbone angle restraints from searching a database for chemical shift and sequence homology. *Journal of Biomolecular NMR*, *13*, 289–302.

Dawson, M. A., Kouzarides, T., & Huntly, B. J. (2012). Targeting epigenetic readers in cancer. *The New England Journal of Medicine*, *367*, 647–657.

Dawson, M. A., Prinjha, R. K., Dittmann, A., Giotopoulos, G., Bantscheff, M., Chan, W. I., et al. (2011). Inhibition of BET recruitment to chromatin as an effective treatment for MLL-fusion leukaemia. *Nature*, *478*, 529–533.

Delaglio, F., Grzesiek, S., Vuister, G. W., Zhu, G., Pfeifer, J., & Bax, A. (1995). NMRPipe: A multidimensional spectral processing system based on UNIX pipes. *Journal of Biomolecular NMR*, *6*, 277–293.

Delmore, J. E., Issa, G. C., Lemieux, M. E., Rahl, P. B., Shi, J. W., Jacobs, H. M., et al. (2011). BET bromodomain inhibition as a therapeutic strategy to target c-Myc. *Cell*, *146*, 903–916.

Dey, A., Chitsaz, F., Abbasi, A., Misteli, T., & Ozato, K. (2003). The double bromodomain protein Brd4 binds to acetylated chromatin during interphase and mitosis. *Proceedings of the National Academy of Sciences of the United States of America*, *100*, 8758–8763.

Dhalluin, C., Carlson, J. E., Zeng, L., He, C., Aggarwal, A. K., & Zhou, M. M. (1999). Structure and ligand of a histone acetyltransferase bromodomain. *Nature*, *399*, 491–496.

Emsley, P., & Cowtan, K. (2004). Coot: Model-building tools for molecular graphics. *Acta Crystallographica. Section D, Biological Crystallography*, *60*, 2126–2132.

Filippakopoulos, P., & Knapp, S. (2014). Targeting bromodomains: Epigenetic readers of lysine acetylation. *Nature Reviews. Drug Discovery*, *13*, 339–358.

Filippakopoulos, P., Picaud, S., Mangos, M., Keates, T., Lambert, J. P., Barsyte-Lovejoy, D., et al. (2012). Histone recognition and large-scale structural analysis of the human bromodomain family. *Cell, 149*, 214–231.

Filippakopoulos, P., Qi, J., Picaud, S., Shen, Y., Smith, W. B., Fedorov, O., et al. (2010). Selective inhibition of BET bromodomains. *Nature, 468*, 1067–1073.

Gallagher, S. J., Mijatov, B., Gunatilake, D., Gowrishankar, K., Tiffen, J., James, W., et al. (2014). Control of NF-κB activity in human melanoma by bromodomain and extra-terminal protein inhibitor I-BET151. *Pigment Cell & Melanoma Research, 27*, 1126–1137.

Grunstein, M. (1997). Histone acetylation in chromatin structure and transcription. *Nature, 389*, 349–352.

Grzesiek, S., & Bax, A. (1992). Correlating backbone amide and side-chain resonances in larger proteins by multiple relayed triple resonance NMR. *Journal of the American Chemical Society, 114*, 6291–6293.

Hassan, A. H., Prochasson, P., Neely, K. E., Galasinski, S. C., Chandy, M., Carrozza, M. J., et al. (2002). Function and selectivity of bromodomains in anchoring chromatin-modifying complexes to promoter nucleosomes. *Cell, 111*, 369–379.

Hou, T., Ray, S., Lee, C., & Brasier, A. R. (2008). The STAT3 NH2-terminal domain stabilizes enhanceosome assembly by interacting with the p300 bromodomain. *The Journal of Biological Chemistry, 283*, 30725–30734.

Jacobson, R. H., Ladurner, A. G., King, D. S., & Tjian, R. (2000). Structure and function of a human TAF(II)250 double bromodomain module. *Science, 288*, 1422–1425.

Jeanmougin, F., Wurtz, J. M., LeDouarin, B., Chambon, P., & Losson, R. (1997). The bromodomain revisited. *Trends in Biochemical Sciences, 22*, 151–153.

Johnson, B. A. (2004). Using NMRView to visualize and analyze the NMR spectra of macromolecules. *Methods in Molecular Biology, 278*, 313–352.

Kasten, M., Szerlong, H., Erdjument-Bromage, H., Tempst, P., Werner, M., & Cairns, B. R. (2004). Tandem bromodomains in the chromatin remodeler RSC recognize acetylated histone H3 Lys14. *The EMBO Journal, 23*, 1348–1359.

Kay, L. E., Xu, G. Y., Singer, A. U., Muhandiram, D. R., & Formankay, J. D. (1993). A gradient-enhanced Hcch Tocsy experiment for recording side-chain H-1 and C-13 correlations in H_2O samples of proteins. *Journal of Magnetic Resonance. Series B, 101*, 333–337.

Kennison, J. A., & Tamkun, J. W. (1988). Dosage-dependent modifiers of polycomb and antennapedia mutations in Drosophila. *Proceedings of the National Academy of Sciences of the United States of America, 85*, 8136–8140.

Kuboniwa, H., Grzesiek, S., Delaglio, F., & Bax, A. (1994). Measurement of H-N-H-alpha J-couplings in calcium-free calmodulin using new 2D and 3D water-flip-back methods. *Journal of Biomolecular NMR, 4*, 871–878.

Laskowski, R. A., Macarthur, M. W., Moss, D. S., & Thornton, J. M. (1993). PROCHECK: A program to check the stereochemical quality of protein structures. *Journal of Applied Crystallography, 26*, 283–291.

Laskowski, R. A., Rullmannn, J. A., MacArthur, M. W., Kaptein, R., & Thornton, J. M. (1996). AQUA and PROCHECK-NMR: Programs for checking the quality of protein structures solved by NMR. *Journal of Biomolecular NMR, 8*, 477–486.

Letunic, I., & Bork, P. (2011). Interactive tree of life v2: Online annotation and display of phylogenetic trees made easy. *Nucleic Acids Research, 39*, W475–W478.

Li, W., Cowley, A., Uludag, M., Gur, T., McWilliam, H., Squizzato, S., et al. (2015). The EMBL-EBI bioinformatics web and programmatic tools framework. *Nucleic Acids Research, 43*, W580–W584.

Li, Y., Wen, H., Xi, Y., Tanaka, K., Wang, H., Peng, D., et al. (2014). AF9 YEATS domain links histone acetylation to DOT1L-mediated H3K79 methylation. *Cell, 159*, 558–571.

Logan, T. M., Olejniczak, E. T., Xu, R. X., & Fesik, S. W. (1993). A general method for assigning NMR spectra of denatured proteins using 3D HC(CO)NH-TOCSY triple resonance experiments. *Journal of Biomolecular NMR*, *3*, 225–231.

Long, F., Vagin, A. A., Young, P., & Murshudov, G. N. (2008). BALBES: A molecular-replacement pipeline. *Acta Crystallographica. Section D, Biological Crystallography*, *64*, 125–132.

Mccoy, A. J., Grosse-Kunstleve, R. W., Adams, P. D., Winn, M. D., Storoni, L. C., & Read, R. J. (2007). Phaser crystallographic software. *Journal of Applied Crystallography*, *40*, 658–674.

Mertz, J. A., Conery, A. R., Bryant, B. M., Sandy, P., Balasubramanian, S., Mele, D. A., et al. (2011). Targeting MYC dependence in cancer by inhibiting BET bromodomains. *Proceedings of the National Academy of Sciences of the United States of America*, *108*, 16669–16674.

Mujtaba, S., He, Y., Zeng, L., Yan, S., Plotnikova, O., Sachchidanand, et al. (2004). Structural mechanism of the bromodomain of the coactivator CBP in p53 transcriptional activation. *Molecular Cell*, *13*, 251–263.

Muller, S., Filippakopoulos, P., & Knapp, S. (2011). Bromodomains as therapeutic targets. *Expert Reviews in Molecular Medicine*, *13*, 1–21.

Murshudov, G. N., Vagin, A. A., & Dodson, E. J. (1997). Refinement of macromolecular structures by the maximum-likelihood method. *Acta Crystallographica. Section D, Biological Crystallography*, *53*, 240–255.

Neri, D., Szyperski, T., Otting, G., Senn, H., & Wuthrich, K. (1989). Stereospecific nuclear magnetic-resonance assignments of the methyl-groups of valine and leucine in the DNA-binding domain of the 434-repressor by biosynthetically directed fractional C-13 labeling. *Biochemistry*, *28*, 7510–7516.

Nicodeme, E., Jeffrey, K. L., Schaefer, U., Beinke, S., Dewell, S., Chung, C. W., et al. (2010). Suppression of inflammation by a synthetic histone mimic. *Nature*, *468*, 1119–1123.

Ott, C. J., Kopp, N., Bird, L., Paranal, R. M., Qi, J., Bowman, T., et al. (2012). BET bromodomain inhibition targets both c-Myc and IL7R in high-risk acute lymphoblastic leukemia. *Blood*, *120*, 2843–2852.

Otwinowski, Z., & Minor, W. (1997). Processing of X-ray diffraction data collected in oscillation mode. *Macromolecular Crystallography, Part A*, *276*, 307–326.

Owen, D. J., Ornaghi, P., Yang, J. C., Lowe, N., Evans, P. R., Ballario, P., et al. (2000). The structural basis for the recognition of acetylated histone H4 by the bromodomain of histone acetyltransferase Gcn5p. *The EMBO Journal*, *19*, 6141–6149.

Pastori, C., Daniel, M., Penas, C., Volmar, C. H., Johnstone, A. L., Brothers, S. P., et al. (2014). BET bromodomain proteins are required for glioblastoma cell proliferation. *Epigenetics*, *9*, 611–620.

Plotnikov, A. N., Yang, S., Zhou, T. J., Rusinova, E., Frasca, A., & Zhou, M. M. (2014). Structural insights into acetylated-histone H4 recognition by the bromodomain-PHD finger module of human transcriptional coactivator CBP. *Structure*, *22*, 353–360.

Rieping, W., Habeck, M., Bardiaux, B., Bernard, A., Malliavin, T. E., & Nilges, M. (2007). ARIA2: Automated NOE assignment and data integration in NMR structure calculation. *Bioinformatics*, *23*, 381–382.

Robert, X., & Gouet, P. (2014). Deciphering key features in protein structures with the new ENDscript server. *Nucleic Acids Research*, *42*, W320–W324.

Sanchez, R., Meslamani, J., & Zhou, M. M. (2014). The bromodomain: From epigenome reader to druggable target. *Biochimica et Biophysica Acta*, *1839*, 676–685.

Sanchez, R., & Zhou, M. M. (2009). The role of human bromodomains in chromatin biology and gene transcription. *Current Opinion in Drug Discovery & Development*, *12*, 659–665.

Shen, Y., Delaglio, F., Cornilescu, G., & Bax, A. (2009). TALOS+: A hybrid method for predicting protein backbone torsion angles from NMR chemical shifts. *Journal of Biomolecular NMR*, *44*, 213–223.

Strahl, B. D., & Allis, C. D. (2000). The language of covalent histone modifications. *Nature*, *403*, 41–45.

Struhl, K. (1998). Histone acetylation and transcriptional regulatory mechanisms. *Genes & Development*, *12*, 599–606.

Syntichaki, P., Topalidou, I., & Thireos, G. (2000). The Gcn5 bromodomain co-ordinates nucleosome remodelling. *Nature*, *404*, 414–417.

Tamkun, J. W., Deuring, R., Scott, M. P., Kissinger, M., Pattatucci, A. M., Kaufman, T. C., et al. (1992). brahma: A regulator of Drosophila homeotic genes structurally related to the yeast transcriptional activator SNF2/SWI2. *Cell*, *68*, 561–572.

Winn, M. D., Ballard, C. C., Cowtan, K. D., Dodson, E. J., Emsley, P., Evans, P. R., et al. (2011). Overview of the CCP4 suite and current developments. *Acta Crystallographica. Section D, Biological Crystallography*, *67*, 235–242.

Wittekind, M., & Mueller, L. (1993). HNCACB, a high-sensitivity 3D NMR experiment to correlate amide-proton and nitrogen resonances with the alpha-carbon and beta-carbon resonances in proteins. *Journal of Magnetic Resonance, Series B*, *101*, 201–205.

Yamazaki, T., Lee, W., Arrowsmith, C. H., Muhandiram, D. R., & Kay, L. E. (1994). A suite of triple-resonance NMR experiments for the backbone assignment of N-15, C-13, H-2 labeled proteins with high-sensitivity. *Journal of the American Chemical Society*, *116*, 11655–11666.

Zeng, L., Zhang, Q., Gerona-Navarro, G., Moshkina, N., & Zhou, M. M. (2008). Structural basis of site-specific histone recognition by the bromodomains of human coactivators PCAF and CBP/p300. *Structure*, *16*, 643–652.

Zeng, L., Zhang, Q., Li, S., Plotnikov, A. N., Walsh, M. J., & Zhou, M. M. (2010). Mechanism and regulation of acetylated histone binding by the tandem PHD finger of DPF3b. *Nature*, *466*, 258–262.

Zeng, L., & Zhou, M. M. (2002). Bromodomain: An acetyl-lysine binding domain. *FEBS Letters*, *513*, 124–128.

Zhang, G. T., Liu, R. J., Zhong, Y. F., Plotnikov, A. N., Zhang, W. J., Zeng, L., et al. (2012). Down-regulation of NF-kappa B transcriptional activity in HIV-associated kidney disease by BRD4 inhibition. *Journal of Biological Chemistry*, *287*, 28840–28851.

Zuber, J., Shi, J. W., Wang, E., Rappaport, A. R., Herrmann, H., Sison, E. A., et al. (2011). RNAi screen identifies Brd4 as a therapeutic target in acute myeloid leukaemia. *Nature*, *478*, 524–528.

CHAPTER FIFTEEN

Preparation, Biochemical Analysis, and Structure Determination of Methyllysine Readers

C.A. Musselman*,[1], T.G. Kutateladze†,[1]
*University of Iowa, Iowa City, IA, United States
†University of Colorado School of Medicine, Aurora, CO, United States
[1]Corresponding authors: e-mail address: catherine-musselman@uiowa.edu; tatiana.kutateladze@ucdenver.edu

Contents

1. Introduction 345
2. Preparation of Methyllysine Readers 347
   2.1 Construct Design 347
   2.2 Overexpression in *E. coli* 348
   2.3 Purification 350
3. Screening for Methyllysine Recognition 355
4. Measuring Affinity for Methylated Substrate 356
5. Analysis of Binding Using NMR Spectroscopy 358
6. Structure Determination by X-ray Crystallography 360
Acknowledgments 361
References 361

Abstract

In-depth in vitro characterization of methyllysine reader domains and their association with cognate methyllysine substrates is essential to better understand fundamental mechanisms of chromatin regulation and to design targeted therapeutics that disrupt these interactions. In this chapter, we summarize commonly used methods for preparation, biochemical characterization, and determination of structures of methyllysine reader domains. We provide a detailed protocol for the preparation of a GST-tagged methyllysine reader domain and for analysis of histone-binding activities using a combination of pull-down, tryptophan fluorescence, and NMR assays, and describe initial steps toward crystallization of the complexes.

1. INTRODUCTION

Methylation of lysine residues has emerged as one of the major post-translational modifications (PTMs) in histone proteins. The side chain of

Methods in Enzymology, Volume 573
ISSN 0076-6879
http://dx.doi.org/10.1016/bs.mie.2015.12.005

lysine undergoes methylation on its ε-amino group yielding mono-, di-, or trimethylated species. In physiological conditions, addition of a methyl group does not eliminate the positive charge on lysine; however, it alters the hydrophobic character, charge distribution, and size of the lysine side chain as well as its ability to act as a hydrogen bond donor. Methylation is a reversible mark, which is generated by lysine methyltransferases (KMTs), whereas lysine demethylases (KDMs) catalyze the removal of this PTM (Black, Van Rechem, & Whetstine, 2012; Del Rizzo & Trievel, 2014). Various histone lysine methylation patterns have been shown to characterize specific genomic regions and chromatin states and are linked to particular biological outcomes such as active gene transcription, repression, and DNA damage response (Bannister & Kouzarides, 2011; Ernst et al., 2011; Kouzarides, 2007; Maunakea, Chepelev, & Zhao, 2010; Zentner & Henikoff, 2013).

Methyllysine marks are recognized by a set of protein effector domains, or readers of these PTMs. The list of identified methyllysine readers is growing very fast and currently includes ADD (ATRX-DNMT3-DNMT3L), Ankyrin, BAH (bromo adjacent homology), chromo-barrel, chromodomain (CD), double chromodomain (DCD), HEAT, MBT (malignant brain tumor), PHD (plant homeodomain), PWWP, SAWADEE, tandem tudor domain (TTD), Tudor, WD40, and zf-CW (zinc finger CW) (reviewed in Musselman, Lalonde, Côté, & Kutateladze, 2012; Patel & Wang, 2013). Comprehensive mechanistic, biochemical, and structural studies reveal a conserved mode for the methyllysine readout, which involves caging of the methylated lysine in an aromatic pocket, typically formed by two to four aromatic residues. Complex formation is driven by cation–π interactions between the methylammonium group of the lysine and the aromatic side chains of the reader, as well as hydrophobic and van der Waals contacts, and release of high energy water molecules (Hughes, Wiggins, Khorasanizadeh, & Waters, 2007; Kamps et al., 2015).

Aberrant catalytic activities of KMTs and KDMs result in alterations in histone methylation patterns and genomic instability. Misregulated functions of these enzymes have been linked to a wide variety of human disorders and there are a growing number of examples of methyllysine readers being implicated in disease (Brookes & Shi, 2014; Chi, Allis, & Wang, 2010; Greer & Shi, 2012; Helin & Dhanak, 2013; Pedersen & Helin, 2010; Portela & Esteller, 2010; Wigle & Copeland, 2013). In-depth characterization of the methyllysine reader domains is vital for a better understanding of fundamental mechanisms of chromatin regulation and is essential in the design of targeted therapeutics. In this chapter we describe our laboratories' overall strategy for preparation, biochemical analysis, and determination of structures of the methyllysine readers.

2. PREPARATION OF METHYLLYSINE READERS

Most methyllysine reader domains can be easily overexpressed as fusion proteins in *Escherichia coli* and purified for biochemical and structural analysis. However, care must be taken to ensure a stable and properly folded protein construct is chosen for monitoring the interaction with methyllysine substrate, and some optimization may be required to select the best expression and purification conditions to ensure high yield and stability. For an excellent in-depth review on optimization of expression in *E. coli* see Gräslund et al. (2008).

2.1 Construct Design

Boundaries of an initial construct can be determined through comparative analysis of the primary sequence homology and the predicted secondary structure, followed by assessment of the best cutoffs to promote stability and solubility. Subsequently the expression plasmid must be selected for the desired purification protocol and ultimate application.

(a) Analyze primary sequence of the domain via a bioinformatics research tool of choice, such as SMART (smart.embl-heidelberg.de; Letunic, Doerks, & Bork, 2015) to determine approximate domain boundaries.

(b) The exact construct boundaries should be chosen such that the N- and C-terminal residues do not truncate a consensus sequence, interrupt predicted secondary structure regions (determined using, eg, JPred; Drozdetskiy, Cole, Procter, & Barton, 2015), and are not hydrophobic.

(c) Once the construct boundaries have been selected, the sequence can either be subcloned out of the full-length gene or custom ordered. If subcloning is chosen, it is possible that the resultant DNA construct harbors "rare" codons and a compensating expression cell line, such as BL21-CodonPlus or Rosetta2 should be used for expression purposes. Custom synthesis of genes can be beneficial as codons can be optimized for bacterial expression.

(d) An inducible expression plasmid should be selected (the most common being isopropyl β-D-1-thiogalactopyranoside (IPTG) inducible) with the desired fusion tag and cleavage site. The most common tags used for the expression of methyllysine reader domains are the glutathione S-transferase (GST) tag, the hexa-Histidine (6 × His) tag, and the maltose-binding protein tag. Tags are typically positioned at the

N-terminus, but can also be placed at the C-terminus. Common cleavage sites, for postexpression cleavage of the fusion tag, include sites for PreScission, thrombin, Factor Xa, or TEV proteases. Care should be taken to ensure that the cleavage site sequence of choice does not occur anywhere in the domain sequence, and that the residual residues after cleavage are acceptable.

(e) The final DNA construct should be sequenced to confirm that the proper coding sequence has been achieved and is in frame in the expression plasmid.

2.2 Overexpression in *E. coli*

Once a proper construct is cloned into an expression plasmid it can be overexpressed in *E. coli*. The choice of growth media depends on the desired application and can also be optimized for the best yield. Rich media such as Luria broth (LB), 2XYT, or terrific broth (TB) will often produce high protein yields and are sufficient for most biochemical and biophysical experiments and X-ray crystallography (unless specific labeling is needed for phasing purposes). For NMR spectroscopy it is often necessary to overexpress proteins in minimal media for ^{15}N and/or ^{13}C (as well as ^{2}H, if desired) isotope enrichment. *E. coli* cells can be grown directly in minimal media, or if desired in rich media to proper OD_{600}, harvested, and transferred to minimal media before induction (Marley, Lu, & Bracken, 2001). Rich broths are commercially available as premixed powders. M9 media can be made as follows:

Recipe for M9 media

Autoclave the following in 1 L of water

12.8 g $Na_2HPO_4 \cdot 7H_2O$

3.0 g KH_2PO_4

0.5 g NaCl

1.0 g ^{15}N-NH_4Cl

2 mL (1 *M*) $MgSO_4$

200 μL (1 *M*) $CaCl_2$

Sterile filter and add the following to the cooled autoclaved media.

5 mL vitamin solution (1 tablet of daily vitamin dissolved in 50 mL water, spin down insoluble portion before filtering), or equivalent of commercially available vitamin solution.

25 mL of 20% (w/v) glucose (^{13}C-glucose should be used if carbon labeling is desired).

Proper antibiotics.

Overexpression

(a) Transform the expression plasmid into cells of choice following the recommended protocol and plate onto LB agar plates supplemented with the proper antibiotics.

(b) Inoculate one, or better two, preculture tubes (10 mL media supplemented with proper antibiotics) with colonies from the plate and incubate in a shaker/incubator at 25–37°C overnight.

(c) Inoculate 1 L of media supplemented with proper antibiotics with one preculture.

(d) Grow cultures in a shaker/incubator at 37°C until $OD_{600} \sim 0.4$–1.0. The choice of OD_{600} value at which to induce should be optimized for each construct. (Save a sample of uninduced cells for SDS-PAGE analysis.)

(e) Induce protein expression with, eg, 0.1–1.0 m*M* IPTG. (For metal-binding domains appropriate salts should also be added at induction. Example for Zn-fingers add 50–100 μ*M* $ZnCl_2$, note that adding too much can be toxic to the cells.)

(f) Incubate induced cultures in a shaker/incubator at 16–37°C for 4–18 h. The choice of induction temperature and time should be optimized for each construct. Generally, it is best to induce at lower temperatures for longer times when overexpressing in minimal media to increase solubility. (Save a sample of induced cells for SDS-PAGE analysis.)

(g) Harvest cells by centrifugation at $\sim 15{,}000 \times g$ for 30 min. Pellets can be stored at −80°C for up to several months.

Troubleshooting

Problem	Potential Cause and Solution
Leaky expression seen in preinduced samples	Switch to a cell line that suppresses background expression such as pLysS or pLysE cells
No expression seen in postinduction sample	The full-length protein is either not expressing or is being degraded in the cell. (1) Rare codons in the gene can lead to severely truncated transcripts. Check for the presence of rare codons and if needed either switch to a compensatory cell line such as BL21-CodonPlus or Rosetta2 or codon optimize the gene. (2) The rate of translation or induction time needs to be altered. Optimize the concentration of IPTG, the induction temperature and length of induction and analyze yield by SDS-PAGE. (3) Expression is low in the chosen media. Switch media.

2.3 Purification

For fusion proteins the first step is affinity purification utilizing the fusion tag. Lysis and wash buffers can be varied considerably to include buffer of choice (eg, Tris, phosphate, HEPES, MOPS), monovalent ion of choice (eg, potassium chloride, sodium chloride) as well as additives such as divalent salts, Arg, glycerol, etc. Below is an example set of buffers that have worked well for purification of a number of PHD fingers and Tudor domains in our hands. These buffers should be optimized for each construct. Many methyllysine-binding domains have a low isoelectric point (pI) and thus a pH of 6.5–8 is appropriate; however, this should be assessed for each construct (using, eg, the ProtParam tool; Gasteiger et al., 2005). A buffer pH that is at least one unit from the pI of the construct is recommended. Note that the pI of the fusion construct should be considered when determining the pH for the lysis buffer.

2.3.1 Affinity-Tag Purification (Per Liter of Culture)

The following protocol is written for purification of a GST-fusion methyllysine reader domain with a PreScission protease cleavage site that is stable in Tris and potassium chloride, but can easily be modified for other constructs. Note that the needed resin, method of elution, and cleavage conditions will differ depending on the choice of tag and cleavage site. It is advisable to follow the resin manufacturer's recommendation as a starting point.

Lysis buffer

20 m*M* Tris, pH 7.5
250 m*M* KCl
0.5% Triton X-100
3 m*M* DTT (freshly made)
Protease inhibitors (freshly made)
1 m*M* EDTA (be cautious in using with metal-binding proteins)

High-salt wash buffer

20 m*M* Tris, pH 7.5
250–500 m*M* KCl
3 m*M* DTT
1 m*M* EDTA (be cautious in using with metal-binding proteins)

Low-salt wash buffer

20 m*M* Tris, pH 8
150 m*M* KCl

3 m*M* DTT

1 m*M* EDTA (be cautious in using with metal-binding proteins)

Note: If EDTA cannot be used due to interference with metal binding, it is best to autoclave water for making buffers to avoid protease degradation during purification.

(a) Resuspend the cells into 50 mL of lysis buffer by running through a 10 mL serological pipette. After resuspension lysozyme can be added if desired at 1 mg/mL and incubated for 30 min to increase lysis efficiency.

(b) Lyse the cells by freeze/thaw, sonication, or homogenization methods.

(c) Clear the lysate by centrifuging at ~25,000 × *g* for 30 min. (Save samples of the pellet and supernatant for SDS-PAGE analysis.)

(d) Incubate the supernatant with glutathione agarose or glutathione sepharose resin for at least 1 h rocking at 4°C. (Lysate can also be loaded onto a prepacked glutathione column using a fast protein liquid chromatography (FPLC) pump.)

(e) Wash the resin with 100–500 mL of high-salt wash buffer (for some domains a high-salt wash is unnecessary) followed by 100–500 mL of low-salt wash buffer. Every 100 mL of flow-through should be analyzed either by UV absorbance or SDS-PAGE to determine the appropriate wash volume. Once no protein is detected in the flow-through washing is complete.

(f) If the cleaved domain is desired, skip to step (g). If the GST-fusion domain is desired, incubate the washed resin with ~20 mL of wash buffer supplemented with 50 m*M* L-glutathione (note that L-glutathione is very acidic, therefore make sure to pH the stock solution). Elute the GST-fusion domain from the column and wash the resin with an additional 50 mL of buffer to collect any residual protein. Be sure to remove glutathione by dialysis or concentration/exchange before further application.

(g) If cleaved protein is desired, incubate washed resin with ~20 mL of low-salt wash buffer supplemented with PreScission Protease, rocking overnight at 4°C. Then elute cleaved protein and wash resin with an additional 50 mL of wash buffer to collect any residual protein.

(h) Analyze all samples by SDS-PAGE (Fig. 1). (Antibodies to fusion tags are commercially available and can also be used for analysis.)

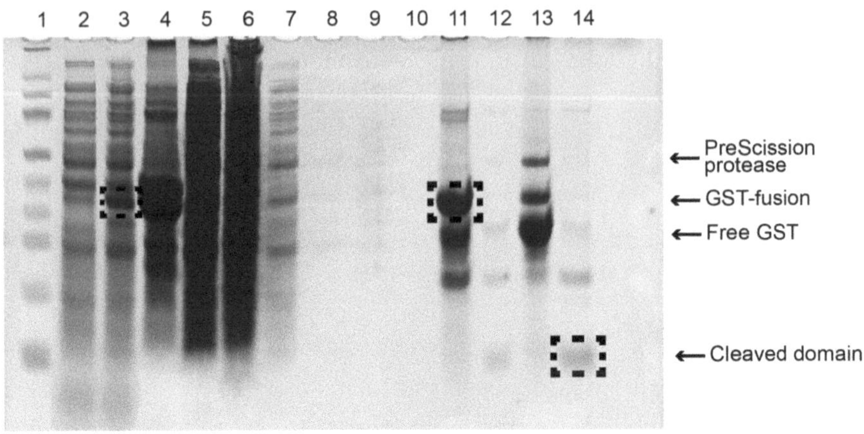

Fig. 1 SDS-PAGE analysis of expression and purification steps for a GST-fusion methyllysine reader domain, stained with Coomassie Blue. Shown are (1) ladder, (2) cells preinduction, (3) cells postinduction, (4) lysate pellet, (5) lysate supernatant, (6) flow-through from glutathione column, (7) wash #1, (8) wash #2, (9) wash #3, (10) wash #4, (11) glutathione beads prior to PreScission protease cleavage, (12) elution, (13) glutathione beads after cleavage with PreScission protease, and (14) concentrated elution. For reference, the fusion domain is boxed in lanes 3 and 11 and the cleaved methyllysine reader domain is boxed in 14.

Troubleshooting

Problem	**Potential Cause and Solution**
The protein expresses but is only in the pellet after lysis	It is wise to first confirm expression and presence of the fusion protein in the pellet through Western analysis. If confirmed, this suggests that the protein is insoluble. There are several potential causes and solutions to this. (1) The lysis buffer could be incompatible with the protein, check the composition of the buffer especially the pH and salt concentration and alter to increase solubility. (2) The cells were not lysed, try relysing, another lysis method or a combination of lysis methods to increase lysis efficiency. (3) The protein misfolds during expression. This can usually be addressed by slowing down the rate of expression. Lower the concentration of IPTG for induction and/or lower induction temperature

The protein does not bind to the resin	The potential cause of this depends on the tag. (1) Confirm that the binding buffer is compatible with the resin. (2) If using a structured fusion tag, it may indicate that the tag unfolds, which could occur either during expression or lysis. Try different induction conditions described above for solubility. Or try altering the lysis method to a less harsh method. (3) The resin-binding site on the tag may be occluded by the domain of interest. Try altering the linker between the fusion tag and domain, switching termini for placement of the tag, or switching tags
There is no cleavage reaction	The cleavage site may be occluded. (1) Try eluting off the resin first and cleaving in solution. (2) Try altering the linker between the fusion tag and protein, switching the termini that the tag is on, or switching tags
The protein cleaves but the tag-free domain does not elute from the resin	It is likely that the cleaved domain is either sticking to the resin or precipitating after cleavage. (1) Try eluting with a higher salt buffer to reduce sticking. (2) Try cleaving after eluting the fusion protein from the resin. (3) Alter cleavage conditions to promote solubility of the tag-free domain. (4) Switch tags

2.3.2 FPLC Purification

For some biochemical assays such as pull-downs, the affinity-tag purification step may yield a sufficiently pure sample. For most biophysical assays and structural determination it is best practice to incorporate further purification steps, for example, utilizing FPLC (Fig. 2). A common procedure would incorporate cation or anion exchange (eg, a Source S or Q column purification) and/or size-exclusion chromatography (eg, Superdex 75 or Sephacryl S-100 column purification) in order to yield high-purity samples.

(a) Concentrate the elution from the affinity resin to 0.5–2.0 mL (depending on the size of the FPLC loading loop). Note that for a cation/anion exchange column a super loop can also be used to load the elution.

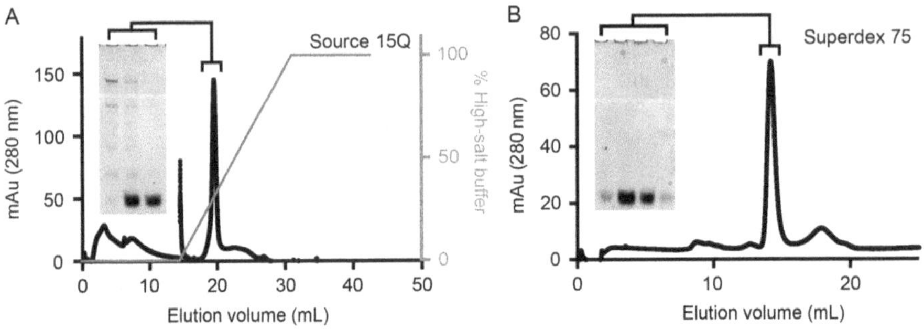

Fig. 2 FPLC purification of a methyllysine reader domain. (A) Chromatogram from Source 15Q purification. The UV trace at 280 nm is shown in black and the percentage of 1 *M* KCl buffer is shown in gray. SDS-PAGE analysis of the major peak is shown in the inset. (B) Chromatogram from the subsequent Superdex 75 purification. The UV trace at 280 nm is shown and the SDS-PAGE analysis of the major peak is shown in the inset.

(b) Spin down the sample at $5000 \times g$ for 3 min to remove any insoluble particles and load onto the appropriate column that has been preequilibrated into the desired buffer. Run according the manufacturer's recommendation.

(c) Check the purity of fractions by SDS-PAGE, pool the clean fractions and concentrate the sample to the desired concentration for further purification or application.

(d) At this point many methyllysine reader domains are compatible with long-term storage at −80°C. It is recommended to flash freeze the protein before storage. Protein stability should be confirmed by SDS-PAGE after thawing to ensure the freeze thaw process has not damaged the domain.

Troubleshooting

Problem	Potential Cause and Solution
Protein does not bind the cation/anion exchange column	(1) The wrong ion exchange column is being used. Assess the pI (and pH of the loading buffer) and confirm that the domain of interest will bind to the chosen column. (2) The salt concentration in the loading buffer is too high. Exchange into a lower salt buffer before loading
The protein has the incorrect retention volume on gel filtration based on expected for the molecular weight	Run standards in the chosen buffer to confirm column performance and where the protein should be eluting. (1) The protein may have degraded. Check

by SDS-PAGE and alter conditions to reduce degradation. (2) The protein may be unfolded. This can be confirmed by another method such as circular dichroism. Construct boundaries may need to be altered to obtain a well-folded domain, or expression/ purification conditions may need to be altered to promote folding. (3) The protein is aggregating. Alter buffer composition to decrease aggregation

3. SCREENING FOR METHYLLYSINE RECOGNITION

A simple way to screen for methyllysine-binding activity is through the pull-down assay. The following protocol assumes a GST-fusion methyllysine reader domain that is stable in Tris and potassium chloride, and biotinylated methylated peptides. Other combinations of fusion tags can also be used. There are a large number of biotinylated methylated peptides commercially available. As lysine can be mono-, di-, or trimethylated it is wise to obtain all possible methylation states of the substrate of interest as well as the unmodified counterpart for screening.

Example binding buffer

20 m*M* Tris, pH 8
150 m*M* KCl
0.5% Triton X-100
3 m*M* DTT (freshly made)

Example wash buffer

20 m*M* Tris, pH 8
150–250 m*M* KCl
0.5% Triton X-100
3 m*M* DTT (freshly made)

(a) Prepare protein and peptide stocks at 1 μg/μL.
(b) Mix 1 μg of GST-fusion domain with 1 μg of biotinylated peptide and bring to 100 μL with binding buffer. Also prepare a negative control of GST-fusion alone. Incubate overnight at 4°C rotating end-over-end.
(c) Add 25 μL of streptavidin sepharose or streptavidin agarose slurry (~50% slurry equilibrated in binding buffer) to each sample and incubate for 1 h at 4°C rotating end-over-end.
(d) Centrifuge for 2 min at 2500 × *g* and remove supernatant.

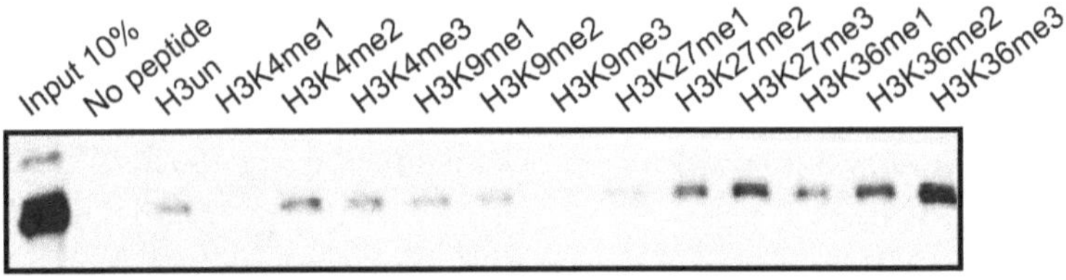

Fig. 3 Pull-down experiment with a GST-fusion methyllysine reader domain using a series of biotinylated histone peptides each containing a single modification as denoted, detected using an anti-GST antibody.

(e) Add 100 μL of wash buffer, invert 5–10 times and centrifuge for 2 min at 2500 × *g*. Repeat this wash step three more times.

(f) After the final wash add 25 μL of SDS loading dye to the resin and boil samples for 10 min. Also prepare a 10% input (0.1 μg fusion protein in 25 μL loading dye).

(g) Run samples (including negative and positive controls) on an acrylamide gel.

(h) Analyze by Western using an anti-GST antibody (Fig. 3).

Troubleshooting

Problem	Potential Cause and Solution
Protein is detected in the negative control	The GST-tagged domain is sticking to the resin. Try increasing the salt concentration of the washes, Triton X-100 concentration and/or the number of washes
Signal is very weak as compared to positive control	The wash conditions may be too stringent. Try lowering the salt concentration or Triton X-100 concentration in the wash buffer

4. MEASURING AFFINITY FOR METHYLATED SUBSTRATE

There are many methods appropriate for determining the affinity of methyllysine reader domains for substrate peptides, including isothermal titration calorimetry (ITC), surface plasmon resonance (SPR), and fluorescence anisotropy. Often there is a tryptophan in the aromatic cage or otherwise in the binding pocket of these domains and a dearth of tryptophans in the peptide, making intrinsic tryptophan fluorescence spectroscopy an excellent method as well. Below is a procedure for affinity determination by tryptophan fluorescence spectroscopy (Fig. 4). (*Note*: intrinsic tyrosine

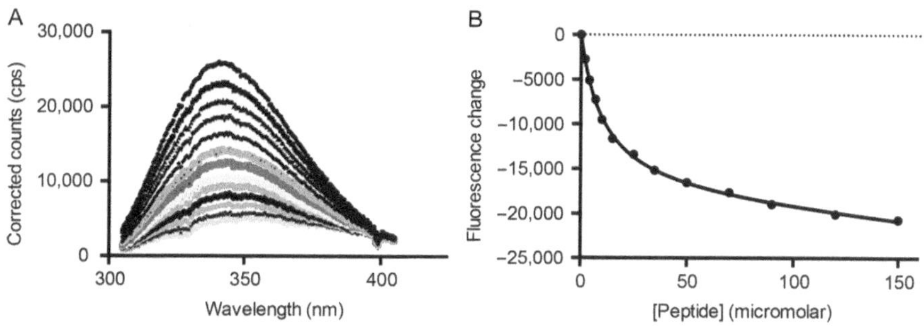

Fig. 4 Monitoring peptide binding of a methyllysine reader domain by intrinsic tryptophan fluorescence. (A) Corrected fluorescence spectra in the presence of increasing concentration of peptide. (B) Nonlinear regression fit of the change in fluorescence from apo as a function of peptide concentration.

or phenylalanine fluorescence can also be used, though the excitation and emission wavelengths should be altered accordingly.)

(a) Freshly prepared protein should be extensively dialyzed into a freshly made appropriate binding buffer and peptide stocks should be prepared in the same buffer.

(b) A quartz cuvette should be extensively washed and dried, then filled with buffer.

(c) The excitation wavelength should be set to 295 nm and emission spectra recorded from 305–405 nm with a 0.5 nm step size and a 1 s integration time. For best results collect and average three scans.

(d) Replace buffer with the protein sample and collect another emission spectrum. The appropriate protein concentration should be determined based on the signal to noise ratio, keeping in mind that fluorescence intensity will likely decrease significantly at saturation. For domains containing a single Trp, concentrations of 0.5–10 μM generally yield good signal for the suggested data collection parameters, however this may take some optimization.

(e) Add progressively increasing concentrations of peptide in a step-wise manner, keeping the total volume increase as small as possible while maintaining good pipetting accuracy. If equipped, setting a low stir speed is helpful to obtain proper mixing of peptide, otherwise the peptide should be mixed by pipetting. Upon each addition of peptide collect another emission spectrum.

(f) Reference titrations should be performed adding buffer to protein alone as well as adding peptide to buffer alone.

(g) Subtract the buffer scan from each sample run and make any other necessary corrections such as for dilution, photobleaching, and inner filter effects (see Kozlov, Galletto, & Lohman, 2012 for details on such corrections). Determine the maximum fluorescence signal for each corrected data set.

(h) Compute the change in fluorescence intensity as compared to apo protein for each titration point and determine the K_d by nonlinear least-squares analysis (Fig. 4).

Troubleshooting

Problem	Potential Cause and Solution
A significant decrease in fluorescence is detected in the buffer reference titration	This can occur for several reasons. (1) The protein may be photo bleaching. Increase the concentration of protein to obtain sufficient signal throughout the titration and adjust for photobleaching in analysis. (2) The protein may be aggregating. Adjust conditions to reduce aggregation. (3) The protein may be sticking to the pipette tip, or to the cuvette. To assess if it is sticking to the cuvette collect several spectra at different concentrations and check for linear dependence of signal. If protein is sticking to the pipette obtain low-retention tips. Pipette tips can also be treated in the lab using a siliconizing agent
An increase in fluorescence is detected in the buffer reference titration	The buffer is likely contaminated. Fresh buffer should be used for all fluorescence experiments
An increase in fluorescence is detected in the peptide reference titration	Confirm that the substrate is not intrinsically fluorescent. If not, the peptide stock may be contaminated

5. ANALYSIS OF BINDING USING NMR SPECTROSCOPY

Nuclear magnetic resonance (NMR) spectroscopy is a very powerful tool for analysis of complex formation. NMR can be used to map binding pockets, determine dissociation constants, examine conformational dynamics on a wide range of timescales, and determine atomic resolution structures. Here we outline procedures for sample preparation and analysis of chemical shift perturbations (CSPs) to map the binding pocket of the methyllysine-binding domains and determine dissociation constants. We do not describe the procedures for NMR data collection and processing or for obtaining resonance assignments here, which should be pursued with the help of an expert.

(a) Samples should be prepared in buffer suitable for NMR spectroscopy. It is generally advisable to have a buffer with a pH $\leq$7.0, total salt concentration of $\leq$150 m*M* and free of any viscous additives, such as glycerol, to obtain the best signal. Deviation from these optimal conditions is possible with careful experimental planning and should be discussed with an expert. Reducing agents, such as DTT, are fine to use here (of note, for some NMR experiments deuterated DTT is required). For the following analysis, the methyllysine recognition domain should be enriched in ^{15}N and the final sample should be ~250–300 μL in volume with a concentration of ~100 μ*M* (assuming the use of a shigemi NMR tube). Lower concentrations can be used if a cryogenic probe is available. Peptide stock should be prepared in matching buffer. This is critical as NMR resonances are sensitive to buffer conditions.

(b) Collect an ^{1}H,^{15}N-heteronuclear single quantum coherence (^{1}H,^{15}N HSQC) spectrum of the apo-state of the protein on a spectrometer of $\geq$500 MHz. Confirm that conditions chosen provide a good signal to noise ratio and that the majority of expected resonances are observed.

(c) Collect additional ^{1}H,^{15}N HSQC spectra after adding increasing concentrations of the unlabeled methyllysine substrate to the NMR sample until saturation is reached. For very weak-binding substrates, it may not be feasible to reach saturation. Overlay of these spectra will reveal resonance perturbations induced by ligand binding (Fig. 5).

(d) CSPs can be determined by computing the change in ^{1}H and ^{15}N chemical shifts, $\Delta\delta_{\mathrm{H}}$ and $\Delta\delta_{\mathrm{N}}$, where $\Delta\delta = \delta_{\mathrm{bound}} - \delta_{\mathrm{apo}}$. These values can be normalized using the equation:

$$\Delta\delta_{\mathrm{norm}} = \left(\Delta\delta_{\mathrm{H}}{}^{2} + (\Delta\delta_{\mathrm{N}}/5)^{2}\right)^{1/2}$$

(e) If resonance assignments have been made, they can be used to determine the residues directly or indirectly involved in binding of the methylated substrate, which can be further confirmed via mutagenesis.

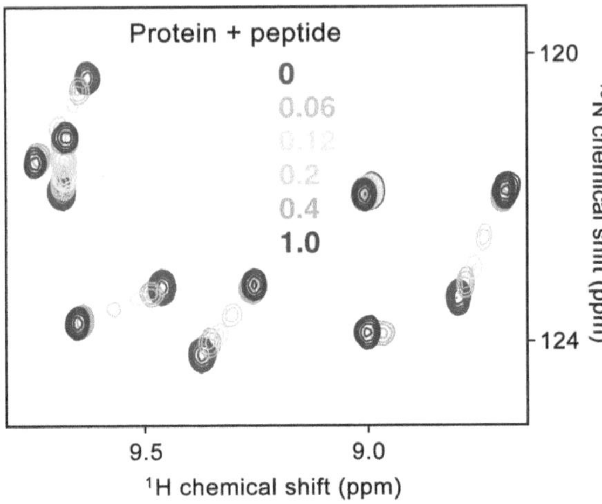

Fig. 5 Overlay of successive $^{1}H,^{15}N$ HSQC spectra of a labeled methyllysine reader domain upon titration of methylated peptide. Arbitrary peptide concentration is shown in legend. (See the color plate.)

(f) Binding constants can be obtained by nonlinear fitting of $\Delta\delta_{norm}$ vs ligand concentration using an equation that accounts for ligand depletion. Note that given the high protein concentration utilized in the NMR experiment, it is advisable to determine binding constants of $K_d < 10\ \mu M$ using an alternative approach as outlined above, as these K_ds will lead to stoichiometric binding conditions in the NMR experiment.

6. STRUCTURE DETERMINATION BY X-RAY CRYSTALLOGRAPHY

X-ray crystallography is the most widely utilized technique for determining the structure of methyllysine reader domains in complex with their cognate substrates. Obtaining well diffracting crystals is an empirically driven process that can take considerable time and effort. Through ample screening one can obtain the appropriate chemical, biochemical, and physical conditions that will produce a crystalline state of the complex as a starting point. From here extensive optimization of conditions might yield diffracting crystals, from which a suitable data set can be obtained and a structure determined. Below is a set of guidelines to aid in obtaining crystals. The process of data collection, processing, and structure determination is not covered here and should be pursued with the help of an expert. Note that the phase problem for most methyllysine reader domains can be determined through molecular replacement using previously solved homologous domains as a model, or in the case of Zn-binding domains can be determined de novo using MAD or SAD.

(a) Beginning with the most minimal construct is critical in the crystallization process and should be considered for both the reader domain and the substrate peptide. If the construct contains significant regions outside those necessary for binding, especially if these regions are not structured, this can preclude crystallization.

(b) A highly pure and homogenous sample is vital in the crystallization process, and thus extensive purification of the protein should be pursued before screening begins. In addition, having freshly purified and not previously frozen samples often yields better results.

(c) The protein should be as concentrated as possible while retaining stability as a starting point.

(d) Cocrystallization is generally the best strategy for methyllysine reader domains. Incubate the protein with a slight molar excess of peptide (1.2–1.5:1.0 molar ratio of peptide:protein) for at least 1 h prior to setting up crystal trays. If the complex has a high enough affinity, it can be repurified using size-exclusion chromatography to obtain a highly homogenous stock mixture. However, weaker affinity complexes will likely partly dissociate during this process, and thus the protein–peptide mixture should be used directly after incubation.

(e) A large number of "shotgun" screens are commercially available and provide the best starting conditions if nothing is known about the crystallization properties of the domain of interest. These will allow for testing a wide range buffers, salts, and precipitants. It is also advisable to screen for different crystallization temperatures.

(f) Once a promising condition is found, systematic screening can be pursued to obtain the best diffracting crystals.

ACKNOWLEDGMENTS

Research in the C.A.M. laboratory is supported by the NSF, CAREER Award 1452411. Research in the T.G.K. laboratory is supported by grants from the NIH, GM101664, GM106416, and GM100907.

REFERENCES

Bannister, A. J., & Kouzarides, T. (2011). Regulation of chromatin by histone modifications. *Cell Research*, *21*(3), 381–395.

Black, J. C., Van Rechem, C., & Whetstine, J. R. (2012). Histone lysine methylation dynamics: Establishment, regulation, and biological impact. *Molecular Cell*, *48*, 491–507.

Brookes, E., & Shi, Y. (2014). Diverse epigenetic mechanisms of human disease. *Annual Review of Genetics*, *48*, 237–268.

Chi, P., Allis, C. D., & Wang, G. G. (2010). Covalent histone modifications—Miswritten, misinterpreted and mis-erased in human cancers. *Nature Reviews. Cancer*, *10*(7), 457–469. Nature Publishing Group.

Del Rizzo, P. A., & Trievel, R. C. (2014). Molecular basis for substrate recognition by lysine methyltransferases and demethylases. *Biochimica et Biophysica Acta*, *1839*(12), 1404–1415.

Drozdetskiy, A., Cole, C., Procter, J., & Barton, G. J. (2015). JPred4: A protein secondary structure prediction server. *Nucleic Acids Research*, *43*(W1), W389–W394.

Ernst, J., Kheradpour, P., Mikkelsen, T. S., Shoresh, N., Ward, L. D., Epstein, C. B., et al. (2011). Mapping and analysis of chromatin state dynamics in nine human cell types. *Nature*, *473*(7345), 43–49.

Gasteiger, E., Hoogland, C., Gattiker, A., Duvaud, S., Wilkins, M. R., Appel, R. D., et al. (2005). Protein identification and analysis tools on the ExPASy server. In John M. Walker (Ed.), *The proteomics protocols handbook* (pp. 571–607). Totowa, NJ: Humana Press.

Gräslund, S., Nordlund, P., Weigelt, J., Bray, J., Gileadi, O., Knapp, S., et al. (2008). Protein production and purification. *Nature Methods*, *5*(2), 135–146. Nature Publishing Group.

Greer, E. L., & Shi, Y. (2012). Histone methylation: A dynamic mark in health, disease and inheritance. *Nature Reviews. Genetics*, *13*(5), 343–357.

Helin, K., & Dhanak, D. (2013). Chromatin proteins and modifications as drug targets. *Nature*, *502*(7472), 480–488.

Hughes, R. M., Wiggins, K. R., Khorasanizadeh, S., & Waters, M. L. (2007). Recognition of trimethyllysine by a chromodomain is not driven by the hydrophobic effect. *Proceedings of the National Academy of Sciences of the United States of America*, *104*(27), 11184–11188.

Kamps, J. J. A. G., Huang, J., Poater, J., Xu, C., Pieters, B. J. G. E., Dong, A., et al. (2015). Chemical basis for the recognition of trimethyllysine by epigenetic reader proteins. *Nature Communications*, *6*, 8911.

Kouzarides, T. (2007). Chromatin modifications and their function. *Cell*, *128*, 693–705.

Kozlov, A. G., Galletto, R., & Lohman, T. M. (2012). *SSB–DNA binding monitored by fluorescence intensity and anisotropy*. Totowa, NJ: Humana Press. pp. 55–83.

Letunic, I., Doerks, T., & Bork, P. (2015). SMART: Recent updates, new developments and status in 2015. *Nucleic Acids Research*, *43*(Database issue), D257–D260.

Marley, J., Lu, M., & Bracken, C. (2001). A method for efficient isotopic labeling of recombinant proteins. *Journal of Biomolecular NMR*, *20*(1), 71–75. Springer.

Maunakea, A. K., Chepelev, I., & Zhao, K. (2010). Epigenome mapping in normal and disease states. *Circulation Research*, *107*(3), 327–339.

Musselman, C. A., Lalonde, M.-E., Côté, J., & Kutateladze, T. G. (2012). Perceiving the epigenetic landscape through histone readers. *Nature Structural & Molecular Biology*, *19*(12), 1218–1227.

Patel, D. J., & Wang, Z. (2013). Readout of epigenetic modifications. *Annual Reviews*, *82*(1), 81–118. http://dxdoiorg/101146/annurev-biochem-072711-165700.

Pedersen, M. T., & Helin, K. (2010). Histone demethylases in development and disease. *Trends in Cell Biology*, *20*(11), 662–671.

Portela, A., & Esteller, M. (2010). Epigenetic modifications and human disease. *Nature Biotechnology*, *28*(10), 1057–1068. Nature Publishing Group.

Wigle, T. J., & Copeland, R. A. (2013). Drugging the human methylome: An emerging modality for reversible control of aberrant gene transcription. *Current Opinion in Chemical Biology*, *17*, 369–378.

Zentner, G. E., & Henikoff, S. (2013). Regulation of nucleosome dynamics by histone modifications. *Nature Structural & Molecular Biology*, *20*(3), 259–266. Nature Publishing Group.

PART IV

DNA Modifications and Nucleic Acid Regulators

CHAPTER SIXTEEN

Quantification of Oxidized 5-Methylcytosine Bases and TET Enzyme Activity

M.Y. Liu, J.E. DeNizio, R.M. Kohli[1]
Perelman School of Medicine, University of Pennsylvania, Philadelphia, PA, United States
[1]Corresponding author: e-mail address: rkohli@upenn.edu

Contents

1. Introduction 366
2. Analysis of Cytosine Modifications in Cellular DNA 366
   2.1 Preparation of Genomic DNA from TET-Transfected Cells 366
   2.2 Qualitative Analysis by Dot Blotting 368
   2.3 Quantitative Analysis by LC-MS/MS 369
3. Analysis of Cytosine Modifications In Vitro 373
   3.1 Purification of TET Enzymes from Sf9 Insect Cells 373
   3.2 Synthesis and Isotopic Labeling of TET Substrates 374
   3.3 Chemoenzymatic Activity Assays on Full-Length Oligonucleotides 378
   3.4 Quantitative Activity Assays on Nucleosides 382
Acknowledgments 383
References 383

Abstract

In eukaryotic DNA, cytosine can be enzymatically modified to yield up to four epigenetic base variants. DNA methyltransferases convert cytosine to 5-methylcytosine (mC), which plays critical roles in gene regulation during development. Ten-eleven translocation (TET) enzymes can sequentially oxidize mC to three products: 5-hydroxymethylcytosine (hmC), 5-formylcytosine (fC), and 5-carboxylcytosine (caC). These oxidized bases have been found in numerous mammalian cell types, where they potentially carry out independent epigenetic functions and aid in DNA demethylation. To gain insight into the mechanisms and functions of TET family enzymes, rigorous approaches are needed to quantify genomic cytosine modifications in cells and track TET enzyme activity in vitro. Here, we present tools developed by our lab and others to report on each of the five forms of cytosine (unmodified, mC, hmC, fC, and caC) with high specificity and sensitivity. We provide detailed protocols for qualitative and quantitative analysis of cytosine modifications in genomic DNA by dot blotting and LC-MS/MS. We then describe methods for generating synthetic oligonucleotide substrates for biochemical studies, provide optimized reaction conditions, and introduce several

Methods in Enzymology, Volume 573
ISSN 0076-6879
http://dx.doi.org/10.1016/bs.mie.2015.12.006

chemoenzymatic assays, as well as HPLC, mass spectrometry, and scintillation counting methods to quantify cytosine modifications in vitro. These approaches enable mechanistic studies of TET activity, which are key to understanding the role of these enzymes in epigenetic regulation.

1. INTRODUCTION

Ten-eleven translocation (TET) enzymes are Fe(II)/α-ketoglutarate-dependent dioxygenases that are increasingly tied to diverse biological and pathological processes, including cellular differentiation, reprogramming, and malignancy (Kohli & Zhang, 2013; Tahiliani et al., 2009). There are three mammalian TET isoforms (TET1, 2, and 3) that are all capable of sequentially oxidizing 5-methylcytosine (mC) to 5-hydroxymethylcytosine (hmC), 5-formylcytosine (fC), and 5-carboxylcytosine (caC) (He et al., 2011; Ito et al., 2011). Stepwise oxidation provides a feasible pathway for DNA demethylation, as the highly oxidized bases fC and caC can be selectively removed by thymine DNA glycosylase (TDG), resulting in an abasic site that can be repaired to regenerate unmodified C (He et al., 2011; Maiti & Drohat, 2011). Emerging evidence also indicates that all three oxidized mC bases (ox-mCs) can exist as stable epigenetic marks with potentially independent functions (Bachman et al., 2014, 2015; Iurlaro et al., 2013; Spruijt et al., 2013). As TET biology has expanded, there has been a need for robust assays to detect, localize, and quantify these rare genomic modifications (Fig. 1). Numerous chemical methods now exist for quantification and base-resolution sequencing of ox-mC bases in a variety of cell types (Booth, Raiber, & Balasubramanian, 2015; Song, Yi, & He, 2012; Yardimci & Zhang, 2015). At the same time, rigorous biochemical assays are needed to address open mechanistic questions. Early approaches to the study of TET enzymes have been reviewed previously (Shen & Zhang, 2012). Here, we present the most current in vivo and in vitro methods developed by our lab and others to distinguish between the modified forms of cytosine and measure TET enzyme activity with high sensitivity.

2. ANALYSIS OF CYTOSINE MODIFICATIONS IN CELLULAR DNA

2.1 Preparation of Genomic DNA from TET-Transfected Cells

One major area of study centers on the comparative roles of the three TET isoforms and their mutants, many of which have been catalogued in human

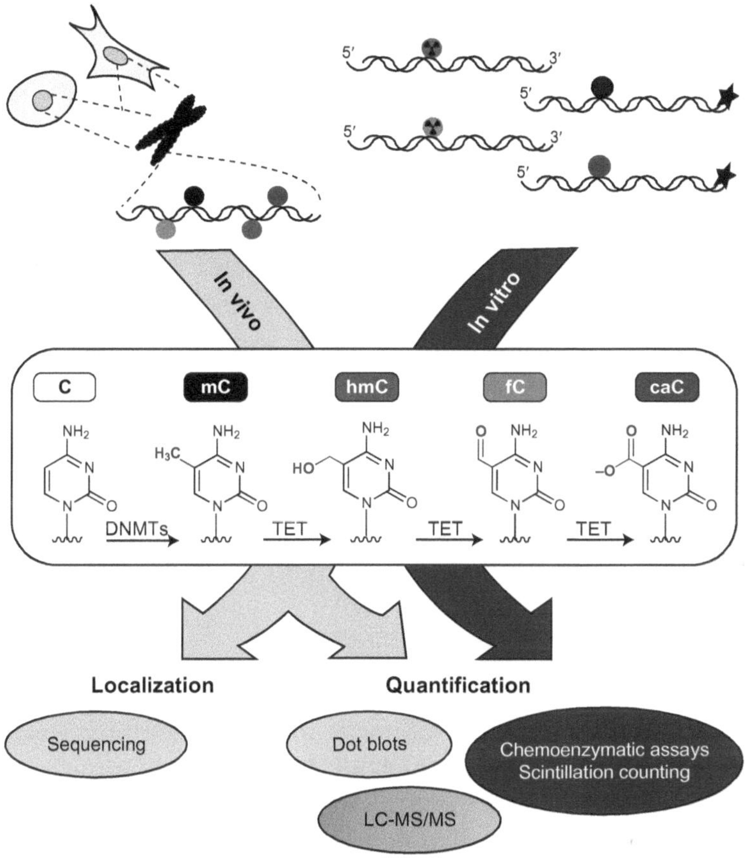

Fig. 1 Current methods for static analysis of TET activity via modified cytosines in vivo and in vitro. (See the color plate.)

diseases and may provide key insights into enzyme mechanisms. HEK293T cells provide a convenient overexpression system for assessing the relative activities of TET constructs. HEK293T cells have low levels of endogenous TET proteins and support efficient transfection and high levels of overexpression. Available constructs include full-length (FL) and catalytic domains (CD) of mouse and human TET1, 2, and 3 cloned into standard mammalian overexpression vectors. Our lab and others have demonstrated activity from the following: hTET1-FL (residues 1–2136), hTET1-CD (1418–2136), hTET2-FL (1–2002), hTET2-CD (1129–2002), hTET3-FL (1–1660), mTet1-FL (1–2007), mTet1-CD (1367–2039), mTet2-FL (1–1912), mTet2-CD (1042–1921), mTet3-FL (1–1668), and mTet3-CD (697–1668) (He et al., 2011; Ito et al., 2010; Tahiliani et al., 2009). In addition, Hu et al. crystallized a truncated form of the hTET2 catalytic domain (hTET2-CS, 1129–1936 Δ1481–1843) and demonstrated activity similar to the full-length construct (Hu et al., 2013). Common negative controls include the corresponding empty expression vector or mutation of the iron-binding HxD motif in the catalytic domain that renders TET inactive.

We use the following protocol to overexpress TET constructs. First, culture HEK293T cells in Dulbecco's Modified Eagle Medium with GlutaMAX (Thermo Fisher Scientific) and 10% fetal bovine serum (Sigma). When cells are 70–90% confluent, transfect using Lipofectamine 2000 (Thermo) according to the manufacturer's protocol. Change media 24 h posttransfection, harvest cells by trypsinization at 48 h, and resuspend pellets in phosphate buffered saline. A sample of transfected cells can be set aside for Western blotting to evaluate protein expression (see Section 2.2). Purify genomic DNA (gDNA) using the DNeasy Blood & Tissue Kit (Qiagen), including addition of RNase A where specified in the manufacturer's protocol. Note that this kit is also adaptable for extracting gDNA from tissue specimens and cells that express endogenous TET enzymes.

2.2 Qualitative Analysis by Dot Blotting

Dot blots are commonly used to probe for modified bases in gDNA. DNA is denatured to expose the bases, spotted onto an absorbent membrane, and probed with antibodies against each of the four cytosine modifications. Dot blots offer a clear visual result and can be performed using either serial dilutions or single concentrations of DNA. We consider the former to be semiquantitative, while the latter is only qualitative but still particularly useful for screening a large number of samples. Dot blotting also works for plasmids but is generally not well suited for short oligonucleotides, likely because these do not adhere consistently to membranes.

The first step is to determine the appropriate amount of DNA for blotting, considering the amount of expected modifications. For gDNA from HEK293T cells overexpressing TET, load 400 ng of gDNA into each well of a Bio-Dot microfiltration apparatus (Bio-Rad). Calculate the total amount of DNA needed (based on number of blots and number of serial dilutions) and dilute to 10 ng/μL in TE buffer (10 m*M* Tris–Cl, pH 8.0, 1 m*M* EDTA). Add 1/4 volume of 2 *M* NaOH/50 m*M* EDTA. Denature the DNA at 95°C for 10 min, transfer quickly to ice, and add 1 volume of ice-cold 2 *M* ammonium acetate to stabilize single strands. Serial dilutions may be performed at this point into TE buffer. Meanwhile, prepare membranes for blotting; we have found that Sequi-Blot PVDF membranes (Bio-Rad) give cleaner results than nitrocellulose. Wet membranes in methanol and equilibrate in TE buffer; then, assemble the dot blotting apparatus, taping off any unused wells. Wash each well with 400 μL TE and draw through with gentle vacuum. Purge any air bubbles in the wells, as these can interfere with washing and spotting DNA, and release the vacuum gently to avoid regurgitation that can cross-contaminate wells. Apply 100 μL of DNA samples at the desired dilutions and wash with another 400 μL of TE. Carefully place the membranes into 50 mL conical tubes for blotting. Note that replicate membranes are needed for each separate mC, hmC, fC, and caC blot.

The blotting procedure begins with blocking for 2 h in TBST buffer (50 m*M* Tris–Cl, pH 7.6, 150 m*M* NaCl, 0.5% Tween 20) with 5% (w/v) milk at room temperature. Then, wash three times with TBST and incubate at 4°C overnight with primary antibodies against each modified cytosine (Active Motif offers mouse monoclonal mC and rabbit polyclonal hmC, fC, and caC antibodies). We use the following antibody dilutions in 5% milk/TBST: 1:5000 mC; 1:10,000 hmC; 1:5000 fC; and 1:10,000 caC. Volumes should be enough to cover the membrane evenly, and solutions should be poured off cleanly between steps. Wash the blots three times with TBST for 5 min each and incubate with secondary 1:2000 goat anti-mouse IgG-HRP or 1:5000 goat anti-rabbit IgG-HRP (Santa Cruz Biotechnology) at room temperature for 2 h. Wash three times again and, just before imaging, apply Immobilon Western Chemiluminescent HRP Substrate (Millipore) evenly over the entire blot. Expose on an imager with chemiluminescent detection capabilities (we use a Fujifilm LAS-1000), taking care to smooth the blot over the imaging surface and remove air bubbles and excess HRP substrate. As positive and negative controls for this optimized protocol, we typically use gDNA from cells transfected with WT hTET2-CD or empty vector, respectively. Fig. 2 shows an example of dot blotting results for select TET constructs.

2.3 Quantitative Analysis by LC-MS/MS

To quantify genomic levels of cytosine modifications more rigorously, we favor liquid chromatography tandem mass spectrometry (LC-MS/MS). Several alternatives exist to detect global cytosine modifications, including antibody-based or chemoenzymatic assays (Booth et al., 2015; Song et al., 2012), but we focus on LC-MS/MS as a direct, reliable, and flexible assay that can quantify diverse modifications simultaneously. A host of sequencing methods have also been developed to localize specific modifications at base resolution; these methods are not covered in this chapter, as

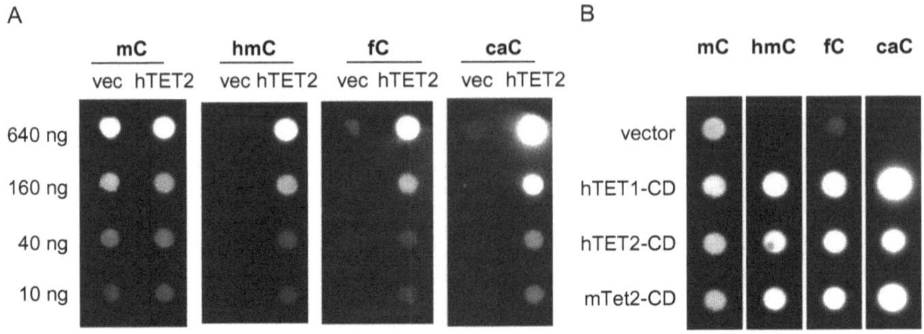

Fig. 2 Representative dot blots from analysis of gDNA from transfected HEK293T cells. Shown are (A) serial dilutions of gDNA from cells transfected with either empty expression vectors or hTET2-CD and (B) dot blots on 400 ng of gDNA from HEK293T cells transfected with empty expression vector, hTET1-CD, hTET2-CD, or mTet2-CD.

they have been thoroughly reviewed elsewhere (Booth et al., 2015; Yardimci & Zhang, 2015). Here, we discuss the protocols we developed for micro- and nanoscale analysis of nucleosides by LC-MS/MS, though we note that multiple methods and instruments have been described (Bachman et al., 2014; Pfaffeneder et al., 2014; Tsuji et al., 2014), and optimization for each system will be unique.

We use an Agilent 6460 triple quadrupole mass spectrometer with Agilent 1200 Series HPLC and Supelcosil LC-18-S reverse phase analytical column (5 μm particle size, 2.1 mm × 25 cm, Sigma) (Fig. 3A). This system offers a good starting platform for nucleoside analysis, is well suited for most applications, and is likely more accessible for most researchers than the nanoscale setup described later. To prepare samples, concentrate up to 10 μg of purified gDNA by ethanol precipitation and, in a total of 20 μL, degrade the DNA to component nucleosides with 10 U DNA Degradase Plus (phosphodiesterase and phosphatase cocktail available from Zymo Research) in 1 × DNA Degradase Buffer (Zymo). Incubate this mixture at 37°C overnight, then dilute 10-fold into 0.1% formic acid, since on-column retention of caC in particular requires pH adjustment.

Equilibrate the HPLC column to 50°C in Buffer A1 (5 m*M* ammonium formate, pH 4.0). Load 2 μL (up to 0.1 μg) of the nucleoside mixture and separate by gradient elution at a flow rate of 0.5 mL/min: 0–5 min, 0% Buffer B1 (4 m*M* ammonium formate, pH 4.0, 20% (v/v) methanol); 5–12 min, 0–10% B1; 12–12.5 min, 10–100% B1; 12.5–20 min, 100% B1; 20–20.5 min, 100–0% B1; and 20.5–28 min, 0% B1. Note that this gradient is optimized for efficient separation of all modified cytosine

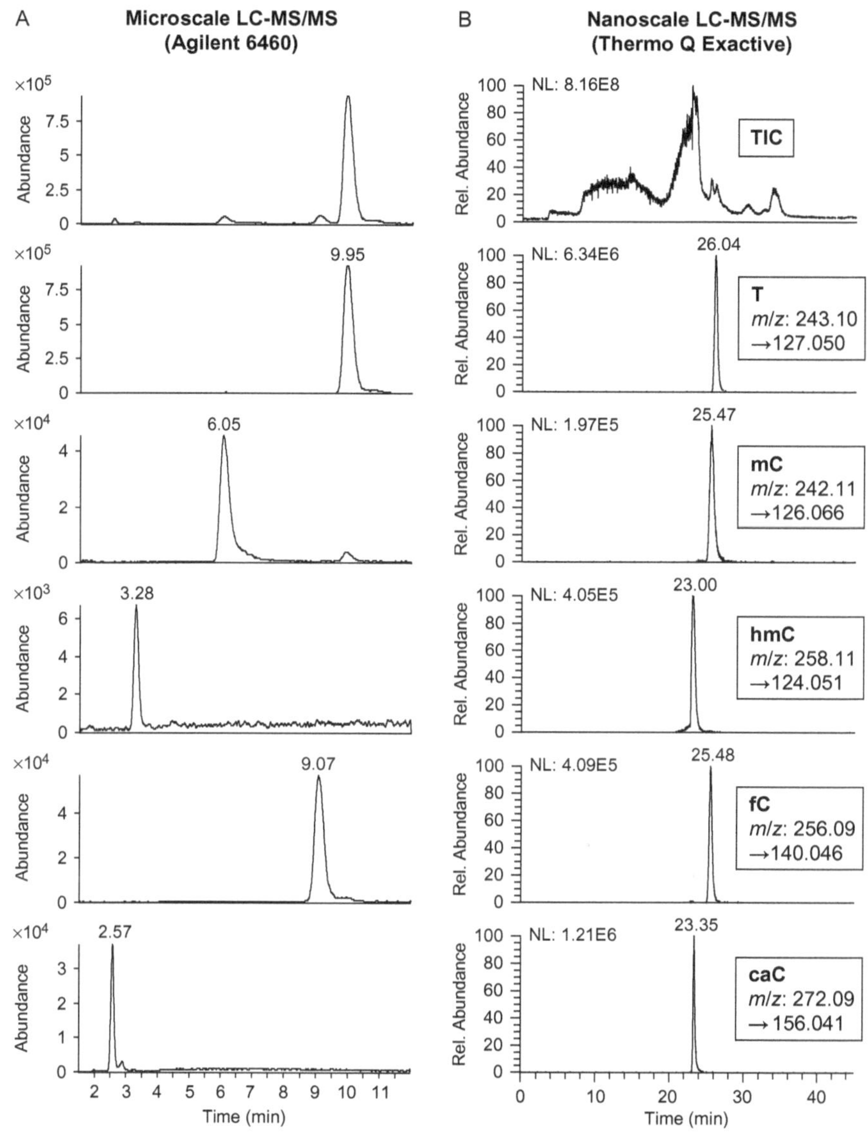

Fig. 3 Examples of LC-MS/MS analysis of DNA nucleosides. Protocols can be applied to genomic DNA or oligonucleotide substrates. Shown are traces from oligonucleotide substrates containing a single mC that were treated with hTET2-CS in vitro, degraded to nucleosides, and analyzed by (A) HPLC with tandem Agilent 6460 triple quadrupole mass spectrometer or (B) nano-LC with tandem Thermo Q Exactive hybrid quadrupole-orbitrap mass spectrometer. *From top to bottom are the total ion current (TIC) and MS2 chromatograms for T, mC, hmC, fC, and caC deoxyribonucleosides, along with retention times and mass transitions.*

nucleosides, but the most hydrophobic bases, such as adenosine, elute at >10% B1. Set downstream positive ion mode electrospray ionization for gas temperature of 175°C, gas flow of 10 L/min, nebulizer at 35 psi, sheath gas temperature of 300°C, sheath gas flow of 11 L/min, capillary voltage of 2000 V, and fragmentor voltage of 70. Collect MS2 scans on the following mass transitions: mC 242.11 → 126.066 *m/z*; hmC 258.11 → 142.061 and 124.051; fC 256.09 → 140.046; caC 272.09 → 156.041 (A, C, G, and, T are optional but can be useful as a loading control: A 252.11 → 136.062; C 228.10 → 112.051; G 268.10 → 152.057; and T 243.10 → 127.050). Optimized collision energies are 10 for mC, fC, and T; 15 for caC; and 25 for hmC. For quantification, generate standard curves from nucleosides (Berry & Associates) ranging from approximately 5 μ*M* to 10 p*M*; many isotopically labeled nucleosides are available as internal standards. Fit all peak areas to the standard curve to determine amounts of each modified cytosine in the gDNA sample.

This method attains low-femtomolar and high-attomolar range detection limits, allowing for quantification of 1 modification in 10^5–10^6 of all cytosines. This provides excellent detection of rare oxidized bases, especially in HEK293T cells overexpressing TET, though quantifying fC and caC in physiological samples remains a challenge. However, the triple quadrupole system has two notable limitations: the need for a large electron multiplier voltage and low mass resolution. At best, the Agilent 6460 differentiates only 0.7 amu, which can present a challenge if analytes are 1 amu or less apart, as might occur in some stable isotope labeling studies.

For analysis of isotopically labeled cytosines, we use nano-LC in tandem with a Q Exactive hybrid quadrupole-orbitrap mass spectrometer (Thermo Scientific) (Fig. 3B). A key advantage to this system is the very high mass resolution, which improves signal-to-noise and clearly distinguishes between isotopes with 1 amu mass difference. Detection limits for modified nucleosides are generally in the low-femtomole range. However, the system is prone to clogging and inconsistent electrospray, so samples and column fittings should be prepared with care.

We make columns from fused-silica tubing (New Objective) with a frit at one end: dip the column into a 400 μL mixture of 1:3 formamide:KASIL 1624 potassium silicate solution (PQ Corporation), let polymerize at 100°C overnight, and trim to ~3 mm. Using a pressure injection cell, pack a 150 μm × ~15 cm precolumn and 100 μm × ~25 cm analytical column with Supelcosil LC-18-S resin (Sigma). Connect columns to an Easy-nLC 1000 (Thermo) with a two-column setup, and add a 10 μm SilicaTip emitter (New Objective). (Alternatively, nanospray tips can be generated with a laser-based micropipette puller (Sutter Instrument), which eliminates the need for frits and significantly reduces dead volume, but this process is delicate and the setup can clog more easily.) Equilibrate the precolumn and analytical column in 5–10 column volumes of Buffer A2 (0.1% formic acid in H_2O) at a constant pressure of 275 bar.

Prepare samples as described earlier and inject 1 μL (0.05 μg) onto the LC-MS system. Set a sample loading step to send 5 μL of Buffer A2 through the sample loop to the precolumn at a constant pressure of 275 bar; this way, the sample is bound to the precolumn and desalted by sending the flow-through to waste. Next, run the gradient at a flow rate of 300 nL/min: 0–2 min, 0% Buffer B2 (0.1% formic acid in acetonitrile); 2–7 min, 0–10% B2; 7–37 min, 10–40% B2; 37–39 min, 40–70% B2; and 39–45 min, 70% B2. Perform tandem mass spectrometry in positive ion mode nanospray ionization with spray voltage of 2.9 kV, capillary temperature of 275°C, and normalized collision energy of 35%. Mass transitions and data analysis are the same as described earlier for the microscale setup. These LC-MS/MS methods, while technically challenging, are broadly applicable for sensitive quantification of cytosine modifications in gDNA. Moreover, in the next section, we will return to these methods as a powerful tool for quantifying the results of TET reactions on oligonucleotides in vitro.

3. ANALYSIS OF CYTOSINE MODIFICATIONS IN VITRO

3.1 Purification of TET Enzymes from Sf9 Insect Cells

To prepare recombinant TET enzymes for biochemical analysis, we and most other groups use a baculovirus expression system in Sf9 insect cells. TET constructs are typically cloned into a pFastBac1 vector containing a FLAG-tag at the N-terminus. Biochemical studies to date have largely relied on the catalytic domain constructs, as these retain the activity of the full-length protein but are generally obtained in higher purity and are more stable in vitro. Bacmid preparation and transfection were previously described in detail (Shen & Zhang, 2012). We express proteins for only 24 h, which we find minimizes the formation of truncation products. Cells are collected by centrifugation and the pellet is stored at −80°C until ready for purification. Alternative strategies have been described for expressing and purifying TET proteins from HEK293T cells and, in the crystal structure studies, from *Escherichia coli* (Hu et al., 2013, 2015); however, we favor Sf9 cells as a relatively straightforward way to obtain pure and active enzyme at intermediate yield.

We adapted a simple protocol for FLAG-based affinity purification of TET enzymes from Sf9 cells; the volumes below are for purification from a 500 mL culture. Resuspend the cell pellet in 10 mL of lysis buffer (50 m*M* HEPES, pH 7.5, 300 m*M* NaCl, 0.2% (v/v) NP-40) with cOmplete, EDTA-free Protease Inhibitor Cocktail (Roche, 1 tablet/10 mL) and 10 U/mL of Benzonase Nuclease (Millipore). To lyse the cells, freeze the cell suspension overnight at −20°C, thaw the following morning on ice, and pass through both a 20-gauge and then a 25-gauge needle. For larger volumes, a microfluidizer or sonicator is a good alternative. Clear the lysate by centrifugation at 20,000 × *g* for 20 min, collect the supernatant, and pass it through a 0.2 μm syringe filter. Next, prepare a 500 μL column of anti-FLAG M2 affinity gel (Sigma) per manufacturer instructions and equilibrate in lysis buffer. Apply the filtered lysate twice over the column under gravity flow. Wash the protein-bound column three times with 10 mL of wash buffer (50 m*M* HEPES, pH 7.5, 150 m*M* NaCl, 15% (v/v) glycerol). To elute the protein, apply 500 μL of wash buffer containing 100 μg/mL 3 × FLAG peptide (Sigma) to the column, incubating for 5–10 min before collecting the fraction. Collect until no protein is detected by Bio-Rad Protein Assay and SDS-PAGE. Pool fractions, add DTT to 1 m*M*, aliquot, and store at −80°C. By this method, we are able to obtain highly pure, active TET proteins with minimal steps.

3.2 Synthesis and Isotopic Labeling of TET Substrates

Our lab employs a variety of techniques to synthesize substrates for in vitro TET reactions and to label the substrates with fluorophores and/or heavy isotopes to enable sensitive detection and quantification. In general, we use DNA oligonucleotides 12–35 nt in length, containing a single TET substrate (mC, hmC, or fC) in a CpG context (although useful HPLC-based assays for substrates as short as 4–6 nt have also been developed; Kizaki & Sugiyama, 2014). TET enzymes exhibit a strong preference for CpGs (Hu et al., 2013) but are thought to be less sensitive to surrounding sequences (Yu et al., 2012). Importantly, the self-complementarity of CpGs requires special considerations, since one or both strands of a DNA duplex can contain a TET substrate. This consideration makes it important to control the identity of the top and bottom strands independently, and if both strands are set up to contain a substrate, assays must be designed to distinguish reactivity on each strand.

For most assays, we synthesize oligonucleotides in-house on an Applied Biosystems 394 DNA/RNA synthesizer. All four modified cytosines are now available (Glen Research) and compatible with standard phosphoramidite synthesis protocol, although fC requires postsynthetic processing to obtain the final formyl group from the precursor. Custom-made oligonucleotides are also available from Integrated DNA Technologies (IDT) and the W.M. Keck Biotechnology Resource Laboratory at Yale University, among other facilities. To facilitate tracking of the designated "top" strand, we typically attach 6-carboxyfluorescein (6-FAM) or alternative fluorophores to either the 5′- or 3′-end of the modified oligonucleotide. This can be done during synthesis or, for 3′-end labeling, can also be done after purification using terminal transferase enzymes and fluorophore-conjugated ddUTP analogs. Importantly, neither the identity nor location of the fluorescent tag alters the reactivity of TET on these substrates. As the first step toward differentiating duplexed strands, we leave the bottom strand unlabeled.

Starting with single strands, the top and bottom strands can be selected to address the experimental goals. In most cases, we use the FAM-labeled top strand containing a single TET substrate and, to ensure that all substrates are double stranded, anneal a 1.1- to 1.5-fold excess of unlabeled bottom strand containing unmodified C, so that the resulting DNA duplex contains only one reactive site. Typically, the CpG is embedded in a restriction site for MspI to allow for downstream analysis (see Section 3.3). Annealing is performed in a thermocycler: mix the top and bottom strands, heat at 95°C for 5 min, then cool slowly by decreasing steps of 5°C for 30 s per step. These duplexes can be added directly to a TET reaction, and reaction products can be visualized by fluorescent detection after separation by denaturing polyacrylamide gel electrophoresis (PAGE), as described in the next sections.

Fluorescence-based assays are well suited for addressing many important mechanistic questions but have limits of detection in the high-femtomole range. To increase the sensitivity of detection for cases of low product formation (eg, in kinetic studies), one previous strategy has been to radiolabel the 5′-end of full-length substrates with T4 polynucleotide kinase (New England Biolabs); however, similar to fluorescent labeling, this method only reports on the activity of a single strand. To measure activity on both strands independently, we developed enzymatic methods to generate substrates where the 5-methyl group of mC is labeled with either $^{13}C^{2}H_3$ or $^{14}CH_3$, which permits several avenues for sensitive and strand-specific detection of product formation (see Section 3.4).

The isotopically labeled substrates are generated enzymatically using the CpG methyltransferase (Fig. 4A). For $^{13}C^{2}H_3$ labeling, we start with [$^{13}C^{2}H_3$]-L-methionine (Sigma) and, in a single reaction mixture, enzymatically generate *S*-[$^{13}C^{2}H_3$-Me]-adenosyl-L-methionine ([$^{13}C^{2}H_3$]-SAM) in situ and simultaneously transfer the methyl moiety from SAM to DNA. For ^{14}C labeling, we simply start with *S*-[^{14}C-Me]-adenosyl-L-methionine ([^{14}C]-SAM) from Perkin Elmer and transfer the methyl group to a CpG-containing oligonucleotide using CpG methyltransferase (NEB).

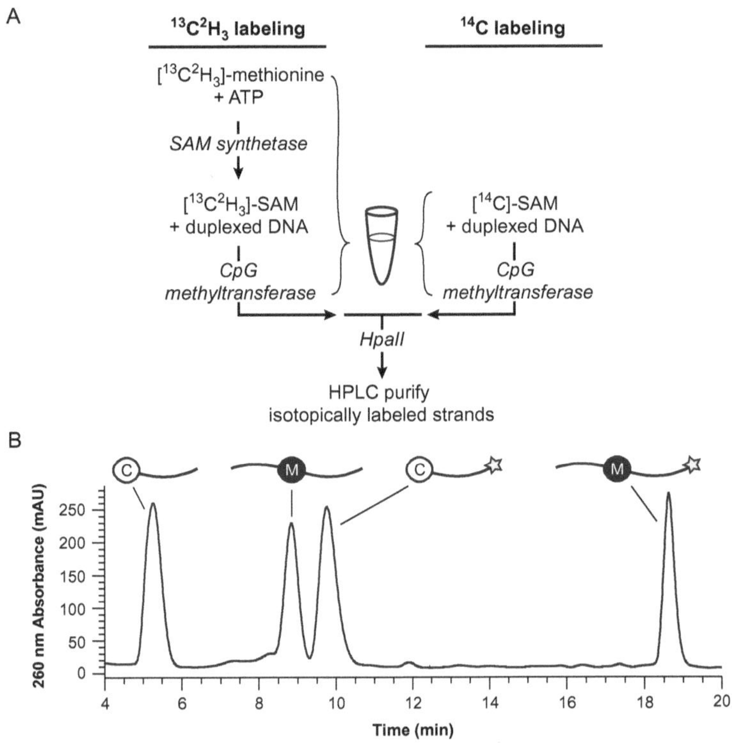

Fig. 4 Synthesis of isotopically labeled substrates. (A) Preparation of oligonucleotide substrates containing isotopically labeled mC. For $^{13}C^{2}H_3$ labeling (*left*), [$^{13}C^{2}H_3$]-*S*-adenosylmethionine is enzymatically generated and simultaneously consumed in a single reaction mixture, yielding oligonucleotide duplexes containing [$^{13}C^{2}H_3$]-mC. For $^{14}CH_3$ labeling (*right*), [^{14}C]-SAM is commercially available, requiring only CpG methyltransferase M.SssI to generate DNA duplexes containing [^{14}C]-mC. Unmethylated strands are digested by the methylation-sensitive restriction enzyme HpaII, and the full-length methylated oligonucleotides are purified by HPLC. (B) Representative HPLC chromatogram showing the HpaII-digested products of $^{13}C^{2}H_3$ or $^{14}CH_3$ labeling. Complementary strands are distinguished by a 3′-FAM label (*star*) on the top strand, which increases on-column retention and is detectable by 490 nm absorbance (not shown).

Notably, CpG methyltransferase acts on double-stranded DNA, so we first anneal complementary oligonucleotides, both containing an unmodified CpG, and purify methylated top and bottom strands separately by HPLC.

For the generation of substrates with a $^{13}C^{2}H_3$-modified mC, first purify recombinant *E. coli* SAM synthetase as described (Ottink, Nelissen, Derks, Wijmenga, & Heus, 2010). The in situ generation of [$^{13}C^{2}H_3$]-SAM is initiated by reacting 1 mg/mL SAM synthetase with 1 m*M* [$^{13}C^{2}H_3$]-L-methionine and 2 m*M* ATP in 1 × NEB Buffer 2 (10 m*M* Tris–Cl, pH 7.9, 50 m*M* NaCl, 10 m*M* $MgCl_2$, 1 m*M* DTT). Incubate this mixture at 37°C for 15 min, then add 4 × volume of duplexed DNA (final concentration 10 μ*M*) and the CpG methyltransferase M.SssI (1:20 total reaction volume, 1000 U/mL) (NEB) in 1 × NEB Buffer 2. The simultaneous SAM synthesis and CpG [$^{13}C^{2}H_3$]-methylation is carried out for an additional 5 h at 37°C.

For the generation of substrates with a [^{14}C]-mC label, with [^{14}C]-SAM supplied at ~350 μ*M* in ~500 μL, prepare a reaction containing 35 μ*M* [^{14}C]-SAM, 5.8 μ*M* duplexed DNA, and 224 U/mL CpG methyltransferase (M.SssI), incubating at 37°C for 4 h. Set aside a small sample of the reaction to determine specific radioactivity for liquid scintillation counting (LSC). Extract DNA from the $^{13}C^{2}H_3$- and ^{14}C-labeling reactions by ethanol precipitation, resuspend the dried pellet in H_2O, and desalt with illustra MicroSpin G-25 Columns (GE Healthcare Life Sciences) equilibrated in water. The resulting oligonucleotide mixture contains methylated products and unmethylated substrate, with and without FAM labels. To digest residual unmethylated substrates, treat with 5000 U/mL of HpaII (NEB) in 1 × CutSmart Buffer (NEB) at 37°C overnight. Ethanol precipitate again and dissolve the dried pellet in Buffer A3 (100 m*M* triethylamine acetate (TEAA), pH 7). Purify by ion-pairing HPLC over an Agilent Zorbax Eclipse Plus C18 reverse phase column (3.5 μm particle size, 4.6 mm × 10 cm); we use an Agilent Infinity 1260 Quaternary Pump VL with 1260 FC-AS fraction collector. Equilibrate the column to 65°C in 65% Buffer A3 and 35% Buffer B3 (50% (v/v) methanol/100 m*M* TEAA, pH 7), and separate over a 20 min gradient from 35% to 45% Buffer B3 at a flow rate of 1 mL/min. Four major peaks should be observed (along with smaller digestion products): (1) HpaII-digested, unmethylated bottom strand without FAM; (2) full-length, methylated bottom strand without FAM; (3) HpaII-digested, unmethylated top strand with FAM; and (4) full-length, methylated top strand with FAM (Fig. 4B). Collect fractions (2) and (4), the methylated products ± FAM, and lyophilize. These purified

[$^{13}C^{2}H_3$]-mC or [^{14}C]-mC oligonucleotides can then be duplexed to complementary strands for use in various TET activity assays. For basic analysis of reactivity at one CpG site, we use a complementary strand with an unmodified CpG. Alternatively, the complementary strand can contain mC or ox-mC bases with natural isotope composition, allowing for strand-specific measurements of activity when both strands contain TET substrates.

Finally, some experiments may benefit from longer substrates with multiple targets, rather than simple oligonucleotides. In this case, we use PCR to generate amplicons containing mC, hmC, or fC at all cytosines using ox-mC dNTPs that are commercially available (Trilink). Set up a 50 μL reaction under the following conditions, which have worked well for the majority of our substrates: 1 × PCR buffer (20 m*M* Tris–Cl, pH 8.4, 50 m*M* KCl), 1.5 m*M* $MgCl_2$, 200 μ*M* of each dNTP (including the desired modified cytosine and no natural cytosine), 1 μ*M* of forward and reverse amplification primers, 1 ng of template DNA (containing unmodified cytosines), and 5 U of Taq polymerase (Invitrogen). When generating mC-containing substrates, add 5% DMSO to aid in denaturation. The thermocycler settings will depend upon the polymerase, nature of the substrate, primer length, etc., but most standard settings translate to these conditions. Amplification of DNA containing caC is not efficient under these conditions, but an alternative strategy using Phusion polymerase has been reported (Neri et al., 2015). Run the entire reaction on an agarose gel and excise the desired band. Use the ZymoClean Gel DNA Recovery Kit (Zymo) to purify the amplicons.

3.3 Chemoenzymatic Activity Assays on Full-Length Oligonucleotides

Despite the diversity of substrates, TET reaction conditions are largely the same for all applications. Our optimized conditions are as follows: 50 m*M* HEPES, pH 6.5, 100 m*M* NaCl, 1 m*M* α-ketoglutarate, 75 μ*M* ammonium iron(II) sulfate (Sigma), 1 m*M* DTT, and 2 m*M* sodium ascorbate. The concentration of DNA substrate and TET protein varies based on the experimental goals (see examples later). Importantly, both the α-ketoglutarate and Fe(II) must be fresh, and Fe(II) should be added immediately prior to the start of the reaction to minimize oxidation to Fe(III). Sodium ascorbate is not essential but increases activity, likely by helping to keep Fe(II) in the reduced state (Blaschke et al., 2013; Yin et al., 2013). We note minor differences between our methods and those used successfully by other groups.

First, we performed a pH titration for our enzymes and found pH 6.5 to be optimal, with decreasing activity at higher pH; however, other groups routinely perform reactions at pH 8.0 (Hu et al., 2015). ATP has also been reported to stimulate TET activity (He et al., 2011), but it made no detectable difference when evaluated under our reaction conditions.

Incubate the reaction at 37°C for the desired time. We specifically tested our purified mTet2-CD protein under low turnover conditions (eg, 500 n*M* DNA, 10 μg/mL enzyme) and showed linear activity through at least 20 min. Under higher turnover conditions (eg, 25 n*M* DNA, 30 μg/mL enzyme), we typically achieve complete conversion of mC and hmC to fC and caC by 30 min; longer reactions are also possible with activity detectable out to 3 h, although turnover slows past ~30 min. Quench the reaction by adding 8× volume of 100% ethanol with 2× volume of Oligo Binding Buffer (Zymo). Purify reactions using the Zymo Oligo Clean & Concentrator kit.

The purified products consist of a mixture of DNA duplexes containing mC, hmC, fC, and/or caC. A number of assays are available to probe for specific bases qualitatively and quantitatively. As noted earlier, Kizaki and Sugiyama reported using 4–6 nt substrates with direct resolution of the reaction products by HPLC (Kizaki & Sugiyama, 2014). However, longer substrates are needed to understand how strand specificity, sequence context, and other factors impact TET activity. Assays on these longer substrates rely on indirect methods involving either enzymatic processing or degradation to nucleosides to deliver rigorous, reliable quantification of TET reaction products in vitro.

There are several bacterial restriction enzymes that display variable capacity to cleave the different cytosine modifications. MspI, the most well known of these enzymes, cleaves the sequence C***X***GG completely when ***X*** is C, mC, or hmC; cleaves partially when ***X*** is fC; and cannot cleave when ***X*** is caC (Fig. 5A and B). There is some evidence that MspI cleavage behavior is affected by the identity of the opposite strand CpG (Pais et al., 2015), but the majority of our experiments use substrates that have unmodified CpG on the complement. We have also observed a similar pattern of discrimination with HaeIII on oligonucleotides containing a GGC***X*** sequence. We exploit these restriction enzymes' behavior further by enzymatically modifying hmC and/or chemically modifying fC to prevent cleavage. The difference in cleavage patterns between treated and untreated samples reflects the fraction of these bases in the total reaction mixture. Notably, by selectively excluding either/both chemical modification steps, this method allows the levels of specific bases to be probed.

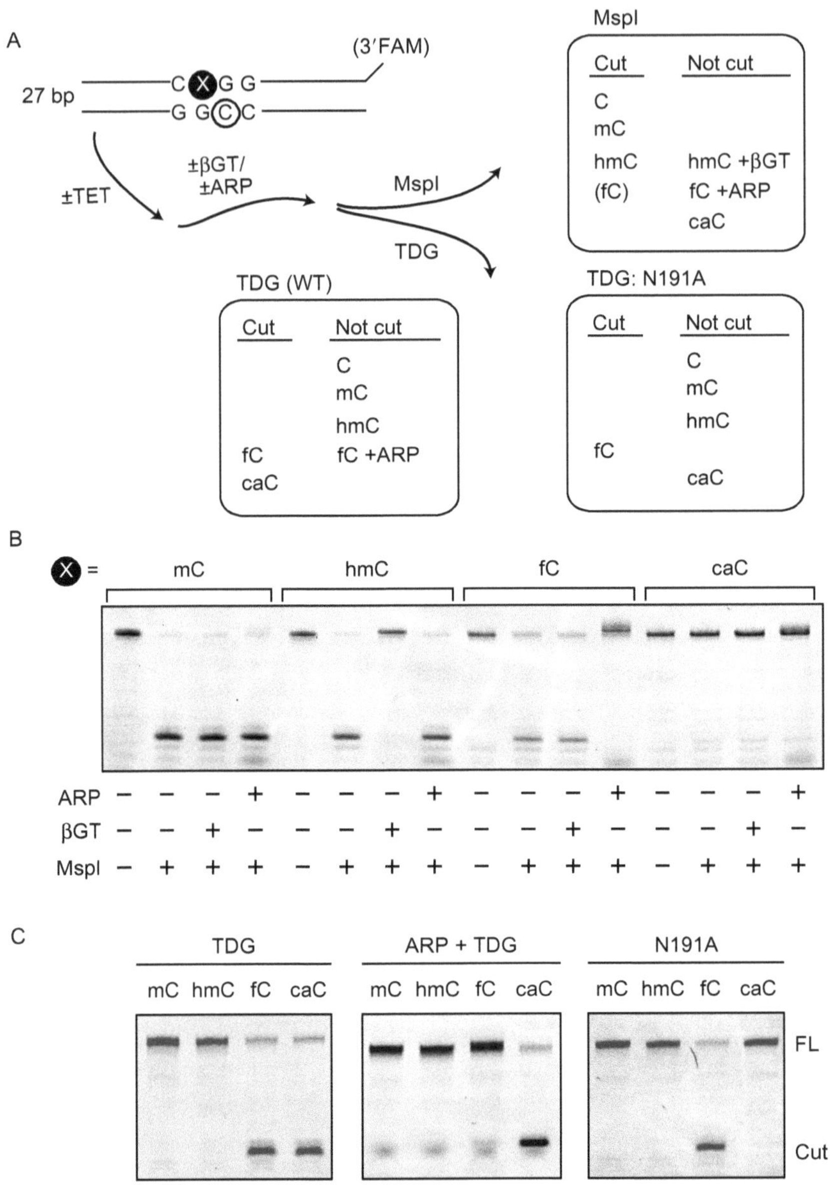

Fig. 5 Enzyme-coupled assays for TET activity. (A) Chemoenzymatic assays indirectly measure TET activity on full-length oligonucleotides. Typically, duplexed DNA containing a FAM-labeled reactive strand, which has a mCpG in the MspI cleavage-sequence context, is treated with recombinant TET protein according to optimized assay conditions. Any resulting hmC and fC among reaction products are chemically protected by glucose (via β-glucosyltransferase) or ARP, respectively. Then, the full-length, duplexed DNA is treated with either MspI or TDG (WT or mutant N191A), which each have a unique cleavage capacity against natural and chemically protected cytosine derivatives. (B and C) Denaturing polyacrylamide gels displaying chemoenzymatic activity assay products. Oligonucleotides containing each cytosine modification at the reactive site (X) were employed in lieu of a TET reaction to display MspI (B) and TDG (C) cleavage signatures, in the presence or absence of various chemical protection steps.

To protect hmC from cleavage, treat the reaction mixture with 2 m*M* UDP-glucose and 1:25 by volume of T4 β-glucosyltransferase (βGT) (NEB) in 1 × CutSmart buffer at 37°C to transfer the glucose moiety from UDP-glucose to hmC (Terragni, Bitinaite, Zheng, & Pradhan, 2012). All hmC bases are glucosylated in 30 min, and MspI/HaeIII does not cleave the resulting glucosyl-hmC. To protect fC, mix DNA substrate with at least 35,000-fold molar excess of aldehyde reactive probe (ARP, Dojindo) in a reaction with 6 m*M* HEPES at pH 5.0 (the lower pH improves the efficiency of the reaction). Incubate at 37°C overnight, or for at least 3 h. MspI/HaeIII does not cleave fC when ARP is covalently attached. Note that ARP and βGT can be used sequentially to protect both hmC and fC in a reaction mixture: perform the ARP reaction first and then dilute into 1 × CutSmart buffer for the βGT reaction, since βGT is more tolerant of buffer conditions. Proceed directly to MspI digestion in 1 × CutSmart buffer at 37°C for at least 2 h, without need for further purification.

In addition to restriction enzymes, we also utilize TDG to recognize fC and caC in TET reaction mixtures (Fig. 5A and C). TDG is expressed and purified as described previously (Morgan, Bennett, & Drohat, 2007). Treat TET reaction products with 25-fold molar excess of TDG (eg, 125 n*M* substrate and 3.125 μ*M* TDG) in TDG buffer (20 m*M* HEPES, pH 7.5, 100 m*M* NaCl, 0.2 m*M* EDTA, 2.5 m*M* $MgCl_2$) for 4 h at 37°C. TDG excises fC and caC, leaving abasic sites, but does not react with mC and hmC. To cleave oligonucleotides that now contain abasic sites, add 1:1 volume of 0.3 *M* NaOH/0.03 *M* EDTA and incubate at 85°C for 15 min. As an added means for discrimination of TET products, we use a mutant form of TDG, N191A, that has been shown to preferentially excise fC but not caC (Maiti, Michelson, Armwood, Lee, & Drohat, 2013); this mutant can be purified and used in the same manner to identify fC specifically. Finally, it is possible to selectively cleave caC-containing DNA by treating with ARP (as described earlier), which protects fC from excision by WT TDG, leaving only caC susceptible. A variant of TDG that shows preferential excision of caC has also been reported and may provide a complementary approach to quantifying caC in reaction products (Hashimoto, Zhang, & Cheng, 2013).

As the final step of all chemoenzymatic assays, mix samples 1:1 with formamide containing bromophenol blue, denature, separate full-length from cleaved oligonucleotides on a 7 *M* urea/20% acrylamide/1 × TBE gel prewarmed to 50°C, and image for FAM fluorescence. Altogether, these chemoenzymatic activity assays provide a complete toolbox to probe for specific ox-mC base modifications in a TET reaction mixture.

3.4 Quantitative Activity Assays on Nucleosides

Our chemoenzymatic assays with fluorescent oligonucleotides are both convenient and quantifiable, but digesting the reaction products to nucleosides greatly enhances the accuracy and sensitivity of quantification. This is particularly true of ^{14}C-labeled reactions, for which we can use HPLC to separate modified nucleosides, collect fractions, and perform liquid scintillation counting (LSC) (Fig. 6A). After reacting [^{14}C]-mC substrate with TET enzymes and purifying as described earlier, degrade the products to nucleosides using DNA Degradase Plus (Zymo). Note that we typically use radioactivity for reactions where very low turnover is desired and product formation would be difficult to detect by any other means. Since these product peaks would be invisible to UV detection during HPLC, spike the samples with 10 μ*M* each of nonradioactive mC, hmC, fC, and caC nucleosides (Berry & Associates), which act as chromatographic markers. Separate the samples by gradient elution over the Supelcosil LC-18-S column, as described in Section 2.3. In lieu of mass spectrometry, collect the peaks into 0.25 mL fractions, mix with Opti-Fluor liquid scintillant (Perkin Elmer),

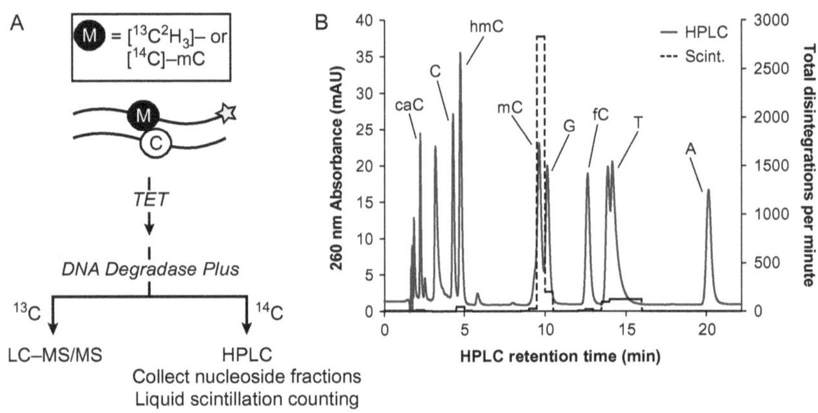

Fig. 6 High-sensitivity in vitro assays. (A) Isotopically labeled oligonucleotides duplexed with an unlabeled complementary strand can be reacted with purified TET enzymes in vitro. Depending on the nature of the label, the reaction products are degraded to nucleosides and analyzed by either LC-MS/MS or HPLC with liquid scintillation counting to quantify oxidized bases with high sensitivity. (B) Representative HPLC chromatogram (*solid line*) of nucleosides from a degraded TET reaction on ^{14}C-labeled DNA. The samples were spiked with unlabeled mC, hmC, fC, and caC to mark these peak positions for fraction collection, and the fractions were subjected to liquid scintillation counting. The disintegrations per minute (*dashed line*) for each fraction are overlaid with the HPLC chromatogram. Very low (<1%) product formation can be reliably detected by this approach.

and subject to LSC (we use a Tri-Carb 2910 TR (Perkin Elmer)), counting each vial for 10 min using the ^{14}C setting (Fig. 6B). Measure specific radioactivity from the small sample that was set aside during preparation of radiolabeled substrate (Section 3.2), and correct for background radiation using fractions collected from a nonradioactive control reaction. The resulting measurements of disintegrations per minute should be normalized to the input volume and known input concentration to yield molar concentrations of each modified cytosine base.

For nonradioactive reactions, LC-MS/MS offers the best method of quantification (Fig. 3). The application of this protocol using one strand that is selectively $^{13}C^{2}H_3$-labeled can also facilitate tracking of oxidation of two strands of DNA simultaneously. For a typical analysis, degrade at least 5 pmol of TET reaction products to component nucleosides with 1 U DNA Degradase Plus (Zymo) in 10 μL at 37°C for at least 2 h. Dilute this mixture 10-fold into 0.1% formic acid, and inject at least 20 μL onto the Agilent LC-MS/MS system as described in Section 2.3. Alternatively, for better resolution of isotopically labeled bases, we start with ~20 pmol of reaction products and inject 1 μL onto the nano LC-MS/MS system.

ACKNOWLEDGMENTS

We are grateful to our colleagues who have previously shared reagents or protocols, including Drs. Guoliang Xu (Shanghai Institutes for Biological Sciences), Yi Zhang (Harvard, HHMI), and Alex Drohat (University of Maryland). We thank Dr. Benjamin Garcia for assistance in development of LC-MS/MS protocols. This work was supported by the Rita Allen Foundation.

REFERENCES

Bachman, M., Uribe-Lewis, S., Yang, X., Burgess, H. E., Iurlaro, M., Reik, W., et al. (2015). 5-Formylcytosine can be a stable DNA modification in mammals. *Nature Chemical Biology*, *11*, 555–557.

Bachman, M., Uribe-Lewis, S., Yang, X., Williams, M., Murrell, A., & Balasubramanian, S. (2014). 5-Hydroxymethylcytosine is a predominantly stable DNA modification. *Nature Chemistry*, *6*, 1049–1055.

Blaschke, K., Ebata, K. T., Karimi, M. M., Zepeda-Martinez, J. A., Goyal, P., Mahapatra, S., et al. (2013). Vitamin C induces Tet-dependent DNA demethylation and a blastocyst-like state in ES cells. *Nature*, *500*, 222–226.

Booth, M. J., Raiber, E. A., & Balasubramanian, S. (2015). Chemical methods for decoding cytosine modifications in DNA. *Chemical Reviews*, *115*, 2240–2254.

Hashimoto, H., Zhang, X., & Cheng, X. (2013). Selective excision of 5-carboxylcytosine by a thymine DNA glycosylase mutant. *Journal of Molecular Biology*, *425*, 971–976.

He, Y. F., Li, B. Z., Li, Z., Liu, P., Wang, Y., Tang, Q., et al. (2011). Tet-mediated formation of 5-carboxylcytosine and its excision by TDG in mammalian DNA. *Science*, *333*, 1303–1307.

Hu, L., Li, Z., Cheng, J., Rao, Q., Gong, W., Liu, M., et al. (2013). Crystal structure of TET2-DNA complex: Insight into TET-mediated 5mC oxidation. *Cell*, *155*, 1545–1555.

Hu, L., Lu, J., Cheng, J., Rao, Q., Li, Z., Hou, H., et al. (2015). Structural insight into substrate preference for TET-mediated oxidation. *Nature*, *527*, 118–122.

Ito, S., D'Alessio, A. C., Taranova, O. V., Hong, K., Sowers, L. C., & Zhang, Y. (2010). Role of Tet proteins in 5mC to 5hmC conversion, ES-cell self-renewal and inner cell mass specification. *Nature*, *466*, 1129–1133.

Ito, S., Shen, L., Dai, Q., Wu, S. C., Collins, L. B., Swenberg, J. A., et al. (2011). Tet proteins can convert 5-methylcytosine to 5-formylcytosine and 5-carboxylcytosine. *Science*, *333*, 1300–1303.

Iurlaro, M., Ficz, G., Oxley, D., Raiber, E. A., Bachman, M., Booth, M. J., et al. (2013). A screen for hydroxymethylcytosine and formylcytosine binding proteins suggests functions in transcription and chromatin regulation. *Genome Biology*, *14*, R119.

Kizaki, S., & Sugiyama, H. (2014). CGmCGCG is a versatile substrate with which to evaluate Tet protein activity. *Organic & Biomolecular Chemistry*, *12*, 104–107.

Kohli, R. M., & Zhang, Y. (2013). Tet, TDG and the dynamics of DNA demethylation. *Nature*, *502*, 472–479.

Maiti, A., & Drohat, A. C. (2011). Thymine DNA glycosylase can rapidly excise 5-formylcytosine and 5-carboxylcytosine: Potential implications for active demethylation of CpG sites. *The Journal of Biological Chemistry*, *286*, 35334–35338.

Maiti, A., Michelson, A. Z., Armwood, C. J., Lee, J. K., & Drohat, A. C. (2013). Divergent mechanisms for enzymatic excision of 5-formylcytosine and 5-carboxylcytosine from DNA. *Journal of the American Chemical Society*, *135*, 15813–15822.

Morgan, M. T., Bennett, M. T., & Drohat, A. C. (2007). Excision of 5-halogenated uracils by human thymine DNA glycosylase. Robust activity for DNA contexts other than CpG. *The Journal of Biological Chemistry*, *282*, 27578–27586.

Neri, F., Incarnato, D., Krepelova, A., Rapelli, S., Anselmi, F., Parlato, C., et al. (2015). Single-base resolution analysis of 5-formyl and 5-carboxyl cytosine reveals promoter DNA methylation dynamics. *Cell Reports*, *10*, 674–683.

Ottink, O. M., Nelissen, F. H., Derks, Y., Wijmenga, S. S., & Heus, H. A. (2010). Enzymatic stereospecific preparation of fluorescent S-adenosyl-L-methionine analogs. *Analytical Biochemistry*, *396*, 280–283.

Pais, J. E., Dai, N., Tamanaha, E., Vaisvila, R., Fomenkov, A. I., Bitinaite, J., et al. (2015). Biochemical characterization of a naegleria TET-like oxygenase and its application in single molecule sequencing of 5-methylcytosine. *Proceedings of the National Academy of Sciences of the United States of America*, *112*, 4316–4321.

Pfaffeneder, T., Spada, F., Wagner, M., Brandmayr, C., Laube, S. K., Eisen, D., et al. (2014). Tet oxidizes thymine to 5-hydroxymethyluracil in mouse embryonic stem cell DNA. *Nature Chemical Biology*, *10*, 574–581.

Shen, L., & Zhang, Y. (2012). Enzymatic analysis of Tet proteins: Key enzymes in the metabolism of DNA methylation. *Methods in Enzymology*, *512*, 93–105.

Song, C. X., Yi, C., & He, C. (2012). Mapping recently identified nucleotide variants in the genome and transcriptome. *Nature Biotechnology*, *30*, 1107–1116.

Spruijt, C. G., Gnerlich, F., Smits, A. H., Pfaffeneder, T., Jansen, P. W., Bauer, C., et al. (2013). Dynamic readers for 5-(hydroxy)methylcytosine and its oxidized derivatives. *Cell*, *152*, 1146–1159.

Tahiliani, M., Koh, K. P., Shen, Y., Pastor, W. A., Bandukwala, H., Brudno, Y., et al. (2009). Conversion of 5-methylcytosine to 5-hydroxymethylcytosine in mammalian DNA by MLL partner TET1. *Science*, *324*, 930–935.

Terragni, J., Bitinaite, J., Zheng, Y., & Pradhan, S. (2012). Biochemical characterization of recombinant beta-glucosyltransferase and analysis of global 5-hydroxymethylcytosine in unique genomes. *Biochemistry*, *51*, 1009–1019.

Tsuji, M., Matsunaga, H., Jinno, D., Tsukamoto, H., Suzuki, N., & Tomioka, Y. (2014). A validated quantitative liquid chromatography-tandem quadrupole mass spectrometry method for monitoring isotopologues to evaluate global modified cytosine ratios in genomic DNA. *Journal of Chromatography*, *953–954*, 38–47.

Yardimci, H., & Zhang, Y. (2015). Charting oxidized methylcytosines at base resolution. *Nature Structural & Molecular Biology*, *22*, 656–661.

Yin, R., Mao, S. Q., Zhao, B., Chong, Z., Yang, Y., Zhao, C., et al. (2013). Ascorbic acid enhances Tet-mediated 5-methylcytosine oxidation and promotes DNA demethylation in mammals. *Journal of the American Chemical Society*, *135*, 10396–10403.

Yu, M., Hon, G. C., Szulwach, K. E., Song, C. X., Zhang, L., Kim, A., et al. (2012). Base-resolution analysis of 5-hydroxymethylcytosine in the mammalian genome. *Cell*, *149*, 1368–1380.

CHAPTER SEVENTEEN

Characterization of How DNA Modifications Affect DNA Binding by C2H2 Zinc Finger Proteins

A. Patel, H. Hashimoto, X. Zhang[1], X. Cheng[1]
Emory University School of Medicine, Atlanta, GA, United States
[1]Corresponding authors: e-mail address: xzhan02@emory.edu; xcheng@emory.edu

Contents

1. Introduction 387
2. Soluble Expression and Purification of ZnF Proteins 391
   2.1 Expression 391
   2.2 Purification 391
3. Fluorescence Polarization Assay for Analysis of DNA Binding 393
4. Crystallization of ZnF Proteins in Complex with DNA 396
Acknowledgments 397
References 398

Abstract

Much is known about vertebrate DNA methylation and oxidation; however, much less is known about how modified cytosine residues within particular sequences are recognized. Among the known methylated DNA-binding domains, the Cys2-His2 zinc finger (ZnF) protein superfamily is the largest with hundreds of members, each containing tandem ZnFs ranging from 3 to >30 fingers. We have begun to biochemically and structurally characterize these ZnFs not only on their sequence specificity but also on their sensitivity to various DNA modifications. Rather than following published methods of refolding insoluble ZnF arrays, we have expressed and purified soluble forms of ZnFs, ranging in size from a tandem array of two to six ZnFs, from seven different proteins. We also describe a fluorescence polarization assay to measure ZnFs affinity with oligonucleotides containing various modifications and our approaches for cocrystallization of ZnFs with oligonucleotides.

1. INTRODUCTION

The control of gene expression in mammals relies significantly on the modification status of DNA cytosine residues. DNA cytosine modification is

Methods in Enzymology, Volume 573
ISSN 0076-6879
http://dx.doi.org/10.1016/bs.mie.2016.01.019

a dynamic process catalyzed by specific DNA methyltransferases (DNMTs) that convert cytosine (C) to 5-methylcytosine (abbreviated 5mC or M; Bestor, Laudano, Mattaliano, & Ingram, 1988; Okano, Xie, & Li, 1998), usually within the sequence context of CpG (Bestor et al., 1988; Okano, Bell, Haber, & Li, 1999; Okano et al., 1998) or CpA (Gowher & Jeltsch, 2001; Kubo et al., 2015; Lister et al., 2013, 2009; Ramsahoye et al., 2000; Vlachogiannis et al., 2015). A subset of 5mC may then be oxidized to 5-hydroxymethylcytosine (5hmC), 5-formylcytosine (5fC), and 5-carboxylcytosine (5caC) by the ten-eleven translocation (Tet) dioxygenases in three consecutive Fe(II) and α-ketoglutarate-dependent oxidation reactions (He et al., 2011; Ito et al., 2010, 2011; Tahiliani et al., 2009).

The best-known modified DNA-recognition domains are two that recognize methylated cytosine: methyl-binding domains (MBDs) recognize fully methylated CpG dinucleotides (Dhasarathy & Wade, 2008; Guy, Cheval, Selfridge, & Bird, 2011), and "SET and RING finger-associated" (SRA) domains that bind hemimethylated CpG sites generated transiently by DNA replication (Hashimoto, Horton, Zhang, & Cheng, 2009; Sharif & Koseki, 2011; reviewed in Hashimoto, Zhang, Vertino, & Cheng, 2015; Liu, Zhang, Blumenthal, & Cheng, 2013). Both MBD and SRA domains have been structurally characterized in complexes with 5mC (Arita, Ariyoshi, Tochio, Nakamura, & Shirakawa, 2008; Avvakumov et al., 2008; Hashimoto et al., 2008; Ho et al., 2008; Ohki et al., 2001; Scarsdale, Webb, Ginder, & Williams, 2011).

A third class of mammalian proteins that can recognize methylated DNA is the Cys2-His2 (C2H2) zinc finger (ZnF) proteins, which can preferentially bind to methylated CpG within a longer specific sequence (Sasai, Nakao, & Defossez, 2010). Kaiso is the first known methyl-binding ZnF protein that belongs to the BTB/POZ family (Prokhortchouk et al., 2001), which also includes ZBTB24, whose mutations are associated with *i*mmunodeficiency, *c*entromeric instability, and *f*acial anomalies (ICF) syndrome (Cerbone et al., 2012; Chouery et al., 2012; de Greef et al., 2011; Nitta et al., 2013), a disease also caused by mutations in a DNA methyltransferase gene, *DNMT3B* (Hansen et al., 1999; Okano et al., 1999; Shirohzu et al., 2002; Xu et al., 1999). Recently, ZnF DNA-binding domains from five proteins, Kaiso, Zfp57, Klf4, Egr1, and WT1, have been structurally analyzed in complex with their respective methylated DNA elements (Buck-Koehntop et al., 2012; Hashimoto et al., 2014; Liu et al., 2014; Liu, Toh, Sasaki, Zhang, & Cheng, 2012; Zandarashvili, White, Esadze, & Iwahara, 2015), allowing comparison to other 5mC-binding proteins. In addition, WT1 binds 5caC DNA, as does a mutant Zfp57 (Hashimoto et al., 2014; Liu, Olanrewaju, Zhang, & Cheng, 2013).

Among the C2H2 ZnF proteins, KRAB-ZnF transcription factors (KRAB-ZnFs) act mostly as chromatin-modulating transcription repressors (Meylan et al., 2011). Of the >300 human or mouse KRAB-ZnF proteins examined, the number of tandem ZnFs ranges from 3 to 35, with a mode of around 11–13 fingers (Liu, Zhang, et al., 2013; Fig. 1A and B). The domain

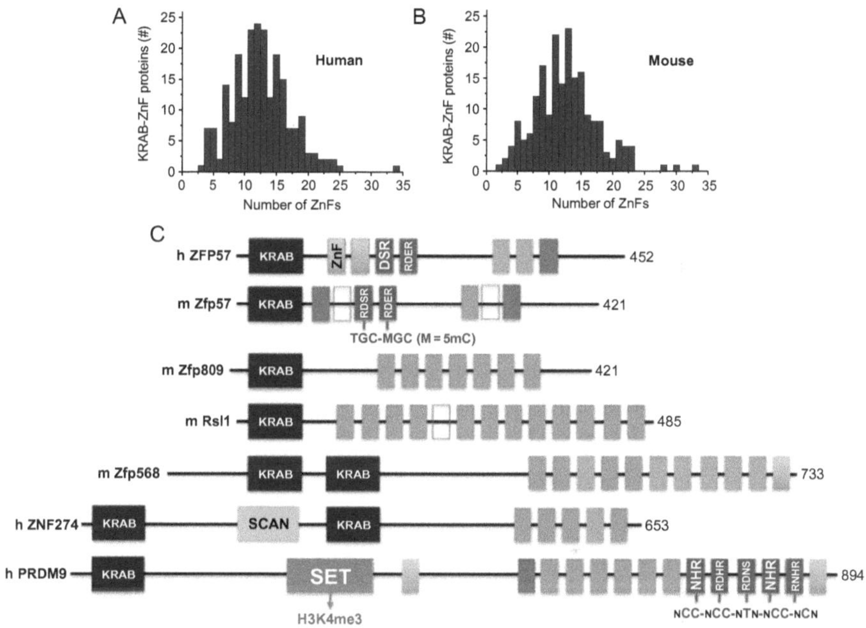

Fig. 1 KRAB-ZnF proteins. The SysZNF database was examined for Krüppel-associated box (KRAB)-ZnF proteins in human (A) or mouse (B). In each case, the distribution of KRAB-ZnF proteins containing a given number of ZnF repeats is shown. (C) Examples of mammalian KRAB-ZnF proteins with known biologic roles. The size of each protein (in amino acids) is shown on the right. The classic C2H2 ZnF motifs are shown in *blue boxes* and the *gray* or *open boxes* indicate degenerate ZnFs that contain mutations affecting zinc coordination. The *red boxes* of mouse Zfp57 and human PRDM9 indicate the structurally characterized ZnFs in complex with their recognition sequences shown below. The amino acids within the *red box* indicate the three or four residues of each ZnF involved in base specific interactions. The SCAN box, a leucine-rich region, was named after *S*RE-ZBP, *C*Tfin51, *A*W-1 (ZNF174), *N*umber 18 cDNA (ZnF20) (Williams, Khachigian, Shows, & Collins, 1995). The SET domain was named after *S*u (var)3-9, *E*nhancer of zeste, *T*rithorax (Jenuwein, Laible, Dorn, & Reuter, 1998). *Modified from Liu, Y., Zhang, X., Blumenthal, R. M., & Cheng, X. (2013). A common mode of recognition for methylated CpG.* Trends in Biochemical Sciences, 38, *177–183.* (See the color plate.)

structures of a few examples of mammalian KRAB-ZnF proteins with known biological roles are shown (Fig. 1C). ZFP57 mutations have been found in patients with transient neonatal diabetes (Mackay et al., 2008). Zfp809 restricts retroviral transposition in embryonic stem cells (Wolf & Goff, 2009), and retroviral silencing has been suggested to be the ancestral role of KRAB-ZnFs (Thomas & Schneider, 2011). *R*egulator of *s*ex *l*imitation (Rsl1) regulates sex- and tissue-specific promoter methylation (Krebs, Schultz, & Robins, 2012). Zfp568 regulates extraembryonic tissue morphogenesis (Garcia-Garcia, Shibata, & Anderson, 2008). Like Zfp57 (Quenneville et al., 2011; Zuo et al., 2012), ZNF274 recruits the histone H3 lysine 9 methyltransferase SETDB1 (SET domain, bifurcated 1; Frietze, O'Geen, Blahnik, Jin, & Farnham, 2010) via the corepressor TRIM28 (tripartite motif-containing 28; also known as KAP1 for Krüppel-associated protein), an essential regulator of genomic imprinting (Messerschmidt et al., 2012). PRDM9 (PR domain zinc finger protein 9), a major determinant of meiotic recombination hotspots, contains a SET domain that methylates histone H3 lysine 4 (Mihola, Trachtulec, Vlcek, Schimenti, & Forejt, 2009). These examples further illustrate the coordinated chromatin controls between DNA methylation and the lysine methylation status of histone H3 (at residues 4 and 9; Cheng & Blumenthal, 2010).

In the last few years, we have biochemically and structurally characterized mouse Zfp57 (2 ZnFs) (Liu, Olanrewaju, et al., 2013; Liu et al., 2012), mouse Krüppel-like factor 4 (Klf4) (3 ZnFs) (Liu et al., 2014), human early growth response protein (Egr1, also known as Zif268) (3 ZnFs) (Hashimoto et al., 2014), human Wilms tumor protein (WT1) (3 ZnFs) (Hashimoto et al., 2014), human PRDM9 allele-A (5 ZnFs) (Patel, Horton, Wilson, Zhang, & Cheng, 2016), human PRDM9 allele-C (6 ZnFs) (Patel et al., unpublished), and human CTCF (4 ZnFs) (Hashimoto et al., unpublished). Rather than following published methods of refolding insoluble proteins, such as WT1 (Laity, Chung, Dyson, & Wright, 2000) and Egr1/Zif268 (Pavletich & Pabo, 1991), we expressed and purified the tandem array of ZnF DNA-binding domains in soluble form as fusion proteins with glutathione S-transferase (GST). One key to our success in obtaining quality ZnF protein suitable for biochemistry analysis and crystallization is the use of *p*oly*e*thylen*i*mine (PEI), a polymer with repeating unit composed of the amine and two carbon aliphatic spacer, and anion exchange column to completely remove bacterial nucleic acids associated with the ZnF proteins.

2. SOLUBLE EXPRESSION AND PURIFICATION OF ZnF PROTEINS

ZnF cDNA fragments were cloned into the *Bam*HI site of pGEX6p-1 vector (GE Healthcare), leaving five extra residues at the N-terminal, Gly-Pro-Leu-Gly-Ser, after PreScission protease cleavage.

2.1 Expression

1. Day 1: Inoculate 15 mL of noninducing MDAG media (Studier, 2005) supplemented with 100 μg/mL sterile ampicillin, with a single colony or glycerol freezer stock of *Escherichia coli* BL21 (DE3) Codon-plus RIL cells containing the expression plasmid. Incubate overnight with shaking at 37°C.
2. Day 2: Inoculate 1 L LB medium supplemented with 100 μg/mL ampicillin with 2 mL of the starter culture. Grow at 37°C with shaking at 200 rpm until the A_{600} reaches ~0.5 when the shaker temperature is reduced to 16°C.

   Note: Monitor the liquid temperature with a thermometer until it has reached 16°C, which can take 0.5–1.5 h depending on shaker. Add 200 μL of 0.5 *M* $ZnCl_2$ (to a final concentration of 100 μ*M*) and induce the expression of protein by adding isopropyl β-D-1-thiogalactopyranoside (IPTG) to 0.2 m*M* final concentration. Incubate overnight with shaking at 16°C.
3. Day 3: Harvest the cells by centrifugation at 3500 rpm for 30 min at 4°C and freeze the pellets at −20°C until purification.

2.2 Purification

The general scheme involves four-column chromatography (Fig. 2A).

Day 1

1. Resuspend cells from 6 L culture into 120 mL of lysis buffer [20 m*M* Tris (pH 7.5), 5% (v/v) glycerol, 25 μ*M* $ZnCl_2$, 0.5 m*M* tris(2-carboxyethyl) phosphine (TCEP), and 0.1 m*M* phenylmethylsulfonyl fluoride (PMSF)] containing 250–700 m*M* NaCl.

   Note: The salt concentration needs to be individually determined to assure maximum solubility.
2. Lyse cells by sonication with 1 s on and 2 s off cycles for 8 min in total.
3. Treat the lysate with PEI (Sigma—408727) neutralized by HCl to pH 7.

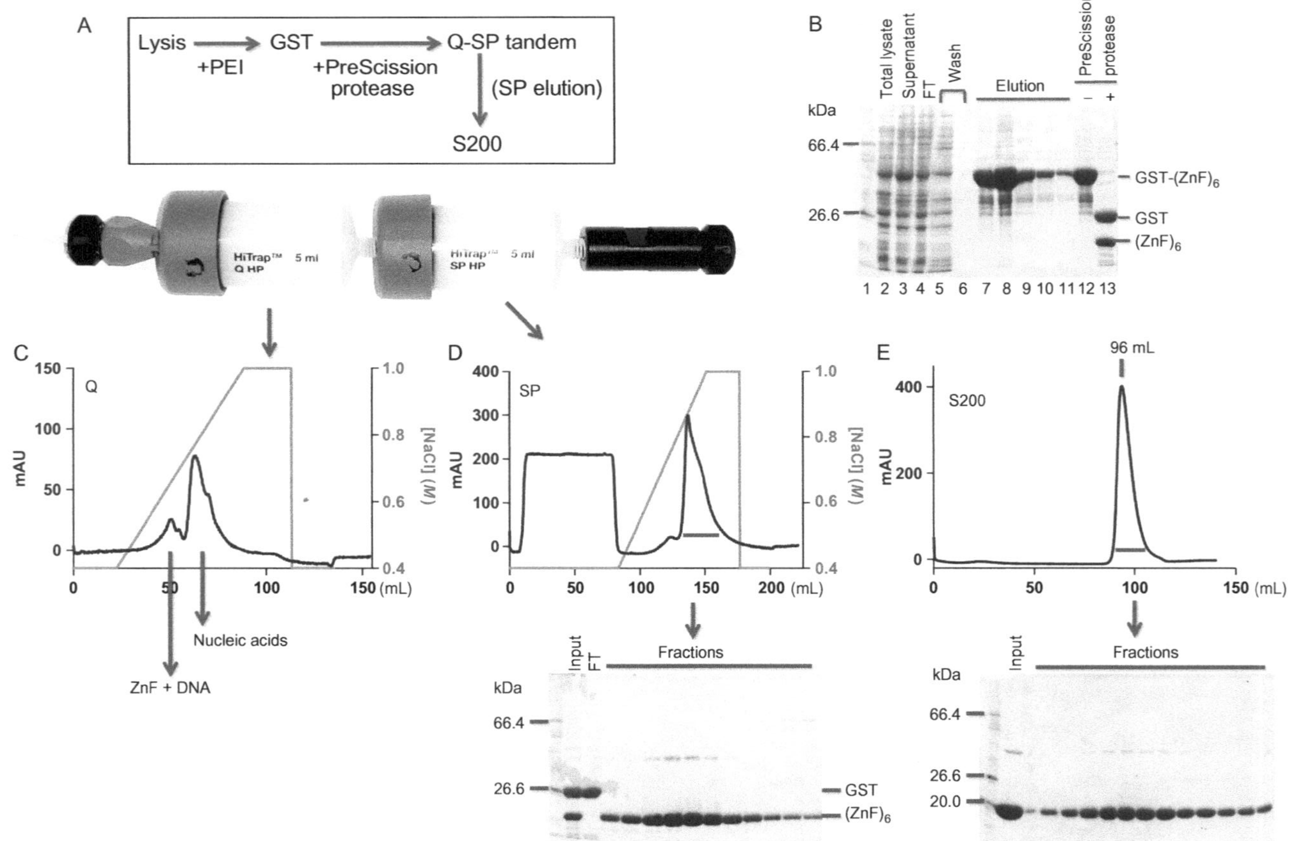

Fig. 2 One example of ZnF protein purification. (A) The general scheme of four-column chromatography. Inserted is a picture of tandem Q-SP columns. (B) A 12% SDS-PAGE showing total lysate (lane 2), supernatant after PEI treatment (lane 3), flowthrough (FT; lane 4), and washing through the GST column (lanes 5 and 6), five elutions by GSH (lanes 7–11), and before and after PreScission protease cleavage (lanes 12 and 13). Note that more soluble protein appears after PEI treatment (comparing lanes 2 and 3). (C) Elution profile of the HiTrap Q column. (D) Elution profile of the HiTrap SP column and an accompanying SDS-PAGE showing the fractions. (E) Elution profile of a Superdex-200 (16/60 GL) column and an accompanying SDS-PAGE showing the fractions.

Note: Slowly add 6 mL of 2% (w/v) PEI solution drop by drop into the lysate to a final concentration of 0.1% while stirring on an ice bath.

4. Clear the lysate by centrifugation at 16,500 rpm for 45 min at 4°C.
5. Load the supernatant onto a Glutathione Sepharose 4B column (GE Healthcare) with 5 mL bed volume equilibrated with 250–700 m*M* NaCl lysis buffer (see Step 1) at ~0.7 mL/min flow rate. Wash the column with 40 mL of lysis buffer followed by 25 mL of washing buffer containing 100 m*M* Tris (pH 8.0), 500 m*M* NaCl, 5% glycerol, 25 μ*M* $ZnCl_2$, and 0.5 m*M* TCEP. Elute the GST-tagged protein with 50 mL of elution buffer (washing buffer + 20 m*M* reduced glutathione) into fractions of 10 mL (Fig. 2B).
6. Remove the GST tag by treating the eluted protein with ~100 μg of PreScission protease (GE Healthcare 27-0843-01 or purified in-house) at 4°C overnight.

Day 2

1. Load the protein onto 5 mL HiTrap Q-SP columns connected in tandem (GE Healthcare; Fig. 2A), equilibrated with column buffer (lysis buffer minus PMSF) with 500 m*M* NaCl. After washing 20 mL with the same buffer, disconnect the Q and SP columns and elute separately with a 50 mL (or 75) linear gradient of 0.5–1 *M* NaCl (Fig. 2C and D).

   Note: DNA-free protein flows through the Q column and binds the SP column, while the DNA-containing protein binds to the Q column along with free DNA (Fig. 2C).
2. Concentrate the protein eluted from the SP column to ~2 mL using a centrifugal concentrator such as Vivaspin. Load onto a Superdex-200 (16/60) column equilibrated with column buffer with 500 m*M* NaCl. Collect the protein eluted as a single peak (Fig. 2E). Concentrate to about 5 mg/mL and flash freeze with liquid nitrogen and store at −80°C in aliquots. Final yields of the protein range from 10 to 15 mg/6 L culture.

3. FLUORESCENCE POLARIZATION ASSAY FOR ANALYSIS OF DNA BINDING

1. Synthesize 5′-FAM (6-carboxyfluorescein)-labeled oligonucleotides containing various cytosine modifications. Only one strand is labeled.

   Note: Generally, blunt-ended DNA duplex is preferred, although sometimes labeling at one (or two) base overhang at the 5′-end yields better signal (Hashimoto et al., 2014; Liu et al., 2014, 2012). The FAM label may not be compatible with 5caC modification within the same strand due to currently available synthesis chemistry.

2. Mix twofold serially diluted protein solutions (1–10 μ*M* starting concentration, 10–15 points) with 1–5 n*M* final concentration of DNA probe in a Corning 3575 plate, using binding buffer of 20 m*M* Tris–HCl, pH 7.5, 5% glycerol, and 0.5 m*M* TCEP with varying NaCl concentration (150–300 m*M*). Incubate the mixture for 10 min at room temperature. Perform at least two duplicate experiments.
3. Measure fluorescence polarization at 25°C on a Synergy 4 Microplate Reader (BioTek) using 485/20 nm and 528/20 nm filters for emission and excitation, respectively.

   Note: The presence of protein should cause no change in fluorescence intensity.
4. Calculate the dissociation constants (K_D) by fitting the experimental data to the following equation using GraphPad Prism software (version 6.0): [mP] = [maximum mP] × [C] / (K_D + [C]) + [baseline mP], then replot the curve using % of saturation calculated as ([mP] − [baseline mP]) / ([maximum mP] − [baseline mP]), where mP is millipolarization and [C] is protein concentration.

   Note: The maximum increase of mP observed is protein and/or oligonucleotide dependent, and is most commonly between 50 and 100 mP. Very large mP change (>200) at high protein concentration (>1 μ*M*) often indicates nonspecific binding.
5. Effect of NaCl concentration: The K_D values are extremely sensitive to the ionic strength of the binding buffer. Fig. 3 illustrated that the WT1 +KTS isoform binds most strongly to 5caC-containing DNA. Affinity is uniformly low in 300 m*M* NaCl (Fig. 3A) but considerably higher (>10-fold) in 200 m*M* NaCl (Fig. 3B; Hashimoto et al., 2014). The effect is even more pronounced for human PRDM9 allele-A: increasing NaCl concentration by 20 m*M* can result in as much as 3-fold increase in K_D value (Fig. 3C), and a 50 m*M* increase of NaCl resulted in an 18-fold reduction in affinity. As documented in previous studies (Jantz & Berg, 2010), the double-logarithmic plot of K_D as a function of NaCl concentration is linear (Fig. 3D).

   Note: Due to extreme salt sensitivity of the assay, different batches of the same buffer formulation can give noticeably different K_D values. When possible, use the same batch of buffer for all assays in one study.

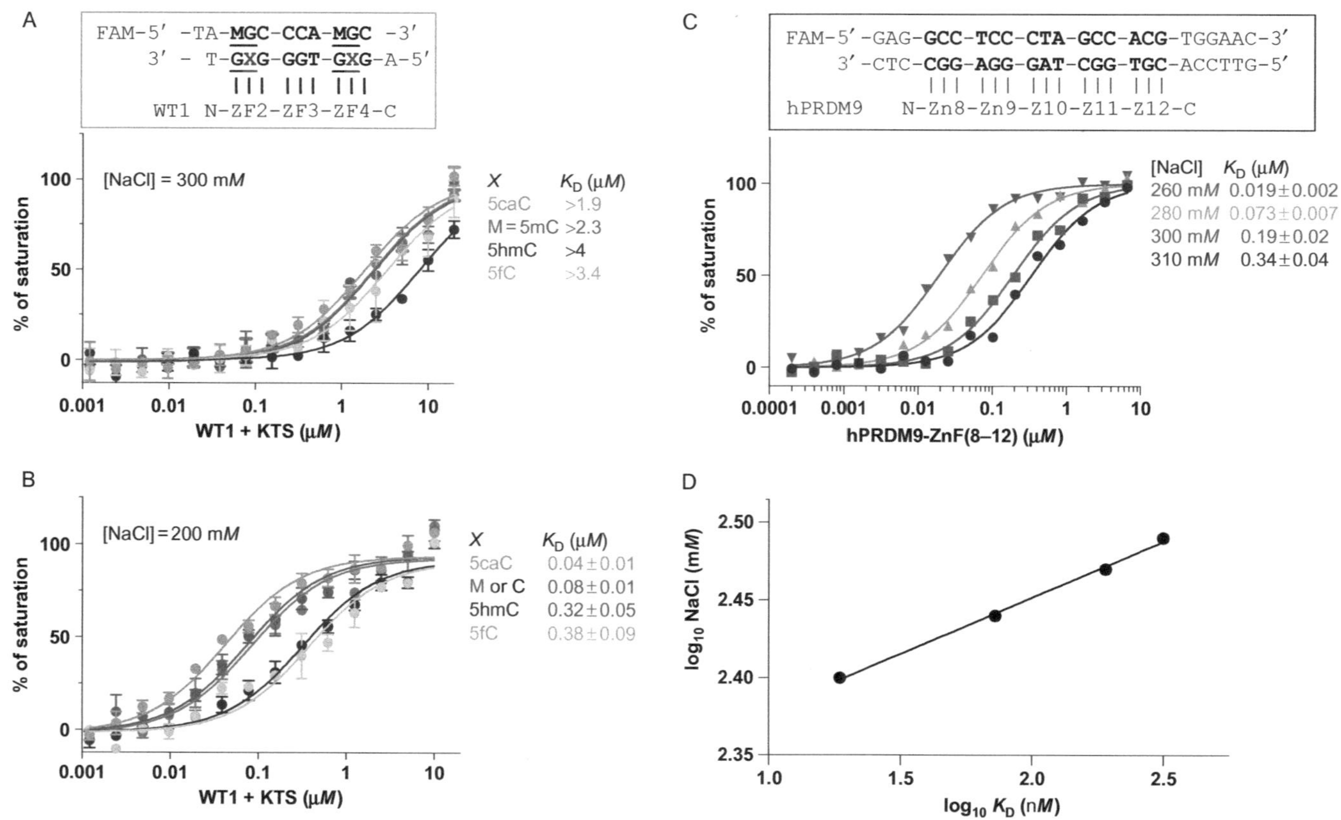

Fig. 3 The effect of NaCl on binding affinity. The DNA-binding affinity of WT1 + KTS isoform is uniformly low in 300 m*M* NaCl (A) but increases markedly in 200 m*M* NaCl (B). (C) The DNA-binding affinity (K_D) of human PRDM9 allele-A under NaCl concentrations from 260 to 310 m*M* with 10–20 m*M* increments. (D) The linear correlation of double-logarithmic plot of K_D values and NaCl concentrations. *Panel (B): Adopted from Hashimoto, H., Olanrewaju, Y. O., Zheng, Y., Wilson, G. G., Zhang, X., & Cheng, X. (2014). Wilms tumor protein recognizes 5-carboxylcytosine within a specific DNA sequence.* Genes & Development, 28, *2304–2313.* (See the color plate.)

4. CRYSTALLIZATION OF ZnF PROTEINS IN COMPLEX WITH DNA

1. Design of oligonucleotide suitable for cocrystallization: both the length and the ends must be considered. Fig. 4 illustrates the process of obtaining diffraction quality crystals for human PRDM9 allele-A ZnF 8–12 by varying oligonucleotides. We started with a 15 + 1 base pairs (bp) double-stranded oligonucleotide (oligo)—the minimum length required for recognition by five ZnFs—plus a 5′-overhanging thymine

```
              hPRDM9A
bp          ZF8--9-10-11-12                 Crystals              Diffraction limit
             |          |
15+1  5'-TGCCTCCCTAGCCACG -3'               NO
      3'- CGGAGGGATCGGTGCA-5'
             |          |
16+1  5'-TGGCCTCCCTAGCCACG -3'              NO
      3'- CCGGAGGGATCGGTGCA-5'
             |          |
17+1  5'-TGGCCTCCCTAGCCACGT -3'             NO
      3'- CCGGAGGGATCGGTGCAA-5'
             |          |
18    5'-GCCTCCCTAGCCACGTGG-3'              Small plates/rods     Poor (~15Å)
      3'-CGGAGGGATCGGTGCACC-5'              (a)(b)
             |          |
18+1  5'-TGCCTCCCTAGCCACGTGG -3'            NO
      3'- CGGAGGGATCGGTGCACCA-5'
             |          |
19    5'-GGCCTCCCTAGCCACGTGG-3'             Needle clusters
      3'-CCGGAGGGATCGGTGCACC-5'
             |          |
19+1  5'-TGCCTCCCTAGCCACGTGGA -3'           Needle-like
      3'- CGGAGGGATCGGTGCACCTA-5'
             |          |
20    5'-GGCCTCCCTAGCCACGTGGA-3'            Needle clusters       Poor (~20Å)
      3'-CCGGAGGGATCGGTGCACCT-5'            (c)
             |          |
20+1  5'-TGGCCTCCCTAGCCACGTGGA -3'          Suitable for data     2 Å
      3'- CCGGAGGGATCGGTGCACCTA-5'          collection (d)(e)
```

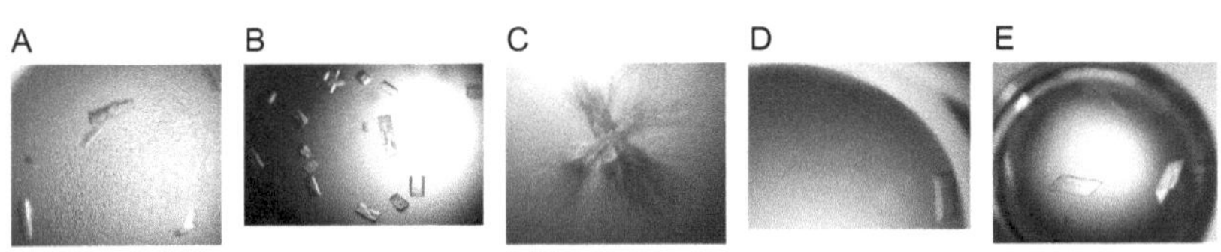

A	B	C	D	E
(Sitting drop) 0.1 *M* Bis–tris (pH 5.5) 25% PEG 3350 0.2 *M* NaCl	(Hanging drop) 0.1 *M* Bis–tris (pH 5.5) 25% PEG 3350 0.2 *M* NaCl	(Sitting drop) 0.1 *M* Bis–tris (pH 5.5) 25% PEG 3350	(Sitting drop) 60 m*M* Bis–tris propane, 40 m*M* citric acid (pH 6.4) 20% PEG 3350	(Hanging drop) 60 m*M* Bis–tris propane, 40 m*M* Citric acid (pH 5.5) 18% PEG 3350

Fig. 4 Examples of DNA oligonucleotide sequences used for cocrystallization with human PRDM9 allele-A, crystals observed, and quality of X-ray diffractions. Five examples of crystals and corresponding conditions were shown (A–E).

or adenine on either strand (Fig. 4). This design was then lengthened 1 or 2 bp at a time to become 16 + 1 (1 bp increase on one end), 17 + 1 (1 bp increase on both ends), 18-, 19-, and 20-bp blunt ends, and 18 + 1, 19 + 1, and 20 + 1 bp with 5′-overhangs. In the end, only the 20 + 1 oligo yielded high-diffraction quality crystals. Alternatively, 3′-overhangings or asymmetric overhanging only on one strand could also be used.

2. Purification of crude oligonucleotides: Due to cost consideration, we use crude unmodified oligos to screen the length and ends. Resuspend each single-stranded DNA into annealing buffer containing 10 m*M* Tris (pH 8.0), 50 m*M* NaCl, and 1 m*M* EDTA (ethylenediaminetetraacetic acid) to final concentration of ~1 m*M*. Mix complimentary stands of DNA in equimolar ratio, heat in a boiling water bath that is slowly cooled overnight to room temperature. Load the annealed double-stranded (ds) DNA to a 5 mL HiTrap Q column with 20 m*M* Tris (pH 8.0) as buffer A and 20 m*M* Tris (pH 8.0) and 1 *M* NaCl as buffer B. Collect DNA eluted as a single large peak between 0.55 and 0.65 *M* NaCl using a linear gradient of NaCl from 0.1 to 1 *M*. Pool peak fractions and measure DNA concentration by absorbance at 260 nm.
3. Protein–DNA complex formation by dialysis: Mix ZnF protein with purified dsDNA in 1:1 molar ratio to a final concentration of 25 μ*M* each in buffer containing 20 m*M* Tris (7.5), 500 m*M* NaCl, 5% glycerol, 25 μ*M* $ZnCl_2$, and 0.5 m*M* TCEP. Dialyze the mixture against low salt buffer (150–250 m*M* NaCl) at 4°C with two 500 mL changes of buffer every 6–8 h. The slow exchange to low salt buffer can potentially reduce nonspecific binding and promote specific complex formation between DNA and protein. [Alternatively, modified oligos are often HPLC or PAGE purified after synthesis, thus can be directly mixed with protein at high concentration (~1 m*M*).] After dialysis, centrifuge the protein–DNA complex at 4000 rpm for 15 min to remove any precipitate. Concentrate the supernatant to desired concentration and centrifuge at 13,000 rpm for 10 min before crystallization trial.

ACKNOWLEDGMENTS

We sincerely thank Yusuf Olatunde Olanrewaju and Yiwei Liu for their early efforts in purifying soluble ZnF proteins. This work was supported by grant from the National Institutes of Health GM049245-22 to X.C. (who is a Georgia Research Alliance Eminent Scholar).

REFERENCES

Arita, K., Ariyoshi, M., Tochio, H., Nakamura, Y., & Shirakawa, M. (2008). Recognition of hemi-methylated DNA by the SRA protein UHRF1 by a base-flipping mechanism. *Nature, 455*, 818–821.

Avvakumov, G. V., Walker, J. R., Xue, S., Li, Y., Duan, S., Bronner, C., et al. (2008). Structural basis for recognition of hemi-methylated DNA by the SRA domain of human UHRF1. *Nature, 455*, 822–825.

Bestor, T., Laudano, A., Mattaliano, R., & Ingram, V. (1988). Cloning and sequencing of a cDNA encoding DNA methyltransferase of mouse cells. The carboxyl-terminal domain of the mammalian enzymes is related to bacterial restriction methyltransferases. *Journal of Molecular Biology, 203*, 971–983.

Buck-Koehntop, B. A., Stanfield, R. L., Ekiert, D. C., Martinez-Yamout, M. A., Dyson, H. J., Wilson, I. A., et al. (2012). Molecular basis for recognition of methylated and specific DNA sequences by the zinc finger protein Kaiso. *Proceedings of the National Academy of Sciences of the United States of America, 109*, 15229–15234.

Cerbone, M., Wang, J., Van der Maarel, S. M., D'Amico, A., D'Agostino, A., Romano, A., et al. (2012). Immunodeficiency, centromeric instability, facial anomalies (ICF) syndrome, due to ZBTB24 mutations, presenting with large cerebral cyst. *American Journal of Medical Genetics. Part A, 158A*, 2043–2046.

Cheng, X., & Blumenthal, R. M. (2010). Coordinated chromatin control: Structural and functional linkage of DNA and histone methylation. *Biochemistry, 49*, 2999–3008.

Chouery, E., Abou-Ghoch, J., Corbani, S., El Ali, N., Korban, R., Salem, N., et al. (2012). A novel deletion in ZBTB24 in a Lebanese family with immunodeficiency, centromeric instability, and facial anomalies syndrome type 2. *Clinical Genetics, 82*, 489–493.

de Greef, J. C., Wang, J., Balog, J., den Dunnen, J. T., Frants, R. R., Straasheijm, K. R., et al. (2011). Mutations in ZBTB24 are associated with immunodeficiency, centromeric instability, and facial anomalies syndrome type 2. *American Journal of Human Genetics, 88*, 796–804.

Dhasarathy, A., & Wade, P. A. (2008). The MBD protein family-reading an epigenetic mark? *Mutation Research, 647*, 39–43.

Frietze, S., O'Geen, H., Blahnik, K. R., Jin, V. X., & Farnham, P. J. (2010). ZNF274 recruits the histone methyltransferase SETDB1 to the 3' ends of ZNF genes. *PloS One, 5*, e15082.

Garcia-Garcia, M. J., Shibata, M., & Anderson, K. V. (2008). Chato, a KRAB zinc-finger protein, regulates convergent extension in the mouse embryo. *Development, 135*, 3053–3062.

Gowher, H., & Jeltsch, A. (2001). Enzymatic properties of recombinant Dnmt3a DNA methyltransferase from mouse: The enzyme modifies DNA in a non-processive manner and also methylates non-CpG [correction of non-CpA] sites. *Journal of Molecular Biology, 309*, 1201–1208.

Guy, J., Cheval, H., Selfridge, J., & Bird, A. (2011). The role of MeCP2 in the brain. *Annual Review of Cell and Developmental Biology, 27*, 631–652.

Hansen, R. S., Wijmenga, C., Luo, P., Stanek, A. M., Canfield, T. K., Weemaes, C. M., et al. (1999). The DNMT3B DNA methyltransferase gene is mutated in the ICF immunodeficiency syndrome. *Proceedings of the National Academy of Sciences of the United States of America, 96*, 14412–14417.

Hashimoto, H., Horton, J. R., Zhang, X., Bostick, M., Jacobsen, S. E., & Cheng, X. (2008). The SRA domain of UHRF1 flips 5-methylcytosine out of the DNA helix. *Nature, 455*, 826–829.

Hashimoto, H., Horton, J. R., Zhang, X., & Cheng, X. (2009). UHRF1, a modular multi-domain protein, regulates replication-coupled crosstalk between DNA methylation and histone modifications. *Epigenetics, 4*, 8–14.

Hashimoto, H., Olanrewaju, Y. O., Zheng, Y., Wilson, G. G., Zhang, X., & Cheng, X. (2014). Wilms tumor protein recognizes 5-carboxylcytosine within a specific DNA sequence. *Genes & Development*, *28*, 2304–2313.

Hashimoto, H., Zhang, X., Vertino, P. M., & Cheng, X. (2015). The mechanisms of generation, recognition, and erasure of DNA 5-methylcytosine and thymine oxidations. *The Journal of Biological Chemistry*, *290*, 20723–20733.

He, Y. F., Li, B. Z., Li, Z., Liu, P., Wang, Y., Tang, Q., et al. (2011). Tet-mediated formation of 5-carboxylcytosine and its excision by TDG in mammalian DNA. *Science*, *333*, 1303–1307.

Ho, K. L., McNae, I. W., Schmiedeberg, L., Klose, R. J., Bird, A. P., & Walkinshaw, M. D. (2008). MeCP2 binding to DNA depends upon hydration at methyl-CpG. *Molecular Cell*, *29*, 525–531.

Ito, S., D'Alessio, A. C., Taranova, O. V., Hong, K., Sowers, L. C., & Zhang, Y. (2010). Role of Tet proteins in 5mC to 5hmC conversion, ES-cell self-renewal and inner cell mass specification. *Nature*, *466*, 1129–1133.

Ito, S., Shen, L., Dai, Q., Wu, S. C., Collins, L. B., Swenberg, J. A., et al. (2011). Tet proteins can convert 5-methylcytosine to 5-formylcytosine and 5-carboxylcytosine. *Science*, *333*, 1300–1303.

Jantz, D., & Berg, J. M. (2010). Probing the DNA-binding affinity and specificity of designed zinc finger proteins. *Biophysical Journal*, *98*, 852–860.

Jenuwein, T., Laible, G., Dorn, R., & Reuter, G. (1998). SET domain proteins modulate chromatin domains in eu- and heterochromatin. *Cellular and Molecular Life Sciences*, *54*, 80–93.

Krebs, C. J., Schultz, D. C., & Robins, D. M. (2012). The KRAB zinc finger protein RSL1 regulates sex- and tissue-specific promoter methylation and dynamic hormone-responsive chromatin configuration. *Molecular and Cellular Biology*, *32*, 3732–3742.

Kubo, N., Toh, H., Shirane, K., Shirakawa, T., Kobayashi, H., Sato, T., et al. (2015). DNA methylation and gene expression dynamics during spermatogonial stem cell differentiation in the early postnatal mouse testis. *BMC Genomics*, *16*, 624.

Laity, J. H., Chung, J., Dyson, H. J., & Wright, P. E. (2000). Alternative splicing of Wilms' tumor suppressor protein modulates DNA binding activity through isoform-specific DNA-induced conformational changes. *Biochemistry*, *39*, 5341–5348.

Lister, R., Mukamel, E. A., Nery, J. R., Urich, M., Puddifoot, C. A., Johnson, N. D., et al. (2013). Global epigenomic reconfiguration during mammalian brain development. *Science*, *341*, 1237905.

Lister, R., Pelizzola, M., Dowen, R. H., Hawkins, R. D., Hon, G., Tonti-Filippini, J., et al. (2009). Human DNA methylomes at base resolution show widespread epigenomic differences. *Nature*, *462*, 315–322.

Liu, Y., Olanrewaju, Y. O., Zhang, X., & Cheng, X. (2013). DNA recognition of 5-carboxylcytosine by a zfp57 mutant at an atomic resolution of 0.97 A. *Biochemistry*, *52*, 9310–9317.

Liu, Y., Olanrewaju, Y. O., Zheng, Y., Hashimoto, H., Blumenthal, R. M., Zhang, X., et al. (2014). Structural basis for Klf4 recognition of methylated DNA. *Nucleic Acids Research*, *42*, 4859–4867.

Liu, Y., Toh, H., Sasaki, H., Zhang, X., & Cheng, X. (2012). An atomic model of Zfp57 recognition of CpG methylation within a specific DNA sequence. *Genes & Development*, *26*, 2374–2379.

Liu, Y., Zhang, X., Blumenthal, R. M., & Cheng, X. (2013). A common mode of recognition for methylated CpG. *Trends in Biochemical Sciences*, *38*, 177–183.

Mackay, D. J., Callaway, J. L., Marks, S. M., White, H. E., Acerini, C. L., Boonen, S. E., et al. (2008). Hypomethylation of multiple imprinted loci in individuals with transient neonatal diabetes is associated with mutations in ZFP57. *Nature Genetics*, *40*, 949–951.

Messerschmidt, D. M., de Vries, W., Ito, M., Solter, D., Ferguson-Smith, A., & Knowles, B. B. (2012). Trim28 is required for epigenetic stability during mouse oocyte to embryo transition. *Science*, *335*, 1499–1502.

Meylan, S., Groner, A. C., Ambrosini, G., Malani, N., Quenneville, S., Zangger, N., et al. (2011). A gene-rich, transcriptionally active environment and the pre-deposition of repressive marks are predictive of susceptibility to KRAB/KAP1-mediated silencing. *BMC Genomics*, *12*, 378.

Mihola, O., Trachtulec, Z., Vlcek, C., Schimenti, J. C., & Forejt, J. (2009). A mouse speciation gene encodes a meiotic histone H3 methyltransferase. *Science*, *323*, 373–375.

Nitta, H., Unoki, M., Ichiyanagi, K., Kosho, T., Shigemura, T., Takahashi, H., et al. (2013). Three novel ZBTB24 mutations identified in Japanese and Cape Verdean type 2 ICF syndrome patients. *Journal of Human Genetics*, *58*, 455–460.

Ohki, I., Shimotake, N., Fujita, N., Jee, J., Ikegami, T., Nakao, M., et al. (2001). Solution structure of the methyl-CpG binding domain of human MBD1 in complex with methylated DNA. *Cell*, *105*, 487–497.

Okano, M., Bell, D. W., Haber, D. A., & Li, E. (1999). DNA methyltransferases Dnmt3a and Dnmt3b are essential for de novo methylation and mammalian development. *Cell*, *99*, 247–257.

Okano, M., Xie, S., & Li, E. (1998). Cloning and characterization of a family of novel mammalian DNA (cytosine-5) methyltransferases. *Nature Genetics*, *19*, 219–220.

Patel, A., Horton, J. R., Wilson, G. G., Zhang, X., & Cheng, X. (2016). Structural basis for human PRDM9 action at recombination hot spots. *Genes & Development*, *30*, 257–265.

Pavletich, N. P., & Pabo, C. O. (1991). Zinc finger-DNA recognition: Crystal structure of a Zif268-DNA complex at 2.1 A. *Science*, *252*, 809–817.

Prokhortchouk, A., Hendrich, B., Jorgensen, H., Ruzov, A., Wilm, M., Georgiev, G., et al. (2001). The p120 catenin partner Kaiso is a DNA methylation-dependent transcriptional repressor. *Genes & Development*, *15*, 1613–1618.

Quenneville, S., Verde, G., Corsinotti, A., Kapopoulou, A., Jakobsson, J., Offner, S., et al. (2011). In embryonic stem cells, ZFP57/KAP1 recognize a methylated hexanucleotide to affect chromatin and DNA methylation of imprinting control regions. *Molecular Cell*, *44*, 361–372.

Ramsahoye, B. H., Biniszkiewicz, D., Lyko, F., Clark, V., Bird, A. P., & Jaenisch, R. (2000). Non-CpG methylation is prevalent in embryonic stem cells and may be mediated by DNA methyltransferase 3a. *Proceedings of the National Academy of Sciences of the United States of America*, *97*, 5237–5242.

Sasai, N., Nakao, M., & Defossez, P. A. (2010). Sequence-specific recognition of methylated DNA by human zinc-finger proteins. *Nucleic Acids Research*, *38*, 5015–5022.

Scarsdale, J. N., Webb, H. D., Ginder, G. D., & Williams, D. C., Jr. (2011). Solution structure and dynamic analysis of chicken MBD2 methyl binding domain bound to a target-methylated DNA sequence. *Nucleic Acids Research*, *39*, 6741–6752.

Sharif, J., & Koseki, H. (2011). Recruitment of Dnmt1: Roles of the SRA protein Np95 (Uhrf1) and other factors. *Progress in Molecular Biology and Translational Science*, *101*, 289–310.

Shirohzu, H., Kubota, T., Kumazawa, A., Sado, T., Chijiwa, T., Inagaki, K., et al. (2002). Three novel DNMT3B mutations in Japanese patients with ICF syndrome. *American Journal of Medical Genetics*, *112*, 31–37.

Studier, F. W. (2005). Protein production by auto-induction in high density shaking cultures. *Protein Expression and Purification*, *41*, 207–234.

Tahiliani, M., Koh, K. P., Shen, Y., Pastor, W. A., Bandukwala, H., Brudno, Y., et al. (2009). Conversion of 5-methylcytosine to 5-hydroxymethylcytosine in mammalian DNA by MLL partner TET1. *Science*, *324*, 930–935.

Thomas, J. H., & Schneider, S. (2011). Coevolution of retroelements and tandem zinc finger genes. *Genome Research*, *21*, 1800–1812.

Vlachogiannis, G., Niederhuth, C. E., Tuna, S., Stathopoulou, A., Viiri, K., de Rooij, D. G., et al. (2015). The Dnmt3L ADD domain controls cytosine methylation establishment during spermatogenesis. *Cell Reports*, *10*, 944–956. http://dx.doi.org/10.1016/j.celrep.2015.01.021. pii: S2211-1247(15)00034-0.

Williams, A. J., Khachigian, L. M., Shows, T., & Collins, T. (1995). Isolation and characterization of a novel zinc-finger protein with transcription repressor activity. *The Journal of Biological Chemistry*, *270*, 22143–22152.

Wolf, D., & Goff, S. P. (2009). Embryonic stem cells use ZFP809 to silence retroviral DNAs. *Nature*, *458*, 1201–1204.

Xu, G. L., Bestor, T. H., Bourc'his, D., Hsieh, C. L., Tommerup, N., Bugge, M., et al. (1999). Chromosome instability and immunodeficiency syndrome caused by mutations in a DNA methyltransferase gene. *Nature*, *402*, 187–191.

Zandarashvili, L., White, M. A., Esadze, A., & Iwahara, J. (2015). Structural impact of complete CpG methylation within target DNA on specific complex formation of the inducible transcription factor Egr-1. *FEBS Letters*, *589*, 1748–1753.

Zuo, X., Sheng, J., Lau, H. T., McDonald, C. M., Andrade, M., Cullen, D. E., et al. (2012). Zinc finger protein ZFP57 requires its co-factor to recruit DNA methyltransferases and maintains DNA methylation imprint in embryonic stem cells via its transcriptional repression domain. *The Journal of Biological Chemistry*, *287*, 2107–2118.

CHAPTER EIGHTEEN

Crystallographic Studies of Telomerase

H. Hoffman, E. Skordalakes[1]
The Wistar Institute, Philadelphia, PA, United States
[1]Corresponding author: e-mail address: skorda@wistar.org

Contents

1. Introduction 404
2. Crystal Structure Determination of the Catalytic Subunit of *Triobolium castaneum* Telomerase (tcTERT) 407
   2.1 tcTERT Protein Isolation 407
   2.2 tcTERT Crystallization 410
   2.3 tcTERT Structure Determination 412
3. Crystal Structure Determination of a Partial Telomerase Elongation Complex 414
   3.1 Nucleic Acid Substrate Design 414
   3.2 Activity Assays 414
   3.3 Complex Crystallization 414
   3.4 Structure Determination 416
4. Conclusions 417
Acknowledgments 417
References 417

Abstract

Telomeres are nucleoprotein complexes that maintain the ends of our chromosomes thus providing genomic stability. Telomerase is a ribonucleoprotein reverse transcriptase that replicates the short tandem repeats of DNA known as telomeres. The telomeric DNA is specifically associated with two major complexes, the shelterin and CST complexes both of which are involved in telomere length regulation and maintenance along with telomerase. Obtaining structural information on these nucleoprotein complexes has been a major bottleneck in fully understanding the mechanism of action of telomeric nucleoproteins for over two decades. The recent advances in molecular and structural biology have enabled us to obtain atomic resolution structures of telomeric proteins alone and in complex with their nucleic acid substrates transforming the field and our understanding and interpretation of this unique biological pathway. Here we report our approach to obtain the structure of the *Triobolium castaneum* catalytic subunit of telomerase TERT (tcTERT) in its apo- and substrate-bound states.

Methods in Enzymology, Volume 573
ISSN 0076-6879
http://dx.doi.org/10.1016/bs.mie.2016.04.006

ABBREVIATIONS

Cdc13 cell division cycle 13
Stn1 suppressor of cdc thirteen
Ten1 telomeric pathways in association with Stn1 number 1
TER telomerase RNA
TERT telomerase reverse transcriptase

1. INTRODUCTION

Telomerase replicates the ends of our chromosomes by adding multiple identical repeats of DNA known as telomeres. Even though telomeres are noncoding parts of the chromosome, they are critically important for the integrity of our genome as indicated by a host of diseases associated with dysfunctional telomerase, telomerase deregulation, and short telomeres. Some of the most common maladies associated with dysfunctional telomeric complexes are cancer, aplastic anemia, and dyskeratosis congenital, just to mention a few (Calado & Young, 2009; Kim et al., 1994; Vulliamy et al., 2002). In fact, telomerase is known to be active in approximately 85% of human cancers where the continuous replication of the chromosome ends is required, a prerequisite for cell proliferation as it allows the cells to overcome their Hayflick limit and to continue dividing indefinitely (Hayflick, 1965; Kim et al., 1994).

The core telomerase holoenzyme is a ribonucleoprotein complex, consisting of a protein subunit (telomerase reverse transcriptase, TERT) that catalyzes the reaction of nucleotide addition at the ends of chromosomes via phosphodiester bond formation. TERT belongs to the family of reverse transcriptases and consists of four or five domains depending on species. These include the TEN (telomerase essential N-terminal), TRBD (telomerase RNA binding domain), fingers, palm, and thumb domains (reverse transcriptase nomenclature) (Gillis, Schuller, & Skordalakes, 2008). The TRBD, fingers, palm, and thumb domains are universally conserved and essential for telomerase function, while the TEN domain, which comprises the N-terminal portion of the protein is absent in insects and possibly in worms (Fig. 1A) (Osanai et al., 2006). The TEN domain contributes to telomerase recruitment to telomeres but does not affect telomerase activity (Schmidt, Dalby, & Cech, 2014). One possibility is that the TEN domain in the later organisms exists as an independent polypeptide. This hypothesis is supported by data showing that the fully active enzyme can be

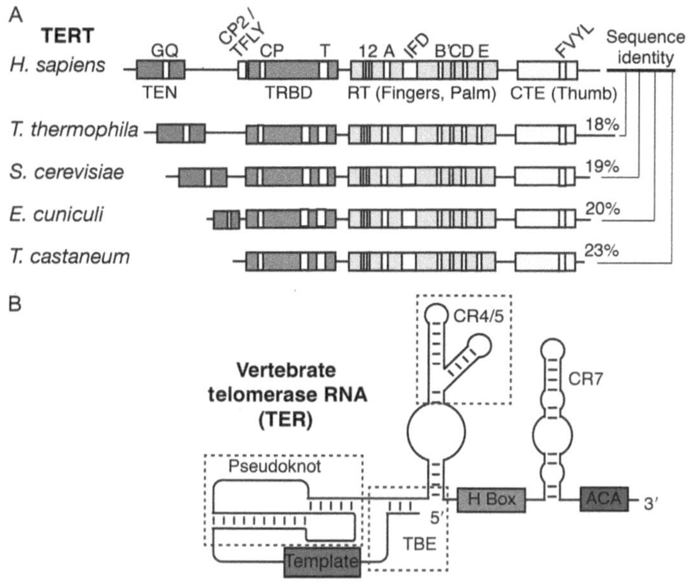

Fig. 1 (A) Primary structure and sequence identity of TERT proteins from a diverse group of organisms (Human, *T. thermophila*, *S. cerevisiae*, *E. cuniculi*, and *T. castaneum*). (B) Schematic of the vertebrate telomerase RNA; conserved motifs are shown in color. The activation domain (CR4/5) and the pseudoknot are highlighted with a blue, dashed line. (See the color plate.)

reconstituted by transiently adding the TEN domain to the TEN-truncated telomerase in vitro (Robart & Collins, 2011).

The telomerase RNA (TER) is an integral component of the holoenzyme and carries the template TERT uses during telomere replication (Fig. 1B). The TER template usually consists of 1.5 telomeric repeats (Lee & Blackburn, 1993; Lingner, Hendrick, & Cech, 1994; Shippen-Lentz & Blackburn, 1990). For example, in humans the telomeric repeat is TTAGGG with the RNA template being CUAACCCU (Lee & Blackburn, 1993; Lingner et al., 1994; Shippen-Lentz & Blackburn, 1990). Interestingly, while TERT is conserved across species, TER is vastly divergent. One of the smallest known TERs belongs to the ciliate *Tetrahymena paravorax* and is 147 nucleotide long (Ye & Romero, 2002), while the fungi TERs are the largest known with *Mycosphaerella graminicola* (2425 dNTPs) (Qi et al., 2013) being the largest identified to date (http://telomerase.asu.edu/sequences_tr.html). To add to the complexity of this system, the proper telomerase holoenzyme assembly requires the enzymatic contribution of a number of proteins with most of them acting on the RNA component of the enzyme. In vertebrates, the enzyme dyskerin modifies the 3′-end of TER, a process required for its stability and proper assembly of telomerase (Mitchell, Wood, & Collins, 1999; Shay & Wright, 1999). In ciliates, the enzyme p65 is required to bind and fold the RNA for proper TERT, TER assembly (Witkin & Collins, 2004). The complexity of the holoenzyme is demonstrated in the recent EM structure of *Tetrahymena thermophila* telomerase (Jiang et al., 2013).

Replicating telomeres is only part of chromosome end protection and overcoming the Hayflick limit. Chromosome end maintenance is carried out by telomere binding complexes such as CST and shelterin (de Lange, 2005; Grandin, Reed, & Charbonneau, 1997; Grandin, Damon, & Charbonneau, 2000, 2001; Lin & Zakian, 1996; Miyake et al., 2009). Both are essential complexes as the deletion of any of their components has deleterious effects for cells, usually leading to telomere length deregulation, chromosome end uncapping, and genomic instability associated with senescence, apoptosis, and in some cases cell immortalization and carcinogenesis. Here, we will focus primarily on the CST named after the three proteins that constitute this complex (cell division cycle 13 (Cdc13)/Ctc1, suppressor of cdc thirteen (Stn1), and Ten1). The CST complex specifically binds the telomeric overhang, and although it is commonly known as a telomere-capping complex, its function at the ends of chromosomes is far more complex. It was believed that this complex was unique to yeast until 2013 when the Ishikawa group identified the human CST complex, which was followed by the discovery of CST in plants, and most recently in ciliates (Lin & Zakian, 1996; Miyake et al., 2009; Wan et al., 2015).

Although the degree of functional conservation of the CST complex across species is currently unclear, there are a few things one should take into consideration when thinking of this nucleoprotein assembly. The trimeric CST complex binds single-stranded telomeric DNA (ssDNA) with high affinity and specificity, and caps the ends of chromosomes. The Stn1 and Ten1 subcomplex structurally resembles the replication protein A complex (containing 32 and 14 kDa subunits), binds DNA nonspecifically, and localizes to various parts of the chromosome where it regulates DNA replication (Bryan et al., 2013; Grandin et al., 2001, 1997; Sun et al., 2009). Yeast (*Saccharomyces cerevisiae*) CST recruits telomerase to telomeres via binding to the Est1 protein (Pennock, Buckley, & Lundblad, 2001), a process that most likely requires significant structural rearrangement of the trimeric complex (Mason & Skordalakes, 2010). In contrast, current data suggest that the human CST complex, and in particular Ctc1, interacts with TPP1, a component of the shelterin complex, which recruits telomerase to telomeres and downregulates telomere replication (Chen, Redon, & Lingner, 2012; Tejera et al., 2010; Wang et al., 2007).

At this point, it is well established that telomeres are replicated and maintained by large nucleoprotein complexes consisting of multiple domains involved in protein–protein and nucleic acid binding. As previously stated, TERT consists of four or five domains containing a large integral RNA, which, as we now know, makes extensive contacts with almost all domains of TERT. Cdc13, the core subunit of the yeast CST complex, consists of five domains, each of which contributes to the function of this complex. Four of these domains are OB (oligonucleotide/oligosaccharide binding) folds with some being involved in Cdc13 dimerization while others are important for Cdc13 ssDNA binding (Bourns et al., 1998; Greetham et al., 2015; Mason & Skordalakes, 2010; Mason et al., 2013; Mitchell et al., 2010a; Nugent et al., 1996; Qi & Zakian, 2000). With this wide array of intricate protein complexes serving multiple essential functions, it is not surprising that structural data were limited to only a domain of Cdc13 (DNA-binding domain, DBD) and the TEN domain of telomerase until the mid-2000s (Jacobs, Podell, & Cech, 2006; Mitton-Fry et al., 2004).

Our understanding of telomeric complexes has been greatly enhanced by the recent structural breakthroughs in the field. In this chapter, we examine strategies utilized in recent years to obtain high-resolution structures of telomerase, breakthroughs that have transformed not only the telomere field and our understanding of telomere biology but also its implications in human disease.

2. CRYSTAL STRUCTURE DETERMINATION OF THE CATALYTIC SUBUNIT OF *TRIOBOLIUM CASTANEUM* TELOMERASE (tcTERT)

2.1 tcTERT Protein Isolation

Considering the size and complexity of telomerase, it is not surprising that efforts to obtain high-resolution atomic data on this enzyme have been met with limited success. NMR structures of fragments of the RNA component of telomerase were accessible early on, mostly because of their small size. One can still glean much information from individual domains, especially when combined with biochemical studies. The concerted effort from a confluence of laboratories resulted in the high-resolution structures of several fragments of the RNA component of telomerase such as the pseudoknot, the TBE, and the activation domain of TER (Theimer & Feigon, 2006). In the case of TERT, the structures of the TEN and TRBD domains became available in 2006 and 2007, respectively (Jacobs et al., 2006; Rouda & Skordalakes, 2007).

The catalytic subunit of telomerase, TERT, posed a different challenge. A major limitation to obtaining structural information on TERT was associated with the lack of accessible active, soluble protein in sufficient quantities for crystallographic studies. Efforts to obtain soluble TERT by a number of prominent labs were met with limited success. These issues were no exception to our lab. Initially, we tested TERT genes from a variety of organisms, primarily focusing on vertebrate, fungi, and ciliates. These groups of organisms had been studied extensively, and hence there was a wealth of biochemical data to draw from when studying these systems. Unfortunately, this approach did not yield any suitable protein for further studies. When expression was possible, the protein was either degraded or insoluble. At this point, we decided that coexpression of TERT with its integral RNA component TER may address this problem. We tried this approach both in *Escherichia coli* and in baculovirus, but the results were no better than with TERT alone. At this point it was clear that we needed to think outside of the box and go after TERT genes from organisms that had not been looked at previously; a comprehensive list of TERT genes can be found in the online available telomerase database (http://telomerase.asu.edu). Although there are significant evolutionary differences among telomerases from different species (Fig. 1), the core mechanism of telomere replication appears to be highly conserved and as such one can still garner a wealth of information from their structural studies. A survey of the telomerase database revealed that the size of the catalytic subunit of telomerase varies considerably in size. TERTs in birds are in the range of 150–160 kDa, in yeasts, around 90–110 kDa, and in worms and insects around 70–80 kDa, while human TERT is 127 kDa. Identifying a TERT gene that lacks all the nonconserved insertion loops could be the answer to addressing this difficult crystallographic problem. With this notion in mind, we decided to test TERT of some of the smaller genes, including the insect *T. castaneum* (70 kDa) and the nematode *Caenorhabditis elegans* (66 kDa).

The next step was to identify the correct expression system and conditions so that we could obtain the protein in large, stable, pure, and active quantities required for crystallographic studies. Expressing proteins in *E. coli* is inexpensive and efficient, plus it eliminates all of the posttranslational modifications that could interfere with structural studies as they can be quite heterogeneous. However, expressing eukaryotic proteins in *E. coli* can be challenging, and therefore the baculovirus system provides a good alternative if necessary. To maximize our yields and quality of protein obtained, we screened protein expression in a range of *E. coli* cell lines (BL21 (pLysS), RIPL, Rosetta (DE3) pLysS, *E. coli* ScarabXpress-1 (T7Lac) cells, and SHuffle T7 Competent *E. coli*), all of which are commercially available (EMD Millipore (Novagen), Billerica, MA, Scarab Genomics, Madison, WI, and New England BioLabs Inc., Ipswich, MA). Many of these cell lines have been modified to be compatible with eukaryotic protein expression by the introduction of rare codons required for mRNA translation. Alternatively, one can obtain the codon-optimized gene that is compatible with the prokaryotic protein expression system being used, although this can be rather expensive, especially when dealing with large genes such as TERT, which spans the length of 1800–3500 nucleotides, depending on organism. For tcTERT overexpression studies, we used the synthetic gene (GeneArt and ThermoFisher Scientific) codon optimized for *E. coli* overexpression cloned into a modified version of the pET28b vector containing a cleavable hexahistidine tag at its N-terminus. tcTERT protein overexpression tests showed that the *E. coli* Rosetta (DE3) pLysS (EMD Millipore (Novagen)) was the ideal cell line for obtaining this protein in large quantities required for structural studies.

Identifying the appropriate cell line is important as it is essential to identify the correct expression conditions, which will define the rate of cell growth and protein expression. Frequently, overexpression of proteins in *E. coli* or baculovirus leads to protein misfolding, which in turn forces the protein into inclusion bodies or to be degraded by the cell. One can improve the protein quality and eliminate aggregation by slowing down the rate and time at which the protein is expressed. One way to achieve this is by varying the temperature at which the cells are grown and the protein expressed. Using temperatures of 16–20°C is common practice, although temperatures ranging from 37°C to 4°C have been used successfully. For tcTERT, constant temperatures in the range of 37–16°C produced mostly small amounts of protein that allowed for a limited number of crystal studies. However, this procedure was expensive and cumbersome so optimization was essential. Interestingly, our overexpression tests indicated that a slow transition in temperature (from high to low) after cell induction with IPTG showed promise worth pursuing. With this in mind, we screened several temperature ranges (37–30°C, 37–20°C, 37–16°C, and 30–20°C). Through this process, we found that a drop in temperature to 30°C from 37°C when the cell density (OD) was 0.4, followed by cell induction when the OD reached 0.6, was ideal for obtaining maximum yields of protein. The period of cell induction/protein expression is also critical. At 37–30°C, 2–4 h overexpressions are commonly used, and this was also true in the case of tcTERT.

The purification of the tcTERT was a three-step process. The protein was overexpressed with a hexahistidine tag at its N-terminus, so we used a Ni-NTA resin for the initial purification of the protein from cell lysate, which resulted in 90% pure protein. However, for crystallographic studies one needs a highly pure and monodisperse protein, so this step alone was not sufficient. At this point, we took advantage of the high pI of the protein (9.37; Wilkins et al., 1999) as it allows for tight binding to cation exchange resin, such as HS poros (Applied Biosystems). Loading the tcTERT to the resin at high salt concentration (500 m*M* KCl) allows for contaminant proteins coming from the host cell (*E. coli* Rosetta (DE3) pLysS (EMD Millipore (Novagen))) to flow through the column while tcTERT binds to the resin tightly. Following loading of the protein to the resin, the column was washed thoroughly (~100 mL) with 50% buffer B (25 m*M* Tris–HCl, pH 7.5, 1 *M* KCl, 1 m*M* DTT, 5% glycerol) to remove any contaminant proteins still associated with the sample. The sample was then eluted off the HS poros using a salt gradient (50–100% buffer B, which translates to 500 m*M*–1 *M* KCl), and the fractions were run on a gel to determine the purity of the sample. The fractions, clean of any contaminants, were pulled together and concentrated to less than 1 mL for the final purification step involving size exclusion chromatography (Sephedex S200). Size exclusion chromatography is an essential step for crystallographic studies as it allows for the separation of protein aggregates generated primarily due to the overexpression of the protein and which can significantly hinder protein crystallization. We eluted the protein from the S200 column in 25 m*M* Tris–HCl, 500 m*M* KCl, 1 m*M* TCEP, and 5% glycerol, pH 7.5, which we refer to as "storage buffer." We found that most of the proteins we work with are relatively stable under these buffer conditions for a few days after purification. At this point, the protein (tcTERT) was >99% pure and therefore ready to be used for crystallographic studies.

2.2 tcTERT Crystallization

For crystallographic studies, there are several points of concern one should consider prior to setting trays. The protein concentration and pI are critical for identifying one or more crystallization conditions. The temperature at which the crystallization is carried out can also be significant as it can affect the stability but most importantly the solubility of the protein. The concentration of the protein should also be taken into account, and our experience indicates that the smaller the protein, the higher the concentration required for crystallization, primarily due to crowding reasons. However, there is no hard rule as to the concentration of the protein sample since the protein solubility factors into this process. Our experience has shown that for large proteins such as ~300 kDa, even 1 mg/mL protein is sufficient in some cases to obtain crystals.

Crystal screening is usually done using a range of sparse matrix conditions commercially available from a number of sources such as Hampton Research (https://www.hamptonresearch.com). The goal of this step is to screen a wide range of precipitants, salts, buffers, pHs, and additives that when added together facilitate the orderly arrangement of the protein molecules leading to a diffracting crystal. For this reason and for the sparse matrix conditions to be effective, the high concentration of buffer, salt, and glycerol of the "storage buffer" needs to be replaced with much lower concentration of buffer, salt, and in most cases glycerol must be removed. The gentlest way to exchange existing buffer conditions of a macromolecule is through dialysis. Dialysis allows for the gradual exchange of buffers, thus providing a window of time for the protein to adjust to the new buffer conditions. There are several devices available for this purpose; we used Slide-A-Lyzer MINI Dialysis Units from ThermoFisher Scientific. We dialyzed the stock tcTERT into 5 m*M* Tris–HCl, 100 m*M* KCl, 1 m*M* TCEP (pH 7.5) prior to crystallization trials. Crystallization trials were carried out at room temperature and 4°C. As mentioned previously, temperature is an important factor as it can affect the solubility and therefore the stability and rate of crystallization of the protein. We found that tcTERT is far more soluble at 4°C than at room temperature, which resulted in a large number of clear crystallization drops even at high concentrations of protein that ultimately did not produce any useful crystals. We addressed this issue by performing all additional crystallization trials at room temperature. However, neither temperature produced any crystal hits. At this point, we decided that cocrystallization of the tcTERT with a natural substrate such as ssDNA or RNA may facilitate crystallization. We prepared the tcTERT–ssDNA sample by mixing the dialyzed (5 m*M* Tris–HCl, 100 m*M* KCl, 1 m*M* TCEP (pH 7.5)) tcTERT protein with 1.2-fold excess ssDNA. The ssDNA consisting of two or three telomeric repeats was purchased from Integrated DNA Technologies (IDT) and prepared in 5 m*M* Tris–HCl and 50 m*M* KCl, pH 7.5. Mixing the dialyzed tcTERT resulted in heavy precipitation of the protein as indicated by a thick white solution; it was unclear at this point, if the protein was useful for crystallization but we proceeded anyway. Interestingly, we found that mixing the protein with various sparse matrix crystallization conditions resulted in resolubilization of the protein and clear crystallization drops. Our approach subsequently resulted in two crystal hits (a) 100 m*M* Tris pH 8.5, 1.5 *M* $NaNO_3$ and (b) 100 m*M* Tris (pH 8.0) and 2 *M* $(NH_4)_2SO_4$.

Crystal optimization is standard and is usually achieved with adjustment of the concentration and pH of the reagents that produced the crystal hit, as well as the addition of a wide range of additives or detergents. Additive and detergent screens, sampling a wide range of metal ions, small organics, and detergents are commercially available and used widely by the crystallographic community. Crystals grown in 100 m*M* Tris, pH 8.5, and 1.5 *M* $NaNO_3$ were small and diffracted to 4 Å resolution at best, so improvement of crystal quality for better resolution was essential. Using the current crystal growth conditions as a guide, we screened a wide range of buffers, pHs, salts, and precipitant ($NaNO_3$) concentrations. Through this process we found that the best quality crystals, diffracting to better than 3 Å, grew in 50 m*M* HEPES (pH 7.0), 1.2–1.3 *M* $NaNO_3$ and 1 m*M* TCEP at room temperature. Crystals grown in the NH_4SO_4 condition did not diffract better than 3.25 Å even after extensive optimization, so we focused solely on the $NaNO_3$-dependent crystals of tcTERT plus ssDNA.

2.3 tcTERT Structure Determination

Data collection was carried out at NSLS, beam line X6A. The best crystals grown in $NaNO_3$, diffracted to 2.71 Å resolution, belonged to the orthorhombic space group $P2_12_12_1$ and contained two molecules in the asymmetric unit. Phasing the structure proved challenging. We initially tried preparing the selenomethionine protein for phasing, but the protein expression levels from minimal media growths were disappointingly low. At this point, we decided that metal derivatization of the crystals may be a good alternative. This can be done either by use of a covalent binder, such as a mercury compound that will coordinate free cysteines or by a noncovalent binder, such as NaBr. We found that soaking the crystals with high concentrations of NaBr caused significant damage to the crystals and impaired diffraction. At this point, we took a step back, examined the protein sequence, found that tcTERT consists of 13 cysteines, and therefore the use of mercury derivatization was a possible option. Soaking the crystals with methyl mercury(II) chloride CH_3HgCl or $HgCl_2$ solution (1–10 m*M*) even for a few minutes resulted in poorly diffracting crystals suggesting that a gentler approach was needed for effective derivatization. To address this issue, we decided to try the following two approaches: the first involved the addition of a very small amount of solid CH_3HgCl into the crystallization drop containing the crystals of interest. The idea behind this approach was to allow for the relatively insoluble mercury to slowly seep into the crystals, thus allowing for slow and specific derivatization of solvent-accessible cysteines without damaging the crystal lattice. The second approach involved the addition of 5 m*M* $MeHgCl_2$ solution incrementally (0.5 m*M* every 5 min). We found that the second method of slowly adding CH_3HgCl to the crystals resulted in derivatized crystals that diffracted to 3.5 Å resolution.

We calculated phases using the method of single isomorphous replacement with anomalous signal (SIRAS). More specifically, we used one native and two isomorphous CH_3HgCl derivative datasets collected at wavelengths Hg1—1.00850 Å and Hg2—1.00800 Å. We first used SOLVE (https://solve.lanl.gov) (Terwilliger, 2000; Terwilliger & Berendzen, 1999) to identify the high occupancy sites (12 heavy atom sites were located), which we subsequently refined and used to calculate new phases with MLPHARE (Collaborative Computational Project No. 4, 1994). We then used the MLPHARE improved phases to calculate an anomalous difference map, which allowed us to identify the remaining heavy atom sites (22 in total) to 3.5 Å resolution. MLPHARE phases obtained using all the heavy atom sites where then used in DM with twofold NCS and phase extension using the high-resolution native data (2.71 Å) to calculate starting experimental maps and build the model.

The structure revealed a ring configuration that closely resembles the polymerase domain of HIV RTs and viral RNA polymerases (Gillis et al., 2008). A striking difference between TERT and RTs as well as viral RNA polymerases is the presence of the N-terminal extension, which comprises the TRBD and is unique to telomerase. The structure reveals that this domain adopts an almost all-helical structure that makes extensive interactions with the C-terminal (thumb domain) extension, which facilitates the formation of the closed ring configuration observed in the TERT structure. Subsequent EM studies of the *T. thermophila* and human TERT (Jiang et al., 2013; Sauerwald et al., 2013) showed that the structure of TERT is conserved across species.

3. CRYSTAL STRUCTURE DETERMINATION OF A PARTIAL TELOMERASE ELONGATION COMPLEX

3.1 Nucleic Acid Substrate Design

We then asked how TERT assembles with TER and the telomeric overhang to form an active elongation complex. While we have the tcTERT protein in large quantities, the RNA component (tcTER) of *T. castaneum* telomerase is still unknown. For this reason, we decided to determine the structure of tcTERT in complex with the RNA template and telomeric DNA. As we mentioned earlier, the TER template is usually 1.5 telomeric repeats. In the case of *T. castaneum* telomerase, the telomeric repeat is TCAGG, so the putative RNA template should be (5′-rCrUrGrArCrCrU-3′) (Mitchell et al., 2010b). We designed the RNA–DNA hairpin so that the RNA and DNA strands are linked together for stability with a stable tetraloop (TTCG). We also designed the hairpin to contain a three-nucleotide overhang (5′-rCrUrGrArCrCrG**TTCG**AGGT-3′) so that we can (a) test the enzymatic activity of tcTERT in the presence of the hairpin and (b) trap the enzyme in its catalytic state by cocrystallizing with dNTPs and Mg^{2+} ions.

3.2 Activity Assays

To ensure tcTERT assembles with the designed RNA–DNA hairpin in a productive manner, we tested its activity in the presence of the hairpin. Since the hairpin does not contain the full-length TER required for repeat addition processivity, we only expected the addition of three nucleotides to the end of the telomeric DNA associated with the hairpin. We carried out a standard reverse transcriptase assay in a buffer containing 50 m*M* Tris–HCl, 100 m*M* KCl, 1.25 m*M* $MgCl_2$, 5 m*M* DTT, 5% (w/v) glycerol, pH 8, 100 μ*M* dNTPs (dATP, dTTP, and dGTP), 10 μ*M* [^{32}P]dCTP (80 Ci/mmol), 5 μ*M* RNA–DNA hairpin, and 1 μ*M* recombinant TERT. The results clearly indicated the enzyme is indeed active under these conditions and allowed us to move forward in attempting to capture its catalytic state.

3.3 Complex Crystallization

Trapping telomerase in its catalytic state can be challenging as the active enzyme can hydrolyze the dNTPs during either the sample preparation or crystallization. To avoid this issue, we purchased nonhydrolyzable dNTPs (dNTPαS—Jenna Biosciences) and prepared the sample by mixing the protein–RNA–DNA hairpin at a 1:1.2 ratio in the presence of 5 m*M* dNTPαs and 5 m*M* $MgCl_2$.

Crystallization of tcTERT in the presence 5′-rCrUrGrArCrCrU**TTCG**AGGT-3′, dNTPαS, and $MgCl^{2+}$ was carried out at room temperature and 4°C using the commercially available sparse matrix conditions we used for the apo-tcTERT. Co-crystallization of tcTERT with the above hairpin, dNTPαS and Mg^{2+} did not produce any diffracting crystals, so we decided to modify the length of the linker that connects the RNA template with the telomeric DNA to facilitate crystallization. To this end, we designed two additional hairpins consisting of the RNA template (5′-rCrUrGrArCrCrU-3′), the complementary telomeric DNA (5′-AGGT-3′), the tetraloop (5′-**TTCG-3**′), and two or six additional nucleotides (Fig. 2). Of all three hairpins only the one with the additional six nucleotides (highlighted in bold-italic, 5′-rCrUrGrArCrCrG***rGrArC*****TTCG*****GTC***AGGT-3′) produced crystals amenable to structural studies. Interestingly, several crystal hits appeared within a week, however, all of them with the exception of one gave the rod-like crystals we observed previously with the tcTERT protein alone, suggesting that these crystals did not contain nucleic acid. We confirmed the absence of RNA–DNA in these crystals by dissolving them and running them on an agarose gel in the presence of ethidium bromide. Inspection of the crystallization conditions revealed that crystals that did not contain the hairpin had high salt concentration as a precipitant, which suggested that the contacts between the protein and the nucleic acid were mostly hydrophilic in nature and involved the backbone of the RNA–DNA hairpin.

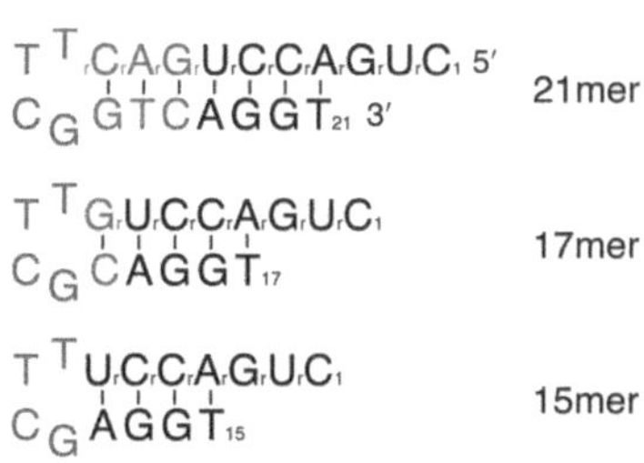

Fig. 2 Primary structure and sequence of the RNA–DNA hairpins used in cocrystallization trials with tcTERT. The RNA template and complementary telomeric DNA are shown in *black letters*; the linker in *blue* and the stable tetraloop in *red*. (See the color plate.)

Crystals containing the hairpin grew in 0.1 *M* HEPES (pH 7.5), 0.2 *M* KCl, and 12% 1,6-hexanediol or PEG 4K. Under these crystallization conditions, small crystals appeared within a couple of days and grew to useful diffracting size (200 × 50 × 20 μm) within 2 weeks. For diffraction studies, we transferred these crystals into cryoprotectant containing 0.1 *M* HEPES (pH 7.5), 15% (w/v) 1,6-hexanediol or PEG 4K, 15% (w/v) glycerol, 0.2 *M* KCl, and 1 m*M* TCEP and harvested them by flash-freezing in liquid nitrogen. The crystals ultimately diffracted to 2.7 Å resolution and belong to the monoclinic space group $P2_1$.

3.4 Structure Determination

We determined the structure using the method of molecular replacement and the apo-tcTERT (PDB ID: 3DU6) structure as a search model. The structure revealed one molecule in the asymmetric unit with clear density for the nucleic acid in the interior cavity of the TERT ring. Close inspection of the nucleic acid density revealed three additional nucleotides at the 3′-end of the telomeric DNA (Fig. 3), which suggested that the active tcTERT had extended the DNA in the crystallization drop. The nucleotide located at the 3′-end of the DNA occupied the active site of the enzyme thus providing partial evidence of the mechanism of dNTP addition by telomerase to the end of our chromosomes.

This structure revealed that during telomere elongation TERT binds the RNA template and telomeric DNA in a similar fashion to that of HIV RTs revealing new insights into the mechanism of telomere elongation by telomerase.

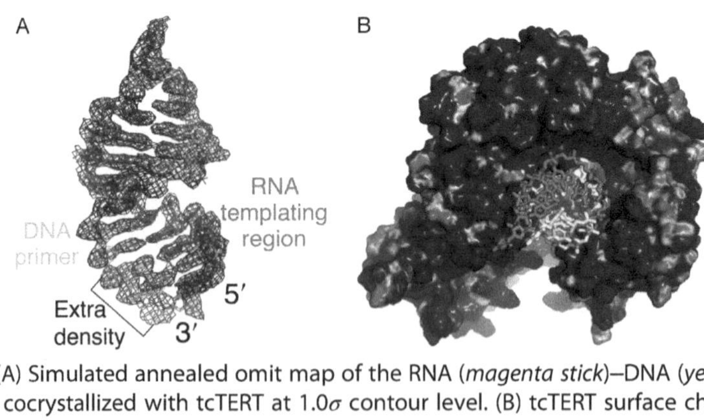

Fig. 3 (A) Simulated annealed omit map of the RNA (*magenta stick*)–DNA (*yellow stick*) hairpin cocrystallized with tcTERT at 1.0σ contour level. (B) tcTERT surface charge representation, showing the RNA–DNA hybrid (*stick*) docked in the interior cavity of the TERT ring. (See the color plate.)

4. CONCLUSIONS

The main obstacle to obtaining the high-resolution structure of telomerase has been the isolation of the active TERT in large quantities required for crystallographic studies. Screening a large number of TERT genes from different organisms paid off in identifying a TERT gene, which lacked most of the nonconserved regions of the protein. However, several other factors contributed to the successful completion of this project. We found that using a synthetic TERT gene, codon-optimized for bacterial overexpression and identifying the appropriate cell strain for protein overexpression were essential to the overall success of this project.

Another finding of note was that using fresh protein and avoiding flash-frozen samples was critical for effective crystallization with this being especially true for the protein and RNA samples, while DNA samples appeared relatively stable. Not only was the design of a substrate important, but an understanding of the biochemistry of the assembly in general is indispensable to the success of any project.

These experiments provide a view of the catalytic subunit of telomerase alone and in complex with its RNA template and telomeric DNA. Telomerase has evolved over time, which explains the diversity observed both in the protein and the RNA component of telomerase across species. It is clear that the core TERT enzyme is highly conserved across species as are the RNA motifs implicated in telomerase-dependent telomere replication. The information obtained from these studies has been instrumental in the interpretation and design of many experiments and has shaped the way we view telomerase in the field.

ACKNOWLEDGMENTS

The research was funded by the NIGMS (5 R01 GM088332) and The Wistar Cancer Center Support Grant (P30 CA10815).

REFERENCES

Bourns, B. D., et al. (1998). Sir proteins, Rif proteins, and Cdc13p bind Saccharomyces telomeres in vivo. *Molecular and Cellular Biology*, *18*(9), 5600–5608.

Bryan, C., et al. (2013). Structure of the human telomeric Stn1-Ten1 capping complex. *PloS One*, *8*(6), e66756.

Calado, R. T., & Young, N. S. (2009). Telomere diseases. *The New England Journal of Medicine*, *361*(24), 2353–2365.

Chen, L. Y., Redon, S., & Lingner, J. (2012). The human CST complex is a terminator of telomerase activity. *Nature*, *488*, 540–544.

Collaborative Computational Project No. 4. (1994). The CCP4 suite: Programs for protein crystallography. *Acta Crystallographica. Section D, Biological Crystallography, 50*(Pt. 5), 760–763.

de Lange, T. (2005). Shelterin: The protein complex that shapes and safeguards human telomeres. *Genes & Development, 19*(18), 2100–2110.

Gillis, A. J., Schuller, A. P., & Skordalakes, E. (2008). Structure of the *Tribolium castaneum* telomerase catalytic subunit TERT. *Nature, 455*(7213), 633–637.

Grandin, N., Damon, C., & Charbonneau, M. (2000). Cdc13 cooperates with the yeast Ku proteins and Stn1 to regulate telomerase recruitment. *Molecular and Cellular Biology, 20*(22), 8397–8408.

Grandin, N., Damon, C., & Charbonneau, M. (2001). Ten1 functions in telomere end protection and length regulation in association with Stn1 and Cdc13. *The EMBO Journal, 20*(5), 1173–1183.

Grandin, N., Reed, S. I., & Charbonneau, M. (1997). Stn1, a new *Saccharomyces cerevisiae* protein, is implicated in telomere size regulation in association with Cdc13. *Genes & Development, 11*(4), 512–527.

Greetham, M., et al. (2015). The telomere binding protein Cdc13 and the single-stranded DNA binding protein RPA protect telomeric DNA from resection by exonucleases. *Journal of Molecular Biology, 427*, 3023–3030.

Hayflick, L. (1965). The limited in vitro lifetime of human diploid cell strains. *Experimental Cell Research, 37*, 614–636.

Jacobs, S. A., Podell, E. R., & Cech, T. R. (2006). Crystal structure of the essential N-terminal domain of telomerase reverse transcriptase. *Nature Structural & Molecular Biology, 13*(3), 218–225.

Jiang, J., et al. (2013). The architecture of Tetrahymena telomerase holoenzyme. *Nature, 496*(7444), 187–192.

Kim, N. W., et al. (1994). Specific association of human telomerase activity with immortal cells and cancer. *Science, 266*(5193), 2011–2015.

Lee, M. S., & Blackburn, E. H. (1993). Sequence-specific DNA primer effects on telomerase polymerization activity. *Molecular and Cellular Biology, 13*(10), 6586–6599.

Lin, J. J., & Zakian, V. A. (1996). The Saccharomyces CDC13 protein is a single-strand TG1-3 telomeric DNA-binding protein in vitro that affects telomere behavior in vivo. *Proceedings of the National Academy of Sciences of the United States of America, 93*(24), 13760–13765.

Lingner, J., Hendrick, L. L., & Cech, T. R. (1994). Telomerase RNAs of different ciliates have a common secondary structure and a permuted template. *Genes & Development, 8*(16), 1984–1998.

Mason, M., & Skordalakes, E. (2010). Insights into Cdc13 dependent telomere length regulation. *Aging (Albany NY), 2*(10), 731–734.

Mason, M., et al. (2013). Cdc13 OB2 dimerization required for productive Stn1 binding and efficient telomere maintenance. *Structure, 21*(1), 109–120.

Mitchell, J. R., Wood, E., & Collins, K. (1999). A telomerase component is defective in the human disease dyskeratosis congenita. *Nature, 402*(6761), 551–555.

Mitchell, M., et al. (2010a). Cdc13 N-terminal dimerization, DNA binding, and telomere length regulation. *Molecular and Cellular Biology, 30*(22), 5325–5334.

Mitchell, M., et al. (2010b). Structural basis for telomerase catalytic subunit TERT binding to RNA template and telomeric DNA. *Nature Structural & Molecular Biology, 17*(4), 513–518.

Mitton-Fry, R. M., et al. (2004). Structural basis for telomeric single-stranded DNA recognition by yeast Cdc13. *Journal of Molecular Biology, 338*(2), 241–255.

Miyake, Y., et al. (2009). RPA-like mammalian Ctc1-Stn1-Ten1 complex binds to single-stranded DNA and protects telomeres independently of the Pot1 pathway. *Molecular Cell, 36*(2), 193–206.

Nugent, C. I., et al. (1996). Cdc13p: A single-strand telomeric DNA-binding protein with a dual role in yeast telomere maintenance. *Science*, *274*(5285), 249–252.

Osanai, M., et al. (2006). Identification and characterization of the telomerase reverse transcriptase of *Bombyx mori* (silkworm) and *Tribolium castaneum* (flour beetle). *Gene*, *376*(2), 281–289.

Pennock, E., Buckley, K., & Lundblad, V. (2001). Cdc13 delivers separate complexes to the telomere for end protection and replication. *Cell*, *104*(3), 387–396.

Qi, H., & Zakian, V. (2000). The Saccharomyces telomere-binding protein Cdc13p interacts with both the catalytic subunit of DNA polymerase alpha and the telomerase-associated est1 protein. *Genes & Development*, *14*(14), 1777–1788.

Qi, X., et al. (2013). The common ancestral core of vertebrate and fungal telomerase RNAs. *Nucleic Acids Research*, *41*(1), 450–462.

Robart, A. R., & Collins, K. (2011). Human telomerase domain interactions capture DNA for TEN domain-dependent processive elongation. *Molecular Cell*, *42*(3), 308–318.

Rouda, S., & Skordalakes, E. (2007). Structure of the RNA-binding domain of telomerase: Implications for RNA recognition and binding. *Structure*, *15*(11), 1403–1412.

Sauerwald, A., et al. (2013). Structure of active dimeric human telomerase. *Nature Structural & Molecular Biology*, *20*(4), 454–460.

Schmidt, J. C., Dalby, A. B., & Cech, T. R. (2014). Identification of human TERT elements necessary for telomerase recruitment to telomeres. *eLife*, *3*.

Shay, J. W., & Wright, W. E. (1999). Mutant dyskerin ends relationship with telomerase. *Science*, *286*(5448), 2284–2285.

Shippen-Lentz, D., & Blackburn, E. H. (1990). Functional evidence for an RNA template in telomerase. *Science*, *247*(4942), 546–552.

Sun, J., et al. (2009). Stn1-Ten1 is an Rpa2-Rpa3-like complex at telomeres. *Genes & Development*, *23*(24), 2900–2914.

Tejera, A. M., et al. (2010). TPP1 is required for TERT recruitment, telomere elongation during nuclear reprogramming, and normal skin development in mice. *Developmental Cell*, *18*(5), 775–789.

Terwilliger, T. C. (2000). Maximum-likelihood density modification. *Acta Crystallographica. Section D, Biological Crystallography*, *56*(Pt. 8), 965–972.

Terwilliger, T. C., & Berendzen, J. (1999). Automated MAD and MIR structure solution. *Acta Crystallographica. Section D, Biological Crystallography*, *55*(Pt. 4), 849–861.

Theimer, C. A., & Feigon, J. (2006). Structure and function of telomerase RNA. *Current Opinion in Structural Biology*, *16*(3), 307–318.

Vulliamy, T., et al. (2002). Association between aplastic anaemia and mutations in telomerase RNA. *Lancet*, *359*(9324), 2168–2170.

Wan, B., et al. (2015). The Tetrahymena telomerase p75-p45-p19 subcomplex is a unique CST complex. *Nature Structural & Molecular Biology*, *22*, 1023–1026.

Wang, F., et al. (2007). The POT1-TPP1 telomere complex is a telomerase processivity factor. *Nature*, *445*(7127), 506–510.

Wilkins, M. R., et al. (1999). Protein identification and analysis tools in the ExPASy server. *Methods in Molecular Biology*, *112*, 531–552.

Witkin, K. L., & Collins, K. (2004). Holoenzyme proteins required for the physiological assembly and activity of telomerase. *Genes & Development*, *18*(10), 1107–1118.

Ye, A. J., & Romero, D. P. (2002). Phylogenetic relationships amongst tetrahymenine ciliates inferred by a comparison of telomerase RNAs. *International Journal of Systematic and Evolutionary Microbiology*, *52*(Pt. 6), 2297–2302.

CHAPTER NINETEEN

Detection and Analysis of Long Noncoding RNAs

F. Lai, E. Blumenthal, R. Shiekhattar[1]
Sylvester Comprehensive Cancer Center, University of Miami Miller School of Medicine, Miami, FL, United States
[1]Corresponding author: e-mail address: rshiekhattar@med.miami.edu

Contents

1. Introduction 422
2. Traditional Methods for lncRNA Detection 423
3. Genome-Wide Detection of lncRNAs Through Direct RNA Sequencing Approaches 423
   3.1 Total RNA Sequencing 423
   3.2 Chromatin-Associated RNA Sequencing 425
   3.3 Mapping Transcription Start Sites by Direct Cap Sequencing 427
4. Indirect Sequencing Methods Coupled with Transcription 428
   4.1 Nascent Transcript Global Run-On Sequencing 428
   4.2 Precision Nuclear Run-On and Sequencing Assay (PRO-seq) 434
5. Protein-Based Approaches for Genomic Characterization of lncRNAs 435
   5.1 RNA Immunoprecipitation and Sequencing (RIP-seq) 435
   5.2 Photoactivatable Ribonucleoside-Enhanced Cross-Linking and Immunoprecipitation 435
   5.3 Individual Nucleotide Resolution CLIP 440
6. Future Directions and Conclusion 441
References 442

Abstract

Recent genome-wide studies have yielded new insights into the biological function of long noncoding RNAs (lncRNAs), predominantly through analysis of their genomic addresses. These studies have revealed that a large number of lncRNAs map to regulatory elements in eukaryotic genome regions known as promoter and enhancer elements. Here, we review the principles of current methodologies for analyzing lncRNAs with high-throughput sequencing approaches. These include (1) direct RNA sequencing, (2) sequencing coupled with transcription, and (3) isolation of protein complexes associated with lncRNAs followed by high-throughput sequencing. Within these categories, we also describe detailed protocols for chromatin-associated RNA sequencing, nascent transcript Global run-on sequencing, and photoactivatable ribonucleoside-enhanced cross-linking and immunoprecipitation.

Methods in Enzymology, Volume 573
ISSN 0076-6879
http://dx.doi.org/10.1016/bs.mie.2016.03.010

1. INTRODUCTION

Beyond the abundant classes of noncoding RNAs, transfer RNAs (tRNAs), and ribosomal RNAs (rRNAs), a large number of messenger RNA-like noncoding transcripts have been described in a wide variety of organisms. The first few functional noncoding RNAs were identified between the 1970s and 1990s, including RNase P (Stark, Kole, Bowman, & Altman, 1978), snRNAs (Yang, Lerner, Steitz, & Flint, 1981), 7SL (Walter & Blobel, 1982), H19 (Brannan, Dees, Ingram, & Tilghman, 1990), Malat1, and Xist (Brockdorff et al., 1992; Brown et al., 1992; Ji et al., 2003). Despite the early identifications of quite a few noncoding RNAs, their prominence was not fully recognized until the discovery of small noncoding RNAs (miRNAs and siRNAs) in the last 20 years.

However, the greatest impact in the expanding world of noncoding RNAs has come from the emergence of high-throughput sequencing technologies, also called next-generation sequencing (NGS) technologies. Numerous transcripts without protein-coding potential have been identified in nearly all species. To distinguish these transcripts from previously known small noncoding RNAs (mi/siRNAs), long noncoding RNAs (lncRNAs) have been defined as RNA species of longer than 200 nucleotides. Nearly 75% of the human genome has been shown to encode RNA transcripts and a large number of these transcripts are classified as lncRNAs (Bertone et al., 2004; Carninci et al., 2005; Djebali et al., 2012; Encode Project Consortium et al., 2007; Okazaki et al., 2002; Ota et al., 2004).

Physical and functional classification of lncRNAs has proven very challenging. We favor classification based on genomic addresses of lncRNAs. In this context, a large number of lncRNAs map to promoters of protein-coding genes in the antisense direction or localize in a bidirectional manner to distal regulatory elements known as enhancers (Sigova et al., 2013; Wilusz, Sunwoo, & Spector, 2009). For the lncRNAs involved in transcriptional regulation, many of them repress or activate mRNA expression in either a *cis* or *trans* mode of action (Ponting, Oliver, & Reik, 2009). Multiple lines of evidence describe lncRNAs acting as decoys, scaffold, or guides for transcription regulators or epigenetic markers (Wang & Chang, 2011). Over the past few years, new discoveries suggest that similar to protein-coding genes, lncRNAs have a broad range of functions and specific expression patterns. To clearly understand their biological function, it is critical to dissect these transcripts with appropriate methods. Here we summarize different approaches for identifying lncRNAs. The main emphasis of this chapter is to focus on current high-throughput techniques for detecting these transcripts at genome-wide levels.

2. TRADITIONAL METHODS FOR lncRNA DETECTION

The traditional methods for the detection of lncRNAs were PCR-based amplification methods or direct Northern blot analysis. These approaches and their modified versions are widely used for both coding and noncoding RNAs. To precisely determine the expression level of lncRNAs, reverse transcription polymerase chain reaction (RT-PCR) approaches have higher sensitivity than Northern blot. However, Northern blot can provide reliable visual evidence for the analysis of transcripts. To their detriment, these methods are often based on the previous knowledge of lncRNA sequences used to target individual lncRNAs. The success of these approaches largely relies on the expression level of the lncRNA transcripts. Fluorescence in situ hybridization is also a commonly used approach for visualizing abundant lncRNAs in cells or tissue samples.

3. GENOME-WIDE DETECTION OF lncRNAs THROUGH DIRECT RNA SEQUENCING APPROACHES

The initial genome-wide studies of transcriptome were based on Sanger sequencing or tiling array technologies. Several Sanger sequencing-based large-scale analyses have already identified thousands of lncRNAs from human and mouse (Carninci et al., 2005; Encode Project Consortium, 2012; Schuler et al., 1996). In the tiling array approaches, a pool of tiled oligonucleotides has been designed against a contiguous region on one chromosome or at whole-genome level. These oligonucleotides are used to probe microarray slides and hybridize cDNA samples for detecting novel transcripts (Guttman et al., 2009; Kapranov et al., 2002; Rinn et al., 2003, 2007). Due to the high rate of weak-binding and nonspecific hybridization events between probes and the cDNA samples, array-based technologies have become less popular for the detection of lncRNAs.

3.1 Total RNA Sequencing

Compared to tiling arrays, high-throughput NGS technologies provide more robust sequencing capabilities with higher read numbers, longer read lengths, and fewer error rates. There are four NGS systems—454, SOLiD,

Ion Torrent, and Illumina sequencing. The basic principle for NGS technologies is to create a library of cDNA from an RNA sample that can be amplified and placed on a two-dimensional surface for the sequencing reaction. Using light-emitting reactions, an image can be constructed from the surface at each cycle. During the analysis of sequential images, specific nucleotides can be determined at these spots. As millions of reactions can be imaged multiple times, the number of reads obtained in a single run can be very large.

The first step of RNA sequencing is to create a DNA library from RNA samples by RNA fragmentation, adapter ligation, cDNA synthesis, size selection, and limited cycles of PCR amplification (Fig. 1). Based on the experimental needs, the library could be based on "single-end" or "paired-end". Single-end libraries allow for the amplification from one end of the transcripts, while paired-end libraries can obtain sequences from the other end of the same molecules. As the paired-end strategy can provide additional information from two directions, it has proven useful for the analysis of splicing events. During the library preparation step, the extra nucleotides of the adapter primer (bar code) will provide information to separate different

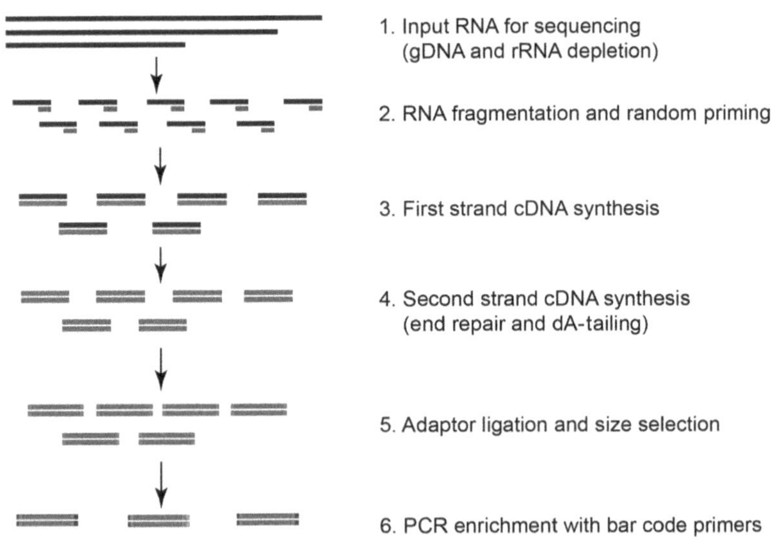

Fig. 1 Principle of RNA sequencing library preparation. Following removal of genomic DNA and ribosomal RNA, the remaining RNA is fragmented, and primers are added for cDNA synthesis. At this stage, polyadenylated transcripts can be enriched using oligo $d(T)_{25}$ magnetic beads and oligo $d(T)_6$ primers in the cDNA synthesis. To select for all RNA species, random hexamer primers are used. The PCR product is then ligated with sequencing adaptors, size-selected, and purified for sequencing. (See the color plate.)

samples in a single run. Thus, libraries from many samples can be mixed and sequenced together to reduce costs. The reads from each sample can be sorted by their unique nucleotide composition (bar code). After the run, the sequenced reads can be mapped to the corresponding genome. The read counts are a measure of the expression level of each individual transcript.

Similar to most mRNAs, many lncRNAs are capped at their 5′-ends, spliced, and polyadenylated at their 3′ ends. However, there are a large number of lncRNAs that do not contain poly (A) tails (Derrien et al., 2012; Yang, Duff, Graveley, Carmichael, & Chen, 2011). To separate these two lncRNA populations, the most convenient way is to use oligo $d(T)_6$ primer during the cDNA synthesis step for the polyadenylated transcripts and random hexamer for total transcripts. An important caveat of this approach is the low annealing specificity between RNA transcripts and RT primers. Whether they are polyadenylated or not, transcripts having stretches of A or adenine-rich regions can be erroneously amplified. Even when using the designed RT primers for a specific transcript, there is still a possibility of annealing with nonspecific transcripts as a result of the low temperature used during the RT step. Thus far, the most reliable approach is to hybridize the RNA sample with oligo $d(T)_{25}$ magnetic beads. The flow-through fraction is the nonpolyadenylated transcripts, and the polyadenylated RNA is attached to the beads. Both RNAs can be adapted into different RNA-seq protocols for sequencing.

Similar to mRNA sequencing, lncRNA transcripts can also be sequenced from both ends. The advantage of sequencing from the ends is to identify the boundaries of transcription units, especially for alternative start and polyadenylation sites. Instead of an RNA fragmentation step, paired-end technologies (PETs) typically use a type IIS restriction endonuclease to cut cDNA a certain distance away from the restriction sites introduced by adaptor sequences. The short DNA fragments can be concatenated for efficient sequencing analysis. There are a few different PET technologies, including RNA-PET for transcriptome, ChIP-PET for transcription and epigenetic regulators, ChIA-PET (Chromatin Interaction Analysis) for chromatin structure, and DNA-PET for genome assembly (Fullwood, Wei, Liu, & Ruan, 2009).

3.2 Chromatin-Associated RNA Sequencing

Within the standard high-throughput sequencing methods, 20–40 million reads is adequate for detecting transcripts. In many cases, low expression of lncRNAs may hinder the detection of these transcripts. There is a tendency to generate more reads per library to capture the low-abundant transcripts. Due to the technical limitations associated with ligation efficiency, only minor increases in read numbers can be generated from low-abundant transcripts. High number of total reads may result in the biased creation of more reads for the highly expressed transcripts, but not for transcripts that are expressed at lower levels. To overcome this issue, the practical approach is to fractionate the RNA samples into cytoplasmic, nuclear, and chromatin fractions. Each individual fraction can then be sequenced with the standard RNA sequencing protocols. Utilizing this approach, the low-abundant lncRNA transcripts from each fraction can be easily detected with 20–40 million reads per sample. The following fractionation protocol was initially established in Stephen Smale's laboratory, which we present following minor modifications (Bhatt et al., 2012):

1. 20–30 million cells are resuspend in 200–400 μL cytoplasmic lysis buffer (0.15% NP-40, 10 m*M* Tris pH 7.5, 150 m*M* NaCl) using wide orifice tips and incubated on ice for 10 min;
2. The lysate is layered on 500 μL cold sucrose buffer (10 m*M* Tris pH 7.5, 150 m*M* NaCl, 24% sucrose w/v) and centrifuged at 13,000 rpm for 10 min at 4°C. The supernatant is the cytoplasmic fraction;
3. The nuclear pellet is gently resuspend into 250 μL cold glycerol buffer (20 m*M* Tris pH 7.9, 75 m*M* NaCl, 0.5 m*M* EDTA, 50% glycerol, 0.85 m*M* DTT) using wide orifice tips;
4. The same volume of cold nuclei lysis buffer (20 m*M* HEPES pH 7.6, 7.5 m*M* $MgCl_2$, 0.2 m*M* EDTA, 0.3 *M* NaCl, 1 *M* urea, 1% NP-40, 1 m*M* DTT) is mixed with the samples;
5. The nuclear sample is quickly vortexed until the solution is mixed well and incubated for 1 min on ice;
6. The sample is centrifuged at 14,000 rpm for 2 min at 4°C;
7. The supernatant is the nucleoplasmic fraction and the pellet is the chromatin fraction;
8. The chromatin pellet is resuspend with 50 μL cold 1 × PBS by vortex.

All the fractions can be suspended into QIAGEN RLT buffer or directly performed with Trizol or RNA extraction (RNeasy Mini kit, QIAGEN; Trizol, ThermoFisher, USA). The chromatin fraction will be extremely viscous and appear sticky. This fraction is resuspending using Trizol and chloroform followed by vortexing. This protocol is easy to use and especially beneficial for the detection of chromatin-associated lncRNAs such as enhancer RNAs by RNA sequencing.

3.3 Mapping Transcription Start Sites by Direct Cap Sequencing

In addition to direct sequencing of lncRNA transcripts, cap analysis of gene expression (CAGE) is a complementary high-throughput technology that allows for the precise mapping of lncRNA transcription start sites (TSSs). Instead of sequencing the entire transcripts, CAGE-seq identifies RNA transcripts from their capped 5′ end. Similar to RNA-seq, CAGE-seq can map and quantify the expression of lncRNAs by sequencing from TSSs. However, as CAGE captures the 5′ end of capped transcripts, the RNAs without a 5′ cap cannot be identified, for example, Pol III transcripts, IRES-dependent transcripts, and circular RNAs.

To perform CAGE sequencing, cDNAs are synthesized from total RNA samples by using random or oligo dT primers. The 5′ end of cDNAs is selected by a cap-trapper method. Then a biotinylated linker is attached to the 5′ end of the cDNA. This linker contains recognition sites for endonuclease. After the second strand cDNA synthesis, a short fragment (20 or 27 nucleotides) will be cut from the 5′ end. Then the 3′ linker can be attached to the 3′ end for PCR amplification and sequencing (Fig. 2).

1. First strand cDNA synthesis: cDNAs are first synthesized from total RNA samples by using random or oligo dT primers. The reverse primer contains the EcoP15I sequence. The reverse transcriptase without RNase H enzymatic activity is the most important step to avoid

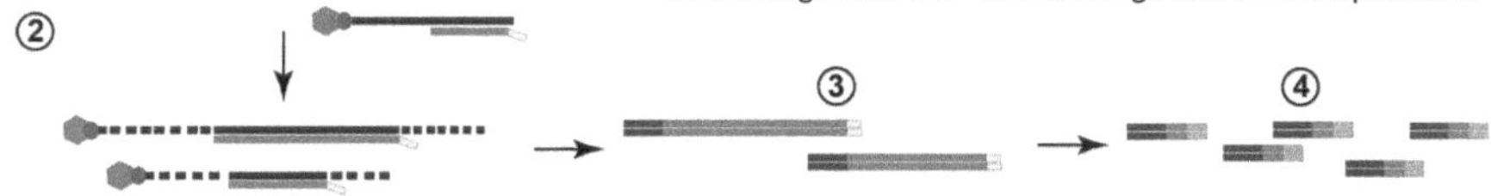

Fig. 2 Overview of CAGE RNA library preparation. First strand cDNA synthesis is performed with random hexamer or oligo d(T) primers containing EcoP15I digestion site. Reverse transcriptase without RNase H activity must be used. 5′ Caps are then biotinylated and captured using strepatividin beads. The resultant RNA is subjected to RNase I digestion and purification. Complete second strand cDNA synthesis with modified 5′ linkers and cleave double-stranded cDNA by EcoP15I. After ligation with 3′ linker, the cDNA library is purified with streptavidin beads. (See the color plate.)

generating nicks on the RNA strand. Random primers (RT-N15-EcoP) are used for capturing all RNAs.

2. Cap trapping: chemical biotinylation and capture of complete cDNAs. After purification of cDNA/RNA from short fragments (<100 nt) and chemical oxidation with $NaIO_4$, the RNA caps are biotinylated. Followed by RNase I treatment and multiple purification steps, biotinylated caps are captured by streptavidin-coated magnetic beads, and 5′ complete cDNAs are enriched for the next step.
3. The second strand cDNA synthesis: A phosphate- and NH2-modified 5′ linkers are ligated to 5′ complete cDNAs. This linker is used for generation of the second strand.
4. Cleavage of the 27 nt CAGE tags and PCR amplification: Before the restriction digestion, the cDNA is treated with Antarctic phosphatase to remove the phosphate modification at the 5′ end of the 5′ linker. The second strand cDNA is then cleaved by EcoP15I, which cleaves 27 nt downstream of the enzyme recognition site. After ligated with 3′ linker, the final tagged double-stranded cDNA is purified again with streptavidin beads. The desired PCR product is 96 bp long. The PCR cycle number normally is under 15 cycles.

4. INDIRECT SEQUENCING METHODS COUPLED WITH TRANSCRIPTION

Like mRNA transcripts, lncRNA transcripts can be identified by tracking transcriptional events. Instead of measuring steady-state levels of lncRNAs by conventional RNA sequencing approaches, it is possible to measure RNA polymerase II (RNAPII)-derived nascent transcripts produced during transcriptional elongation. Analysis of nascent transcription allows for additional mechanistic insights into the transcriptional cycle as the nascent transcript represents the dynamic movements of RNAPII molecules.

4.1 Nascent Transcript Global Run-On Sequencing

The original intent of global run-on sequencing (GRO-seq) was to identify/distinguish transcriptionally engaged RNAPII (transcriptional elongation step) from the paused RNAPII (Core, Waterfall, & Lis, 2008). In principle, there are different stages of RNAPII at individual genomic loci in the cells. Low temperature induces polymerase pausing and nascent transcripts can be reinitiated by brief incubation at 37°C with nucleoside triphosphates

(NTPs) and Bromouridine-triphosphate (BrUTP). As the newly synthesized transcripts incorporate BrUTP, they can be purified and identified by high-throughput sequencing (Fig. 3). Due to its high sensitivity and resolution for sequencing nascent transcript genome-wide, GRO-seq has been used for the identification of transcriptionally active lncRNAs.

GRO-seq protocol

1. Extraction of nuclei:
   - Wash cells with cold 1 × PBS twice;
   - Scrape cells into 50-mL Falcon tubes and add 10 mL cold swelling buffer for 5 min;
   - Spin at 400 × *g* for 10 min;
   - Remove supernatant and resuspend cell pellet in 10 mL swelling buffer with 10% glycerol;
   - Add an equal amount of fresh swelling buffer with 10% glycerol and 1% NP-40;

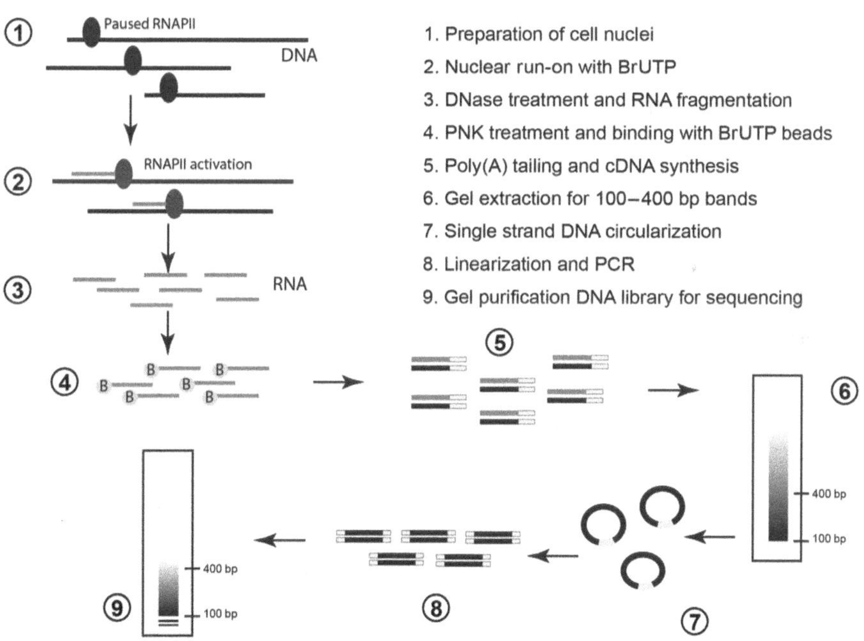

Fig. 3 Steps of GRO-seq RNA library preparation. Nuclear run-on is performed by incubating prechilled nuclei at 37°C with NTPs and BrUTP. Following DNAse treatment, RNA fragmentation, and purification, nascent transcripts are treated with PNK and isolated with BrUTP beads. Addition of poly(A) tails provides a sequence for reverse primers used during cDNA synthesis. 100–400 base pair transcripts are then extracted, circularized, linearized, and amplified prior to sequencing. (See the color plate.)

- Incubate on ice for 10 min and spin at 600 × *g* for 5 min;
- Remove supernatant and resuspend pellet in 10 mL freezing buffer;
- Centrifuge at 900 × *g* for 6 min and resuspend pellet in freezing buffer at 10–15 million cells per 100 μL freezing buffer.

40–60 million cells are needed per library. The nuclei can be stored in −80°C for a few months. Before using, SUPERase In RNase Inhibitor is added to all buffers at 2 U/mL.

Swelling buffer: 10 m*M* Tris–HCl pH 7.5, 2 m*M* $MgCl_2$, 3 m*M* $CaCl_2$;

Freezing buffer: 50 m*M* Tris–HCl pH 7.5, 0.1 m*M* EDTA, 5 m*M* $MgCl_2$, 40% glycerol.

2. Nuclear run-on reaction:
- Mix 100 μL nuclear extract with 100 μL 2 × nuclear run-on buffer (NRO buffer) with wide-orifice tips;
- Incubate the reaction at 30°C for 7.5 min;
- Add 400 μL Trizol and 100 μL chloroform for RNA extraction;
- Shake on vortex for 15 s and incubate for 5 min at RT;
- Centrifuge at 16,000 × *g* for 15 min at 4°C;
- Take the supernatant to a new eppendorf tube and precipitate with 2 volumes of ethanol and 2 μL of glycogen;
- Incubate at −20°C a few hours or −80°C overnight;
- Centrifuge at 16,000 × *g* for 15 min at 4°C;
- Wash the pellet with 1 mL 75% EtOH;
- Centrifuge at 16,000 × *g* for 10 min at 4°C, carefully remove the supernatant, and air dry the pellet;
- Resuspend pellet in 20 μL of UltraPure water (DNAse, RNAse free);
- Incubate at 60°C for 10 min.

2 × NRO buffer: 10 m*M* Tris–HCl pH 8.0, 5 m*M* $MgCl_2$, 300 m*M* KCl, 1 m*M* DTT, 500 μ*M* ATP, 500 μ*M* GTP, 500 μ*M* BrUTP, 2 μ*M* CTP, 200 U/mL Superase-In, 1% Sarkosyl (*N*-lauroylsarcosine sodium salt solution).

3. DNAse treatment (TURBO DNA-free Kit, Ambion AM1907):
- Add 1.4 μL of DNAse buffer and 2 μL DNase with 10 μL of RNA sample;
- Incubate at 37°C for 30 min;
- Stop the reaction with 2 μL DNAse inactivation reagent;
- Incubate for 5 min at RT;
- Centrifuge at 10,000 × *g* for 2 min;
- Transfer the supernatant to a new tube.

4. RNA fragmentation (Ambion fragmentation kit AM8740):
   - Add 4 μL water to 5 μL of RNA sample in a PCR tube;
   - Add 1 μL of fragmentation reagent to each sample;
   - Incubate at 70°C for 13 min;
   - Add 1 μL of stop solution and put on ice;
   - Add water to a final volume of 20–25 μL.

   RNA Purification is carried out through a P-30 RNAse free spin column (Bio-Rad #732-6250).
5. PNK treatment:
   - Add 1 μL of Superase, 3 μL of 10 × PNK buffer, and 1.5 μL PNK;
   - Incubate at 37°C for 60 min;
   - Add additional 1 μL of PNK and 3 μL 100 m*M* EDTA and incubate at 37°C for another 60 min;
   - Incubate at 70°C for 5 min to inactive PNK;
   - Incubate the samples on ice for 2 min and combine two of the same samples;
   - Bring up to 200 μL with binding buffer.

   Binding buffer: 0.25 × SSPE, 37.5 m*M* NaCl, 1 m*M* EDTA, 0.05% Tween-20.
6. Binding with BrUTP beads:
   - Wash 50 μL beads with 500 μL of binding buffer twice at RT;
   - Block beads with 250 μL of blocking buffer and rotate at RT for 60 min;
   - Wash beads with 500 μL of binding buffer twice at RT;
   - Resuspend beads in 500 μL of binding buffer and mix with 200 μL of PNK-treated RNA;
   - Rotate at RT for 60 min;
   - Wash with 500 μL binding buffer twice;
   - Wash with 500 μL low salt buffer twice;
   - Wash with 500 μL high salt buffer;
   - Wash with 500 μL TET buffer twice;
   - Elute RNA with 100 μL of elution buffer (heated at 42°C) four times, each time incubate 10 min at RT;
   - Precipitate RNA with 300 m*M* NaCl, 2 μL glycogen, and 1 mL ethanol;
   - Incubate at −20°C for a few hours or −80°C overnight;
   - Centrifuge at 16,000 × *g* for 15 min at 4°C;
   - Wash the pellet with cold 75% ethanol;
   - Centrifuge again at 16,000 × *g* at 4°C;

- Resuspend the pellet in 5 μL UltraPure water with 0.05% Tween-20.

Blocking buffer: Binding buffer with 0.1% polyvinylpyrrolidone, 0.1% BSA;

Low salt wash buffer: 0.2× SSPE, 1 m*M* EDTA, 0.05% Tween-20;

High salt wash buffer: 137.5 m*M* NaCl, 0.2× SSPE, 1 m*M* EDTA, 0.05% Tween-20;

Elution buffer: 150 m*M* NaCl, 50 m*M* Tris–HCl pH 7.5, 0.1% SDS, 1 m*M* EDTA, fresh 20 m*M* DTT;

TET buffer: 10 m*M* Tris–HCl pH 7.5, 1 m*M* EDTA, 0.05% Tween-20.

7. Poly(A) tailing reaction (Epicentre, PAP5104H):
   - 5 μL of RNA incubated for 30 min at 37°C with the following in the kit:
     a. 0.7 μL of 10× poly(A) polymerase buffer;
     b. 0.25 μL of 4 m*M* ATP;
     c. 1 μL poly(A) polymerase (4 U);
8. cDNA synthesis (Superscript III kit, ThermoFisher Scientific #18080-051):
   - Add 0.9 μL of dNTP mix (10 m*M* each) and 0.9 μL of oNTI223 primer (5′-pGATCGTCGGACTGTAGAACTCT;CAAGCAGAAGACGGCATACGATTTTTTTTTTTTTTTTTTTTVN-3′);
   - Heat at 75°C for 3 min;
   - Chill on ice;
   - Add the following to the reaction:
     a. 1.7 μL of 10× RT buffer;
     b. 3 μL of 25 m*M* $MgCl_2$;
     c. 1.7 μL 100 m*M* DTT;
     d. 0.5 μL Superase (20 U/μL);
     e. 1 μL Superscipt IIIRT (200 U/μL);
   - Incubate at 48°C for 40 min;
   - Add 3 μL exonuclease I (20 U/μL);
   - Incubate at 37°C for 15 min;
   - Add 2 μL of 1 *M* NaOH;
   - Incubate at 98°C for 20 min;
   - Neutralize the reaction with 1 μL of 2 *M* HCl.
9. Gel extraction with 10% TBE-urea gel:
   - Add 1 volume of Novex 2× sample buffer to the cDNA samples;

- Denature samples at 70°C for 3 min and chill on ice;
- Load samples with 1 empty lane between each samples;
- Run gel at 170 V with low range ssRNA ladder;
- Stain gel with 25 mL SYBER-GOLD for 3 min and visualize over light box;
- Extract all transcripts between 100 and 400 bp in size;
  a. Puncture the bottom of a sterile, nuclease free 0.5-ml tube multiple times with a 21-g needle;
  b. Place the 0.5-mL tube into 1.5-mL tube;
  c. Cut the slices of gel into 0.5-mL tube;
  d. Spin at 14,000 × *g* for 3 min.
- Add 500 μL TE with 0.1% Tween-20 to gel debris;
- Rotate 4 h at RT;
- Transfer elute and gel debris to a Millipore MC column;
- Spin at 12,000 × *g* for 2 min;
- Add 2 μL glycogen, 30 μL 5 *M* NaAc, and 1 mL cold ethanol to precipitate at −80°C overnight;
- Centrifuge at 16,000 × *g* for 15 min at 4°C;
- Carefully discard the supernatant;
- Wash the pellet with 1 mL cold 75% ethanol;
- Centrifuge for 10 min at 4°C;
- Remove the supernatant and dry the cDNA pellet;
- Resuspend into 7.5 μL water.

10. Circularization with ssDNA ligase (CircLigase™ ssDNA Ligase, Epicentre CL4111K):
- For each reaction, add the following:
  a. 1 μL of 10 × reaction buffer;
  b. 0.5 μL of 1 m*M* ATP;
  c. 0.5 μL of 50 m*M* $MnCl_2$;
  d. 0.5 μL CircLigase (100 U/μL).
- Incubate at 60°C for 60 min;
- Heat inactivate samples at 80°C for 10 min.

11. Relinearization with APE1:
- Add 3.5 μL relinearization buffer (100 m*M* KCl, 2 m*M* DTT) and 1 μL APE1;
- Incubate at 37°C for 45 min;
- Heat inactivate samples at 65°C for 20 min;
- Dilute the volume to 30 μL with water.

12. PCR amplification:
- Each PCR reaction should contain:
  a. 5 μL of cDNA;
  b. 2.5 μL oNTI200 (10 μ*M*);
  c. 2.5 μL oNTI201 (10 μ*M*);
  d. 10 μL of 5 × HF buffer;
  e. 1 μL of dNTP (10 m*M*);
  f. 5 μL Betaine;
  g. 0.5 μL DNA polymerase;
  h. 23.5 μL water.
- PCR program:
  a. 98°C for 30 s;
  b. 98°C for 10 s;
  c. 57°C for 30 s;
  d. 72°C for 15 s and back to b step for 19 cycles;
  e. 72°C for 5 min and hold at 4°C.

13. Gel extraction from 10% polyacrylamide TBE gel:
- Purify and concentrate cDNA with ChIP DNA Clean and Concentrator kit (ZYMO research D5201);
- Elute to 15 μL for loading into one lane;
- Run samples at 120 V for 2 h;
- Gel extract 150–300 bp PCR product;
- Elute DNA with 400 μL elution buffer (TE with 0.1% Tween-20 and 150 m*M* NaCl);
- Precipitate DNA with ethanol overnight at −80°C;
- Resuspend DNA in 10–20 μL TE with 0.05% Tween-20;
- DNA samples are ready for sequencing.

4.2 Precision Nuclear Run-On and Sequencing Assay (PRO-seq)

Recently, GRO-seq methods have been improved to monitor nascent RNA transcripts with single-nucleotide resolution (Kwak, Fuda, Core, & Lis, 2013). These techniques are particularly powerful for identifying regulatory lncRNAs for their function in transcriptionally engaged RNAPII at TSS sites, splicing events and 3′ polyadenylation sites. This technique enriches for nascent short RNAs, offering the advantage of a high number of reads at lncRNA sites, especially at enhancer RNA loci.

5. PROTEIN-BASED APPROACHES FOR GENOMIC CHARACTERIZATION OF lncRNAs

An important strategy, which can shed further light into the biological function of lncRNAs, is the analysis of lncRNAs' interacting protein partners. There are a number of established methods that have been successfully applied to the analysis of lncRNA–protein interactions. Understanding the protein complexes that associate with lncRNAs provides an important avenue for further functional analysis of lncRNAs (Kretz et al., 2013; Lai et al., 2013; Rinn et al., 2007; Tsai et al., 2010; Zhao, Sun, Erwin, Song, & Lee, 2008).

5.1 RNA Immunoprecipitation and Sequencing (RIP-seq)

Interaction between RNA and protein partners can be analyzed following UVC cross-linking of RNA–protein complexes and treatment with nucleases to minimize indirect interaction. The cross-linked RNA–rotein complexes are lysed and isolated via immunoprecipitation. After the purification of lncRNA transcripts, the specific lncRNAs can be detected by ligating the lncRNA with linkers for RT-PCR or NGS technologies. The combination of UV-irradiated RIP followed by purification of the cross-linked protein complexes after fractionation on an SDS-PAGE is known as CLIP (UV cross-linking and immunoprecipitation (Ule, Jensen, Mele, & Darnell, 2005; Ule et al., 2003; Zhao et al., 2008). To combine CLIP with high-throughput technologies, cross-linking and immunoprecipitation sequencing (CLIP-seq) is used to identify protein associated RNAs genome-wide. This method is also known as HITS-CLIP (High-throughput sequencing of Clip cDNA library) (Licatalosi et al., 2008).

5.2 Photoactivatable Ribonucleoside-Enhanced Cross-Linking and Immunoprecipitation

Photoactivatable ribonucleoside-enhanced cross-linking and immunoprecipitation (PAR-CLIP) is a method that utilizes photoactivatable nucleoside 4-thiouridine (4SU) or 6-thioguanosine (6SG) incorporation into the nascent RNA transcripts (Hafner et al., 2010a). It is a modified approach from HITS-CLIP that improves cross-linking efficiency and shows a higher nucleotide resolution along with decreased signal-to-noise ratio. In this method, photoreactive ribonucleoside analogs (4SU or 6SG) are applied to living cells to facilitate stronger cross-linking between RNAs and proteins (Fig. 4). The photoreactive analogs undergo a structural change in response

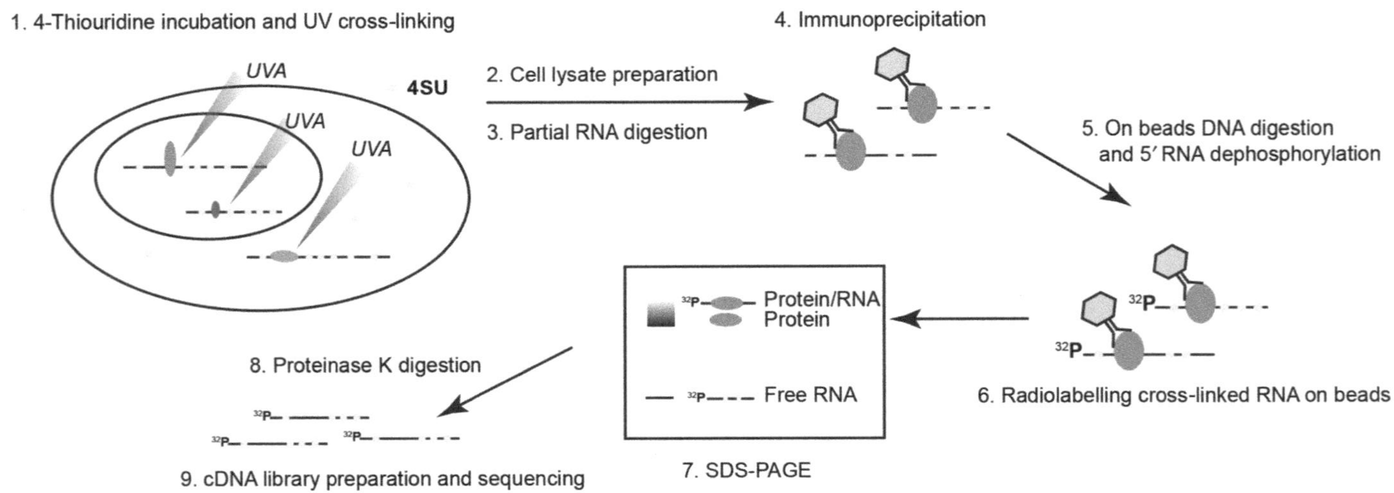

Fig. 4 PAR-CLIP workflow. Incubating cells with 4-thiouridine and UV radiation efficiently cross-links RNA to proteins. Cells are lysed, and the extracted RNA is partially digested with RNase. The protein of interest is immunoprecipitated and pulled down using magnetic beads. DNA digestion, 5′ RNA dephosphorylation, and radiolabeling occur on the magnetic beads. Separation of protein-bound RNA from free RNA is achieved with SDS-PAGE electrophoresis. Proteinase K digestion isolates radiolabeled RNA from its immunoprecipitated target for analysis by sequencing. (See the color plate.)

to cross-linking with ultraviolet A light (wavelength 365 nm) and develop specific sequence mutations: T to C in 4SU, and G to A in 6SG (Spitzer et al., 2014).

PAR-CLIP protocol

1. Incubate cells with 4SU and carry out UV cross-linking:
   - Add 4-SU to cell growth media to achieve 100 μ*M* final concentration;
   - Incubate 14–16 h at 37°C;
   - Wash cells with cold 15 mL PBS twice;
   - Aspirate PBS;
   - Cross-link cells on a Stratalinker 2400 at 365 nm UV light, 150 mJ/cm^2;
   - Scrape cells in 10 mL cold PBS with 2 m*M* EDTA and 0.2 m*M* P*M*SF;
   - Centrifuge at 400 × *g* for 5 min to pellet cells;

     Cell pellet can be stored at -80°C for a few months.
2. Cell lysate preparation:
   - Thaw the cross-linked cell pellet on ice;
   - Resuspend cell pellet in two volumes of CLIP lysis buffer with RNase inhibitor (1:200) and Turbo DNase (1:200);
   - Sonicate lysate with bioruptor for 5 min at 4°C, 30 s on/off to disrupt chromatin;
   - Incubate the lysate for 10 min on ice;
   - Centrifuge at 20,000 × *g* for 30 min at 4°C and transfer the clear lysate to a clean tube;
   - Clear the cell lysate further by filtering through a 5-μm membrane syringe filter;
   - Add CLIP buffer to reduce the concentration of SDS to 0.25%;
   - Retain 100 μL for an input sample;

   CLIP buffer: 150 m*M* KCl, 5 m*M* EDTA, 2% EMPIGEN, 0.1% Na-deoxycholate, 0.5% *N*-lauroylsarcosine, 10 m*M* Na-butyrate, 2 m*M* DTT (fresh), proteinase inhibitors (fresh).

   CLIP lysis buffer: 150 m*M* KCl, 5 m*M* EDTA, 20 m*M* HEPES pH 8.0, 2% EMPIGEN, 0.1% Na-deoxycholate, 0.5% *N*-lauroylsarcosine, 0.5% SDS, 2 m*M* DTT (fresh), proteinase inhibitors (fresh), 0.2 U/μL murine RNase Inhibitor, 0.02 U/μL Turbo DNase I.
3. Initial RNase digestion:
   - Add 0.02 U/μL TURBO DNase to every sample;
   - Add RNase cocktail (Ambion, AM2286) to the cell lysate (1:4000);

- Incubate RNase-treated samples at RT for 5 min with low vortex speed (900 rpm).

4. Preparation of magnetic beads:
   - Aliquot 15 μL of Protein A or G dynabeads per mL cell lysate to an eppendorf tube;
   - Wash beads three times with cold PBS, 1% BSA;
   - Add 500 μL of 0.1% BSA and 2 μg of primary antibody (for 15 μL of beads);
   - A corresponding amount of IgG can be added into the control samples;
   - Incubate at 4°C for 1 h with rotation;
   - Dynabeads are washed once with CLIP buffer and resuspended into the original volume.
5. Immunoprecipitation:
   - Add 20 μL of freshly prepared antibody-conjugated beads to the RNase-treated samples;
   - Incubate for 2 h at 4°C;
   - Wash beads with 1 mL CLIP wash buffer twice at 4°C;
   - Wash beads with 1 mL IP50 buffer twice at 4°C;
   - Resuspend beads with CLIP buffer;
   - Add RNase cocktail (Ambion, AM2286) to the samples (RNase A 0.01 U/μL, RNase T1 0.4 U/μL);
   - Incubate for 15 min at 37°C, 900 rpm and cool on ice for 5 min;
   - Aliquot 10 μL of samples as a control for the size of protein during immunoblotting.

   CLIP wash buffer: 400 m*M* KCl, 5 m*M* EDTA, 0.5 m*M* EGTA, 2% EMPIGEN, 0.7% Na-deoxycholate, 0.5% *N*-lauroylsarcosine, 2 *M* urea, 2 m*M* DTT (fresh), proteinase inhibitors (fresh).

   IP50 buffer: 50 m*M* KCl, 20 m*M* Tris–HCl pH 8.0, 0.2 m*M* EDTA, 10% glycerol, 2 m*M* DTT (fresh), proteinase inhibitors (fresh).
6. TURBO DNase digestion:
   - Resuspend beads in 43 μL of IP50 buffer;
   - Add 5 μL 10 × Turbo DNase buffer and 2 μL of TURBO DNase (Ambion, AM1907);
   - Incubate for 15 min at 37°C with low vortex speed (900 rpm);
   - Wash beads with 1 mL cold CLIP wash buffer twice;
   - Wash beads once with 1 mL cold IP50 buffer.
7. Dephosphorylate ends of RNA on beads:
   - Resuspend beads in 40 μL IP50 buffer;

- Add 5 μL 10 × cutsmart buffer;
- Add recombinant Shrimp Alkaline Phosphatase (rSAP) to a final concentration of 0.1 U/μL and resuspend beads by pipetting;
- Incubate at 37°C for 15 min with 900 rpm rotation;
- Wash beads with 1 mL phosphatase wash buffer twice;
- Wash beads with 1 mL T4 PNK buffer (without DTT) twice and resuspend in 47 μL T4 PNK buffer (with DTT).

  Phosphatase wash buffer: 400 m*M* KCl, 20 m*M* Tris–HCl pH 8.0, 0.05% NP-40, 10 m*M* EDTA, 10 m*M* EGTA, 2 m*M* DTT (fresh), proteinase inhibitors (fresh).

  T4 PNK buffer: 50 m*M* NaCl, 10 m*M* $MgCl_2$, 50 m*M* Tris–HCl pH 7.5, 5 m*M* DTT (fresh).

8. Radiolabelling cross-linked RNA:
- Add 2 μL (20U) of T4 PNK;
- Add 1 μL of 10 uCi/gamma-^{32}P-ATP to the reaction;
- Incubate for 30 min at 37°C with 900 rpm rotation;
- Add 100 μ*M* nonradioactive ATP and incubate for another 5 min at 37°C;
- Wash beads with 1 mL PNK buffer without DTT twice;
- Wash beads with 1 mL CLIP wash buffer twice;
- Wash with 1 mL IP50 buffer twice;
- Collect beads on the magnetic rack.

9. SDS-PAGE and transfer:
- Add 5 μL LDS loading buffer to 20 μL input samples and heat at 65°C for 15 min;
- Resuspend beads in 60 μL 1 × LDS loading buffer and heat at 65°C for 15 min;
- Load 30 μL per sample into two adjacent wells on a Novex 10-well 1.5 mm 3–8% Tris–acetate gel;
- Run gel at 100 V for 2 h in the cold room until the loading dye reaches the bottom of the gel;
- Disassemble the gel chamber and cut away all the front dye at the bottom of the gel;
- For the control samples: transfer gel on a 0.2 μ*M* PVDF membrane overnight at 30 V in the cold room and blot with the specific protein antibody.

10. Imaging and recovering the cross-linked RNA–protein complex:
- Wash gel with distilled water to remove the remaining buffer;

- Wrap gel with a plastic wrap and expose to a phosphor screen for 30 min to 1 h;
- Align the gel with the scanned image to cut out the bands that correspond to the expected protein size;
- Transfer to a D-Tube Dialyzer Midi Tube and add 800 μL 1 × SDS running buffer;
- Electro-elute the cross-linked RNA at 100 V for 2 h;
- Reverse the current for 2 min to release any protein attached on the membrane;
- Transfer the solution into two siliconized tubes according to the manufacturer instructions.

11. Proteinase K digestion:

- Add 1 volume of 2 × proteinase K buffer and Proteinase K (Roche) to a final concentration of 1.2 mg/mL;
- Incubate at 55°C for 30 min;
- Recover the RNA by acidic phenol/chloroform/IAA extraction;
- Chloroform extraction with equal volume of chloroform;
- Add 1/10 volume of 3 *M* NaAc, 2 μL of glycogen (10 mg/mL stock) to precipitate the RNA with 3 volumes of ethanol;
- Dissolve the pellet in 10 μL of water.

12. cDNA library preparation and high-throughput sequencing.

The recovered RNA is carried through a standard cDNA library preparation protocol for cloning of small regulatory RNAs. The RNA fragments are ligated with different Solexa sequencing adapters at 5′ and 3′. After the cDNA library preparation and PCR amplification, the right size of PCR product is excised from a 2.5% low melt agarose gel. After the gel purification step, the purified cDNA is ready for sequencing. The detailed protocol refers to papers from the Thomas Tuschl laboratory (Hafner et al., 2010a, 2010b).

5.3 Individual Nucleotide Resolution CLIP

Similar to HITS-CLIP and PAR-CLIP, individual nucleotide resolution cross-linking and immunoprecipitation (iCLIP) can identify protein-binding sites at single-nucleotide resolution (Konig et al., 2011). In this method, Protein–RNA complexes are UV cross-linked, partial RNase digested, and immunoprecipitated with specific protein antibody. After 3′ linker ligation, Protein–RNA complexes can be purified and subjected to proteinase K digestion. As proteinase K leaves cross-linked polypeptide together with the RNA fragment, the truncated cDNA can be generated during the reverse transcription step. By comparing truncated products against read-through products, it is possible to determine protein-binding sites at single-nucleotide resolution. The detailed protocol can be seen in papers from Jernej Ule's laboratory (Huppertz et al., 2014; Konig et al., 2011).

6. FUTURE DIRECTIONS AND CONCLUSION

A large body of literature has pointed to regulatory roles for lncRNAs in epigenetic regulation and gene expression. In this chapter, we focused on the principles to detect and analyze lncRNAs using high-throughput approaches. We characterize these NGS approaches based on capturing the steady state of lncRNA transcripts, or analysis of lncRNA nascent transcripts. We also described methods that could provide further insight through analysis of lncRNAs protein interaction networks. Additionally, there are emerging technologies and new tools that have been utilized for investigating lncRNAs. Particularly, the recent advances in the CRISPR/Cas9 system have been suggested as a powerful tool for the analysis of the function of lncRNAs. The nuclease-deficient Cas9 mutant has also been used as a RNA-binding partner to detect lncRNA interaction within specific chromatin regions (O'Connell et al., 2014).

Current studies mainly focus on the expression patterns for lncRNA detection. To further address the molecular bases of lncRNA function, information on lncRNA structure and functional motifs becomes critical, especially for the recognition of lncRNA with RNA-binding partners or interacting with specific chromatin regions. By using short antisense biotinylated oligonucleotides tiling along the target lncRNA, RNA antisense purification, capture hybridization analysis of RNA targets, and chromatin isolation by RNA purification (ChIRP) methods have been developed for the identification of lncRNA and DNA interactions using genome-wide sequencing approaches (Chu, Qu, Zhong, Artandi, & Chang, 2011; Engreitz et al., 2014; Simon et al., 2011). These approaches can also be optimized for the identification of lncRNA-binding proteins (Chu et al., 2015). The advantage of these methods is to circumvent the secondary structure of lncRNAs by using short affinity-tagged oligonucleotides. Such high-throughput sequencing technologies have generated large amounts of data from a variety of biological samples. As these methods mature, there will be a greater need for development of bioinformatics pipelines for analysis of such large data sets. Importantly, it would become critical to integrate experimental results using a variety of these approaches to arrive at a cohesive understanding of biological function and mechanism of action of different classes of lncRNAs.

REFERENCES

Bertone, P., Stolc, V., Royce, T. E., Rozowsky, J. S., Urban, A. E., Zhu, X., … Snyder, M. (2004). Global identification of human transcribed sequences with genome tiling arrays. *Science*, *306*(5705), 2242–2246. http://dx.doi.org/10.1126/science.1103388.

Bhatt, D. M., Pandya-Jones, A., Tong, A. J., Barozzi, I., Lissner, M. M., Natoli, G., … Smale, S. T. (2012). Transcript dynamics of proinflammatory genes revealed by sequence analysis of subcellular RNA fractions. *Cell*, *150*(2), 279–290. http://dx.doi.org/10.1016/j.cell.2012.05.043.

Brannan, C. I., Dees, E. C., Ingram, R. S., & Tilghman, S. M. (1990). The product of the H19 gene may function as an RNA. *Molecular and Cellular Biology*, *10*(1), 28–36.

Brockdorff, N., Ashworth, A., Kay, G. F., McCabe, V. M., Norris, D. P., Cooper, P. J., … Rastan, S. (1992). The product of the mouse Xist gene is a 15 kb inactive X-specific transcript containing no conserved ORF and located in the nucleus. *Cell*, *71*(3), 515–526.

Brown, C. J., Hendrich, B. D., Rupert, J. L., Lafreniere, R. G., Xing, Y., Lawrence, J., & Willard, H. F. (1992). The human XIST gene: Analysis of a 17 kb inactive X-specific RNA that contains conserved repeats and is highly localized within the nucleus. *Cell*, *71*(3), 527–542.

Carninci, P., Kasukawa, T., Katayama, S., Gough, J., Frith, M. C., Maeda, N., … Genome Science, Group. (2005). The transcriptional landscape of the mammalian genome. *Science*, *309*(5740), 1559–1563. http://dx.doi.org/10.1126/science.1112014.

Chu, C., Qu, K., Zhong, F. L., Artandi, S. E., & Chang, H. Y. (2011). Genomic maps of long noncoding RNA occupancy reveal principles of RNA-chromatin interactions. *Molecular Cell*, *44*(4), 667–678. http://dx.doi.org/10.1016/j.molcel.2011.08.027.

Chu, C., Zhang, Q. C., da Rocha, S. T., Flynn, R. A., Bharadwaj, M., Calabrese, J. M., … Chang, H. Y. (2015). Systematic discovery of Xist RNA binding proteins. *Cell*, *161*(2), 404–416. http://dx.doi.org/10.1016/j.cell.2015.03.025.

Core, L. J., Waterfall, J. J., & Lis, J. T. (2008). Nascent RNA sequencing reveals widespread pausing and divergent initiation at human promoters. *Science*, *322*(5909), 1845–1848. http://dx.doi.org/10.1126/science.1162228.

Derrien, T., Johnson, R., Bussotti, G., Tanzer, A., Djebali, S., Tilgner, H., … Guigo, R. (2012). The GENCODE v7 catalog of human long noncoding RNAs: Analysis of their gene structure, evolution, and expression. *Genome Research*, *22*(9), 1775–1789. http://dx.doi.org/10.1101/gr.132159.111.

Djebali, S., Davis, C. A., Merkel, A., Dobin, A., Lassmann, T., Mortazavi, A., … Gingeras, T. R. (2012). Landscape of transcription in human cells. *Nature*, *489*(7414), 101–108. http://dx.doi.org/10.1038/nature11233.

Encode Project Consortium. (2012). An integrated encyclopedia of DNA elements in the human genome. *Nature*, *489*(7414), 57–74. http://dx.doi.org/10.1038/nature11247.

Encode Project Consortium, Birney, E., Stamatoyannopoulos, J. A., Dutta, A., Guigo, R., Gingeras, T. R., … de Jong, P. J. (2007). Identification and analysis of functional elements in 1% of the human genome by the ENCODE pilot project. *Nature*, *447*(7146), 799–816. http://dx.doi.org/10.1038/nature05874.

Engreitz, J. M., Sirokman, K., McDonel, P., Shishkin, A. A., Surka, C., Russell, P., … Lander, E. S. (2014). RNA-RNA interactions enable specific targeting of noncoding RNAs to nascent Pre-mRNAs and chromatin sites. *Cell*, *159*(1), 188–199. http://dx.doi.org/10.1016/j.cell.2014.08.018.

Fullwood, M. J., Wei, C. L., Liu, E. T., & Ruan, Y. (2009). Next-generation DNA sequencing of paired-end tags (PET) for transcriptome and genome analyses. *Genome Research*, *19*(4), 521–532. http://dx.doi.org/10.1101/gr.074906.107.

Guttman, M., Amit, I., Garber, M., French, C., Lin, M. F., Feldser, D., … Lander, E. S. (2009). Chromatin signature reveals over a thousand highly conserved large non-coding RNAs in mammals. *Nature*, *458*(7235), 223–227. http://dx.doi.org/10.1038/nature07672.

Hafner, M., Landthaler, M., Burger, L., Khorshid, M., Hausser, J., Berninger, P., … Tuschl, T. (2010a). Transcriptome-wide identification of RNA-binding protein and microRNA target sites by PAR-CLIP. *Cell*, *141*(1), 129–141. http://dx.doi.org/10.1016/j.cell.2010.03.009.

Hafner, M., Landthaler, M., Burger, L., Khorshid, M., Hausser, J., Berninger, P., … Tuschl, T. (2010b). PAR-CliP—A method to identify transcriptome-wide the binding sites of RNA binding proteins. *Journal of Visualized Experiments*, (41). http://dx.doi.org/10.3791/2034.

Huppertz, I., Attig, J., D'Ambrogio, A., Easton, L. E., Sibley, C. R., Sugimoto, Y., … Ule, J. (2014). iCLIP: Protein-RNA interactions at nucleotide resolution. *Methods*, *65*(3), 274–287. http://dx.doi.org/10.1016/j.ymeth.2013.10.011.

Ji, P., Diederichs, S., Wang, W., Boing, S., Metzger, R., Schneider, P. M., … Muller-Tidow, C. (2003). MALAT-1, a novel noncoding RNA, and thymosin beta4 predict metastasis and survival in early-stage non-small cell lung cancer. *Oncogene*, *22*(39), 8031–8041. http://dx.doi.org/10.1038/sj.onc.1206928.

Kapranov, P., Cawley, S. E., Drenkow, J., Bekiranov, S., Strausberg, R. L., Fodor, S. P., & Gingeras, T. R. (2002). Large-scale transcriptional activity in chromosomes 21 and 22. *Science*, *296*(5569), 916–919. http://dx.doi.org/10.1126/science.1068597.

Konig, J., Zarnack, K., Rot, G., Curk, T., Kayikci, M., Zupan, B., … Ule, J. (2011). iCLIP—Transcriptome-wide mapping of protein-RNA interactions with individual nucleotide resolution. *Journal of Visualized Experiments*, (50). http://dx.doi.org/10.3791/2638.

Kretz, M., Siprashvili, Z., Chu, C., Webster, D. E., Zehnder, A., Qu, K., … Khavari, P. A. (2013). Control of somatic tissue differentiation by the long non-coding RNA TINCR. *Nature*, *493*(7431), 231–235. http://dx.doi.org/10.1038/nature11661.

Kwak, H., Fuda, N. J., Core, L. J., & Lis, J. T. (2013). Precise maps of RNA polymerase reveal how promoters direct initiation and pausing. *Science*, *339*(6122), 950–953. http://dx.doi.org/10.1126/science.1229386.

Lai, F., Orom, U. A., Cesaroni, M., Beringer, M., Taatjes, D. J., Blobel, G. A., & Shiekhattar, R. (2013). Activating RNAs associate with Mediator to enhance chromatin architecture and transcription. *Nature*, *494*(7438), 497–501. http://dx.doi.org/10.1038/nature11884.

Licatalosi, D. D., Mele, A., Fak, J. J., Ule, J., Kayikci, M., Chi, S. W., … Darnell, R. B. (2008). HITS-CLIP yields genome-wide insights into brain alternative RNA processing. *Nature*, *456*(7221), 464–469. http://dx.doi.org/10.1038/nature07488.

O'Connell, M. R., Oakes, B. L., Sternberg, S. H., East-Seletsky, A., Kaplan, M., & Doudna, J. A. (2014). Programmable RNA recognition and cleavage by CRISPR/Cas9. *Nature*, *516*(7530), 263–266. http://dx.doi.org/10.1038/nature13769.

Okazaki, Y., Furuno, M., Kasukawa, T., Adachi, J., Bono, H., Kondo, S., … Team, I. I. (2002). Analysis of the mouse transcriptome based on functional annotation of 60,770 full-length cDNAs. *Nature*, *420*(6915), 563–573. http://dx.doi.org/10.1038/nature01266.

Ota, T., Suzuki, Y., Nishikawa, T., Otsuki, T., Sugiyama, T., Irie, R., … Sugano, S. (2004). Complete sequencing and characterization of 21,243 full-length human cDNAs. *Nature Genetics*, *36*(1), 40–45. http://dx.doi.org/10.1038/ng1285.

Ponting, C. P., Oliver, P. L., & Reik, W. (2009). Evolution and functions of long noncoding RNAs. *Cell*, *136*(4), 629–641. http://dx.doi.org/10.1016/j.cell.2009.02.006.

Rinn, J. L., Euskirchen, G., Bertone, P., Martone, R., Luscombe, N. M., Hartman, S., … Snyder, M. (2003). The transcriptional activity of human chromosome 22. *Genes & Development*, *17*(4), 529–540. http://dx.doi.org/10.1101/gad.1055203.

Rinn, J. L., Kertesz, M., Wang, J. K., Squazzo, S. L., Xu, X., Brugmann, S. A., … Chang, H. Y. (2007). Functional demarcation of active and silent chromatin domains in human HOX loci by noncoding RNAs. *Cell*, *129*(7), 1311–1323. http://dx.doi.org/10.1016/j.cell.2007.05.022.

Schuler, G. D., Boguski, M. S., Stewart, E. A., Stein, L. D., Gyapay, G., Rice, K., … Hudson, T. J. (1996). A gene map of the human genome. *Science*, *274*(5287), 540–546.

Sigova, A. A., Mullen, A. C., Molinie, B., Gupta, S., Orlando, D. A., Guenther, M. G., … Young, R. A. (2013). Divergent transcription of long noncoding RNA/mRNA gene pairs in embryonic stem cells. *Proceedings of the National Academy of Sciences of the United States of America*, *110*(8), 2876–2881. http://dx.doi.org/10.1073/pnas.1221904110.

Simon, M. D., Wang, C. I., Kharchenko, P. V., West, J. A., Chapman, B. A., Alekseyenko, A. A., … Kingston, R. E. (2011). The genomic binding sites of a noncoding RNA. *Proceedings of the National Academy of Sciences of the United States of America*, *108*(51), 20497–20502. http://dx.doi.org/10.1073/pnas.1113536108.

Spitzer, J., Hafner, M., Landthaler, M., Ascano, M., Farazi, T., Wardle, G., … Tuschl, T. (2014). PAR-CLIP (photoactivatable ribonucleoside-enhanced crosslinking and immunoprecipitation): A step-by-step protocol to the transcriptome-wide identification of binding sites of RNA-binding proteins. *Methods in Enzymology*, *539*, 113–161. http://dx.doi.org/10.1016/B978-0-12-420120-0.00008-6.

Stark, B. C., Kole, R., Bowman, E. J., & Altman, S. (1978). Ribonuclease P: An enzyme with an essential RNA component. *Proceedings of the National Academy of Sciences of the United States of America*, *75*(8), 3717–3721.

Tsai, M. C., Manor, O., Wan, Y., Mosammaparast, N., Wang, J. K., Lan, F., … Chang, H. Y. (2010). Long noncoding RNA as modular scaffold of histone modification complexes. *Science*, *329*(5992), 689–693. http://dx.doi.org/10.1126/science.1192002.

Ule, J., Jensen, K., Mele, A., & Darnell, R. B. (2005). CLIP: A method for identifying protein-RNA interaction sites in living cells. *Methods*, *37*(4), 376–386. http://dx.doi.org/10.1016/j.ymeth.2005.07.018.

Ule, J., Jensen, K. B., Ruggiu, M., Mele, A., Ule, A., & Darnell, R. B. (2003). CLIP identifies Nova-regulated RNA networks in the brain. *Science*, *302*(5648), 1212–1215. http://dx.doi.org/10.1126/science.1090095.

Walter, P., & Blobel, G. (1982). Signal recognition particle contains a 7S RNA essential for protein translocation across the endoplasmic reticulum. *Nature*, *299*(5885), 691–698.

Wang, K. C., & Chang, H. Y. (2011). Molecular mechanisms of long noncoding RNAs. *Molecular Cell*, *43*(6), 904–914. http://dx.doi.org/10.1016/j.molcel.2011.08.018.

Wilusz, J. E., Sunwoo, H., & Spector, D. L. (2009). Long noncoding RNAs: Functional surprises from the RNA world. *Genes & Development*, *23*(13), 1494–1504. http://dx.doi.org/10.1101/gad.1800909.

Yang, L., Duff, M. O., Graveley, B. R., Carmichael, G. G., & Chen, L. L. (2011). Genomewide characterization of non-polyadenylated RNAs. *Genome Biology*, *12*(2), R16. http://dx.doi.org/10.1186/gb-2011-12-2-r16.

Yang, V. W., Lerner, M. R., Steitz, J. A., & Flint, S. J. (1981). A small nuclear ribonucleoprotein is required for splicing of adenoviral early RNA sequences. *Proceedings of the National Academy of Sciences of the United States of America*, *78*(3), 1371–1375.

Zhao, J., Sun, B. K., Erwin, J. A., Song, J. J., & Lee, J. T. (2008). Polycomb proteins targeted by a short repeat RNA to the mouse X chromosome. *Science*, *322*(5902), 750–756. http://dx.doi.org/10.1126/science.1163045.

CHAPTER TWENTY

Identifying Centromeric RNAs Involved in Histone Dynamics In Vivo

D. Quénet[1,2], D. Sturgill, Y. Dalal[1]
Center for Cancer Research, National Cancer Institute, National Institutes of Health, Bethesda, MD, United States
[1]Corresponding authors: e-mail address: delphine.quenet@med.uvm.edu; dalaly@mail.nih.gov

Contents

1. Introduction 446
2. Materials 446
   2.1 Tissue Cell Culture Elements 447
   2.2 RNA Immunoprecipitation Components 447
   2.3 Immunofluorescence/RNA FISH Buffers 449
3. Methods 450
   3.1 Tissue Cell Culture: Maintenance of HeLa Cells 450
   3.2 RIP-Seq 451
   3.3 Immunofluorescence/RNA FISH 455
   3.4 Computational Analysis 457
Acknowledgments 464
References 464

Abstract

Over the last decade, the long accepted dogma that heterochromatin is silent has been challenged by increasing evidence of active transcription in these apocryphally annotated quiescent regions of the genome. The recent discovery of noncoding RNAs (ncRNAs) originating from, or localizing to, centromeres, pericentromeres, and telomeres (ie, constitutive heterochromatin) suggest a potential role for ncRNAs in genome integrity. This new paradigm suggests that ncRNAs may recruit chromatin-binding factors, stabilize the higher order folded state of the chromatin fiber, and participate in regulation of processes such as transcription-mediated nucleosome assembly. Thus, identifying, purifying, and elucidating the function of ncRNAs has the potential to provide key insights into genome organization and is currently a topic of intense experimental investigation.

[2] Current address: Department of Biochemistry, College of Medicine, University of Vermont, Burlington, VT.

Methods in Enzymology, Volume 573
ISSN 0076-6879
http://dx.doi.org/10.1016/bs.mie.2016.01.010

1. INTRODUCTION

Constitutive heterochromatin, including centromeres, pericentromeres, and telomeres, forms condensed chromatin structure and is associated with transcriptional silencing. In humans, centromeres consist of AT-rich 171 bp α-satellite repeats lacking any known genes (Miga, 2015; Waye & Willard, 1987). This alpha-satellite repeat sequence is not conserved across species; however, centromeres do play a conserved function in guiding correct chromosome segregation during mitosis, suggesting an epigenetic basis for centromere identity (Verdaasdonk & Bloom, 2011). Despite the heterogeneity of α-satellite repeats, recent studies suggest that transcription at centromeres is essential for maintaining centromere structure and function (Chan & Wong, 2012; Hall, Mitchell, & O'Neill, 2012).

In a recent study, we reported on a potential role for cell-cycle specific centromeric transcription and the resulting transcripts in the maintenance of centromere integrity in HeLa cells (Quenet & Dalal, 2014). Our data supported the hypothesis that centromeric noncoding RNAs (ncRNAs) may be involved in the targeting of centromeric factors to the centromere, including the histone variant CENP-A. By their nature, centromeric transcripts are presumed to be repetitive, and consequently may originate from regions underrepresented in their entirety in the current database of the human genome project (http://www.ncbi.nlm.nih.gov/genome/guide/human/). Therefore, we adapted protocols described in the literature to better characterize centromeric RNAs. In this chapter, we provide a detailed description of tools currently being used by others and us, in the investigation of ncRNAs associated with chromatin from cultured cells. We also include suggestions for controls, as well as cautionary notes for the interpretation of data generated by such studies. In Sections 3.2 and 3.3, we describe strategies for performing chromatin-associated RNA immunoprecipitation sequencing (RIP-Seq) and immunofluorescence/RNA FISH. In Section 3.4, we discuss bioinformatics approaches to analyze RIP-Seq data.

2. MATERIALS

RNAs are easily deteriorated by nucleases and heat. To protect samples from degradation, extremely cautious laboratory practices are essential, including a dedicated "RNA" bench decontaminated with nuclease-inhibitor solutions like RNase AWAY (#10328-011, Thermo Fisher

Scientific/Ambion, Grand Island, NY) or equivalent, wearing clean gloves and lab coat, and covering up hair if necessary. Buffers must be prepared fresh with pure and nuclease-free reagents in nuclease-free reaction tubes using decontaminated pipettes and nuclease-free filter tips. Samples must be kept at 4°C or on ice as much as possible.

2.1 Tissue Cell Culture Elements

2.1.1. Human HeLa cell line (ATCC®, Manassas, VA).

2.1.2. 10× Phosphate-buffered saline, pH 7.4 (PBS, #70011-044, Thermo Fisher Scientific/Gibco, Grand Island, NY).

2.1.3. Trypsin (#25300, Thermo Fisher Scientific/Gibco).

2.1.4. Cell growth medium: Dulbecco/Vogt modified Eagle's minimal essential medium (DMEM, #11995-081, Thermo Fisher Scientific/Gibco); 10% fetal bovine serum (FBS, #10437-028, Thermo Fisher Scientific/Gibco); and 1% penicillin/streptomycin (#15140-163, Thermo Fisher Scientific/Gibco).

2.2 RNA Immunoprecipitation Components

2.2.1. 1× PBS; 0.1% Tween® 20 (#P2287, Sigma-Aldrich, St Louis, MO).

2.2.2. Formaldehyde (#15680, Electron Microscopy Sciences, Hatfield, PA).

2.2.3. Glycine (#50046, Sigma-Aldrich).

Note: A 1 *M* solution is prepared in RNase–DNase free water (#AM9932, Thermo Fisher Scientific/Ambion), filtered under a tissue culture hood using a 50 mL sterile filter unit (#09-741-88, Nalgene, Cleveland, OH), and stored at 4°C.

2.2.4. TM2: 20 m*M* Tris–HCl, pH 8.0 (#15568-025, Thermo Fisher Scientific); 2 m*M* magnesium chloride ($MgCl_2$, #AM9530G, Thermo Fisher Scientific); 0.5 m*M* phenylmethylsulfonyl fluoride (PMSF, #78830, Sigma-Aldrich); 1 × complete protease inhibitor cocktail (#05892953001, Roche, Indianapolis, IN).

Note: A 200 m*M* stock solution of PMSF diluted in ethanol is prepared in advance and stored at −20°C.

Note: A 50× stock solution of complete protease inhibitor cocktail is prepared by dissolving one tablet in 1 mL RNase–DNase free water and stored for up to 12 weeks at −20°C.

2.2.5. TM2 ± 0.5% Nonidet™ P 40 Substitute (#74385, Sigma-Aldrich).

2.2.6. Ribonucleoside Vanadyl Complex (RVC, #1402, New England Biolabs—NEB, Ipswich, MA).

Note: RVC or other RNase inhibitor is used in order to avoid RNA degradation. RVC is added in buffers which do not contained ethylenediaminetetraacetic acid (EDTA) or ethylene glycol tetraacetic acid (EGTA) (eg, TM2), as they have the property to dissociate and inactivate the complex.

2.2.7. TE: 0.1 *M* sodium chloride (NaCl, #24740-011, Thermo Fisher Scientific); 10 m*M* Tris–HCl, pH 8.0; 0.2 m*M* EGTA (#03777, Sigma-Aldrich).

Note: A 0.5 *M* stock solution of EGTA (pH 7.5) is prepared in RNase–DNase free water, filtered under a tissue culture hood using a 50 mL sterile filter unit, and stored at room temperature.

2.2.8. Micrococcal Nuclease (MNase, #N3755, Sigma-Aldrich).

Note: A stock solution at 0.5 unit/μL is prepared in 10 m*M* Tris–HCl, 50 m*M* NaCl, 1 m*M* EDTA, 50% glycerol, aliquoted, and stored at −20°C. Thawed aliquot is kept at 4°C for up to 1 week.

2.2.9. $CaCl_2$ (#746495, Sigma-Aldrich).

Note: A 100 m*M* stock solution of $CaCl_2$ is prepared in RNase–DNase free water, filtered under a tissue culture hood using a 50 mL sterile filter unit, and stored at room temperature.

2.2.10. EGTA (see step 2.2.7).

2.2.11. Low-salt buffer: 0.5× PBS, 5 m*M* EGTA, 0.5 m*M* PMSF, 1× complete protease inhibitor cocktail.

2.2.12. Murine RNase inhibitor (#M0314S, NEB).

Note: Murine RNase inhibitor is a replacement for RVC in buffers containing EDTA or EGTA (eg, low-salt buffer).

2.2.13. Protein A/G PLUS agarose beads (#sc-2003, Santa Cruz Biotechnology, Dallas, TX).

2.2.14. Anti-CENP-A antibody (#ab13939, Abcam, Cambridge, MA).

2.2.15. Elution buffer: 1% sodium dodecyl sulfate (SDS); 0.1 *M* sodium bicarbonate ($NaHCO_3$, #S6014, Sigma-Aldrich).

Note: A 1 *M* stock solution of $NaHCO_3$ is prepared in RNase–DNase free water, filtered under a tissue culture hood using a 50 mL sterile filter unit, and stored at room temperature.

2.2.16. NaCl (#24740-011, Thermo Fisher Scientific).

2.2.17. Proteinase K (#AM2548, Thermo Fisher Scientific/Ambion).

2.2.18. Proteinase K buffer: 10 m*M* Tris–HCl, pH 6.5; 1 m*M* EDTA.

2.2.19. DNaseI (RNase free, #AM2222, Thermo Fisher Scientific/Ambion).

2.2.20. Phenol:chloroform:isoamylalcohol (25:24:1, #77617, Sigma-Aldrich).
2.2.21. Chloroform (#C7559, Sigma-Aldrich).
2.2.22. Sodium acetate 3 *M*, pH 5.2 (#S7899, Sigma-Aldrich).
2.2.23. Absolute Ethanol (#459844, Sigma-Aldrich).

2.3 Immunofluorescence/RNA FISH Buffers

2.3.1. HANKS buffer (#14170112, Thermo Fisher Scientific).
2.3.2. Fixation solution: 4% Paraformaldehyde (#15714, Electron Microscopy Sciences); 1 × PBS; RNase–DNase free water.
2.3.3. 1 × PBS (#10010-023, Thermo Fisher Scientific).
2.3.4. Permeabilization solution: 0.1% Triton X-100 (#T8787, Sigma-Aldrich); 1 × PBS.
2.3.5. IF buffer: 1 × PBS; 1% normal goat serum (#005-000-121, Jackson ImmunoResearch, West Grove, PA); 50 units of murine RNase inhibitor (see Section 2.2; step 2.2.12).
2.3.6. Anti-CENP-A antibody (#ab13939, Abcam, Cambridge, MA).
2.3.7. Goat anti-Mouse IgG (H + L) secondary antibody, Alexa Fluor® 488 conjugate (#A-11001, Thermo Fisher Scientific).
2.3.8. RNase A (#12091-021, Thermo Fisher Scientific).
2.3.9. Wash solution A: 2 × saline-sodium citrate buffer (SSC, #AM9763, Thermo Fisher Scientific/Ambion); 10% deionized formamide (# 15745, Electron Microscopy Sciences); RNase–DNase free water.

Note: Deionized formamide has to be stored at 4°C. Before using it, let the solution warm up at room temperature.

2.3.10. Hybridization solution: 10% dextran sulfate (w:v, #S4030, EMD Millipore, Billerica, MA), 2 × SSC nuclease-free, 10% deionized formamide, RNase–DNase free water.

Note: For a final volume of 10 mL, 1 g of dextran sulfate is dissolved in 7 mL of RNase–DNase water on an end-over rotator for 30–60 min. Then, 1 mL of 20 × SSC and 1 mL of deionized formamide are added, and the volume is completed to 10 mL with RNase–DNase free water. Aliquots of 500 μL are stored at −20°C for future use.

2.3.11. RNA FISH Probes labeled with Quasar dyes (ie, Quasar®570—Cy3™ Replacement or Quasar®670—Cy5™ Replacement) (Biosearch Technologies, Petulama, CA).

Note: The desiccated probe set is redissolved in TE buffer (10 m*M* Tris–HCL, 1 m*M* EDTA, pH 8.0) to obtain a 25 μ*M* stock solution, aliquoted, and stored at −20°C. A 2.5 μ*M* working solution is prepared by diluting 10 times the stock solution.

Note: The robustness of RNA FISH depends critically on probe quality. Several companies offer access to software which can automatically design probes specifically recognizing a transcript of interest (Baker, 2012). For our assays, we used RNA FISH Probes designed by LGC Biosearch Technologies' Stellaris RNA FISH Probe Designer (https://www.biosearchtech.com/support/tools/design-software/stellaris-probe-designer). Probe sets can contain up to 48 labeled oligonucleotides modified with a single fluorophore. The probe set targeting the consensus centromeric sequence (ie, α-satellite repeats) was designed against the 171-bases consensus sequence first described by Waye and Willard (1987). The short α-satellite sequence allows the generation of only 11 probes (5′-gcacacatcacaaagaagtt-3; 5′-ggttcaactctgtgagttga-3′; 5′-gagtgtttcaaaactgctct-3′; 5′-caaagcgctccaaatatcca-3′; 5′-agatattccgtttccaacga-3′; 5′-gagaatgcttctgtctagtt-3′; 5′-gcacacatcacaaagaagtt-3′; 5′-ggttcaactctgtgagttga-3′; 5′-gagtgtttcaaaactgctct-3′; 5′-caaagcgctccaaatatcca-3′; 5′-agatattccgtttccaacga-3′). Chosen control probe set targets human NEAT1 (#SMF-2036-1, Biosearch Technologies).

2.3.12. Hybridization mix: 50 μL of hybridization solution; 2 μL of 2.5 μ*M* probe set (per 18 mm coverslip).

Note: No competitor DNA (eg, sheared salmon sperm DNA, human Cot1, *Escherichia coli* tRNA) was added to the hybridization mix. Competitor DNA is used primarily to decrease background and reduce binding of probes to nonspecific loci. However, if the hybridization is limited to 4 h, these potential issues can be limited. If desired, add 5 μg of sheared salmon sperm DNA and 5 μg of *E. coli* tRNA to the hybridization mix per slide.

2.3.13. Wash solution B: 2× SSC; RNase–DNase free water.

2.3.14. Antifade mounting medium Prolong Gold with DAPI (#P36935, Thermo Fisher Scientific).

3. METHODS

3.1 Tissue Cell Culture: Maintenance of HeLa Cells

HeLa cells are typically harvested twice a week, and a new F175 flask is prepared to maintain the cell line at each passage. Passage numbers should be noted for replicate experiments regardless of which cells in culture are used.

3.1.1. Wash cells two times with sterile 1 × PBS.

3.1.2. Incubate for 2–3 min at 37°C with 1 mL of 0.05% trypsin, or until cells start to be round.

Note: Check under an inverted microscope every minute to assess the extent of cell trypsinization. A long incubation or presence of high levels of trypsin can damage cell membrane, leading to cell death and potential artifacts.

3.1.3. Stop trypsin activity by addition of 10 volumes of cell growth medium.

3.1.4. Resuspend cells very gently, pipetting up and down using a cut-off pipet tip.

3.1.5. Transfer 1 mL of trypsinized cells in a new F175 flask containing 19 mL of cell growth medium.

3.1.6. Mix gently.

3.1.7. Incubate in CO_2 incubator (37°C, 5% CO_2, high relative humidity level).

3.2 RIP-Seq

This protocol was previously described in Quenet and Dalal (2014). Notes are added below to discuss steps that can be modified or where caution should be used.

Day 1

3.2.1. Spread cells in F175 flask using the same protocol as described in Section 3.1.

Note: Five F175 flasks at a final confluency of 80% are used per RIP-Seq. The number of flasks depends on the abundance of the studied histone variants, the efficacy of the immunoprecipitation and the specificity of the antibody.

Day 2

3.2.2. Trypsinize cells with 1 mL trypsin (0.05% final concentration).

3.2.3. Wash cells two times with cold 1 × PBS; 0.1% Tween 20.

3.2.4. Fix cells with 0.1% formaldehyde for 10 min at room temperature.

Note: Formaldehyde fixation allows study of direct and indirect bindings. To analyze direct interactions only, UV-crosslinking is preferred (150–200 mJ/cm^2 at 254 nm using a Stratalinker® UV crosslinker).

3.2.5. Quench by addition of glycine (125 m*M* final concentration).

3.2.6. Wash cells twice with cold 1 × PBS, 0.1% Tween 20. Between washes, collect cells by 5 min centrifugation at 800 rpm at 4°C.

Note: A low speed of centrifugation is preferred, so as not to damage cells.

3.2.7. Isolate nuclei in TM2 buffer complemented with 0.5% Nonidet™ P 40 Substitute and 10 m*M* of RVC.

3.2.8. Wash once with TM2 buffer containing 10 m*M* of RVC.

3.2.9. Collect chromatin by centrifugation at 800 rpm for 5 min at 4°C.

3.2.10. Digest chromatin with MNase in 0.1 *M* TE buffer complemented with 2 m*M* $CaCl_2$.

3.2.11. Stop the MNase action by addition of 10 m*M* EGTA and transfer on ice.

Note: The time of incubation and the MNase concentration depends on the desired size of the chromatin array. In our hands, 2 min and 8 min incubation with 0.2 unit/mL of MNase (Sigma) results in chromatin arrays containing 3–5 and 1–2 nucleosomes, respectively.

3.2.12. Collect chromatin by gentle centrifugation at 800 rpm for 5 min at 4°C.

Note: It is critical that all buffers be pure and nuclease/protease free for all subsequent handling of chromatin. Gentle handling of chromatin can reduce degradation and dissociation, which can be monitored by taking a small sample for DNA (agarose gel) and histone (Western blot) analyses respectively, or by analyzing a small portion of the chromatin under an atomic force microscope for "beads on a string" appearance.

3.2.13. Resuspend the nuclear pellet in 1 mL low-salt buffer complemented with 50 units of murine RNase inhibitor.

Note: Salt concentration can be increased to a total of 0.35 *M* NaCl, which will lead to the disassociation of chromatin-binding proteins strongly associated to chromatin, including most linker histone H1 species.

3.2.14. Extract chromatin overnight at 4°C in an end-over-end rotator.

Day 3

3.2.15. Centrifuge sample for 5 min at 8000 rpm at 4°C.

3.2.16. Save an aliquot of the supernatant, and label it "input."

3.2.17. Preclear the supernatant with 30 μL of protein A/G Plus agarose beads for 30 min at 4°C in an end-over-end rotator.

Note: To avoid RNAs and protein degradation, fresh murine RNase inhibitor and complete protease inhibitor cocktail are added at this step.

3.2.18. Centrifuge at 800 rpm at 4°C for 5 min.

3.2.19. Incubate the supernatant with the anti-CENP-A (or protein of choice) primary antibody (5 μL for 5 flasks) for a minimum of 4 h and a maximum of 12 h.

Note: The incubation with the primary antibody can be extended up to a maximum of 12 h, but shorter incubation periods are preferred to minimize background. Incubating fewer than 4 h results in low yields, although this depends on the avidity of the antibody for its epitope in the antigen.

3.2.20. Immunoprecipitate CENP-A/primary antibody complex with 50 μL of protein A/GPLUS agarose beads for 2 h at 4°C on the end-over-end rotator.

3.2.21. Centrifuge at 800 rpm at 4°C for 5 min.

3.2.22. Save an aliquot of the supernatant, label it "unbound."

3.2.23. Wash three times CENP-A/primary antibody/beads complex with 1 mL low-salt buffer containing murine RNase inhibitor and complete protease inhibitor cocktail.

Note: To validate the efficacy of the immunoprecipitation technique and the purification of the protein of interest, Western blot analysis must be performed. For this purpose, samples can be divided into two, if the quantity of material is high enough to perform the protein and RNA study with five F175 flasks. Otherwise the number of initial flasks has to be increased.

Note: At this point, samples can be stored at −80°C. RNA purification described in the following steps should be performed no later than the following week of the immunoprecipitation to avoid potential degradation.

Day 4

3.2.24. Elute RNA–protein complex in 250 μL elution buffer two times by incubation in an end-over-end rotator for 15 min at room temperature.

3.2.25. Denature all samples (input, unbound, and beads) in 200 m*M* NaCl for 2 h at 65°C.

3.2.26. Digest proteins with 20 μg of proteinase K in proteinase K buffer at 42°C for 45 min.

3.2.27. To avoid genomic DNA contamination, treat sample with DNaseI for 30 min at 37°C.

3.2.28. Stop the reaction by adding 5 m*M* EDTA.

3.2.29. Purify RNAs by the phenol:chloroform:isoamylalcohol method.

Note: Mix equal volume of sample (~500 μL) with phenol: chloroform:isoamylalcohol solution and shake by hand for 1 min. Centrifuge for 15 min at 10,000 rpm at 4°C. Transfer the top aqueous phase to a new tube, mix with equal volume of chloroform and hand shake for 1 min. Centrifuge for 15 min at 10,000 rpm at 4°C. Transfer the top aqueous phase to a new tube, add 1/10 volume of sodium acetate (final concentration ~0.3 *M*) and 3 volumes of absolute ethanol. Incubate at −20°C overnight or −80°C for 30 min. Centrifuge for 30 min at 10,000 rpm at 4°C. Resuspend dry pellet in 30 μL RNase–DNase free water. Samples can be stored at −80°C until further analysis.

3.2.30. Determine RNA concentration by UV-spectra and quality using a Bioanalyzer instrument (#5067-1511, Agilent Technologies, Santa Clara, CA).

Day 5

3.2.31. Prepare DNA library and sequence by the Illumina approach or other method of your choice.

Note: The library was prepared and sequenced using Illumina TruSeq® Stranded Total RNA Sample Preparation and TruSeq v4 chemistry (Illumina, San Diego, CA).

Note: An important decision at the library preparation stage is whether enrichment for poly-A transcripts or depletion of ribosomal transcripts is necessary. This choice could greatly increase the final signal-to-noise ratio. Ribosomal transcripts represent the majority of molecules in a total RNA sample, potentially reducing RNA signal (Zhao et al., 2014). However, targeting ncRNA would abrogate the use of oligo-dT beads (ie, ncRNA can either be poly-A or not), and RiboZero depletion may not work efficiently if those transcripts are fragmented by the chromatin-immunoprecipitation protocol. Since a robust and specific immunoprecipitation should effectively purify only targets of interest, ribosomal depletion is commonly not performed. Still, in practice, RIP samples can be composed of up to 90% rRNA (Lu, Guan, Schmidt, & Matera, 2014). The ribosomal component can be estimated before sequencing (Schroeder et al., 2006). Once a library is sequenced, ribosomal reads can also be easily removed computationally post hoc. The protocol described in Section 3.4 presents this approach.

3.2.32. Analysis of data (see Section 3.4).

Note: Since nucleic acids can bind to various surfaces, we suggest including a number of controls:

i. Testing the specificity of each new lot of antibody (confirm by Western blot or mass spectrometry that the protein of choice is the primary captured target);

ii. Using a heterologous protein to ensure specificity of identified RNAs bound to the protein of interest;

iii. Including mock immunoprecipitation with beads alone or IgG immunoprecipitation;

iv. Performing at least two (preferably more) biological replicates spread out over the course of a few months to sample cell passage numbers;

v. Where possible, performing the "gold standard" test, namely, depleting the protein of interest to ensure that any RNAs identified in the subsequent RIP-Seq are specific to that protein, and do not bind indiscriminately to the antibody or beads;

vi. Confirming the localization at the same locus by an alternative approach, like RNA FISH coupled to immunofluorescence (see Section 3.3);

vii. Analyzing functional consequences of the downregulation of identified ncRNAs with a focus on the heterochromatic domain of interest (eg, centromeres for CENP-A).

3.3 Immunofluorescence/RNA FISH

The goal of RNA FISH is to visualize the localization of centromeric RNAs in human cells. In order to define their subnuclear localization, the RNA signal can be compared to the localization of proteins of interest, like centromeric proteins, by an experiment of immunofluorescence combined to RNA FISH (IF/RNA FISH).

Several published protocols were tested (eg, combined immunofluorescence/RNAFISH described by Chaumeil et al., and simultaneous immunofluorescence/RNAFISH described in Raj et al. (Chaumeil, Augui, Chow, & Heard, 2008; Raj, van den Bogaard, Rifkin, van Oudenaarden, & Tyagi, 2008)). For centromeric ncRNA detection, the clearest and most reproducible results were obtained from the following modified protocol of sequential immunofluorescence/RNA FISH from Raj et al. (2008).

Day 1

3.3.1. Spread cells on poly-L lysine coverslip in 6-well plate at ~75% confluency using the same protocol than described in Section 3.1.

Day 2

3.3.2. Wash cells three times with HANKS buffer.

3.3.3. Fix cells with the fixation solution for 10 min at room temperature.

3.3.4. Wash out fixation solution with 1 × PBS for 5 min two times at room temperature.

3.3.5. Permeabilize cells for 5 min at room temperature.

3.3.6. Wash coverslip with 1 × PBS for 5 min at room temperature.

3.3.7. Incubate with diluted anti-CENP-A primary antibody ($d = 1/20{,}000$) in IF buffer overnight at 4°C.

Note: Several kinetochore proteins are present at human centromeres all over cell cycle. Consequently, other kinetochore proteins can be stained, such as CENP-B (dilution 1/400, Anti-CENP-B antibody, #ab25734, Abcam).

3.3.8. Wash cells with 1× PBS for 10 min at room temperature three times.

3.3.9. Incubate with secondary antibody ($d = 1/750$) in IF buffer for 1 h at room temperature in the dark.

3.3.10. Wash with 1 × PBS for 10 min at room temperature in the dark three times.

Note: At this step, a couple of slides can be washed with 1 × PBS complemented with 1 mg/mL of RNase A, to run out potentially nonspecific recognition of the probe set with DNA. It is essential to separate RNase A treated coverslips from RNase A not treated coverslips, to avoid contamination, by using two sets of 6-well plates and pipettes.

3.3.11. Fix with the fixation solution for 10 min at room temperature in the dark.

3.3.12. Wash twice with 1× PBS for 10 min at room temperature in the dark.

3.3.13. Wash with wash solution A for 5 min at room temperature in the dark.

3.3.14. Incubate with hybridization mix for 4 h at 37°C in the dark.

Note: Hybridization can proceed for up to 16 h, however, with increasing risk of higher background, nonspecific binding, and RNA degradation.

Note: Keeping the coverslips in a humidified atmosphere is essential for the efficacy of the hybridization. Several options exist to achieve this, including the use of Petri dish sealed twice with parafilm "M" and containing a wet tissue. Additionally, hybridization is performed on parafilm "M" with cells facing down to assure the equal hybridization mix on the surface of the coverslip and avoid bubbles.

Note: It is crucial to set the temperature of the incubator/hybridization oven at precisely 37°C. Any change (lower or higher) will affect the efficiency of the hybridization.

3.3.15. Wash twice with wash solution A for 30 min at 37°C in the dark.

Note: Wash solution A needs to be warmed up before use.

3.3.16. Wash with solution B for 5 min at room temperature in the dark.

3.3.17. Mount coverslips onto the slide with Prolong Gold or other mounting medium containing DAPI.

Note: Slides can be kept at room temperature for up to 1 h and then stored at 4°C.

3.3.18. Acquire and analyze pictures in the following 3 days.

Note: Slides were observed with a DeltaVision RT microscopy imaging system (GE Healthcare, Pittsburgh, PA) controlling an interline charge-coupled device camera (Coolsnap, Tucson, AZ) mounted on an inverted microscope (IX-70; Olympus, Center Valley, PA). Images were captured by using a 60 × objective at 0.2 μm z-sections and analyzed with Image J (1.50e; Java 1.6.0_20, NIH, Bethesda, MD). One z-stack is represented in Figs. 1 and 2.

3.4 Computational Analysis

Although a standardized unified method for RIP-Seq analysis is currently lacking, one can treat RIP-Seq data similarly to RNA-Seq or ChIP-Seq data, which have more well-established tools. The former approach aims to identify and quantify transcripts that are associated with your protein of interest. The latter is a peak-calling approach that identifies a genomic region and indirectly identifies the originating transcript (Zambelli & Pavesi, 2015). This chapter will focus primarily on the former approach.

The overall analysis strategy is presented in Fig. 3. Individual steps labeled in this figure are described later.

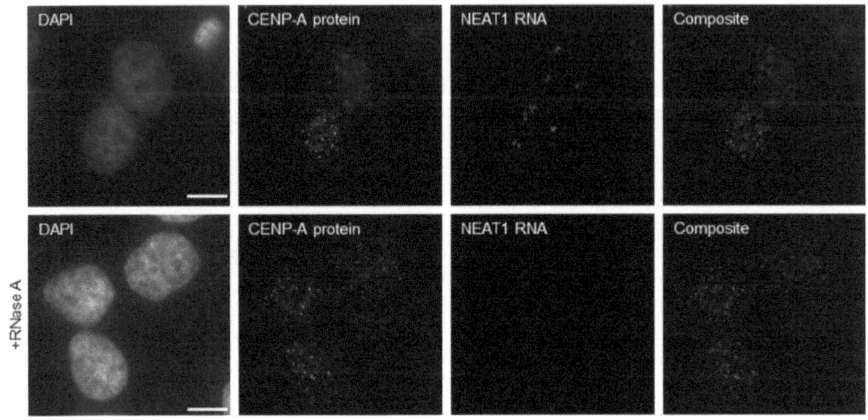

Fig. 1 Control immunofluorescence/RNA FISH on nonsynchronized HeLa cells. CENP-A protein is stained by immunofluorescence, followed by NEAT1 RNA hybridization. *Bottom panels* present cells treated with RNase A prior to RNA FISH. Only one z-stack is shown. Scale bar represents 10 μm.

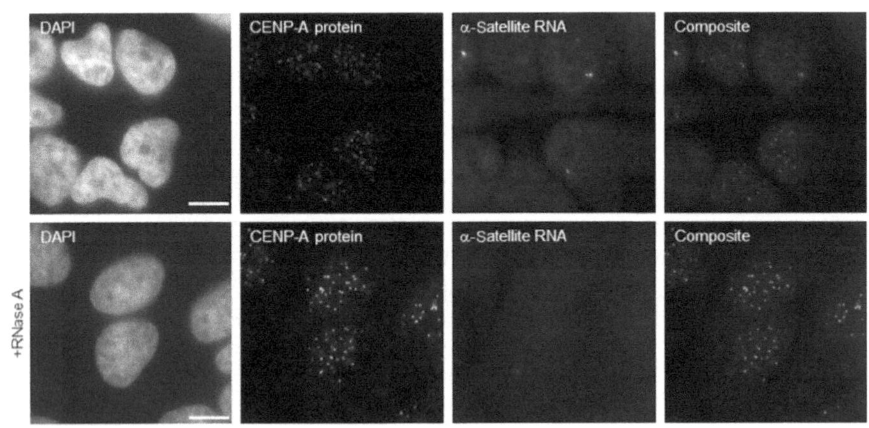

Fig. 2 Immunofluorescence/RNA FISH on nonsynchronized HeLa cells. CENP-A protein is stained by immunofluorescence, followed by α-satellite RNA hybridization. *Bottom panels* present cells treated with RNase A prior to RNA FISH. Only one z-stack is shown. Scale bar represents 10 μm.

3.4.1 Sequencing Strategy

The sequencing strategy for the RIP-Seq experiment involves settling the read length and depth for studied samples. These decisions are important to maximize the inferences you can make about the source molecules in your original pool. For RIP-Seq, the strategy should be similar to an

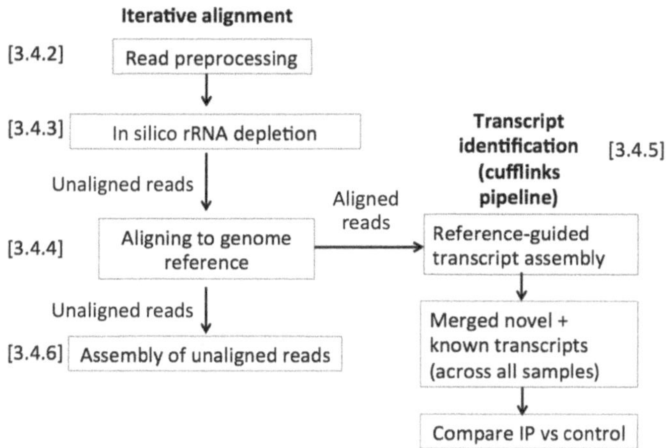

Fig. 3 Example of RIP-Seq analysis pipeline.

RNA-Seq experiment, with long paired end reads. In the example dataset for which this protocol was developed, libraries were sequenced on an Illumina HiSeq 2500 in paired end mode for 125 cycles for each end, using Illumina TruSeq v4 chemistry. To analyze the outcome, reads are aligned to a genomic reference following a multistep process described in Sections 3.4.2–3.4.4.

Note: The choice of sequencing depth is based on the complexity of the library you wish to sequence, the enrichment you hope to obtain, the amount of replication you will perform, and cost of the experiment (Sims, Sudbery, Ilott, Heger, & Ponting, 2014). Early ENCODE guidelines for ChIP-Seq of "point source" peaks, such as transcription factors, suggest a modest requirement of 10 million uniquely mapped reads per experiment to find enriched regions (Landt et al., 2012). For RNA-Seq, 30 million reads is an often-cited minimum for mammalian gene expression profiling, with 100 million needed for novel isoform discovery, and >200 million for saturated detection (Sims et al., 2014). For RIP-Seq experiments, where a less diverse set of transcript molecules is expected to be sequenced, 20 million reads is a reasonable minimum for a protein with a small number of expected targets.

3.4.2 Read Preprocessing

Reads should be preprocessed to remove errors and artifacts that could lead to inaccurate mapping. Raw reads commonly display systematic errors, including reduced base-calling accuracy and adapter sequence contamination. A first pass quality assessment can help identify these systematic errors.

An effective tool to achieve this is FastQC (Simon Andrews, http://www.bioinformatics.babraham.ac.uk/projects/fastqc/). This tool provides several graphical results including base-calling quality by position, base composition, and overrepresented sequences, which allow the identification of adapter contamination or other types of error.

Once the error profile is established, a read preprocessing strategy is chosen that preserves read length while maximizing alignability. The two major options are adaptive trimming, where each read is processed individually and trimmed to variable lengths; or fixed cropping, where an equal number of bases is trimmed from each read. Unless the read mapping percentage is very low (<75%), fixed cropping is recommended as it provides the maximum compatibility with downstream analysis tools (ie, some count-based algorithms assume all reads are of equal length, such as MISO (Katz, Wang, Airoldi, & Burge, 2010)). We used the Trimmomatic tool that can perform either strategy (Bolger, Lohse, & Usadel, 2014). Preprocessed reads were then further refined in order to deplete ribosomal RNA (rRNA).

3.4.3 In Silico rRNA Depletion

Immunoprecipitated samples may include a significant fraction of rRNA transcripts, which are not the binding partners of interest, particularly if no oligo-dT selection or ribosomal depletion is performed. The proportion of rRNA can vary between experiments, leading to the spurious detection of novel transcripts and inaccurate abundance estimations when transcript abundances are normalized by total depth. In silico rRNA depletion mitigates these problems.

In silico rRNA depletion consists of aligning reads to a reference of rDNA sequence and extracting the unaligned reads for downstream analysis. There are many alignment algorithms available to perform this step (Fonseca, Rung, Brazma, & Marioni, 2012). The most frequently cited are the TopHat and Bowtie programs, which are components of a software suite called the Tuxedo tools (Trapnell et al., 2012). These programs also perform well in systematic comparisons with simulated data (Engstrom et al., 2013). The instructions below perform rRNA depletion using the Bowtie2 aligner (Langmead & Salzberg, 2012):

i. Build a reference database.

  a. Obtain or compile rDNA sequence(s).

Note: A 43 kb complete rDNA repeating unit in human can be obtained from GenBank (accession U13369.1) in FASTA format.

b. Using the following command, build a reference database from the rDNA sequence to map reads to:

```
bowtie2 index bowtie2-build rDNA.fa rDNA
```

where "rDNA.fa" contains rDNA sequence in FASTA format. Bowtie2 will compile a set of index files using the prefix "rDNA."

ii. Align reads to the rDNA reference database.

a. Align preprocessed reads to the rDNA reference.

b. Save nonmapping reads to a separate file, so that they can be realigned to the genome later (see Section 3.4.4):

```
bowtie2 -p 8 -un-concr DNA_unaligned -sensitive -x rDNA \
  -1 processed_reads_1.fastq.gz -2 processed_reads_2.fastq.gz \
  -S rDNA_alignments.sam
```

where "rDNA_unaligned" is the file name prefix for your rDNA-filtered reads; "rDNA" is the Bowtie2 index from the previous step (ie, Section 3.4.3.i.b); "processed_reads_x.fastq.gz" is the RIP-Seq reads; and "rDNA_alignments.sam" is the destination file for the alignments to the rDNA sequence.

Note: A similar strategy was recently described by Miga et al. to reduce the number of false positive alignments with an expanded set of repetitive and "blacklisted" regions (Miga, Eisenhart, & Kent, 2015).

3.4.4 Aligning to a Reference Genome

To determine the origin of transcript molecules from the rDNA-filtered reads, they must be aligned to the complete reference genome. Since reads likely derive from processed spliced transcripts, alignments to the genomic reference may be split to indicate removed introns. The TopHat alignment program performs splice-aware read mapping, producing spliced alignments over splice junctions (Kim et al., 2013). This tool performs alignments to transcript sequence first, and then to genomic DNA. To do this, transcript models must be formatted in gene transfer format (GTF) with chromosome nomenclature consistent with the genomic sequence. To ensure these components are correctly arranged, standardized reference sequences and annotations used to perform the alignment are compiled in a collection called iGenomes (Illumina, https://support.illumina.com/sequencing/sequencing_software/igenome.html). Genome sequence can be obtained from this source as a preformatted index for use with Bowtie2. Once the appropriate reference (ie, the human genome in our study) is downloaded and saved to the proper location, the following command is applied to align reads to the human genome and find splice junctions:

```
tophat -G genes.gtf -g 1 GRCh38.p2.genome \
   rDNA_unaligned.1 rDNA_unaligned.2
```

where "genes.gtf" contains transcript coordinates; "GRCh38.p2.genome" is the Bowtie2 index for the build 38 genome sequence; and "rDNA_unaligned.1" and "rDNA_unaligned.2" are the rDNA filtered reads. By default, results will be saved into a folder called "tophat_out." This can be changed with the "—output-dir" option.

Note: Handling read alignments that map to more than one location in the genome is an important parameter to set. For example, paralogs are genes with a common evolutionary origin that are similar in sequence, such that they may not be uniquely mappable. For accurate transcriptional profiling in paralogs, it is appropriate to allow multiple alignments for a single read. For RIP-Seq application, the more conservative setting of "-g 1" limits alignments to only a single location. This is recommended to limit detection of spurious transcripts from repetitive sequence.

Note: Other sources of reference annotation and sequence may be more comprehensive or current than iGenomes, such as Ensembl (Yates, Akanni, et al., 2016) or GENCODE (Harrow et al., 2012). Additionally, cell line specific assemblies may be used for mapping to avoid errors from polymorphisms and structural variants. In these cases, follow the instructions in Section 3.4.3.i.b. to generate a Bowtie2 index from your sequence. To find out what has changed with successive assembly versions, a useful resource is the Genome Reference Consortium (GRC, http://www.ncbi.nlm.nih.gov/projects/genome/assembly/grc/human/).

3.4.5 Transcript Abundance Estimation

To identify enrichment within the RIP sample relative to control, the abundances of the transcripts need to be estimated. For this purpose, the alignments generated in the previous step (ie, Section 3.4.4) are used to reconstruct the transcript molecules that were present in the original sample and estimate their abundances. The Cufflinks program includes algorithms for performing both of these tasks (Trapnell et al., 2012).

Transcript abundance is calculated in units of fragments per kilobase of transcript per million mapped reads (FPKM), which normalize for sequencing depth and transcript length. A significantly higher FPKM in the RIP sample relative to input or control is evidence of enrichment of the target by immunoprecipitation.

This command compares transcript abundances between an immunoprecipitation and control sample:

```
cuffdiff -o cuffdiff_out genes.gtf -L IP,Cntrl \
   rip_alignments.bam input_alignments.bam
```

where "cuffdiff_out" is the name of an output directory; "IP,Cntrl" are labels to identify the immunoprecipitation and control samples in the results; and "rip_alignments.bam" and "input_alignments.bam" are the alignment files (from Section 3.4.4).

Note: The "genes.gtf" can be a reference annotation or de novo assembled transcripts (Trapnell et al., 2012). The latter is preferred if putative targets are unannotated or unknown.

3.4.6 Assembly of Unaligned Reads

The human genome is incompletely sequenced and assembled, particularly within centromeres and heterochromatic regions (Miga, 2015). Even though the most recent reference adds megabases of new centromeric DNA and alternative loci (Rosenbloom et al., 2015), transcript sequence purified by chromatin-immunoprecipitation can still be unalignable to the reference. These unalignable reads may derive from unassembled human genome sequence, or from other sources such as technical artifacts and contaminants. To investigate the source of unaligned reads, they can be assembled into longer fragments. For this purpose, the bam2fastq program (http://gsl.hudsonalpha.org/information/software/bam2fastq) is used to extract unaligned reads from TopHat results (Section 3.4.4) following this command:

```
bam2fastq -no-aligned -o top_unaligned#.txt unmapped.bam
```

where "top_unaligned_#.txt" is the name given to the result file, and "unmapped.bam" is the file where TopHat stores the unmapped reads. The pound sign will be replaced with the mate pair number (1 or 2).

Then, reads are assembled in a two-step process using the Velvet program (Zerbino & Birney, 2008):

i. Step of hashing:

```
velvethmysteryseqs 21 -fastq -short -separate \
   top_unaligned_1.txt top_unaligned_2.txt >& out1
```

ii. Step of contig assembly:

```
velvetg mystery1 -min_contig_lgth 150 -ins_length 50
```

where "mysteryseqs" is your output directory, and the "contig_lgth" and "ins_length" are estimates based on the sequenced fragment length.

The remaining contigs can be mined for additional evidence of transcripts, using BLAST. In the case of contaminants, BLAST will identify significant similarity to bacterial DNA and vector sequence. For technical artifacts, the FastQC program can be used to identify adapter sequence (see Section 3.4.2). Any remaining high-abundance contigs can be mined for evidence of human transcriptional origin or analyzed via RNA FISH (see Section 3.3).

ACKNOWLEDGMENTS

We thank Daniel R. Larson, Murali Palangat, and Joseph Rodriguez for thoughtful discussion, and Tatiana Karpova for the access to microscopy facility. The biochemical protocols used in our study were influenced by prior work on crasiRNAs published by Rachel O'Neill (University of Connecticut-Storrs). For additional resources on lncRNAs, we suggest reading recent papers by Anindya Dutta's research group (University of Virginia). All authors on this study are supported by Intramural Research Program of the Center for Cancer Research at the National Cancer Institute/National Institutes of Health.

Disclosures: The authors declare they have no competing financial interests.

REFERENCES

Baker, M. (2012). RNA imaging in situ. *Nature Methods*, *9*, 787–790.

Bolger, A. M., Lohse, M., & Usadel, B. (2014). Trimmomatic: A flexible trimmer for illumina sequence data. *Bioinformatics (Oxford, England)*, *30*, 2114–2120.

Chan, F. L., & Wong, L. H. (2012). Transcription in the maintenance of centromere chromatin identity. *Nucleic Acids Research*, *40*, 11178–11188.

Chaumeil, J., Augui, S., Chow, J. C., & Heard, E. (2008). Combined immunofluorescence, RNA fluorescent in situ hybridization, and DNA fluorescent in situ hybridization to study chromatin changes, transcriptional activity, nuclear organization, and X-chromosome inactivation. *Methods in Molecular Biology (Clifton, NJ)*, *463*, 297–308.

Engstrom, P. G., Steijger, T., Sipos, B., Grant, G. R., Kahles, A., Ratsch, G., et al. (2013). Systematic evaluation of spliced alignment programs for RNA-seq data. *Nature Methods*, *10*, 1185–1191.

Fonseca, N. A., Rung, J., Brazma, A., & Marioni, J. C. (2012). Tools for mapping high-throughput sequencing data. *Bioinformatics (Oxford, England)*, *28*, 3169–3177.

Hall, L. E., Mitchell, S. E., & O'Neill, R. J. (2012). Pericentric and centromeric transcription: A perfect balance required. *Chromosome Research*, *20*, 535–546.

Harrow, J., Frankish, A., Gonzalez, J. M., Tapanari, E., Diekhans, M., Kokocinski, F., et al. (2012). GENCODE: The reference human genome annotation for The ENCODE Project. *Genome Research*, *22*, 1760–1774.

Katz, Y., Wang, E. T., Airoldi, E. M., & Burge, C. B. (2010). Analysis and design of RNA sequencing experiments for identifying isoform regulation. *Nature Methods*, *7*, 1009–1015.

Kim, D., Pertea, G., Trapnell, C., Pimentel, H., Kelley, R., & Salzberg, S. L. (2013). TopHat2: Accurate alignment of transcriptomes in the presence of insertions, deletions and gene fusions. *Genome Biology*, *14*, R36.

Landt, S. G., Marinov, G. K., Kundaje, A., Kheradpour, P., Pauli, F., Batzoglou, S., et al. (2012). ChIP-seq guidelines and practices of the ENCODE and modENCODE consortia. *Genome Research*, *22*, 1813–1831.

Langmead, B., & Salzberg, S. L. (2012). Fast gapped-read alignment with Bowtie 2. *Nature Methods*, *9*, 357–359.

Lu, Z., Guan, X., Schmidt, C. A., & Matera, A. G. (2014). RIP-seq analysis of eukaryotic Sm proteins identifies three major categories of Sm-containing ribonucleoproteins. *Genome Biology*, *15*, R7.

Miga, K. H. (2015). Completing the human genome: The progress and challenge of satellite DNA assembly. *Chromosome Research*, *23*, 421–426.

Miga, K. H., Eisenhart, C., & Kent, W. J. (2015). Utilizing mapping targets of sequences underrepresented in the reference assembly to reduce false positive alignments. *Nucleic Acids Research*, *43*, e133.

Quenet, D., & Dalal, Y. (2014). A long non-coding RNA is required for targeting centromeric protein A to the human centromere. *eLife*, *3*, e03254.

Raj, A., van den Bogaard, P., Rifkin, S. A., van Oudenaarden, A., & Tyagi, S. (2008). Imaging individual mRNA molecules using multiple singly labeled probes. *Nature Methods*, *5*, 877–879.

Rosenbloom, K. R., Armstrong, J., Barber, G. P., Casper, J., Clawson, H., Diekhans, M., et al. (2015). The UCSC Genome Browser database: 2015 update. *Nucleic Acids Research*, *43*, D670–D681.

Schroeder, A., Mueller, O., Stocker, S., Salowsky, R., Leiber, M., Gassmann, M., et al. (2006). The RIN: An RNA integrity number for assigning integrity values to RNA measurements. *BMC Molecular Biology*, *7*, 3.

Sims, D., Sudbery, I., Ilott, N. E., Heger, A., & Ponting, C. P. (2014). Sequencing depth and coverage: Key considerations in genomic analyses. *Nature Reviews. Genetics*, *15*, 121–132.

Trapnell, C., Roberts, A., Goff, L., Pertea, G., Kim, D., Kelley, D. R., et al. (2012). Differential gene and transcript expression analysis of RNA-seq experiments with TopHat and Cufflinks. *Nature Protocols*, *7*, 562–578.

Verdaasdonk, J. S., & Bloom, K. (2011). Centromeres: Unique chromatin structures that drive chromosome segregation. *Nature Reviews. Molecular Cell Biology*, *12*, 320–332.

Waye, J. S., & Willard, H. F. (1987). Nucleotide sequence heterogeneity of alpha satellite repetitive DNA: A survey of alphoid sequences from different human chromosomes. *Nucleic Acids Research*, *15*, 7549–7569.

Yates, A., Akanni, W., Amode, M. R., Barrell, D., Billis, K., Carvalho-Silva, D., et al. (2016). Ensembl 2016. *Nucleic Acids Research*, *44*(D1), D710–D716.

Zambelli, F., & Pavesi, G. (2015). RIP-Seq data analysis to determine RNA-protein associations. *Methods in Molecular Biology (Clifton, NJ)*, *1269*, 293–303.

Zerbino, D. R., & Birney, E. (2008). Velvet: Algorithms for de novo short read assembly using de Bruijn graphs. *Genome Research*, *18*, 821–829.

Zhao, W., He, X., Hoadley, K. A., Parker, J. S., Hayes, D. N., & Perou, C. M. (2014). Comparison of RNA-Seq by poly (A) capture, ribosomal RNA depletion, and DNA microarray for expression profiling. *BMC Genomics*, *15*, 419.

AUTHOR INDEX

Note: Page numbers followed by "*f*" indicate figures, and "*t*" indicate tables.

A

Aasland, R., 211–213
Abbasi, A., 323
Abelson, J.N., 268–269
Abmayr, S.M., 304, 309
Abou-Ghoch, J., 388–389
Aburatani, H., 287
Acerini, C.L., 389–390
Adachi, J., 422
Adam, S., 100, 108–111
Adamo, A., 243–245, 248–249, 263–264
Adams, M.M., 213
Adams, P.D., 204–205, 227–228, 335–336, 338
Admirand, E.A., 267
Afonine, P.V., 227–228, 336
Aggarwal, A.K., 322–323
Agger, K., 287–288
Ai, H.W., 288–289
Aik, W., 281–282
Aikawa, Y., 141–142
Airoldi, E.M., 460
Akanni, W., 462
Aladini, F., 186–187, 190*t*
Alam, H., 281–282
Albaugh, B.N., 140
Albeck, S., 218–219
Albert, M., 217–218
Aldape, K., 273
Alekseyenko, A.A., 441
Alexander, C., 57
Allain, C.J., 217–218
Allali-Hassani, A., 205
Allis, C.D., 106–108, 210–213, 242, 322–323, 346–347
Almouzni, G., 98–114
Almouzni-Pettinotti, G., 113–114
Alonso, A., 83–84
Al-Sady, B., 288–289
Altman, S., 422
Alzayady, K., 264
Ambrosini, G., 389–390
Amit, I., 423
Amode, M.R., 462
An, J., 217–218
Anderson, D.E., 283–284
Anderson, K.A., 187–188
Anderson, K.V., 389–390
Anderson, M.E., 243–244
Ando, R., 273–274
Andrade, J., 211–213
Andrade, M., 389–390
Andres, M.E., 243–244
Andrews, F.H., 8
Anfinsen, C.B., 16
Anselmi, F., 378
Antal, T., 287–288
Antonietta Vanoni, M., 263–265
Antonysamy, S., 213–216
Antoshenko, T., 205
Aparicio, S.A., 228–229, 313
Appel, R.D., 350
Arcand, M., 287–288
Ardehali, M.B., 211–213
Argyrou, A., 287–288
Aris, A., 221
Arita, K., 388
Ariyoshi, M., 388
Arkin, M., 144–145
Armache, K.-J., 44, 57
Armstrong, C.M., 167
Armstrong, J., 463
Armwood, C.J., 381
Arnaudo, N., 47–48
Arrowsmith, C.H., 336
Artandi, S.E., 441
Ascano, M., 435–440
Ashworth, A., 422
Ateeq, B., 217–218
Atouf, F., 243–244
Attig, J., 440–441
Augui, S., 455–457
Auld, D.S., 144–145, 149–150, 287–288
Austin, C.P., 144–145, 149–152, 156–157

Avalos, J.L., 186–187, 190*t*, 197
Avaritt, N.L., 264–265, 267
Avdic, V., 216
Avvakumov, G.V., 388
Avvedimento, E.V., 243–244

B

Baaske, P., 201–202
Baba, T., 243–244, 248–249, 252
Babaoglu, K., 151–152, 156–157
Bachman, M., 366, 369–370
Baek, S.H., 141–142
Baell, J., 155–156
Bai, S.K., 109
Bai, Y., 44, 68–69
Bailey, A.O., 113–114
Bailey, D., 210–211
Baker, M., 449
Balasubramanian, S., 323, 366, 369–370
Ballard, C.C., 204–205, 295, 335
Ballario, P., 322–323
Ballas, N., 243–244
Ballestar, E., 141–142
Balog, J., 388–389
Bampos, N., 289–290
Banaszynski, L.A., 106–108
Bandukwala, H., 366–368, 387–388
Bannister, A.J., 345–346
Bantscheff, M., 323
Bao, Y., 4–8, 11–12, 15–16, 20, 68–69, 79, 84–85, 87–88, 121, 125, 229
Barber, G.P., 463
Barbera, A.J., 44, 263–264
Barcena, J., 57
Bardiaux, B., 338
Barker, K., 146
Barlev, N.A., 140, 145–146, 151
Bar-Or, D., 289–290
Barozzi, I., 211–213, 425–426
Barrea, R.A., 34
Barrell, D., 462
Barrette, T.R., 217–218
Barsyte-Lovejoy, D., 323
Bartholomew, B., 4–6, 46, 48, 263–264
Barton, G.J., 220*f*, 347
Barton, M.C., 273
Bash, R., 29–30
Basrur, V., 216
Bassett, E.A., 68–69, 113–114
Basso, M., 269, 272–273
Battaglioli, E., 243–245, 248–249, 251, 256, 263–265, 282–283, 292–293
Battye, T.G.G., 335
Batzoglou, S., 459
Bauer, C., 366
Bauman, D., 288–289
Bax, A., 332, 336, 338
Beach, B.M., 211
Beard, C., 309
Beaudet, L., 153–154
Beaufils, F., 113–114
Becker, C.F., 190*t*
Becker, P.B., 119–120, 129–131
Bédard, S., 68–69
Beguelin, W., 217–218
Beinke, S., 323
Bekiranov, S., 423
Bell, D.W., 387–389
Bello, S.H., 282
Ben Jelloul, M., 218–219
Bender, W., 211–213
Bennett, M.T., 381
Bensimon, E., 211–213
Bentley, D.L., 211–213
Berber, E., 184–186
Berdasco, M., 211–213
Berendzen, J., 413
Berezovska, O., 141–142
Berg, J.M., 394
Berg, T., 228–229, 313
Bergamin, E., 210–231
Bergbrede, T., 186–187, 190*t*
Berger, S.L., 140, 145–146, 151, 242, 257
Bergeron, S., 4–37
Berglund, H., 218–219
Beringer, M., 435
Bernard, A., 338
Bernatavichute, Y.V., 211–213
Berndsen, C.E., 140
Berninger, P., 435–440
Berns, M., 100–102
Bertone, P., 422–423
Bertoni, A., 243–244
Bestor, T.H., 387–389
Betzel, C., 57
Bever, K.M., 186–187

Beyermann, M., 288–289
Bhagwat, A.S., 289–290
Bharadwaj, M., 441
Bhat, K.P., 273
Bhatt, D.M., 425–426
Biggar, K.K., 280
Billis, K., 462
Bilodeau, S., 243–244, 257
Bilsel, O., 33–34
Binda, C., 243–245, 248–249, 251, 256, 262–265, 269–271, 282–283, 292–293
Biniszkiewicz, D., 387–388
Bird, A.P., 387–388
Bird, G.H., 217–218
Bird, L., 323
Birney, E., 422, 464
Bitinaite, J., 379, 381
Black, B.E., 4–6, 20, 68–95, 99–102, 113–114
Black, J.C., 280–282, 345–346
Blackburn, E.H., 405–406
Blackburn, G.M., 218–219
Blahnik, K.R., 389–390
Blair, L.P., 264–265, 267
Blanc, E., 227–228
Blaschke, K., 378–379
Blessington, P., 217–218
Blewitt, M.E., 211–213
Blobel, G.A., 248–249, 422, 435
Block, J.N., 227–228
Bloom, K., 446
Blouin, J., 287–288
Blower, M.D., 119–120
Blumenthal, E., 422–441
Blumenthal, R.M., 388–390, 393
Blumer, J., 113–114
Boanca, G., 190*t*
Bodor, D.L., 100
Boeke, J.D., 184–186, 190*t*, 197
Boguski, M.S., 423
Boing, S., 422
Boix-Chornet, M., 141–142
Bolger, A.M., 460
Bolotina, N., 218–219
Bomsztyk, K., 174
Bonanno, J.B., 213–216
Bonday, Z., 100–102
Bone, J.R., 211–213
Bono, H., 422
Bonte, E.J., 119–120, 129–131
Boonen, S.E., 389–390
Booth, M.J., 366, 369–370
Borchers, C.H., 255–256, 280–282, 287–288
Borek, D., 243–244, 248–249, 252, 263–264
Bork, P., 210–211, 327*f*, 347
Borra, M.T., 189
Boskovic, A., 114
Bostick, M., 388
Botrugno, O.A., 269
Botvinick, E., 100–102
Bourc'his, D., 388–389
Bourns, B.D., 407
Bowler, M.G., 61–62
Bowler, M.W., 61–62
Bowman, E.J., 422
Bowman, T., 323
Boyarchuk, E., 112–114
Bozicevic, A., 288–289
Bracken, A.P., 216
Bracken, C., 348–350
Braida, D., 264
Brandmayr, C., 369–370
Brannan, C.I., 422
Brasier, A.R., 323
Braun, D., 201–202
Brautigam, C.A., 263–264, 269–271
Bravo, F.J., 264, 272
Bray, J., 347
Brazma, A., 460–461
Brehove, M., 4–6
Brennan, F.M., 164
Brent, M., 188–189
Bricogne, G., 227–228
Briggs, S.D., 211–213
Brock, M.A., 68–69
Brockdorff, N., 422
Bronner, C., 388
Brookes, E., 27, 346–347
Brooks, H.B., 144–145
Brothers, S.P., 323
Brown, C.J., 422
Brown, P.J., 287–288
Brown, R.S., 57
Brudno, Y., 366–368, 387–388

Brugmann, S.A., 423, 435
Brünger, A.T., 338
Brunzelle, J.S., 213–219, 225, 282–285, 289, 292–294
Bryan, C., 406–407
Bryant, B.M., 323
Bryant, C., 217–218
Bryson, K., 219–221
Buchan, D.W., 219–221
Buchwald, G., 218–219
Buck-Koehntop, B.A., 388–389
Buckley, K., 406–407
Buehler, L., 27
Buesa, C., 281–282
Bugge, M., 388–389
Bulfer, S.L., 154
Bullock, A.N., 218–219
Bunick, G.J., 57, 83–84
Bunkoczi, G., 336
Bunting, K.L., 217–218
Burdick, D., 287–288
Burge, C.B., 460
Burger, L., 435–440
Burgess, H.E., 366
Burgess, R.R., 47
Burgess-Brown, N., 266, 287–288
Burlingame, A.L., 288–289
Busso, D., 218–219
Bussotti, G., 425
Buszczak, M., 272
Butler, D., 282–283, 293
Buxton, R., 287–288

C

Cai, J., 186–187
Cairns, B.R., 119–120, 323
Calabrese, J.M., 441
Calado, R.T., 404
Caldwell, J.A., 211–213
Callaway, J.L., 389–390
Callen, E., 213
Campbell, J.E., 267
Canfield, T.K., 388–389
Cantaloube, S., 112–114
Canzio, D., 44, 57
Cao, F., 216
Cao, H., 46
Cao, Q., 217–218
Cao, R., 216, 228
Cao, W., 27
Cao, X., 141–142
Cappa, A., 269
Caputo, V., 323
Carceller, E., 281–282
Cardin, R.D., 264, 272
Carlin, S.M., 272
Carlino, L., 264–267
Carlson, J.E., 322–323
Carmen, A.A., 167
Carmichael, G.G., 425
Carmo-Fonseca, M., 211–213
Carninci, P., 422–423
Caron, C., 12
Caron, M., 211–213, 287–288
Carpenter, P.B., 211–213
Carrington, J.C., 218–219
Carroll, C.W., 68–69
Carrozza, M.J., 323
Carvalho, S., 211–213
Carvalho-Silva, D., 462
Case, A.W., 190*t*, 204
Casero, R.A., 242–244, 255–256, 280–281, 287–288
Casper, J., 463
Cavalli, G., 211–213
Cawley, S.E., 423
Cech, T.R., 404–408
Celic, I., 190*t*, 197
Celie, P.H., 218–219
Cen, Y., 184–187
Cerbone, M., 388–389
Cereseto, A., 140
Cesaroni, M., 435
Chacón, P., 35
Chai, X., 184–186, 190*t*, 204
Chaidos, A., 323
Chakravarthy, S., 4–37, 68–69, 71, 79, 84–85, 87–88, 121, 125, 229
Chakravarty, D., 243–244
Chambon, P., 322–323
Chan, C., 211
Chan, F.L., 446
Chan, W.I., 323
Chandy, M., 323
Chang, C.S., 140–142
Chang, H.Y., 422–423, 435, 441

Chang, K.H., 287–288
Chapman, B.A., 441
Charbonneau, M., 406–407
Chase, A.J., 217–218
Chatterjee, N., 4–6, 46, 48, 263–264
Chaturvedi, C.P., 216
Chaudhry, M.S., 323
Chaumeil, J., 455–457
Che, K., 287, 293
Cheeseman, I.M., 68–69
Chekliarova, I., 29–30
Chellappan, S.P., 264–268
Cheloufi, S., 114
Chen, F., 263–264
Chen, H., 282–283
Chen, H.W., 211–213
Chen, H.Y., 167
Chen, K., 211–213
Chen, L.L., 425
Chen, L.Y., 406–407
Chen, T., 257, 264
Chen, V.B., 227–228, 336
Chen, X., 8, 211–213, 228–229
Chen, Y., 216, 248–249
Chen, Z., 282, 285–287, 293
Cheng, C.S., 243–244
Cheng, F., 164, 177–178
Cheng, J., 366–368, 373–374, 378–379
Cheng, S.W., 228–229, 313
Cheng, X., 387–395
Cheng, X.D., 211–215, 218–219, 228, 280–282, 287, 293, 296, 381, 388–390, 393–394
Cheng, Z., 287, 293, 295
Chenoweth, J., 243–244
Chepelev, I., 345–346
Cherry, S., 283–284
Cheung, P., 211–213, 287, 293, 295
Cheval, H., 388
Chhoy, P., 187–188
Chi, P., 346–347
Chi, S.W., 435
Chiang, A., 211–213
Chijiwa, T., 388–389
Chikaoka, Y., 287
Chin, J.W., 288–289
Chinnaiyan, A.M., 217–218
Chinwalla, V., 210–211
Chitsaz, F., 323
Chodaparambil, J.V., 4–6, 8, 44
Chong, Z., 378–379
Chouery, E., 388–389
Chow, J.C., 455–457
Chowdhury, R., 262, 281–282
Christensen, J., 287–288
Chu, C., 435, 441
Chu, F., 288–289
Chu, J., 228–229, 313
Chua, K.F., 184–186
Chuikov, S., 211
Chung, C.W., 287–288, 293, 323
Chung, J., 390
Chung, T.D., 149–150
Cierpicki, T., 216
Ciossani, G., 269
Cirilli, R., 269–271
Clamp, M., 220*f*
Clapier, C.R., 4–6, 119–120
Claridge, T.D., 287–288
Clark, E.A., 263–264
Clark, N., 4–6
Clark, N.J., 4–6, 5*f*, 8, 12, 14, 24–26, 32–33
Clark, V., 387–388
Clawson, H., 463
Clayton, A.L., 211–213
Clément, C., 98–114, 264, 272
Cleveland, D.W., 4–6, 68–69, 71, 99–102, 113–114
Cloos, P.A., 287–288
Clore, G.M., 336, 338
Cokus, S., 211–213
Cole, C., 347
Cole, P.A., 242–244, 255–256, 262–275, 280–283, 287–288, 292–293
Coleman, M.L., 287–288
Cölfen, H., 27
Coligan, J.E., 47
Collaborative Computational Project No. 4, 413
Collazo, E., 211, 213–215, 218–219, 225, 282–285, 286*t*, 288–295
Collins, K., 404–406
Collins, L.B., 366, 387–388
Collins, T., 389*f*
Colson, P., 57
Coma, I., 149–150

Condon, B., 213–216
Conery, A.R., 323
Considine, T., 190*t*, 204
Cooch, N., 252, 263–264
Cook, R.G., 210–211
Cooper, P.J., 422
Copeland, R.A., 155–156, 346–347
Copeland, T.D., 283–284
Corbani, S., 388–389
Core, L.J., 428–434
Cornett, E.M., 287–288
Cornilescu, G., 338
Corona, D.F., 119–120, 129–131
Corpet, A., 98–99
Correa, I.R., 113–114
Corsinotti, A., 389–390
Cortez, V., 243–244
Cosgrove, M.S., 216
Costa, M., 282–283
Côté, J., 140, 242, 304–317, 346
Cotsworth, S., 211–213
Coussens, N.P., 144–145
Couture, J.F., 210–231, 280, 282–285, 289, 292–294
Cowley, A., 327*f*
Cowtan, K.D., 204–205, 227–228, 295, 335–336
Crane-Robinson, C., 29–30
Crawford, F., 282, 293
Creasy, C.L., 217–218
Cribbs, A.P., 164
Cross, N.C., 217–218
Cuatrecasas, M., 141–142
Cuatrecasas, P., 16
Culhane, J.C., 262–264, 268–271, 282–283, 292–293
Cullen, D.E., 389–390
Cuomo, A., 211–213
Cuozzo, C., 243–244
Curk, T., 440–441
Curth, U., 188–189, 190*t*, 196–197, 205
Curtis, G., 289–290

D

D'Arcy, S., 8
da Rocha, S.T., 441
Da, G., 268–269
Dabin, J., 109
D'Agostino, A., 388–389
Dai, H., 185*f*, 190*t*, 204
Dai, L., 187–188
Dai, N., 379
Dai, Q., 366, 387–388
Dalal, Y., 446–464
Dalby, A.B., 404–405
D'Alessio, A.C., 366–368, 387–388
Dalke, A., 35
Dall'Aglio, A., 243–244, 263–264
Dalvai, M., 305, 307–309, 313, 315, 317
D'Ambrogio, A., 440–441
D'Amico, A., 388–389
Damon, C., 406–407
Dang, W., 140, 142–144
Daniel, M., 287–288, 323
D'Arcy, S., 48
Darnell, R.B., 435
Davey, C.A., 37, 46
Davie, J.R., 167
Davis, C.A., 422
Davis, I.W., 227–228, 336
Davis, J.S., 288–289
Davrazou, F., 285–287
Dawson, M.A., 323
de Almeida, S.F., 211–213
de Greef, J.C., 388–389
de Jong, P.J., 422
De Koning, L., 98–99
de la Cruz, C.C., 288–289
de Lange, T., 406
de Rooij, D.G., 387–388
de Vries, W., 389–390
Debreczeni, J.E., 218–219
Dechassa, M.L., 4–6, 8, 20
Dees, E.C., 422
DeFeo, K., 141–142
Defossez, P.A., 388–389
DeKelver, R.C., 309
Del Rizzo, P.A., 211–213, 345–346
Delaglio, F., 332, 336, 338
DeLano, W.L., 338
Delcuve, G.P., 167
Delmore, J.E., 323
Demeler, B., 27
den Dunnen, J.T., 388–389
Denisenko, O., 174
DeNizio, J.E., 366–383

Dent, S., 242
Dent, S.Y., 211
Denu, J.M., 140, 189, 190*t*, 198–199, 199*t*, 204
Derewenda, Z., 221
Derijck, A.A., 211–213
Derks, Y., 377
Derrien, T., 425
Dessanti, P., 269
Deuring, R., 322–323
Deutscher, M.P., 47
Devries, M.K., 190*t*, 204
Dewell, S., 323
Dey, A., 323
Dhalluin, C., 322–323
Dhanak, D., 346–347
Dhanasekaran, S.M., 217–218
Dharmarajan, V., 216
Dhasarathy, A., 388
Diaz-Fuertes, M., 141–142
Dickson, B.M., 287–288
Diederichs, S., 422
Diekhans, M., 462–463
Dillon, S.C., 213–215
Dimitrova, E., 281–282
Ding, E.C., 188–189
Dingli, F., 113–114
Dirk, L.M., 211, 216–217
Dittmann, A., 323
Djebali, S., 422, 425
Dobeli, H., 218–219
Dobin, A., 422
Dodson, E.J., 204–205, 295, 335–336
Doerks, T., 210–211, 347
Dong, A., 190*t*, 204–205, 213, 216, 218–219, 346
Dong, Z., 263–264
Donoghue, M.T., 211–215, 218–219
Dorn, R., 210–211, 389*f*
Dorsey, J.A., 257
Dou, Y., 216
Doudna, J.A., 441
Dougherty, W.G., 218–219
Dowen, R.H., 387–388
Doyon, Y., 304–317
Dreher, S.J., 4–6
Drenkow, J., 423
Drohat, A.C., 366, 381
Drozdetskiy, A., 347
Drury, W.J., 211–213
Druzina, Z., 213–216
Du, J., 187–188, 204
Du, Y., 264
Duan, M.R., 228–229
Duan, S., 388
Duan, Z.J., 211–213
Dubendorff, J.W., 218–219
Duff, M.O., 425
Duhr, S., 201–202
Dulac, C., 98
Dunleavy, E.M., 100–102, 113–114
Dunn, B.M., 47
Dunn, J.J., 218–219
Durand, D., 263–264, 269
Dutta, A., 422
Duvaud, S., 350
Dyer, P.N., 4–37, 68–69, 79, 84–85, 87–88, 121, 125, 229
Dyson, H.J., 388–390

E

Easton, L.E., 440–441
East-Seletsky, A., 441
Ebata, K.T., 378–379
Ebert, A., 213
Echols, N., 336
Edayathumangalam, R.S., 4–8, 11–12, 15–16, 68–69, 79, 84–85, 87–88, 121, 125, 229
Edelhoch, H., 16
Edgar, R.C., 220*f*
Edmondson, D.E., 262–263
Edmunds, J.W., 211–213
Edwards, A.M., 190*t*, 204, 218–219
Eisen, D., 369–370
Eisenhaber, F., 211–213
Eisenhart, C., 461
Eissenberg, J.C., 211–213
Ekiert, D.C., 388–389
El Ali, N., 388–389
Eliazer, S., 272
Elling, U., 114
Elsaesser, S.J., 106–108
Emkey, R., 287–288
Emsley, P., 204–205, 227–228, 295, 335–336
Encode Project Consortium, 422–423

England, J.R., 44, 46, 57, 63
England, K.S., 262, 281–282
Engreitz, J.M., 441
Engstrom, A., 296
Engstrom, P.G., 460–461
Enriquez-Gasca, R., 114
Epstein, C.B., 345–346
Eram, M.S., 266, 287–288
Erdjument-Bromage, H., 255–256, 280–282, 287–288, 323
Erickson, B., 211–213
Ernst, J., 345–346
Ernst, T., 217–218
Erwin, J.A., 435
Esadze, A., 388–389
Espada, J., 141–142
Espejo, A.B., 257
Esteller, M., 141–142, 211–213, 346–347
Etani, T., 273–274
Euskirchen, G., 423
Evans, P.R., 204–205, 295, 322–323, 335

F

Fak, J.J., 435
Falandry, C., 211–213
Falk, S.J., 20, 68–69, 79–80, 83–84, 93
Fan, H.Y., 119–120, 128
Fang, J., 255–256, 264–268, 280–282, 287–288
Fang, R., 263–264
Farazi, T., 435–440
Farnham, P.J., 389–390
Farooq, A., 213, 215
Feather, D., 217–218
Federowicz, K.E., 269–271, 274
Fedorov, O., 323
Feigon, J., 407–408
Feldman, J.L., 190*t*, 204
Feldser, D., 423
Felisaz, F., 61–62
Feng, B.Y., 151–152, 156–157
Feng, H., 44, 68–69
Feng, S., 211–213
Fenouil, R., 211–213
Ferguson, A.D., 211
Ferguson-Smith, A., 389–390
Fernandez, I.S., 47–48
Fernandez-Tornero, C., 4–6
Ferrao, R., 243–245, 253, 257, 264
Ferrari, K.J., 211–213
Fesik, S.W., 336
Fick, R.J., 280
Ficz, G., 366
Fierke, C.A., 282–283
Fierz, B., 4–6, 8
Figeys, D., 217–218, 280
Filipescu, D., 98, 112–114
Filippakopoulos, P., 323
Finch, J.T., 61, 98–99
Fisch, T., 287–288
Fischer, F., 186–188, 190*t*, 198, 200–204, 202*f*, 203*t*
Fischle, W., 210
Fish, A., 263–264, 269
Fittipaldi, A., 140
Fitzpatrick, P.F., 282–283, 292–293
Flaus, A.J., 44–47
Flensburg, C., 227–228
Fletterick, R.J., 287, 295
Flint, S.J., 422
Flynn, E.M., 211
Flynn, R.A., 441
Fodor, S.P., 423
Foltz, D.R., 4–6, 71, 99–102, 113–114
Fomenkov, A.I., 379
Fong, N., 211–213
Fonseca, N.A., 460–461
Forejt, J., 389–390
Formankay, J.D., 336
Forneris, F., 243–245, 248–249, 251, 256, 263–265, 282–283, 292–293
Forsberg, G., 218–219
Forsberg, L.J., 4–6
Foster, C.A., 151
Foster, C.T., 243–244, 257
Fourel, G., 211–213
Fox, J.D., 283–284
Fraaije, M.W., 243–244
Fraajie, M.W., 263–264
Fraga, M.F., 141–142, 211–213
Frampton, G.M., 243–244, 257
Francis, P., 287–288
Franke, D., 35
Frankish, A., 462
Frants, R.R., 388–389
Franzel, B., 187–188, 190*t*, 204

Frasca, A., 333*f*, 336
Fraser, M.J., 218–219
Fredericq, E., 57
French, C., 423
Frenkiel, T.A., 213–215, 228
Freund, C., 228, 288–289
Frietze, S., 389–390
Frith, M.C., 422–423
Fry, C.J., 288–289
Fuda, N.J., 434
Fujimori, D.G., 287–289, 295
Fujimoto, D., 167, 179–180
Fujita, N., 388
Fukuzawa, S., 248–249
Fuller, C.J., 68–69
Fullwood, M.J., 425
Furnari, F., 100–102
Furuno, M., 422
Fyodorov, D.V., 119–120

G

Gaillard, P.H., 98–99
Gajda, M., 35
Galasinski, S.C., 323
Gale, M., 266–267
Gallagher, S.J., 323
Gallardo, D., 141–142
Galletto, R., 358
Galy, V., 211–213
Gamblin, S.J., 211–215, 218–219
Ganji, G., 217–218
Gantt, S.L., 282–283
Gao, Q., 309
Garber, M., 423
Garcia Alai, M.M., 288–289
Garcia, B.A., 267
Garcia-Bassets, I., 243–244
Garcia-Garcia, M.J., 389–390
Gardner, J.E., 211–213
Garlick, J.D., 44, 57
Garrard, S., 221
Garruss, A., 211, 228–229
Gassmann, M., 454
Gasteiger, E., 350
Gately Luz, J., 213–216
Gattiker, A., 350
Gattis, S.G., 282–283
Gauthier, N., 287–288
Gautier, A., 113–114
Gaweska, H., 282–283, 292–293
Gdula, D.A., 119–120, 129–131
Geeganage, S., 144–145
Gelbart, W.M., 210–211
Geltinger, B., 288–289
Gendreizig, S., 99–100
Genome Science, Group, 422–423
Georgiev, G., 388–389
Gerona-Navarro, G., 338
Gerth, F., 288–289
Gertz, M., 184–189, 190*t*, 198, 201–205, 202*f*, 203*t*
Gewiess, A., 83–84
Gheyi, T., 213–216
Ghirlando, R., 44, 68–69
Ghosh, R.P., 13, 29, 32–33
Giacca, M., 140
Gibson, M.D., 8
Gierasch, L.M., 13, 29, 32–33
Gileadi, C., 281–282, 287
Gileadi, O., 347
Gillis, A.J., 404–405, 413
Gilmour, S.K., 141–142
Ginder, G.D., 388
Gingeras, T.R., 422–423
Giotopoulos, G., 323
Giri, N.C., 282–283
Givens, J., 269, 272–273
Glass, C.K., 141–142
Glozak, M.A., 163
Gnerlich, F., 366
Gobbo, A., 61–62
Gocke, C.B., 243–244, 248–249, 252, 263–264, 269–271
Goda, S., 287
Godek, K.M., 68–69
Goff, L., 460–463
Goff, S.P., 389–390
Goldberg, A.D., 106–108
Gong, W., 366–368, 373–374
Gonzalez, J.M., 462
Gonzalez-Montalban, N., 221
Gouet, P., 327*f*
Gough, J., 422–423
Gouvedenou, K., 323
Gowher, H., 387–388
Gowrishankar, K., 323

Goya, R., 217–218
Goyal, P., 378–379
Gozani, O., 211–213, 280, 287, 293, 295
Graceffa, R., 34
Grande, C., 211
Grandin, N., 406–407
Grant, G.R., 460–461
Grant, P.A., 211–213
Gräslund, S., 347
Graveley, B.R., 425
Greco, T.M., 179–180
Green, I., 211
Greer, E.L., 217–218, 346–347
Greetham, M., 407
Gregory, B., 164
Grishin, N.V., 219–221
Gronemeyer, T., 99–100
Gronenborn, A.M., 336
Groner, A.C., 389–390
Gronlund, J.L., 288–289
Gros, P., 338
Grosse-Kunstleve, R.W., 204–205, 227–228, 335–336, 338
Grosso, A.R., 211–213
Groulx, A., 216
Grunstein, M., 167, 184–186, 322–323
Grzesiek, S., 332, 336, 338
Gu, B., 281–282
Gu, Y., 4–37
Guan, X., 454
Guarente, L., 167
Guenther, M.G., 422–423
Guigo, R., 422, 425
Guise, A.J., 179–180
Gunatilake, D., 323
Gundimella, S.K.Y., 12
Guo, G., 217–218
Guo, L.Y., 20, 34, 68–69, 79–80, 83–84, 93, 211–213
Gupta, S., 422–423
Gur, T., 327*f*
Gurard-Levin, Z.A., 98–99
Guschin, D.Y., 307
Guse, A., 68–69
Gutierrez, M.I., 140
Guttman, M., 423
Guy, J., 388
Gyapay, G., 423

H

Habeck, M., 338
Haber, D.A., 387–389
Haber, J.E., 98–99
Hacha, R., 57
Hafner, M., 435–440
Haigis, M.C., 184–186
Hajian, T., 266, 287–288
Haley, R.H., 140, 145–146, 151
Hall, L.E., 446
Hall, M.A., 8, 20
Hallows, W.C., 198–199, 199*t*
Hallson, G., 211–213
Hamed, R.B., 287–288
Hamiche, A., 119–120, 129–131
Hamkalo, B.A., 83–84
Hammarstrom, M., 218–219
Hamuro, Y., 190*t*, 204
Hansen, J.C., 4–6, 13, 24–27, 32–33, 35
Hansen, R.S., 388–389
Hansen, S.K., 83–84
Hanson, B.L., 57
Hard, T., 218–219
Hare, S., 44
Harp, J.M., 57, 83–84
Harrow, J., 462
Harte, P.J., 210–211
Hartman, S., 423
Hartmanis, M., 218–219
Hasegawa, M., 273–274
Hashimoto, H., 381, 387–395
Hassan, A.H., 323
Hassig, C.A., 167–168
Hasson, D., 83–84
Hatanaka, Y., 114
Hattori, T., 264
Hauk, G., 211
Hausinger, R.P., 281–282
Hausser, J., 435–440
Hawkins, R.D., 387–388
Hayes, C.S., 141–142
Hayes, D.N., 454
Hayes, J.J., 84–85
Hayflick, L., 404
Hayward, D., 262–275
He, B., 184–186
He, C., 322–323, 366, 369–370
He, H., 213, 216, 218–219

He, S., 288–289
He, X., 119–120, 128, 243–245, 253, 257, 264, 454
He, Y., 322–323
He, Y.F., 366–368, 378–379, 387–388
Heard, E., 455–457
Hegenauer, J., 289–290
Heger, A., 459
Heinis, C., 113–114
Heinrich, N., 288–289
Heisi, C., 264
Helfand, B.T., 156–157
Helin, K., 216–218, 242, 287–288, 346–347
Hellgren, N., 218–219
Hempei, K., 264
Hempel, K., 243–245, 253, 257
Henderson Pozzi, M., 282–283, 292–293
Hendrich, B., 388–389
Hendrich, B.D., 422
Hendrick, L.L., 405–406
Hendzel, M.J., 167, 213
Henikoff, S., 98, 345–346
Henrici, R.C., 44, 46, 48
Hepler, M.R.D., 4–6, 8, 12, 14, 32–33
Herranz, J., 149–150
Herrmann, H., 323
Herst, P., 305, 307–309, 313, 315, 317
Herz, H.M., 211–213, 228–229
Hesman, T., 211–213
Hesse, O., 218–219
Heus, H.A., 377
Hidalgo-Curtis, C.E., 217–218
Hieb, A.R., 4–6, 8, 12, 14, 32–33, 48
Hierl, M., 287–288
Hill, J.M., 264, 272
Hill, W.A., 287–288
Hilton, D.J., 211–213
Hite, K., 13, 29, 32–33
Ho, K.L., 388
Hoadley, K.A., 454
Hobbs, C.A., 141–142
Hochedlinger, K., 114
Hochuli, E., 218–219
Hockemeyer, D., 309
Hoelz, A., 248–249
Hofmann, A., 210–211
Hoffman, H., 404–417
Hoke, H.A., 243–244, 257
Hollebakken, R.E., 211–213
Hollis, T., 56
Holmes, M.C., 307
Holshouser, S.L., 271–272
Holz, C., 218–219
Hon, G., 387–388
Hon, G.C., 374
Honda, B.M., 98–99, 211–213
Hong, K., 366–368, 387–388
Hong, X., 282, 285–287, 293
Hoogland, C., 350
Hopfgartner, B., 114
Hopkinson, R.J., 262, 281–282, 287–288
Horard, B., 211–213
Horowitz-Scherer, R.A., 13, 29, 32–33
Horton, J.R., 211–215, 218–219, 280–282, 287, 293, 296, 388, 390
Hosoda, F., 141–142
Hou, H., 243–244, 262–263, 373, 378–379
Hou, T., 323
Houssier, C., 57
Houston, S.I., 213
Houtz, R.L., 211, 216–217
Howard, C.J., 228–229
Howard, T., 211
Howell, S.A., 218–219
Hsieh, C.L., 388–389
Hu, D., 263–264
Hu, L., 264, 366–368, 373–374, 378–379
Hua, X., 217–218
Huang, B., 268–269, 271–272
Huang, H., 267
Huang, J., 257, 346
Huang, R., 264–265, 267
Huang, Y., 271–272, 280
Huard, C.C., 305, 307–309, 313, 315, 317
Hubalek, F., 262–263
Hudson, T.J., 423
Huet, J., 61–62
Hughes, A.L., 37
Hughes, R.M., 346
Humphrey, W., 35
Hung, L.W., 227–228
Huntly, B.J., 323
Huppertz, I., 440–441
Hurley, J.H., 211
Hutt, K.R., 243–244
Hwang, K.K., 211–213
Hyman, B.T., 141–142

I

Ichiyanagi, K., 388–389
Iida, K., 273–274
Iizuka, M., 140
Ikegami, T., 388
Ilatovskiy, A.V., 29–30
Ilott, N.E., 459
Imai, S., 167
Inagaki, K., 388–389
Incarnato, D., 378
Inglese, J., 149–152, 156–157
Ingram, R.S., 422
Ingram, V., 387–388
Inokuchi, J., 216
Inoue, A., 167, 179–180
Inoue, K., 114
Ioerger, T.R., 227–228
Iovino, N., 211–213
Irie, R., 422
Irving, T.C., 4–37
Isagawa, T., 287
Ishiuchi, T., 114
Issa, G.C., 323
Ito, M., 389–390
Ito, S., 366–368, 387–388
Iurlaro, M., 366
Ivanov, G.S., 211
Iwahara, J., 388–389
Iwase, S., 243–244, 248–249, 252

J

Jackson, B., 186–187, 203
Jackson, J., 211–213
Jacob, Y., 211–215, 218–219
Jacobs, H.M., 323
Jacobs, S.A., 407–408
Jacobsen, S.E., 388
Jacobson, R.H., 322–323
Jacobson, S.J., 140–142
Jacquet, K., 305, 307–309, 313, 315, 317
Jadhav, A., 151–152, 156–157, 290–292
Jaenisch, R., 309, 387–388
Jakobsson, J., 389–390
James, W., 323
Jammula, S., 211–213
Jansen, L.E., 99–102, 113–114
Jansen, L.E.T., 68–69
Jansen, P.W., 280, 366
Jantz, D., 394
Jeanmougin, F., 322–323
Jee, J., 388
Jeffrey, K.L., 323
Jeltsch, A., 228, 387–388
Jennings, M.J., 46, 48, 263–264
Jensen, K.B., 435
Jensen, M.R., 216
Jenuwein, T., 210–213, 242, 389*f*
Jenuwien, T., 242
Ji, P., 422
Jiang, F., 185*f*
Jiang, H., 187–188, 204
Jiang, J., 44, 68–69, 405–406, 413
Jiang, Y., 186–187, 217–218
Jiao, L., 213–216, 218–219
Jin, C., 243–245, 253, 257, 264
Jin, L., 186–187
Jin, V.X., 389–390
Jing, C., 211–215, 218–219, 228
Jinno, D., 369–370
Joberty, G., 287, 293
Johansson, C., 281–282, 287
Johnson, B.A., 332, 338
Johnson, J., 140, 142–144
Johnson, K.S., 218–219
Johnson, N.A., 217–218
Johnson, N.D., 387–388
Johnson, O., 335
Johnson, R., 425
Johnson, R.L., 149–150
Johnsson, K., 99–100, 113–114
Johnston, M., 211–213
Johnston, P.A., 151
Johnstone, A.L., 323
Johnstone, S.E., 174
Jones, A.V., 217–218
Jones, D.T., 219–221
Jones, P.L., 167
Jones, R.S., 210–211
Jorgensen, H., 388–389
Joshi, P., 179–180
Joshua-Tor, L., 218–219
Joukov, V., 44
Juillerat, A., 113–114
Jung, Y.L., 114
Justin, N., 211

K

Kachholz, B., 190*t*
Kadonaga, J.T., 119–120
Kaeberlein, M., 167
Kaehlcke, K., 272
Kagawa, W., 68–69
Kagey, J.D., 213
Kahl, S.D., 144–145
Kahles, A., 460–461
Kakazu, N., 141–142
Kalashnikova, A.A., 13, 26–27
Kalin, J.H., 269, 272–273
Kamada, M., 272
Kamat, C.D., 269–271, 274
Kamimura, S., 114
Kamps, J.J.A.G., 346
Kang, Z., 287–288
Kapadnis, P.B., 288–289
Kaplan, M., 441
Kapoor-Vazirani, P., 213
Kapopoulou, A., 389–390
Kappler, J., 282, 293
Kapral, G.J., 227–228
Kapranov, P., 423
Kaptein, R., 338
Kapust, R.B., 283–284
Kar, S.R., 26
Karaman, B., 190*t*, 204
Karimi, M.M., 378–379
Karpen, G.H., 100–102, 119–120
Karytinos, A., 269–271
Kasten, M., 323
Kasukawa, T., 422–423
Katagiri, Z., 211–213
Katayama, S., 422–423
Kathuria, S.V., 33–34
Katibah, G.E., 307
Kato, H., 44, 68–69
Katz, Y., 460
Kauer, M., 213
Kaufman, P.D., 98–99
Kaufman, T.C., 322–323
Kavanagh, K.L., 282–283, 293
Kawahashi, K., 211–213
Kawamura, A., 262, 281–282, 287–288
Kawamura, T., 287
Kay, G.F., 422
Kay, L.E., 336
Kayatekin, C., 34
Kayikci, M., 435, 440–441
Ke, Q., 282–283
Keates, T., 323
Keefe, L.J., 213–215, 218–219
Keen, N., 211
Kelley, D.R., 460–463
Kelley, R., 461–462
Kelley-Clarke, B., 44
Kelly, G., 211–215, 218–219, 228
Kelly, M.J., 287, 295
Kennedy, A., 164
Kennison, J.A., 322–323
Kent, W.J., 461
Keppler, A., 99–100
Kerenyi, M.A., 217–218
Kertesz, M., 423, 435
Kessler, B.M., 287–288
Khachigian, L.M., 389*f*
Khan, S.I., 187–188, 204, 211–215, 218–219, 280–281
Kharchenko, P.V., 441
Khatri, P., 280
Khavari, P.A., 435
Kheradpour, P., 345–346, 459
Khochbin, S., 12, 184–186
Khorasanizadeh, S., 280, 346
Khorshid, M., 435–440
Kibbe, W.A., 12
Kikhney, A.G., 35
Kilic, S., 4–6, 8
Kim, A., 374
Kim, B.H., 219–221
Kim, D., 460–463
Kim, H., 211–213
Kim, I., 211–213
Kim, N.W., 404
Kim, S., 211–213
Kim, S.-A., 46, 48, 263–264
Kim, W., 217–218
Kindermann, M., 113–114
King, D.S., 322–323
King, O.N., 287–288
Kingsbury, J.S., 26
Kingston, R.E., 44, 57, 119–120, 128, 216, 441

Kinoshita, A., 141–142
Kioi, M., 211–213
Kioussis, D., 211–213
Kissinger, M., 322–323
Kitabayashi, I., 141–142
Kittler, R., 272
Kitz, R., 268–269
Kizaki, S., 374, 379
Klare, K., 113–114
Klaus, C.R., 217–218
Kleer, C.G., 217–218
Klein, B.J., 8
Klose, R.J., 281–282, 388
Klug, A., 57, 61
Kluz, T., 282–283
Knapp, S., 218–219, 323, 347
Knowles, B.B., 389–390
Knutson, S.K., 217–218
Kobayashi, H., 387–388
Koch, F., 211–213
Kodama, E.N., 114
Koh, K.P., 366–368, 387–388
Kohli, R.M., 366–383
Kokocinski, F., 462
Kokura, K., 264–268
Kole, R., 422
Kollipara, R.K., 272
Kondo, S., 422
Kondrashkina, E., 34
Konesky, K., 4–6, 20
Kong, F., 264
Konig, J., 440–441
Kontogiannis, L., 335
Koo, E.H., 141–142
Kopp, N., 323
Koppetsch, K.J., 185*f*
Korban, R., 388–389
Korenchuk, S., 217–218
Korge, G., 210–211
Kornberg, R.D., 134
Koseki, H., 388
Kosho, T., 388–389
Kouskouti, A., 211
Kouzarides, T., 242, 323, 345–346
Kozlov, A.G., 358
Kramer, H.B., 287–288
Kramer, M., 4–6, 5*f*, 8, 12, 14, 24–26, 32–33
Kramer, M.A., 48
Krauss, V., 210–211
Krebs, C.J., 389–390
Krepelova, A., 378
Kretz, M., 435
Krishnan, S., 280–296, 286*t*
Kristie, T.M., 264
Kruidenier, L., 287, 293
Kubicek, S., 211–213
Kubista, M., 46
Kubo, N., 387–388
Kuboniwa, H., 336
Kubota, T., 388–389
Kuchenbecker, K.M., 287, 295
Kumar-Sinha, C., 217–218
Kumazawa, A., 388–389
Kundaje, A., 459
Kuntz, K.W., 217–218
Kuo, A.J., 211–213, 287, 293, 295
Kurash, J.K., 211
Kurumizaka, H., 68–69
Kusch, T., 211–213
Kutateladze, T.G., 8, 280, 285–287, 345–361
Kutz, C.J., 271–272
Kwak, H., 434
Kwon, H.-S., 272
Kwon, Y.S., 243–244

L

Labonte, A., 287–288
Lachner, M., 213
Lacoste, N., 98–102, 105–108, 112–114
Ladurner, A.G., 322–323
Ladwein, K.I., 264–267
Lafon, A., 140–142
Lafreniere, R.G., 422
Lai, F., 422–441
Laible, G., 210–211, 389*f*
Laity, J.H., 390
Lakshminarasimhan, M., 186–189, 190*t*, 196–198, 201–205, 202*f*, 203*t*
Lalonde, M.-E., 8, 346
Lambert, J.P., 323
Lan, F., 242–244, 248–249, 252, 255–256, 262–264, 280–281, 287–288, 435
Lander, E.S., 423, 441
Landry, J., 211–213
Landt, S.G., 459

Landthaler, M., 435–440
Lang, C., 218–219
Langmead, B., 460–461
Langst, G., 119–120, 129–131
Laniel, M.A., 210
Lanouette, S., 216–218, 280
Lapunzina, P., 211–213
Larsen, M.J., 154
Larsen, N.A., 211
Laskey, R.A., 98–99
Laskowski, R.A., 336, 338
Lassmann, T., 422
Latham, J.A., 211
Lau, E., 289–290
Lau, H.T., 389–390
Laube, S.K., 369–370
Laudano, A., 387–388
Laue, T.M., 26
Laurent, B., 242–257, 264
Lauring, J., 211–213
Lavault, B., 61–62
Lawrence, J., 422
Laxman, B., 217–218
Lazo, J.S., 151
Lazzerini Denchi, E., 216
Lea, S.M., 227–228
Leavens, B., 287–288
Leaver-Fay, A., 227–228
LeBlanc, C.A., 211–215, 218–219
LeDouarin, B., 322–323
Lee, C., 323
Lee, H., 216
Lee, J., 68–69
Lee, J.E., 190*t*, 204
Lee, J.H., 287–288
Lee, J.K., 381
Lee, J.T., 435
Lee, J.W., 288–289
Lee, K.K., 162–163
Lee, K.M., 84–85
Lee, M.G., 252, 257, 263–264, 268–269, 281–282
Lee, M.S., 405–406
Lee, T.H., 68–69
Lee, T.I., 174
Lee, W., 336
Legault, M., 153–154
Lehotzky, A., 190*t*, 204
Lei, M., 216, 248–249
Leiber, M., 454
Lemieux, M.E., 323
Leonard, J.D., 128
Lerner, M.R., 422
Lesbats, P., 44
Leslie, A.G.W., 335
Letunic, I., 210–211, 327*f*, 347
Leveridge, M., 287–288
Lewis, M.S., 26
Lewis, P.W., 106–108
Li, B.Z., 366–368, 378–379, 387–388
Li, E., 387–389
Li, F., 213, 216, 218–219
Li, F.-Y., 154
Li, L., 167
Li, M., 8, 20, 142, 262–263
Li, Q., 282–283
Li, S., 322–323
Li, S.S., 280
Li, T., 211–213
Li, W., 243–245, 253, 257, 264, 327*f*
Li, X., 269, 272–273
Li, X.S., 287–288
Li, Y., 211–213, 271–272, 322–323, 388
Li, Z., 217–218, 366–368, 373–374, 378–379, 387–388
Liang, Y., 264
Licatalosi, D.D., 435
Liddle, J., 287, 293
Lienard, B.M., 282–283, 293
Lienlaf, M., 164, 177–178
Lilyestrom, W., 4–6
Lin, H., 184–186
Lin, J.J., 406
Lin, K., 213–215
Lin, M.F., 423
Lin, S., 267
Lin, W., 211
Lin, Y.-Y., 140
Lindsay, S., 29–30
Lingner, J., 405–407
Lipinski, M., 98–99
Lis, J.T., 211–213, 428–434
Lissner, M.M., 425–426
Lister, R., 387–388
Liu, B., 323
Liu, E.T., 425

Liu, M.Y., 366–383
Liu, P., 366–368, 378–379, 387–388
Liu, Q., 289–290
Liu, R.J., 323
Liu, S., 280
Liu, X., 213–216, 218–219, 268–269, 271–272, 296
Liu, Y., 216, 388–390, 393
Liu, Z., 217–218
Lizcano, J.M., 290–292
Lleonart, M., 141–142
Loehr, J., 305, 307–309, 313, 315, 317
Loew, D., 113–114
Logan, T.M., 336
Logsdon, G.A., 20, 68–69, 79–80, 83–84, 93
Lohkamp, B., 227–228, 295
Lohman, T.M., 358
Lohr, D., 29–30
Lohse, M., 460
Long, A.M., 287–288
Long, F., 335–336
Lorch, Y., 134
Losson, R., 211–213, 322–323
Lowary, P., 4–6
Lowary, P.T., 46, 68–69, 121, 229
Lowe, N., 322–323
Loyola, A., 288–289
Lu, J., 373, 378–379
Lu, J.-Y., 140
Lu, M., 348–350
Lu, X., 4–6, 13, 32–33
Lu, Z., 454
Lucio-Eterovic, A.K., 211–213
Luger, K., 4–37, 5*f*, 44–48, 57, 68–69, 71, 74, 79, 83–88, 121, 125, 210, 228–229
Lundblad, V., 406–407
Luo, H., 264
Luo, P., 388–389
Luo, X., 243–244, 248–249, 252, 263–264
Luo, Y., 4–6, 10
Luscombe, N.M., 423
Lusic, M., 140
Lyko, F., 387–388
Lyubchenko, Y.L., 29–30

M

MacArthur, M.W., 336, 338
Machius, M., 243–244, 248–249, 252, 263–264
Mackay, D.J., 389–390
Mackeen, M.M., 287–288
Madden, S.K., 281–282, 287
Mäder, A.W., 7–8, 57, 83–84, 210
Madrid, T.S., 120–121
Maeda, N., 422–423
Maeder, A.W., 46
Maes, T., 281–282
Magdinier, F., 211–213
Mahadevan, L.C., 211–213
Mahajan, M., 211–213
Mahapatra, S., 378–379
Mai, A., 269
Maiolica, A., 287–288
Maiti, A., 366, 381
Majewski, I.J., 211–213
Majumdar, A., 268–269, 282–283, 292–293
Makde, R.D., 44–63
Malaby, A.W., 33–34
Malani, N., 389–390
Malin, S., 213
Malliavin, T.E., 338
Malvaez, M., 272
Mandel, G., 243–244
Manganaro, L., 140
Mangos, M., 323
Mani, R.S., 217–218
Mani, T., 20, 68–69, 79–80, 83–84, 93
Manohar, M., 4–6
Manor, O., 435
Manton, C.A., 273
Manzur, K., 213, 215
Mao, C., 185*f*, 186–187
Mao, S.Q., 378–379
Margueron, R., 211–213, 304
Marheineke, K., 113
Marigo, I., 323
Marinov, G.K., 459
Marioni, J.C., 460–461
Markolovic, S., 281–282
Marks, S.M., 389–390
Marley, J., 348–350
Marmorstein, R., 140–158, 184–189, 190*t*, 203–204, 268–269
Maroli, A., 264
Maroney, M., 282–283
Marra, M.A., 217–218
Marshall, B.L., 189
Martienssen, R.A., 211–215, 218–219

Martin, C., 228
Martin, D.M., 220*f*
Martin, J., 149–150
Martin, K.W., 8
Martin, S.R., 218–219
Martinez, A.M., 211–213
Martinez-Yamout, M.A., 388–389
Martini, E.M., 98–99, 113
Martino, F., 47–48
Martone, R., 423
Maskell, D.P., 44
Mason, M., 406–407
Masuoka, J., 289–290
Matadeen, R., 44
Matera, A.G., 454
Matson, C., 242–244, 248–249, 252, 255–256, 262–264, 280–281, 287–288
Matsunaga, H., 369–370
Mattaliano, R., 387–388
Mattevi, A., 243–244, 248–249, 251, 256, 262–265, 282–283, 292–293
Mattiroli, F., 4–37
Maunakea, A.K., 345–346
Mazur, P.K., 280
McBryant, S.J., 13, 32–33
McCabe, M.T., 217–218
McCabe, V.M., 422
McCafferty, D.G., 268–269, 282–283, 288–289, 292–293
Mccammon, J.A., 35
McCoy, A.J., 204–205, 227–228, 335–336
McCullough, C.E., 140–158
McDonald, C.M., 389–390
McDonel, P., 441
McDonough, M.A., 281–283, 293
McGinty, R.K., 44–63
McGovern, S.L., 156–157
McGrath, J., 281–282
McKinley, K.L., 68–69
McLaughlin, S.H., 47–48
McMahon, S.B., 140, 184–186
McManus, K.J., 213
McManus, M., 164
McNae, I.W., 388
McQuade, T.J., 154
McWilliam, H., 327*f*
Mechtler, K., 211–213
Mecinovic, J., 282–283
Mehra, R., 217–218
Mei, A., 211–213
Mele, A., 435
Mele, D.A., 323
Meleshin, M., 200–201
Mellert, H.S., 140, 142–144
Melnick, A.M., 217–218
Merkel, A., 422
Mertz, J.A., 323
Meslamani, J., 322–323
Messerschmidt, D.M., 389–390
Metzger, E., 243–244, 264–267, 269
Metzger, R., 422
Meyer, A., 57
Meylan, S., 389–390
Michaels, S.D., 211–213
Michan, S., 184–186, 188–189
Michelson, A.Z., 381
Miga, K.H., 446, 461, 463
Mihola, O., 389–390
Mijatov, B., 323
Mikkelsen, T.S., 345–346
Milks, K.J., 68–69
Miller, J.C., 307
Milligan, R.A., 35
Mills, A.D., 98–99
Mimasu, S., 248–249
Min, J.H., 205, 211–213, 228–229
Ming, S.A., 269, 272–273
Minneci, F., 219–221
Mino, K., 273–274
Minor, W., 227–228, 295, 335
Minucci, S., 163
Misteli, T., 323
Mitalipova, M., 309
Mitchell, J.R., 405–406
Mitchell, M., 407, 414
Mitchell, S.E., 446
Mitton-Fry, R.M., 407
Miya, Y., 68–69
Miyake, Y., 406
Mizusawa, Y., 272
Mizutani, E., 114
Mohammad, H., 269–271, 274
Mohan, M., 211–213
Moks, T., 218–219
Molinie, B., 422–423
Mongeon, V., 211–215, 217–219, 280
Moniot, S., 186–189, 190*t*, 196–197, 203–205

Montes de Oca, R., 113–114
Montrose, C., 144–145
Moore, K.E., 280
Moradian, A., 228–229, 313
Moree, B., 68–69
Moreno, N., 100
Morgan, M.T., 381
Morin, R.D., 217–218
Morris, M., 57
Mortazavi, A., 422
Mosalaganti, S., 188–189, 190*t*, 196–197, 205
Mosammaparast, N., 280–282, 296, 435
Moshkina, N., 338
Moss, D.S., 336
Moustacchi, E., 98–99
Mueller, L., 332
Mueller, O., 454
Mueller, U., 61–62
Muhammad, S., 190*t*, 197
Muhandiram, D.R., 336
Muir, T.W., 4–6, 8
Mujtaba, S., 322–323
Mukai, T., 228–229
Mukamel, E.A., 387–388
Mullen, A.C., 422–423
Muller, C.W., 4–6
Müller, J.M., 243–244, 269
Muller, S., 98, 113–114, 323
Muller-Tidow, C., 422
Mulligan, P., 242–244, 255–256, 262–264, 280–281, 287–288
Mum, J., 264
Mungall, A.J., 217–218
Murn, J., 114, 243–245, 253, 257
Murray, K., 210–211
Murrell, A., 366, 369–370
Murshudov, G.N., 295, 335–336
Musacchio, A., 113–114
Muskett, F.W., 213–215, 228
Musselman, C.A., 280, 345–361
Muthurajan, U.M., 4–37, 5*f*, 68–69, 79, 84–85, 87–88, 121, 125, 229

N

Nagai, M., 141–142
Naiki, T., 273–274
Naiki-Ito, A., 273–274
Nair, B.C., 243–244
Nair, S.S., 243–244
Nakagawa, H., 273–274
Nakamura, Y., 113–114, 388
Nakao, M., 388–389
Nakatani, Y., 98–99
Nakayama, Y., 211–213
Narayanan, A., 264
Narlikar, G.J., 44, 57, 119–134
Natoli, G., 425–426
Nazarov, I., 29–30
Neelamegam, R., 272
Neely, K.E., 323
Neff, T., 217–218
Nelissen, F.H., 377
Nelson, H., 144–145
Nelson, J.D., 174
Nelson, S., 210–211
Neri, D., 336
Neri, F., 378
Nery, J.R., 387–388
Neumann, H., 288–289
Ng, J., 57
Ng, S.S., 282–283, 287–288, 290–293
Nguyen, D.P., 288–289
Nguyen, G.T., 186–188, 190*t*, 201–205, 202*f*, 203*t*
Nicholson, T.B., 257, 264
Nicodeme, E., 323
Niederhuth, C.E., 387–388
Nikitina, T., 13, 29, 32–33
Nilges, M., 338
Nilsson, B., 218–219
Ninova, M., 114
Nishikawa, T., 422
Nishioka, K., 288–289
Nitta, H., 388–389
Nnadi, C.I., 287, 295
Nobrega, R.P., 33–34
Noh, K.M., 106–108
Nordlund, P., 347
Norris, D.P., 422
North, B.J., 189, 190*t*, 204
North, J.A., 4–6, 10
Norton, S., 272
Norvell, A., 184–186
Nowak, T., 187–188, 190*t*, 204
Nowotarski, S.L., 271–272

Nugent, C.I., 407
Nugent, T.C., 219–221

O

O'Neill, R.J., 446
Oakes, B.L., 441
O'Carroll, D., 211–213
O'Connell, M.R., 441
Odho, Z., 213–216, 218–219
Offner, S., 389–390
O'Geen, H., 389–390
Ogonuki, N., 114
Ogura, A., 114
Ohba, R., 210–211
Ohgi, K.A., 141–142
Ohki, I., 388
Ohki, M., 141–142
Ohta, H., 267
Oikawa, M., 114
Oishi, K.K., 83–84
Okano, M., 387–389
Okazaki, Y., 422
Olanrewaju, Y.O., 388–390, 393–394
Oldenburg, K.R., 149–150
Olejniczak, E.T., 336
Oleskie, A.N., 216
Olia, A.S., 146
Oliver, P.L., 422–423
Ombra, M.N., 243–244
Oppermann, U., 281–283, 287–288, 293
Orkin, S.H., 217–218
Orlando, D.A., 243–244, 257, 422–423
Ornaghi, P., 322–323
Orom, U.A., 435
Ortega, A., 281–282
Ortiz Torres, I., 288–289
Ortiz-Tello, P.A., 282–285, 286*t*, 288–295
Osakabe, A., 68–69
Osanai, M., 404–405
O'Sullivan, R.J., 211–213
Ota, I., 217–218
Ota, T., 422
Otsuki, T., 422
Ott, C.J., 323
Ott, H.M., 217–218
Ottesen, J.J., 228–229
Otting, G., 336
Ottink, O.M., 377
Otwinowski, Z., 227–228, 295, 335
Owen, D.J., 322–323
Owen-Hughes, T., 119–120
Oxley, D., 366
Ozato, K., 323

P

Pabo, C.O., 390
Pachaiyappan, B., 271–272
Paech, C., 268
Paganini, L., 264
Pagans, S., 272
Pais, J.E., 379
Palacios, V., 272
Palmer, E.L., 83–84
Pan, M., 188–189
Pan, P.W., 190*t*, 204
Pan, Y., 264
Panchenko, T., 8, 83–84
Pandya-Jones, A., 425–426
Pannek, M., 187–188, 190*t*, 204
Panning, B., 288–289
Papatheodorou, P., 190*t*
Paranal, R.M., 323
Park, Y.-J., 4–6, 8, 20
Parker, J.S., 454
Parlato, C., 378
Parmely, T., 309
Pasini, D., 211–213, 216
Pastor, W.A., 366–368, 387–388
Pastori, C., 323
Pate, K.T., 8
Patel, A., 216, 387–395
Patel, D.J., 287, 293, 295, 346
Patnaik, D., 272
Pattatucci, A.M., 322–323
Paul, B., 141–142
Pauli, F., 459
Pavesi, G., 457
Pavletich, N.P., 390
Payne, R.M., 184–186
Peak-Chew, S.Y., 47–48
Pedersen, M.T., 242, 346–347
Pedro, L., 287–288
Pehrson, J.R., 12
Pei, J., 219–221
Pelicci, P.G., 163
Pelizzola, M., 387–388

Pellentz, C., 99–102, 105–108, 112
Penas, C., 323
Peng, C., 187–188
Peng, D., 322–323
Peng, H., 186–187, 264
Pennock, E., 406–407
Perche, P.-Y., 12
Perez-Burgos, L., 211–213
Perez-Villarroel, P., 164, 177–178
Perillo, B., 243–244
Perou, C.M., 454
Perrakis, A., 218–219
Pertea, G., 460–463
Peters, A.H., 211–213, 243–244
Peters, A.H.F.M., 269
Peterson, C., 4–6
Peterson, C.L., 210, 288–289
Petosa, C., 4–6
Petoukhov, M.V., 35
Pfaffeneder, T., 366, 369–370
Pfeifer, J., 332, 338
Pflugrath, J.W., 227–228
Picaud, S., 323
Pick, H., 99–100
Pieters, B.J.G.E., 346
Pijnappel, W.W., 211–213
Pilka, E.S., 282–283, 293
Pillus, L., 140–142
Pilotto, S., 263–264, 269–271
Pimentel, H., 461–462
Piotukh, K., 288–289
Plotnikov, A.N., 322–323, 333*f*, 336
Plotnikova, O., 322–323
Poater, J., 346
Podell, E.R., 407–408
Poirier, M.G., 4–6, 10, 228–229
Pollard, H., 211
Polo, S.E., 100, 108–111
Ponting, C.P., 422–423, 459
Ponzoni, L., 264
Popovic, R., 217–218
Portela, A., 346–347
Porter-Goff, M.E., 13, 26–27
Powell, H.R., 335
Pradhan, S., 381
Prasad, R., 4–6
Prefontaine, G.G., 243–244
Prinjha, R.K., 323
Probst, A.V., 114
Prochasson, P., 323
Procter, J., 347
Procter, J.B., 220*f*
Prokhortchouk, A., 388–389
Prusevich, P., 269, 272–273
Pryor, E.E., 56
Puddifoot, C.A., 387–388
Puri, A., 99–102, 105–108, 112
Purich, D.L., 268–269
Pusey, M., 57

Q

Qi, H., 407
Qi, H.H., 281–282, 287, 293
Qi, J., 323
Qi, X., 405–406
Qian, C., 213, 215
Qian, D., 185*f*
Qiu, J., 146
Qu, K., 435, 441
Quenelle, D.C., 264, 272
Quenet, D., 446–464
Quenneville, S., 389–390
Quinn, A.M., 290–292
Quivy, J.P., 98–99

R

Racki, L.R., 288–289
Rahl, P.B., 323
Raiber, E.A., 366, 369–370
Raj, A., 455–457
Ramon y Cajal, S., 141–142
Ramsahoye, B.H., 387–388
Rando, O.J., 37
Rangasamy, D., 4–6, 20
Rao, N., 289–290
Rao, Q., 366–368, 373–374, 378–379
Rapelli, S., 378
Rappaport, A.R., 323
Rappsilber, J., 287–288
Rashidi, H., 27
Rastan, S., 422
Rathert, P., 228
Ratsch, G., 460–461
Rauh, D., 186–187, 190*t*
Raunser, S., 188–189, 190*t*, 196–197, 205
Rauth, D., 205
Ray, S., 323
Ray-Gallet, D., 98–114

Rea, S., 211–213
Read, R.J., 204–205, 227–228, 335–336
Rebar, E.J., 307
Rechsteiner, T.J., 6–7, 44–47, 68–69, 74, 84–86, 121, 125, 228–229
Redon, S., 406–407
Reed, S.I., 406–407
Reik, W., 366, 422–423
Reinberg, D., 211–213, 304
Ren, C., 322–340
Renault, L., 44
Resch, M., 4–6
Reuter, G., 210–211, 213, 389*f*
Reynoird, N., 280
Rezai-Zadeh, N., 167–168
Rhodes, D., 47–48, 57, 61
Rice, J.C., 211–213
Rice, K., 423
Richardson, D.C., 227–228
Richmond, R.K., 7–8, 57, 83–84, 210
Richmond, T.J., 6–8, 44–47, 57, 61, 68–69, 74, 83–86, 121, 125, 210, 228–229
Richter, M., 257
Ricq, E.L., 272
Riefler, G.M., 211
Rieping, W., 338
Riera, T.V., 190*t*, 204
Riester, D., 167
Rifkin, S.A., 455–457
Rinn, J.L., 423, 435
Ristriani, T., 211–213
Robart, A.R., 404–405
Roberson, A.E., 83–84
Robert, X., 327*f*
Roberts, A., 460–463
Roberts, V.A., 4–6
Robins, D.M., 389–390
Roby, P., 153–154
Roche, D.M., 98–99, 113–114
Rodriguez, M.G., 100
Rodriguez-Suarez, R., 287–288
Rodriguez-Terrones, D., 114
Roe, S.M., 213–216, 218–219
Roessler, C., 186–188, 190*t*, 198, 200–201, 204
Rogers, D.J., 68–69
Rogge, R.A., 13, 26–27
Roguev, A., 211–213
Rojo, F., 141–142
Romanenghi, M., 269–271
Romano, A., 388–389
Romero, D.P., 405–406
Romier, C., 218–219
Ropero, S., 211–213
Roques, C., 305, 307–309, 313, 315, 317
Rose, D.W., 141–142
Rose, N.R., 282–283, 287–288, 290–292
Rosen, M., 217–218
Rosenberg, A.H., 218–219
Rosenbloom, K.R., 463
Rosenfeld, M.G., 141–142, 243–245, 253, 257
Rossetto, D., 140, 142–144, 304
Rosu, S., 4–6, 20
Rot, G., 440–441
Rothbart, S.B., 287–288
Rothbauer, U., 201–202
Rouda, S., 407–408
Rouleau, N., 153–154
Rousseaux, S., 184–186
Roversi, P., 227–228
Rowe, C.E., 134
Roy, T.W., 289–290
Royce, T.E., 422
Rozendaal, M., 44, 68–69
Rozowsky, J.S., 422
Ruan, Y., 425
Ruggiu, M., 435
Ruitu, L., 243–245, 253, 257, 264
Rullmannn, J.A., 338
Rumpf, T., 190*t*, 204
Rundlett, S.E., 167
Rung, J., 460–461
Rupert, J.L., 422
Rupp, B., 203–205
Rusconi, F., 264
Rushton, B., 61
Rusinova, E., 333*f*, 336
Russell, P., 441
Russi, S., 61–62
Ruzov, A., 388–389
Rychkov, G., 29–30
Ryder, O.A., 83–84

S

Sacchetti, S., 243–244
Sachchidanand, 213, 215, 322–323
Sado, T., 388–389

Sahakian, E., 162–180
Saito, K., 68–69
Sakaguchi, A., 213
Sakane, C., 267
Sakane, N., 272
Sakurai, M., 287–288, 290–292
Salach, J.I., 268
Salas, J., 281–282
Salem, N., 388–389
Salimian, K.J., 83–84
Salman, M.U., 83–84
Salowsky, R., 454
Saltman, P., 289–290
Salzberg, S.L., 460–462
Sanchez, R., 322–323
Sanchez-Weatherby, J., 61–62
Sanda, M.G., 217–218
Sandaltzopoulos, R., 119–120, 129–131
Sanders, B.D., 186–187, 203
Sandy, P., 323
Santoro, S.W., 98
Sapountzi, V., 140
Sargent, D.F., 7–8, 46, 57, 83–84, 210
Sarma, K., 213
Sarvan, S., 216
Sasai, N., 388–389
Sasaki, H., 388–390, 393
Sasazuki, T., 216
Sasso, A., 243–244
Sato, S., 228–229
Sato, T., 387–388
Sauerwald, A., 413
Sauve, A.A., 184–187
Scamps, C., 98–99
Scarsdale, J.N., 388
Scelfo, A., 211–213
Schacher, A., 218–219
Schaefer, S., 204–205
Schaefer, U., 323
Schaft, D., 211–213
Schapira, M., 213, 216, 218–219, 228–229, 313
Scharfe, M., 187–188, 190*t*, 204
Scheer, E., 211
Schiedel, M., 190*t*, 204
Schierhorn, A., 184–186
Schimenti, J.C., 389–390
Schlicker, C., 187–188, 190*t*, 204
Schmid, M., 211–213
Schmidt, C.A., 454
Schmidt, D.M.Z., 268–269, 282–283, 292–293
Schmidt, J.C., 404–405
Schmiedeberg, L., 388
Schmitt, M.L., 264–267
Schneider, J., 281–282
Schneider, P.M., 422
Schneider, R., 243–244, 269
Schneider, S., 389–390
Schnitzler, G.R., 228–229
Schofield, C.J., 262, 281–283, 287–288
Scholle, M.D., 267
Schotta, G., 141–142, 213
Schramm, V.L., 184–186
Schreiber, S.L., 167–168
Schroeder, A., 454
Schuck, P., 26
Schuettengruber, B., 211–213
Schuetz, A., 205
Schüle, R., 243–244
Schuler, G.D., 423
Schuller, A.P., 404–405, 413
Schulten, K., 35
Schultz, D.C., 389–390
Schultz, L., 290–292
Schultz, P.G., 288–289
Schulz-Fincke, J., 264–267
Schumacher, J.M., 211
Schutkowski, M., 184–188, 190*t*, 198, 200–205, 202*f*, 203*t*
Schwarzer, D., 288–289
Schwienhorst, A., 167
Score, J., 217–218
Scott, E.M., 140–142
Scott, I., 184–186
Scott, J.E., 149–151
Scott, M.P., 322–323
Scott, W.G., 227–228, 295
Sekirnik, R., 282–283
Sekulic, N., 20, 68–95
Selfridge, J., 388
Selker, E.U., 211–215, 218–219, 280–281
Sengoku, T., 248–249, 287, 293
Sengupta, N., 163
Sengupta, R., 213, 257
Senisterra, G., 213, 216, 218–219

Senn, H., 336
Sennett, M.A., 68–69
Serrao, E., 44
Setien, F., 211–213
Seto, E., 162–180
Sevastopoulos, E., 120–121
Severson, T.M., 217–218
Shah, J.V., 100–102
Shahbazian, M.D., 184–186
Shankar, S., 217–218
Shanks, J.R., 296
Sharif, J., 388
Shay, J.W., 405–406
Sheffield, P., 221
Shen, D., 289–290
Shen, L., 366, 373, 387–388
Shen, R., 217–218
Shen, Y., 323, 338, 366–368, 387–388
Sheng, J., 389–390
Sherrill, J., 267
Shevchenko, A., 211–213
Shi, J.W., 323
Shi, X., 280
Shi, Y., 217–218, 242–257, 262–264, 280–282, 287–288, 293, 296, 346–347
Shi, Y.J., 243–244, 248–249, 252
Shiau, C., 288–289
Shibata, M., 389–390
Shidoji, Y., 267
Shiekhattar, R., 252, 263–264, 268–269, 422–441
Shiga, T., 68–69
Shigemura, T., 388–389
Shilatifard, A., 211–213, 216, 228–229, 281–282
Shim, Y., 228–229
Shimko, J.C., 4–6, 228–229
Shimotake, N., 388
Shinde, S.N., 151
Shinkai, Y., 211–213
Shippen-Lentz, D., 405–406
Shirakawa, M., 388
Shirakawa, T., 387–388
Shirane, K., 387–388
Shirasawa, S., 216
Shirohzu, H., 388–389
Shirouzu, M., 228–229
Shishkin, A.A., 441
Shkumatov, A.V., 35
Shlyakhtenko, L.S., 29–30
Shogren-Knaak, M.A., 288–289
Shoichet, B.K., 151–152, 156–157
Shokat, K.M., 4–6, 228–229, 288–289
Shoresh, N., 345–346
Shows, T., 389*f*
Shun, T.Y., 151
Sibley, C.R., 440–441
Sigova, A.A., 422–423
Silva, M.C.C., 68–69
Silverman, R.B., 268–269
Simeonov, A., 149–152, 156–157, 287–288, 290–292
Simic, Z., 184–186
Simon, J., 211–213
Simon, M.D., 4–6, 228–229, 288–289, 441
Simon, M.I., 268–269
Sims, D., 459
Sims, J.K., 213
Sinclair, D.A., 184–186, 188–189
Singer, A.U., 336
Singer, T.P., 268
Singh, M.M., 211–213, 273
Singhal, A., 57
Sipos, B., 460–461
Siprashvili, Z., 435
Sirokman, K., 441
Sison, E.A., 323
Sittampalam, G.S., 144–145
Sittampalam, S., 144–145
Skiniotis, G., 216
Skordalakes, E., 404–417
Slattery, M., 264
Smale, S.T., 425–426
Smerdon, M.J., 228–229
Smith, B.C., 198–199, 199*t*
Smith, D.B., 218–219
Smith, E.R., 211–213
Smith, G.E., 218–219
Smith, J.S., 190*t*, 197
Smith, M.C., 144–145
Smith, S., 98–99
Smith, W.B., 323
Smitherman, K.N., 269–271, 274
Smits, A.H., 366
Smoak, E.M., 20, 68–69, 79–80, 83–84, 93
Sneeringer, C.J., 217–218

Snyder, M., 422–423
Soares, K.M., 151
Soldner, F., 309
Solter, D., 389–390
Song, C.X., 366, 369–370, 374
Song, J.J., 216, 435
Sonoda, M., 216
Soong, D., 269–271, 274
Sorensen, T.L.-M., 61–62
Southall, S.M., 213–216, 218–219
Sowers, L.C., 366–368, 387–388
Spada, F., 369–370
Spector, D.L., 422–423
Speicher, D.W., 47, 146
Spencer, V.A., 167
Speranzini, V., 263–264, 269
Spitzer, J., 435–440
Spruijt, C.G., 366
Squazzo, S.L., 423, 435
Squizzato, S., 327*f*
Srinivasan, M., 140, 142–144
Stadler, S., 106–108
Stahl, U., 218–219
Stamatoyannopoulos, J.A., 422
Stanek, A.M., 388–389
Stanfield, R.L., 388–389
Stargell, L.A., 8
Stark, B.C., 422
Stassen, M.J., 210–211
Stathopoulou, A., 387–388
Staub, A., 211
Stavropoulos, P., 248–249
Steegborn, C., 184–205, 190*t*
Steijger, T., 460–461
Stein, L.D., 423
Steitz, J.A., 422
Sternberg, S.H., 441
Sternglanz, R., 211–213
Steunou, A.-L., 304
Steward, R., 213
Stewart, A.F., 211–213
Stewart, E.A., 423
Stewart, R., 264
Stillman, B., 98–99, 140
Stocker, S., 454
Stolc, V., 422
Storoni, L.C., 204–205, 227–228, 335–336
Straasheijm, K.R., 388–389
Strahl, B.D., 210–213, 242, 287–288, 322–323
Straight, A.F., 68–69
Strausberg, R.L., 423
Stroud, H., 211–213
Struck, M.M., 61
Struhl, K., 322–323
Studier, F.W., 218–219, 391
Sturgill, D., 446–464
Stutzer, A., 211–213
Su, X., 184–188, 204
Sudbery, I., 459
Sudhof, T.C., 141–142
Suenkel, B., 184–205, 190*t*
Sugano, S., 422
Sugimoto, Y., 440–441
Sugiyama, H., 374, 379
Sugiyama, T., 422
Summers, M.D., 218–219
Sun, B.K., 435
Sun, G., 264
Sun, J., 406–407
Sun, J.M., 167
Sun, L., 264–268
Sun, Z., 282
Sun, Z.W., 211–213
Sundaramoorthy, R., 119–120
Sundstrom, M., 218–219
Sunwoo, H., 422–423
Surka, C., 441
Suto, R.K., 7–8
Sutton, A., 211–213
Suzuki, N., 369–370
Suzuki, T., 268–269, 271–274
Suzuki, Y., 422
Svergun, D.I., 35
Swenberg, J.A., 366, 387–388
Swinger, K.K., 267
Swist, K.M., 264
Syntichaki, P., 323
Syrzycka, M., 211–213
Szczepankiewicz, B.G., 185*f*
Szenker, E., 98
Szerlong, H., 323
Szewczuk, L.M., 263–264, 268–271, 282–283, 292–293
Szulwach, K.E., 374
Szykowska, A., 266, 287–288
Szyperski, T., 336

T

Taatjes, D.J., 435
Tachibana, M., 211–213
Tachiwana, H., 68–69, 113–114
Tackett, A.J., 264–265, 267
Taft, J.M., 264
Tagami, H., 98–99, 113–114
Tahiliani, M., 366–368, 387–388
Takahashi, H., 388–389
Takahashi, M., 228–229
Takahashi, Y.H., 216
Takyi, H.K., 151
Talbert, P.B., 98
Talianidis, I., 211
Tamanaha, E., 379
Tamaru, H., 211–215, 218–219, 280–281
Tamkun, J.W., 322–323
Tan, M., 187–188
Tan, S., 37, 44–63, 140, 263–264
Tan, Y., 243–245, 253, 257, 264
Tanaka, K., 322–323
Tanaka, Y., 211–213
Tang, H.Y., 146
Tang, J., 287–288
Tang, Q., 366–368, 378–379, 387–388
Tanzer, A., 425
Tapanari, E., 462
Taranova, O.V., 366–368, 387–388
Taunton, J., 167–168
Taverna, S.D., 264–265, 267
Taylor, W.R., 213–215
Team, I.I., 422
Teater, M., 217–218
Tejera, A.M., 406–407
Telese, F., 243–245, 253, 257, 264
Tempst, P., 255–256, 280–282, 287–288, 323
Tennen, R.I., 184–186
Tepper, C.G., 211–213
Terragni, J., 381
Terreni, M., 140
Terwilliger, T.C., 413
Thalhammer, A., 281–282
Thåström, A., 46
Theimer, C.A., 407–408
Thiel, A.T., 217–218
Thiele, C.J., 217–218
Thinnes, C.C., 262, 281–282
Thireos, G., 323
Thomas, J.H., 389–390
Thompson, C., 217–218
Thornton, J.M., 336, 338
Thunnissen, M.M.G.M., 61–62
Tiffen, J., 323
Tilghman, S.M., 422
Tilgner, H., 425
Tipton, K.F., 290–292
Tjian, R., 322–323
Tochio, H., 388
Toffolo, E., 244–245, 264
Toh, H., 387–390, 393
Tolia, N.H., 218–219
Tomchick, D.R., 243–244, 248–249, 252, 263–264, 269–271
Tomilin, A., 29–30
Tomioka, Y., 369–370
Tomlins, S.A., 217–218
Tommerup, N., 388–389
Tong, A.J., 425–426
Tonti-Filippini, J., 387–388
Topalidou, I., 323
Tora, L., 211
Torres, I.O., 287, 295
Torres-Padilla, M.E., 114
Tortorici, M., 263–264, 269
Tozser, J., 283–284
Trachtulec, Z., 389–390
Trapnell, C., 460–463
Tremblay, V., 216
Tremethick, D.J., 4–6, 8
Tria, G., 35
Trievel, R.C., 154, 211–219, 225, 280–296, 286*t*, 345–346
Trnka, M.J., 288–289
Trojer, P., 211–213, 281–282
Trounson, E., 211–213
Tsai, B.P., 8
Tsai, M.C., 435
Tsai, S.C., 167–168
Tsai, W.-W., 273
Tschiersch, B., 210–211
Tsuji, M., 369–370
Tsukada, Y., 255–256, 280–282, 287–288
Tsukamoto, H., 369–370

Tsumoto, H., 273–274
Tumber, A., 287–288
Tuna, S., 387–388
Turberfield, A.H., 281–282
Tuschl, T., 435–440
Tuting, C., 200–201
Tyagi, S., 455–457

U

Uberbacher, E.C., 83–84
Ueda, R., 273–274
Uhlen, M., 218–219
Ule, A., 435
Ule, J., 435, 440–441
Uludag, M., 327*f*
Umehara, T., 248–249
Unoki, M., 388–389
Unzeta, M., 290–292
Upadhyay, A.K., 281–282, 287, 293
Urban, A.E., 422
Uribe-Lewis, S., 366, 369–370
Urich, M., 387–388
Usadel, B., 460
Utley, R.T., 316
Uversky, V.N., 13, 29, 32–33

V

Vadlamudi, R.K., 243–244
Vagin, A.A., 295, 335–336
Vaisvila, R., 379
Valente, S., 263–264, 269–271
Van Aller, G.S., 217–218, 269–271, 274
van Den Berg, S., 218–219
van den Bogaard, P., 455–457
Van der Maarel, S.M., 388–389
van der Woerd, M.J., 24–26, 35
Van Dyke, B.R., 289–290
Van Holde, K.E., 27
van Oudenaarden, A., 455–457
Van Rechem, C., 280–281, 345–346
Vanoni, M.A., 251, 256, 282–283, 292–293
Varambally, S., 217–218
Vasisht, N., 211–215, 218–219
Vassias, I., 98–114
Vaughan, R.M., 287–288
Vedadi, M., 266, 287–288
Veerappan, C.S., 211–213
Vera, A., 221
Verdaasdonk, J.S., 446
Verde, G., 389–390
Verdin, E., 189
Vermaak, D., 167
Verpelli, C., 244–245, 264
Verreault, A., 113
Vertino, P.M., 213, 388
Vianello, P., 269
Vidal, F., 141–142
Viiri, K., 387–388
Villagra, A., 162–180
Villar-Garea, A., 141–142
Villaverde, A., 221
Virdee, S., 288–289
Vlachogiannis, G., 387–388
Vlcek, C., 389–390
Vogel, H., 99–100
Vogel, J.L., 264, 272
Voigt, P., 211–215, 218–219
Vollmar, M., 281–282, 287
Volmar, C.H., 323
Vonrhein, C., 227–228
Vought, V.E., 216
Vuister, G.W., 332, 338
Vulliamy, T., 404

W

Wade, P.A., 167, 388
Wagner, E.J., 211–213
Wagner, G.R., 184–186
Wagner, M., 369–370
Waite, A.J., 307
Wakamori, M., 228–229
Wakefield, M.J., 211–213
Walensky, L.D., 217–218
Walker, J.R., 388
Walker, John M., 350
Walker, P.A., 211–215, 218–219, 228
Walkinshaw, M.D., 388
Walport, L.J., 281–282, 287
Walsh, C.T., 268–269
Walsh, M.J., 322–323
Walter, J.C., 44
Walter, P., 422
Walter, W., 140
Walters, M.A., 155–156
Wan, B., 406
Wan, J., 140

Wan, K., 248–249
Wan, Y., 435
Wandrey, C., 27
Wang, B., 142
Wang, C., 217–218, 287–288
Wang, C.I., 441
Wang, C.L., 205
Wang, D., 269–271
Wang, E., 323
Wang, E.T., 460
Wang, F., 248–249, 406–407
Wang, G.G., 346–347
Wang, H., 29–30, 322–323
Wang, H.W., 178
Wang, J., 142, 243–245, 253, 257, 264, 388–389
Wang, J.K., 423, 435
Wang, K.C., 422–423
Wang, L., 217–218
Wang, M.D., 8, 20
Wang, Q., 282, 293
Wang, T., 4–6
Wang, W., 422
Wang, X., 164, 213, 215, 217–218, 227–228, 268–269, 271–272
Wang, Y., 210, 366–368, 378–379, 387–388
Wang, Z., 272, 346
Ward, L.D., 345–346
Wardle, G., 435–440
Warholic, N.M., 217–218
Warren, M.E., 255–256, 280–282, 287–288
Washburn, M.P., 211–213
Watanabe, S., 4–6
Waterfall, J.J., 428–434
Waterhouse, A.M., 220*f*
Waters, M.L., 346
Waugh, D.S., 283–284
Waye, J.S., 446, 449
Waye, M.M., 44–47
Webb, H.D., 388
Webster, D.E., 435
Weemaes, C.M., 388–389
Wegener, D., 167
Wei, C.L., 425
Wei, G., 141–142
Wei, J.S., 217–218
Wei, W., 186–187
Weidner, J.R., 144–145
Weigelt, J., 347
Weis, W.I., 8
Weiss, M.S., 61–62
Weiwad, M., 184–186
Wen, H., 322–323
Wen, Y.D., 167–168
Wendong, L., 264
Werner, M., 323
West, A., 287–288
West, J.A., 441
Westfield, G.H., 216
Weyand, M., 186–188, 201–205, 202*f*, 203*t*
Wheeler, M.J., 61–62
Whelan, C.M., 141–142
Whetstine, J., 285–287
Whetstine, J.R., 242–244, 255–256, 262–264, 280–282, 287–288, 345–346
White, A.E., 4–6, 8, 48
White, C.L., 4–8, 11–12, 15–16, 68–69, 79, 84–85, 87–88, 121, 125, 229
White, H.E., 389–390
White, M.A., 388–389
Whyte, W.A., 243–244, 257
Widlund, H.R., 46
Widom, J., 4–6, 46, 68–69, 121, 229
Wienken, C.J., 201–202
Wiggins, K.R., 346
Wigle, T.J., 217–218, 267, 346–347
Wijmenga, C., 388–389
Wijmenga, S.S., 377
Wilkins, M.R., 350, 410
Wilkins, S.E., 281–282
Wilkinson, A.W., 280
Willard, H.F., 422, 446, 449
Williams, A.J., 389*f*
Williams, D.C., 388
Williams, K.P., 149–151
Williams, M., 366, 369–370
Willmann, D., 264–267
Willson, T.A., 211–213
Wilm, M., 211–213, 388–389
Wilmot, C.M., 287, 293, 295
Wilson, D.M., 287, 293
Wilson, G.G., 388–390, 393–394
Wilson, I.A., 388–389
Wilson, I.B., 268–269
Wilson, J.R., 211–216, 218–219, 228

Wilson, K.S., 295
Wilusz, J.E., 422–423
Wingfield, Paul T., 47
Winkler, D.D., 8, 14, 48
Winn, M.D., 204–205, 227–228, 295, 335–336
Wirsching, F., 167
Wissmann, M., 243–244, 269
Witkin, K.L., 405–406
Witt, H., 264
Wittekind, M., 332
Wolberger, C., 184–187, 190*t*, 197
Wolf, D., 389–390
Wolffe, A.P., 167
Wong, L.H., 446
Wong, P.S., 213–216, 218–219
Woo, C.W., 217–218
Wood, E., 405–406
Wood, S., 113–114
Woodcock, C.L., 13, 29, 32–33
Woods, V.L., 68–69, 71
Woods, V.L.J., 4–6
Woolfe, A., 99–102, 105–108, 112
Workman, J.L., 162–163, 304, 309, 316
Worman, H.J., 211–213
Wozniak, D.J., 56
Wriggers, W., 35
Wright, P.E., 390
Wright, W.E., 405–406
Wu, C., 119–120, 129–131
Wu, H., 213, 216, 218–219, 289–290
Wu, J., 142, 264
Wu, S.C., 366, 387–388
Wu, X., 217–218
Wu, Z., 142
Wurtz, J.M., 322–323
Wuthrich, K., 336
Wynder, C., 252, 263–264, 268–269
Wyns, K., 8, 20

X

Xi, Y., 322–323
Xia, M., 149–150
Xiang, Y., 243–245, 253, 257, 264
Xiao, B., 211–215, 218–219, 228
Xiao, T.S., 44
Xie, N., 142
Xie, S., 387–388
Xie, Z., 187–188
Xing, Y., 422
Xu, C., 346
Xu, C.F., 211–213
Xu, G.L., 388–389
Xu, G.Y., 336
Xu, R.M., 211–213
Xu, R.X., 336
Xu, X., 289–290, 423, 435
Xue, L., 211–213
Xue, S., 388

Y

Yamamoto, K., 216
Yamane, K., 248–249, 282–283
Yamazaki, T., 336
Yan, J., 217–218
Yan, Q., 266–267
Yan, S., 322–323
Yan, Y., 140, 145–146, 151
Yanagisawa, T., 228–229
Yang, C., 24–26, 35
Yang, J.C., 322–323
Yang, J.G., 120–121
Yang, L., 425
Yang, M., 243–244, 248–249, 252, 263–264, 268–271, 282–283, 292–293
Yang, P., 211–213
Yang, S., 264, 333*f*, 336
Yang, T.P., 83–84
Yang, V.W., 422
Yang, W.M., 167–168
Yang, X., 366, 369–370
Yang, X.J., 140
Yang, Y., 142, 248–249, 378–379
Yang, Z., 211–215, 280–281
Yao, T., 4–6, 8, 12, 14, 32–33, 309
Yao, Y.L., 167–168
Yap, D.B., 228–229, 313
Yardimci, H., 366, 369–370
Yates, A., 462
Yates, J.R., 113–114
Ye, A.J., 405–406
Ye, P., 264
Yen, P.M., 269–271
Yennawar, H.P., 44, 46, 57, 63
Yi, C., 366, 369–370
Yin, N., 243–244, 269

Yin, R., 378–379
Yodh, J., 29–30
Yokoyama, A., 141–142
Yokoyama, S., 228–229, 248–249, 287, 293
Youn, D.Y., 184–187
Young, M.D., 211–213
Young, N.S., 404
Young, P., 335–336
Young, R.A., 174, 243–244, 257, 422–423
Yu, H., 243–244, 248–249, 252, 262–263, 268–269, 282–283, 292–293
Yu, J., 217–218
Yu, J.J., 187–188, 204
Yu, M., 374
Yu, V., 287–288
Yu, W., 266, 287–288
Yuan, H., 140, 142–144, 188–189
Yuan, Q., 57
Yukl, E.T., 287, 293, 295

Z

Zakian, V.A., 406–407
Zambelli, F., 457
Zandarashvili, L., 388–389
Zang, J., 282, 285–287, 293
Zangger, N., 389–390
Zarnack, K., 440–441
Zee, B.M., 211–213
Zehnder, A., 435
Zeng, H., 213, 216, 218–219
Zeng, L., 213, 215, 322–340
Zentner, G.E., 345–346
Zepeda-Martinez, J.A., 378–379
Zerbino, D.R., 464
Zhan, P., 268–269, 271–272
Zhang, F., 213–216
Zhang, G., 282, 285–287, 293
Zhang, G.T., 323
Zhang, J., 140
Zhang, J.H., 149–150, 287–288
Zhang, K., 211
Zhang, L., 289–290, 374
Zhang, M., 134
Zhang, P., 216
Zhang, Q., 322–323, 338
Zhang, Q.C., 441
Zhang, R., 282–283
Zhang, W.J., 323
Zhang, X., 163, 211–215, 218–219, 228, 280–282, 287, 293, 296, 381, 387–395
Zhang, Y., 142, 211–213, 216, 228, 242, 248–249, 255–256, 280–283, 287–288, 366–370, 373, 387–388
Zhao, B., 378–379
Zhao, C., 378–379
Zhao, J., 435
Zhao, K., 184–186, 190*t*, 204, 345–346
Zhao, W., 454
Zhao, Y., 267
Zheng, W., 149–150
Zheng, Y., 381, 388–390, 393–394
Zheng, Y.G., 142
Zhong, F.L., 441
Zhong, Y.F., 323
Zhou, B.-R., 44, 68–69
Zhou, C.Y., 119–134
Zhou, K., 4–37
Zhou, M.M., 213, 215, 217–218, 322–340, 333*f*
Zhou, T.J., 333*f*, 336
Zhou, Y., 187–188, 204
Zhu, G., 332, 338
Zhu, H., 140
Zhu, X., 422
Zibetti, C., 244–245, 264
Ziegler-Birling, C., 114
Zobeck, K.L., 211–213
Zoeller, E.L., 296
Zou, T., 217–218
Zuber, J., 323
Zuo, X., 389–390
Zupan, B., 440–441
Zwart, P.H., 227–228

SUBJECT INDEX

Note: Page numbers followed by "*f*" indicate figures, and "*t*" indicate tables.

A

AAVS1 ectopic expression system, 313
Acetylation, 184–186
  chemical control, 146
  histone lysine, 322–323
  lysine, 140, 262
Acetyl-lysine binding, 323
Affinity chromatography, 194–195
Affinity-tag purification, 350–353
AFM. *See* Atomic force microscopy (AFM)
Analytical protein purification
  electrophoresis, 48–52
  PRC1 ubiquitylation module/NCP complex, 48–49
  size exclusion chromatography, 47–48, 50, 51*f*
  stoichiometric preparations, 48–52
Analytical ultracentrifugation (AUC), 27–29, 28*f*
  analysis of chromatin by, 29, 30*f*
  buffers and reagents, 27
  data, 26
  equipment, 27
  list of institutes, 26–27
Antibiotics, 325
Antibody-based assays, LSD1, 266–267
Anti-FLAG M2 affinity purification, 310–312
Artifacts, 187–188
Atomic force microscopy (AFM), 29–33
  buffers and reagents, 30
  data processing, 32
  equipment, 30, 31*f*
  to visualize trinucleosomes, 33*f*
ATP-dependent chromatin remodeling enzymes, 119–120, 128–133
  FRET-based remodeling assay, 131–133, 133*f*
  gel-based remodeling assay, 128–133, 131*f*
ATXR5
  activity of, 231
  characterization of, 230
  Coomassie brilliant blue, 221–224, 223*f*
  crystallization, 225–226, 226*f*
  different homologs, 219–221
  H3.1 peptide substrate selection for, 226–227, 227*f*
  ribbon diagrams, 214*f*
  sequence alignment of, 219–221, 222*f*
AUC. *See* Analytical ultracentrifugation (AUC)

B

Bacteria growth medium, 324
Bacterial protein expression vector, 324
β-glucosyltransferase (bGT), 381
Bizine, 273
BL21(DE3), 218–219
Bromodomain (BrD), 322–323
  ^{1}H–^{15}N-HSQC to detect ligand binding, 331–332
  inhibitors, 323
  LC/MS to determine protein molecular weight, 331
  preparation, 326
    expression, 329
    transformation, 326–329
  protein concentration and purity, 330–331
  protein–ligand binding determination
    competition binding assay, 334
    saturation binding assay, 332–333
  protein sequence alignment, 327*f*
  purification
    affinity chromatography, 329
    bacteria disruption, 329
    gel filtration/size-exclusion chromatography, 330
  role, 323
  SDS-PAGE to check protein purity and molecular weight, 331
  structural determination using NMR
    backbone assignment, 336

Bromodomain (BrD) (*Continued*)
restraints collection, 336–337
side-chain assignment, 336
structure calculations, 338
structural determination using x-ray crystallography, 335–336
crystallization, 334
crystal mounting, 334–335
data collection, 335

C

Cap analysis of gene expression (CAGE), 427–428
CENP-A nucleosome assembly, 68–69, 84–85
DNA preparation, 79
by PCR, 79–83, 80–81*f*
from plasmid excision, 83–84, 84*f*
histones preparation for
His-H2A, H2A, and H2B monomers purification, 74–77
His-H2A/H2B monomers and heterotetramers expression, 69–71
histone H2A/H2B refolding, 77–79
(CENP-A/H4)$_2$ tetramer purification, 71–74, 72*f*
nucleosome
purification, 88–89
quality, 86–88
reconstitution, 85–86
repositioning, 88
CENP-A nucleosome with CENP-C
expression and purification of CENP-C, 90–93, 90*f*
stoichiometry determination, 93–95, 94*f*
Centromere, 68–69
Chemoenzymatic activity assays
bacterial restriction enzymes, 379
βGT reaction, 381
enzyme-coupled assays, 380*f*
TDG, 381
Chromatin, 322–323, 338–340
assembly, 4–6, 5*f*
definition, 4
fundamental unit of, 210
gene transcriptional activation in, 323
interaction, 6
roles of BrD in, 323
templates, 6–7
Chromatin complex
analytical purification
electrophoresis, 48–52
PRC1 ubiquitylation module/NCP complex, 48–49
size exclusion chromatography, 47–48, 50, 51*f*
stoichiometric preparations, 48–52
concentratation, 54–56
crystallization, 56–57
crystal structure determination, 44, 45*f*
fluorescently labeled chromatin proteins, 57–61
postcrystallization soaks, 61–63
preparative purification, 52–54
Chromatin immunoprecipitation assay, 177. *See also* Histone deacetylases (HDACs)
cross-linking, 174–175
preparation, 175
sample wash and elution, 177–178
shear sample, 176
sonication, 176
Chromatin-modifying complexes
endogenous tagging of EZH2 N-terminus, 313
mass spectrometry analysis, 313
Chromatin protein
fluorescently labeled, 57–61
preparation, 47
Chromatin reconstitution strategy
DNA labeling, 10–11
fluorescently labeled histones, 8–10
histone proteins, 7
"microscale"/step dilution method, for nucleosome reconstitution, 15–16
nucleosome assembly, 11–13
trinucleosome assembly, 13–15
X. laevis histones, 7
Chromatin remodeling enzymes, ATP-dependent, 119–120, 128–133
FRET-based, 131–133, 133*f*
gel-based, 128–133, 131*f*
C2H2 ZnF proteins, 388–390
Clustered regularly interspaced short palindromic repeats (CRISPR)/Cas9, 305–308

Computational analysis, RIP-Seq
genome alignment, 461–462
overall analysis strategy, 457, 459*f*
read preprocessing, 459–460
sequencing strategy, 458–459
in silico rRNA depletion, 460–461
transcript abundance estimation, 462–463
unaligned reads, assembly of, 463–464
Constitutive centromere associated network (CCAN) of proteins, 68–69
Counts per minute (CPM), 168
Coupled enzyme assays, LSD1, 265–266
Crystallization
ATXR5, 225–226, 226*f*
Divalent ions, 57
initial screens, 56
KDM, 293–295, 294*f*
orthologous proteins, 57
reagents and tools, 325
SET domain proteins, 225–227
sirtuin
catalytic core constructs, 204
crystals, 204–205
data collection and processing, 204–205
NMR and X-ray measurements, 202–203
of Sirt1, 204
of Sirt3, 203, 203*t*
tcTERT
optimization, 412
protein sample concentration, 410–411
screening, 411–412
trial, 56
ZnF protein, 396–397, 396*f*
Cyclopropylamine analogs, 269–271, 270*f*
Cytosine modifications analysis
cellular DNA
genomic DNA preparation, 366–368
qualitative analysis by dot blotting, 368–369
quantitative analysis by LC-MS/MS, 369–373
chemoenzymatic activity assays
bacterial restriction enzymes, 379
βGT reaction, 381
enzyme-coupled assays, 380*f*
TDG, 381
quantitative activity assays, 382–383
TET enzymes purification
FLAG-based affinity, 374
from Sf9 insect cells, 373–374
TET substrates
CpG, 374–375
custom-made oligonucleotides, 375
fluorescence-based assays, 375
generation of, 377–378
isotopically labeled substrates, 376–377, 376*f*
modified cytosines, 375
thermocycler settings, 377–378

D

DELFIA. *See* Dissociation-enhanced lanthanide fluorescent immunoassay (DELFIA)
Demethylation, 264
De novo histone deposition
with histone chaperone depletion
and replicat cells label, 106–108
systematically monitor variations, 113
UV-DNA damage, 108–111
imaging and quantifying SNAP labeling, 111–112
SNAP-tag
cell imaging, 105
cell lines expression, 100–102
cell-permeable fluorescent, 100
chase assay, 103
pulse assay, 103
quench assay, 103
triton extraction, 103
validation, 104*f*
DIM-5, 214*f*
Dissociation-enhanced lanthanide fluorescent immunoassay (DELFIA), 266–267
Divalent ions, 57
DNA
CENP-A nucleosome assembly, 79
PCR, 79–83, 80–81*f*
plasmid excision, 83–84, 84*f*
crystallization of ZnF proteins in complex with, 396–397, 396*f*
cytosine modification, 387–388

DNA (*Continued*)
fluorescence polarization assay for, 393–395, 395*f*
labeling, 124–125
methylation, 389–390
recognition domains, 388
Dot blotting, qualitative analysis by, 368–369
Dox-inducible expression, of EZH2, 315

E

Enhancer of Polycomb homolog 1 (EPC1), 305
C-terminus, endogenous tagging of, 305–308
design and cloning
of sgRNA targeting, 305–307
of tagging donor, 307–308
gene targeting in K562 cells, 308
isogenic cell lines expressing TAP-tagged, 308–309
SDS–polyacrylamide gel electrophoresis and silver staining, 312–313
TAP of, 310–312, 310*f*
Enhancer of zeste homolog 2 (EZH2), 214*f*, 216–218
Dox-inducible expression of, 315
N-terminus, endogenous tagging, 313
TAp of, 315
Enzyme inactivators, mechanism-based, 268–269
EPC1. *See* Enhancer of Polycomb homolog 1 (EPC1)
Escherichia coli (E. coli), 348–349

F

Fast protein liquid chromatography (FPLC) purification, 353–355
FDH from *Pseudomonas putida* (PpFDH), 289–290
Flavin adenine dinucleotide (FAD) cofactor, 263*f*
Fluorescein thioisocyanate (FITC), 201–202
Fluorescence polarization (FP) assay
to determine protein–ligand binding, 332–334
for DNA binding, 393–395, 395*f*
Fluorescence resonance energy transfer (FRET) pairs, 8
Formaldehyde dehydrogenase (FDH)
from *Pseudomonas putida*, 289–290
purification, 289–290
FRET-based method
labeling
DNA, 124–125
histone, 121–122
on SDS-PAGE gel, 123–124, 123*f*
octamer and nucleosome assembly, 125
purification of nucleosomes, 125–127
quantification of nucleosomes, 127–128, 128*f*
reconstituting nucleosomes for, 121–128, 121*f*
FRET-based remodeling assay, 131–133, 133*f*
FRET-labeled nucleosomes, 121

G

Gel-based assay, 120–121, 134
Gel-based remodeling assay, 128–133, 131*f*
Gel electrophoresis, 20
Gene expression, control, 387–388
Gene targeting, 308
Gene transcription, 322–323, 338–340
Genome editing, 304–305
Global run-on sequencing (GRO-seq), 428–434
Glutamate dehydrogenase (GDH), 198–199, 199*f*
Glutathione S-transferase (GST), 390
GSK2879552, 274

H

HATs. *See* Histone acetyltransferases (HATs)
HDACs. *See* Histone deacetylases (HDACs)
HeLa cells, 450–451
High-throughput inhibitor screening
acetonyl-CoA, 149–150
buffer considerations, 151–152
colloidal aggregators, 156–157
eliminating promiscuous inhibitors, 155
ELISA, 152–153
PAINS, 155–156
radioactivity-based assay, 152

thermofluor binding assay, 154–155
Z-factor, 150
His-H2A and H2B monomers expression, 69–71
His-tagged human Sirt3, 191
His-tagged sirtuin
construct, 193
purification, 191, 192*t*, 193*f*
uses, 198
Histone(s)
code features, 262
demethylases classes, 242–243
labeling, 121–124, 123*f*
LSD1, in vitro enzymatic assays, 251
lysine acetylation, 322–323
lysine methylation, 211–213
methyltransferase
EZH2, 304
PRC2, 317
preparation for CENP-A nucleosome assembly
His-H2A, H2A, and H2B monomers purification, 74–77
His-H2A/H2B monomers and heterotetramers expression, 69–71
histone H2A/H2B refolding, 77–79
$(CENP-A/H4)_2$ tetramer purification, 71–74, 72*f*
proteins
assembling chromatin, 7
fluorescently labeled, 8–10
octamer, 7
Histone acetyltransferases (HATs), 262, 304, 317, 322–323
Histone chaperone depletion
HMT assays, 315–316
and replicat cells label, 106–108
systematically monitor variations, 113
UV-DNA damage, 108–111
Histone deacetylases (HDACs), 322–323
chromatin immunoprecipitation assay, 177
cross-linking, 174–175
preparation, 175
sample wash and elution, 177–178
shear sample, 176
sonication, 176
conventional activity assay
advantage, 167
guarente laboratory, 167
H4 peptide label, 168
definition, 162–163
protein substrates identification, 179–180
role, 163
selective inhibition assessment
deacetylase activity quantification, 171–173
T-cell isolation, 170–171
ultraselective inhibitors, 169–170
stable cell lines, generation of
antibiotic resistance, 164–166
lentivirus, 164
monoclonal populations, 166
Histone lysine methyltransferases (HKMTs)
crystal structure of SET domain, 214*f*
subdivision, 211–213
transfer, 228
H3K9 demethylation, 264
H3K4 methylation, 263–264
H4K20 methyltransferases, 213
HKMTs. *See* Histone lysine methyltransferases (HKMTs)
Host strain, 324
Hydrazine analogs, 271*f*
Hydrogen peroxide (HOOH) detection assay, 263*f*

I

Immobilized metal ion affinity chromatography (IMAC), 191
Immunofluorescence/RNA FISH, 449–450, 455–457
Inhibitors
BrD, 323
LSD1, 268–272
of LSD1, 268–272
MAO, 269
suicide, 268–269
Integrated DNA Technologies (IDT), 375
In vitro reconstituted chromatin quality control strategy
AFM, 29–33
buffers and reagents, 30
data processing, 32
equipment, 30, 31*f*
to visualize trinucleosomes, 33*f*

In vitro reconstituted chromatin quality control strategy (*Continued*)
AUC, 26–29, 28*f*, 30*f*
MNase digestion, 16–20
analyzing products, 20–24
of mono- and trinucleosomes, 20, 21*f*
quality control measures, 16
SAXS, 34, 36*f*
SEC-MALS, 24–26
training resources, 16, 17*t*
Ion exchange chromatography, 196
Ionic strength, 394
Isogenic cell lines, 308–309
ISWI, 119–120

J

JmjC domaincontaining histone demethylases (JHDMs), 242–243
JMJD2 lysine demethylases (KDM)
assay methodology
FDH-coupled fluorescent assay, 290–293
FDH purification, 289–290
histone peptide substrates design, 289
kinetic properties, 288–289
methods, 287–288
catalytic mechanism, 281–282, 281*f*
expression and purification
hexahistidine-tagged, 283–284
metal content analysis, 285–287
strep(II)-tagged, 284–285
family, 280–281
genomic processes, implication, 281–282
nonhistone nuclear proteins, 280
regulatory modification, 280
sites of, 280
structure determination
crystallization, 293–295, 294*f*
crystal structures, 293
peptide complexes, 295
subfamily, 281–282
Jumonji C (JmjC) domain, 242–243

K

Kaiso, 388–389
K562 cells
gene targeting in, 308
large-scale expansion of, 309
KDM1A. *See* Lysine-specific demethylase 1 (LSD1)
Kinetic analysis
chemical acetylation control, 146
MYST acetyltransferases, 145–146
radioactive acetyl-CoA, 147, 148*f*
steady-state, 144–145
Kinetochore, 68–69
Krüppel-associated box (KRAB)-ZnF proteins, 389–390, 389*f*

L

LC/MS reagents and tools, 326
Liquid chromatography tandem mass spectrometry (LC-MS/MS)
Agilent 6460 triple quadrupole mass spectrometer, 370
cytosine modifications, 369–370
DNA nucleosides, 370, 371*f*
fused-silica tube, 372–373
isotopically labeled cytosines, 372
sample preparation, 373
sequencing methods, 369–370
triple quadrupole system, 372
Long noncoding RNAs (lncRNAs)
detection, traditional methods, 423
genome-wide detection
CAGE, 427–428
chromatin-associated RNA sequencing, 425–426
RNA sequencing, 423–425
genomic characterization
individual nucleotide resolution CLIP, 440–441
PAR-CLIP, 435–440
RIP-seq, 435
indirect sequencing methods
nascent transcript global run-on sequencing, 428–434
PRO-seq, 434
physical and functional classification, 422–423
LSD1. *See* Lysine-specific demethylase 1 (LSD1)
Lysine acetylation, 140, 262, 322–323
Lysine demethylases (KDMs), 345–347
Lysine methyltransferases (KMTs), 280–281, 345–347

Lysine-specific demethylase 1 (LSD1)
  assay, 264–265
    antibody-based assays, 266–267
    coupled enzyme assays, 265–266
    mass spectrometry-based assays, 267–268
    radiolabeled assays, 266
  biochemical activity, 243–244
  chemical mechanism, 243*f*
  chromatin-interacting proteins, 244*f*
  description, 262–263
  enzymatic activity
    detection, 255–256
    mass spectrometry, 253–254
    in vivo analysis, 256–257
    Western blot, 254–255
  expression and purification
    bacteria, 245–246
    mammalian cells, 250
    Sf9 insect cells, 246–248
    structure determination, 248–250
  function, 245
  gene, 244–245
  HOOH detection assay for, 263*f*
  inactivation
    by cyclopropylamine analogs, 270*f*
    by hydrazine analogs, 271*f*
  inhibitors, 268–272
    applications, 272–274
    structure, 272*f*
  role on H3K9 demethylation, 264
  in vitro enzymatic assays, 253
    nucleosomes, 252
    peptides and histones, 251
Lys methylation, 262

M

MAO inhibitors, 269
Mass spectrometry-based assay, LSD1, 267–268
MassSQUIRM, 267
Matrix-assisted laser desorption ionization-time-of-flight (MALDI-TOF), 267
MBDs. *See* Methyl-binding domains (MBDs)
Mechanism-based enzyme inactivators, 268–269
Methylamine oxidase (MAO), 290–292
Methyl-binding domains (MBDs), 388
Methyllysine analogs (MLAs), 288–289
Methyllysine reader domains
  affinity of, 356–358
  KMT and KDM, 345–347
  list of identification, 346
  NMR spectroscopy, 358–360
  preparation
    affinity-tag purification, 350–353
    construct design, 347–348
    FPLC purification, 353–355
    overexpression in *E. coli*, 348–349
  screening, 355–356
  X-ray crystallography, structure determination by, 360–361
Methyltransferase assay (MTA)
  histone octamer, 229
  HKMTs, 228
  native nucleosome, 228–229
  peptide substrates, 228–229
Methyltransferases, sequence alignment of, 219–221, 220*f*
MFP-3D software, 32
Micrococcal nuclease (MNase) digestion, 16–20
  advantage, 21*f*, 24
  analyzing products, 20–24
  buffers and reagents, 19
  DNA, 16
  equipment, 19
  of mononucleosomes assembly, 20, 21*f*
  of nucleosomes assembly, 20, 21*f*
  Nuc207 via Bioanalyzer 2100, 21*f*, 24
  of trinucleosomes, 20, 21*f*
Microscale thermophoresis (MST), 201–202
Mitosis, 68–69
MLL1, 214*f*
Monoamine oxidases (MAO), 256–257
Mounting tools and shipping dewar, 326
Multiangle light scattering (MALS), 24–25
MYST family proteins. *See also* High-throughput inhibitor screening
  Kinetic analysis, 145–146
  Ping-pong catalytic mechanism, 141*f*
  preparation, 142–144

N

Native PAGE
  nucleosome quality by native, 86–88
  purification of nucleosomes by, 88–89

NCL1, 273–274
NEs. *See* Nuclear extract (NEs)
Next-generation sequencing (NGS), 422
Nicotinamidase (NCA), 198–199, 199*f*
NMR reagents and tools, 326
Noncoding RNAs (ncRNAs), 422, 446
NuA4 complex, 304–305
Nuclear extract (NEs), 309
Nuclear magnetic resonance (NMR) spectroscopy, 358–360
Nucleosides, quantitative activity assays, 382–383
Nucleosome(s)
  assembly
    on 165 bp DNA, 5*f*, 12–13
    by salt gradient, 11–13, 37
  CENP-A, 68–69
  FRET-labeled, 121
  LSD1, in vitro enzymatic assays, 252
  purification, 88–89
  purification of, 125–127
  quality, 86–88
  quantification of, 127–128
  reconstitution, 85–86
  repositioning, 88
  repositioning by thermal shifting, 88
Nucleosome core particles (NCPs), 44–47, 229
Nucleosome reconstitution, 15–16

O

Octamer, 8–10
Oligonucleotide, 396, 396*f*

P

Pan-assay interference compounds (PAINS), 155–156
PEI. *See* Polyethylenimine (PEI)
Peptides
  LSD1, in vitro enzymatic assays, 251
  propargyl-Lys-containing, 269
Phenylmethylsulfonyl fluoride (PMSF), 197–198
Photoactivatable ribonucleoside-enhanced cross-linking and immunoprecipitation (PAR-CLIP), 435–440
Plasmid excision, DNA preparation, 83–84, 84*f*
Polyacrylamide gel electrophoresis (PAGE)
  nucleosome quality by native, 86–88
  purification of nucleosomes by, 88–89
Polycomb repressive complex-2 (PRC2), 304–305
  purification of, 315
  TAP of, 314*f*
Polyethylenimine (PEI), 390
Polymerase chain reaction (PCR), and DNA preparation, 79–83, 80–81*f*
Posttranslational modifications (PTMs), 184–186, 304, 345–346
PpFDH. *See* FDH from *Pseudomonas putida* (PpFDH)
PRC2. *See* Polycomb repressive complex-2 (PRC2)
PRDM9. *See* PR domain zinc finger protein 9 (PRDM9)
PR domain zinc finger protein 9 (PRDM9), 389–390
Propargyl-Lys-containing peptides, 269
Protein(s)
  C2H2 ZnF, 388–390
  complexes, 305
  expression, 193–194
  inductions, 69*f*
  KRAB-ZnF, 389–390, 389*f*
  purification buffers, 325
Protein–ligand binding determination
  competition binding assay, 334
  saturation binding assay, 332–333
Pseudomonas putida, 289–290
PTMs. *See* Posttranslational modifications (PTMs)

Q

Quantitative activity assays
  cytosine modifications analysis, 382–383
  on nucleosides, 382–383

R

Radiolabeled assays, LSD1, 266
Remodeling assay
  FRET-based, 131–133, 133*f*
  gel-based, 128–133, 131*f*
  reactions, 134

Resin, 325
RNA immunoprecipitation sequencing (RIP-Seq)
components, 447–449
computational analysis
genome alignment, 461–462
overall analysis strategy, 457, 459*f*
read preprocessing, 459–460
sequencing strategy, 458–459
in silico rRNA depletion, 460–461
transcript abundance estimation, 462–463
unaligned reads, assembly of, 463–464
lncRNAs, genomic characterization, 435
methods, 451–455
RNA FISH, 449–450, 455–457
tissue cell culture
elements, 447
HeLa cells, 450–451
RNA sequencing
chromatin-associated, 425–426
library, 424*f*
lncRNAs, 423–425

S

SAMDI, 267
SAXS. *See* Small-angle X-ray scattering (SAXS)
SDS-PAGE
analysis of protein inductions, 69*f*
labeling on, 123–124, 123*f*
SEC-MALS. *See* Size-exclusion chromatography coupled with multiangle light scattering (SEC-MALS)
Selenomethionyl-labeled RcATXR5, preparation, 224–225
SET7/9, 214*f*
SET8, 214*f*
SET and RING finger-associated (SRA) domains, 388
SET domain proteins
biological processes, 210
characteristics, 210–211
crystal structures of, 213, 214*f*
DIM-5, 215
experimental procedures
BL21(DE3), 218–219
construct design strategies, 219–221
data collection and structural determination, 227–228
fusion tag, 218–219
protein crystallization, 225–227
protein sample preparation, 221–224, 223*f*
selenomethionyl-labeled RcATXR5 preparation, 224–225
small-scale test expression, 218–219
in vitro MTAs, 228–231
EZH2, 216–218
histone lysine methylation, 211–213
HKMTs, 210–213
implication of chromatin modifying enzymes, 218
KMT, 216–217
methyltransferase activity, 211
MLL1, 216
roles, 217–218
striking feature, 215
structural features, 213–215
Sf9 insect cells, 246–248
Single isomorphous replacement with anomalous signal (SIRAS), 413
Sirt2, 184–186
Sirt5, 187–188
Sirtuin
activity
coupled continuous activity assay, 198–200, 199*f*, 199*t*
enzyme master mix, 200
fitting equations, 200
fluorescently labeled substrate peptides, 200–201
substrate master mix, 200
applications, 205
architecture and catalytic mechanism, 185*f*
binding affinity measurements, 201–202
complexity of, 187–188
crystallization
catalytic core constructs, 204
crystals, 204–205
data collection and processing, 204–205
NMR and X-ray measurements, 202–203

Sirtuin (*Continued*)
of Sirt1, 204
of Sirt3, 203, 203*t*
defined, 184–186
expression systems, 188–189
human isoforms, 186–187
isoforms feature, 184–186
prominent desuccinylase activity, 187–188
purification, of mammalian
affinity chromatography, 194–195
affinity tag removal, 195–196
cleared lysate preparation, 194
His-tagged human Sirt3 catalytic core, 191, 192*t*, 193*f*
instruments and reagents, 191, 192*t*, 193
ion exchange chromatography, 196
modifications to standard protocol, 197–198
protein expression, 193–194
protocols and chromatography steps, 189, 191*f*
size-exclusion chromatography, 196–197
recombinant preparation— constructs, 188–189, 190*t*
unique mechanism, exploit, 186–187
Size-exclusion chromatography (SEC), 47–48, 50, 51*f*, 196–197
Size-exclusion chromatography coupled with multiangle light scattering (SEC-MALS), 24–26, 26*f*
Small-angle X-ray scattering (SAXS), 34, 36*f*
SNAP-tag
cell imaging, 105
cell lines expression, 100–102
cell-permeable fluorescent, 100
chase assay, 103
pulse assay, 103
quench assay, 103
triton extraction, 103
validation, 104*f*
SNF2h enzyme, 119–120
Suicide inhibitors, 268–269
Suppressor of zeste 12 (SUZ12), 216

T

TAL effector nucleases (TALENs), 305, 313
Tandem affinity purification (TAP)
of EPC1 complex, 310, 310*f*
anti-FLAG M2 affinity purification, 310–312
strep-tactin affinity purification, 312
of EZH2, 315
of EZH2/PRC2, 314*f*
TAP. *See* Tandem affinity purification (TAP)
TAP-tagged EZH2 variants expression, 313–314
TcTERT. *See Triobolium castaneum* telomerase (tcTERT)
Telomerase elongation complex
activity assays, 414
complex crystallization, 414–416
nucleic acid substrate design, 414
structure determination, 416
Telomerase reverse transcriptase (TERT), 405*f*
Telomerase RNA (TER), 405–406
Telomere(s)
definition, 404
holoenzyme, 404–405
replicating, 406
Ten-eleven translocation (TET) enzymes
analysis of cytosine modifications
cellular DNA, 366–373
in vitro, 373–383
biological and pathological processes, 366
mammalian isoforms, 366
purification
FLAG-based affinity, 374
from Sf9 insect cells, 373–374
substrates
CpG, 374–375
custom-made oligonucleotides, 375
fluorescence-based assays, 375
generation of, 377–378
isotopically labeled substrates, 376–377, 376*f*
modified cytosines, 375
thermocycler settings, 377–378
Thermal shifting, 88
Thermofluor assay, 147–149
Thymine DNA glycosylase (TDG), 381

TIP60, 304–305, 317
Tissue cell culture
  elements, 447
  HeLa cells, 450–451
Tranylcypromine analogs, 269
Trinucleosomes, 6
  AFM, 29–33
    buffers and reagents, 30
    data processing, 32
    equipment, 30, 31*f*
    to visualize trinucleosomes, 33*f*
  assembly, 13–15
  EcoRI digestion of, 5*f*, 15
  MNase digestion, 20, 21*f*
  SAXS analysis of, 35, 36*f*
Triobolium castaneum telomerase (tcTERT)
  crystallization
    optimization, 412
    protein sample concentration, 410–411
    screening, 411–412
  protein isolation
    catalytic subunit, 408
    cell line identification, 409–410
    expression system identification, 408–409
    purification, 410
    size and complexity, 407–408
  structure determination
    data collection, 412–413
    ring configuration, 413
    SIRAS, 413

V

vSET, 214*f*

X

Xenopus laevis histones, 7
X-ray crystallography, 360–361

Z

ZFNs. *See* Zinc-finger nucleases (ZFNs)
ZFP57 mutation, 389–390
Zinc-finger nucleases (ZFNs), 305
ZnF protein
  C2H2, 388–390
  crystallization, 396–397, 396*f*
  expression, 391
  purification, 391–393, 392*f*

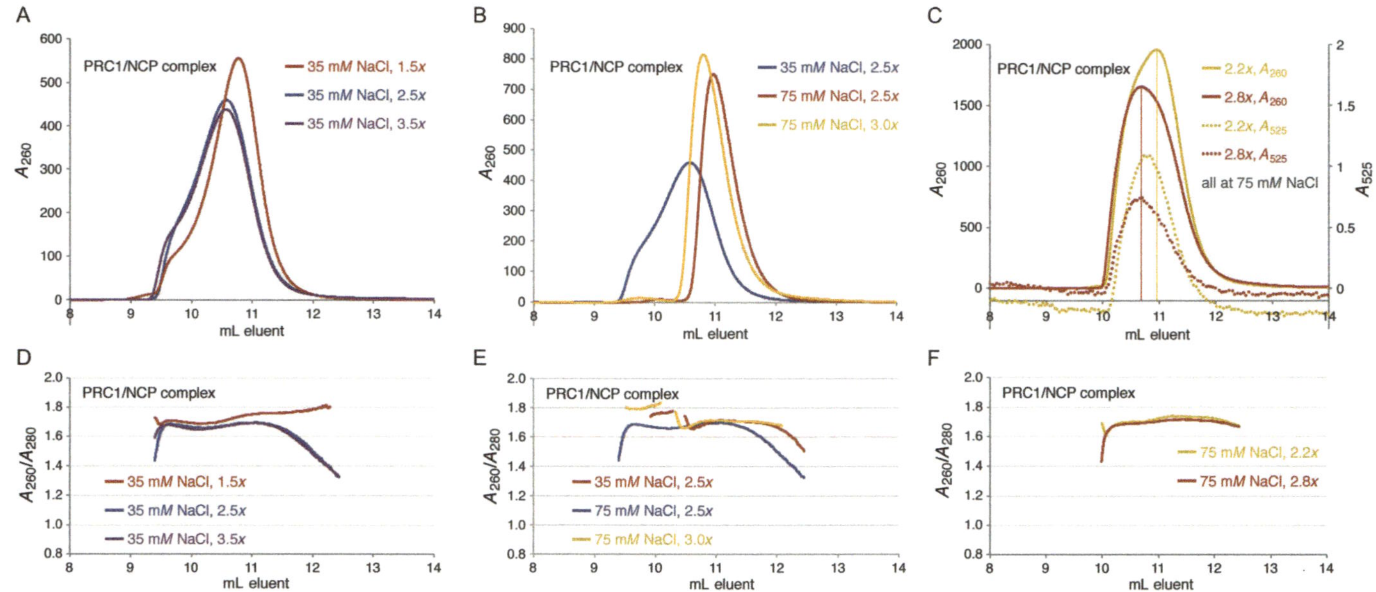

R.K. McGinty *et al.*, Fig. 2 Size exclusion chromatography for the purification and analysis of a reconstituted chromatin complex. (A) and (B) A_{260} chromatograms for reconstituted chromatin complexes on an analytical scale using specified salt concentrations and molar equivalents of chromatin protein relative to NCP. (C) A_{260} and A_{525} for fluorescently labeled chromatin complex reconstituted on a preparative scale. (D–F) A_{260}/A_{280} ratio for chromatograms in top panels.

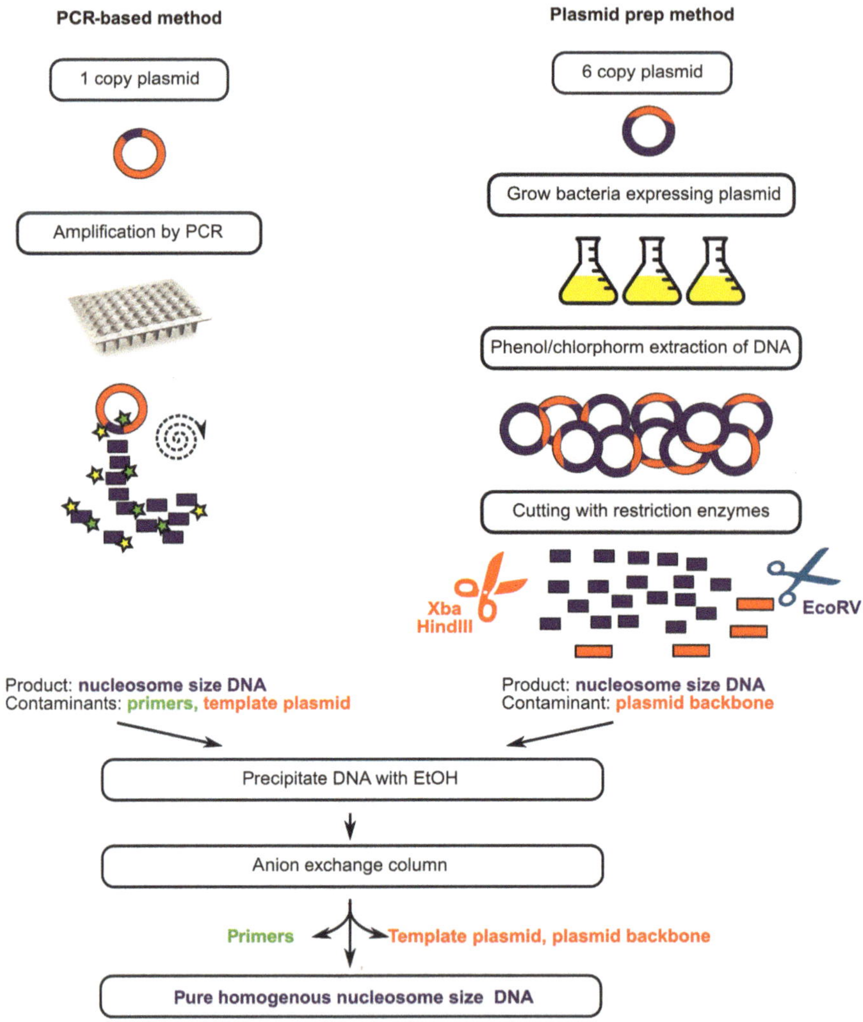

N. Sekulic and B.E. Black, Fig. 3 Overview of DNA preparation for CENP-A nucleosome assembly by PCR method and by mega plasmid prep method.

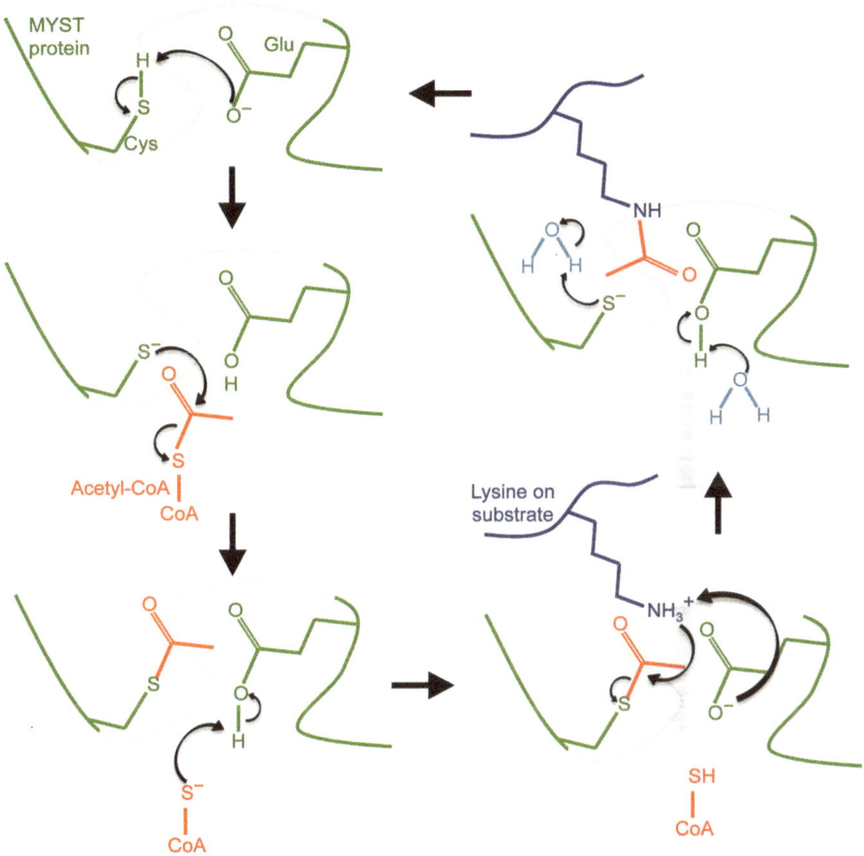

C.E. McCullough and R. Marmorstein, Fig. 1 "Ping-pong" catalytic mechanism employed by MYST family acetyltransferases. A glutamate in the active site acts as the general base to first deprotonate the active site cysteine. Next the acetyl group from acetyl CoA is transferred to the cysteine. The glutamate then protonates the leaving cofactor. The glutamate is then able to deprotonate the substrate lysine so that the cysteine can transfer the acetyl group to the lysine. Finally both the glutamate and cysteine react with water to return to their initial state.

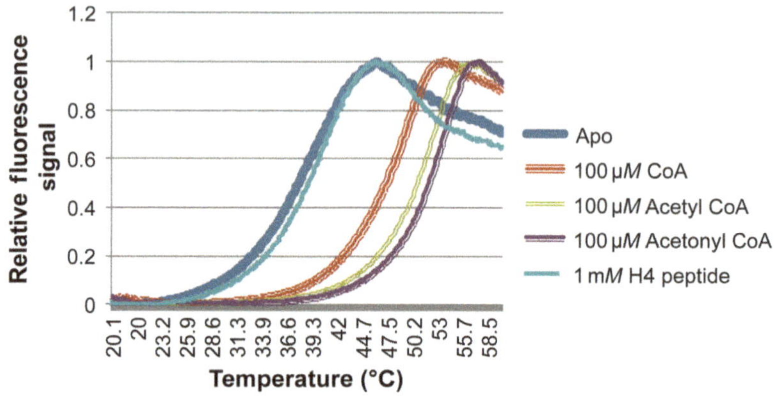

C.E. McCullough and R. Marmorstein, Fig. 6 Thermofluor assay as a readout of hMOF ligand binding. Data for apo hMOF, and hMOF in the presence of 1 m*M* H4 peptide, and 100 μ*M* CoA, acetyl CoA, and acetonyl CoA, respectively. T_m is the inflection point for each curve: ~37°C for apo hMOF, ~47°C for hMOF with CoA, ~50°C for hMOF with acetyl CoA, ~52°C for hMOF with acetonyl CoA, and ~38°C for hMOF with H4 peptide.

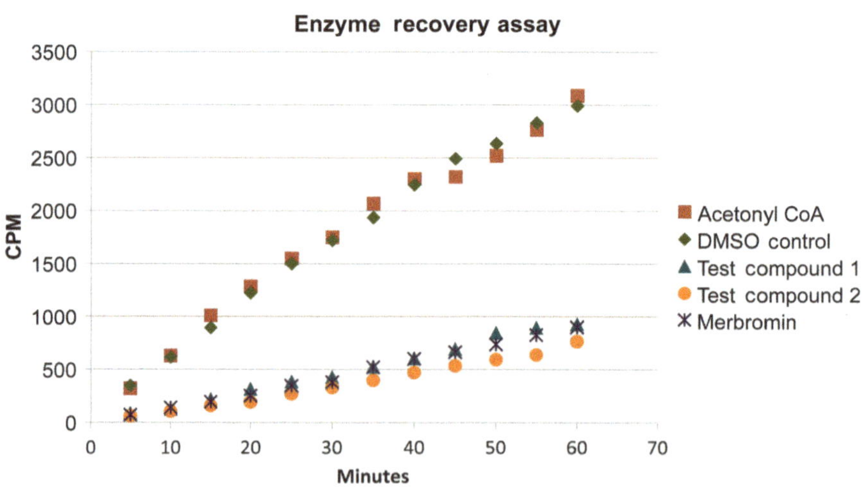

C.E. McCullough and R. Marmorstein, Fig. 8 Sample data for an enzyme recovery assay with a MYST protein. Note that both of the test compounds display irreversible inhibition, as they inhibit enzyme activity after dilution, aligning with the control irreversible inhibitor merbromin, a mercury-containing compound that interacts with the active site cysteine. Also note that acetonyl CoA, a reversible inhibitor, aligns very closely to the DMSO control.

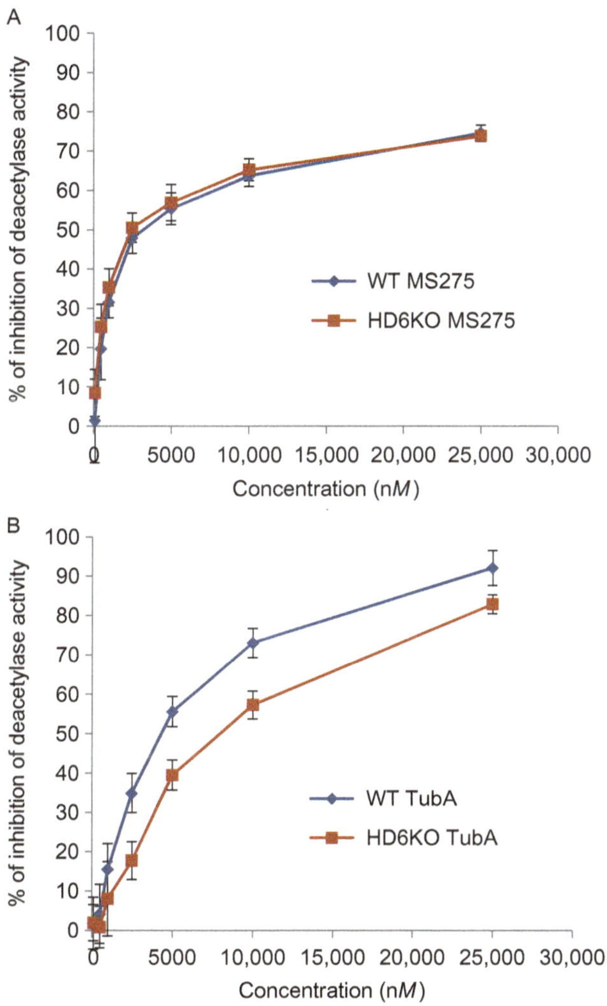

A. Villagra *et al.*, Fig. 2 Quantification of deacetylase activity using HDAC-GLO assay. T-cells isolated from either HDAC6 knockout or wild-type mice were treated with varying concentrations of Entinostat (A) or Tubastatin A (B) and deacetylase activity was measured by the addition of HDAC-GLO reagent and detected by Cytation multimode plate reader.

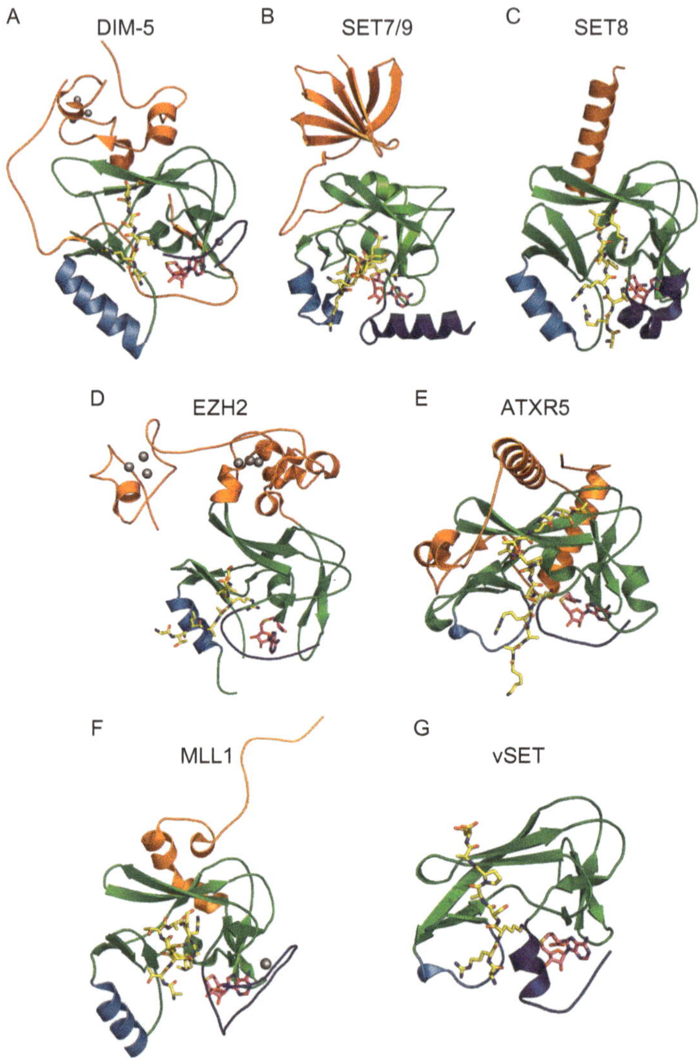

E. Bergamin and J.F. Couture, Fig. 1 Crystal structure of SET domain HKMTs in ternary complexes. Ribbon diagrams of (A) DIM-5 (PDB code 1PEG), (B) SET7/9 (PDB code 1O9S), (C) SET8 (PDB code 1ZKK), (D) EZH2 (PDB code 5CH1), (E) ATXR5 (PDB code 4O30), (F) MLL1 (PDB code 2W5Z), (G) vSET (PDB code 3KMT) are represented with the nSET (*orange*), SET (*green*), iSET (*light blue*), and cSET (*dark blue*) regions highlighted. The product cofactor AdoHcy and the substrate histone peptide are shown in stick representation and colored *magenta* and *yellow*, respectively. The Zn^{2+} ions in the nSET and cSET regions are illustrated as gray spheres.

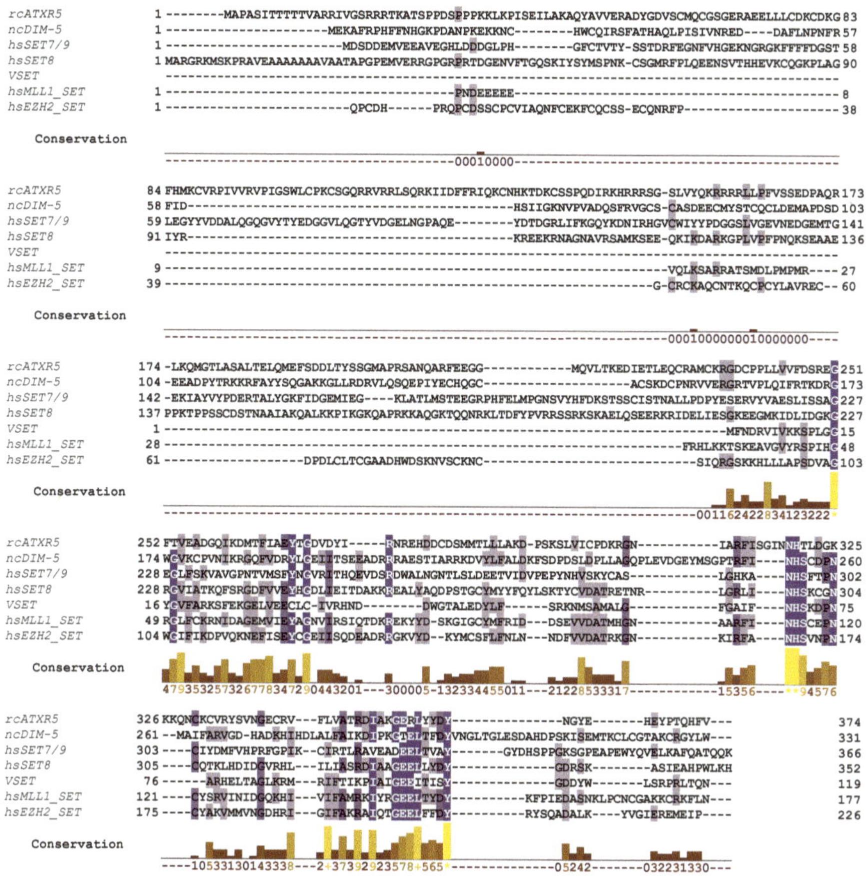

E. Bergamin and J.F. Couture, Fig. 2 Sequence alignment of selected protein methyltransferases. All the sequences are from human with the exception for *Neurospora crassa* DIM-5, *Ricinus communis* ATXR5, and *Paramecium bursaria* Chlorella virus 1 vSET. Conserved residues are highlighted in *blue* with darker shades associated with the higher degree of conservation. The program used was JalView (Waterhouse, Procter, Martin, Clamp, & Barton, 2009), and the alignment was produced using Muscle (Edgar, 2004) with default parameters.

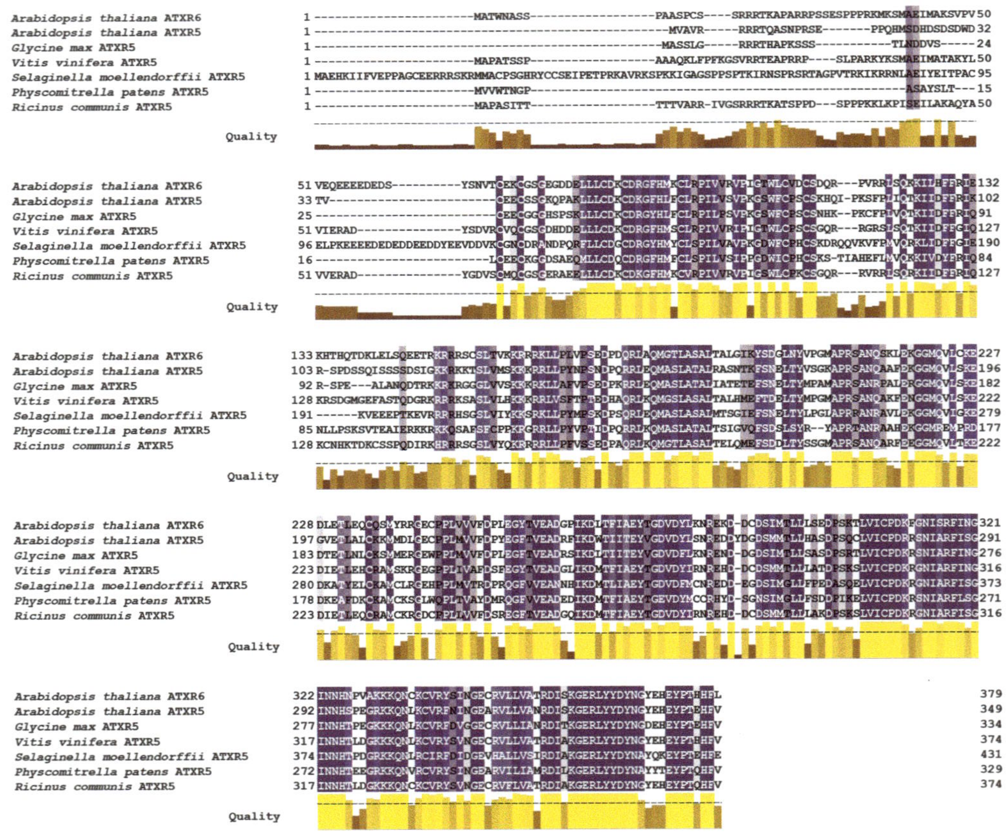

E. Bergamin and J.F. Couture, Fig. 3 Sequence alignment of ATXR5 homologs. Conserved residues of aligned sequences of ATXR5/6 homologs are highlighted in *blue* with *darker shades* associated with the higher degree of conservation. The programs used are the same as for Fig. 2.

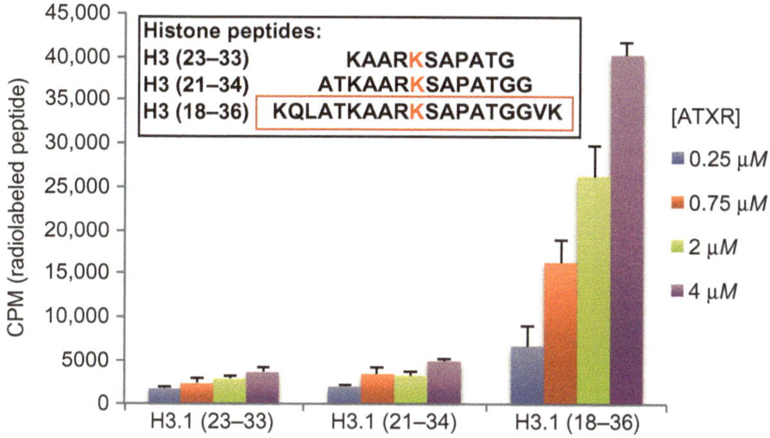

E. Bergamin and J.F. Couture, Fig. 6 H3.1 peptide substrate selection for ATXR5. Methyltransferase assay of four different concentrations of ATXR5 (0.25, 0.75, 2, 4 μ*M*, colored in *blue*, *red*, *green*, and *purple*, respectively) incubated with three different constructs of histone H3.1 peptide (their sequence is illustrated in the inset panel). The peptide chosen for crystallization trials is boxed in *red*.

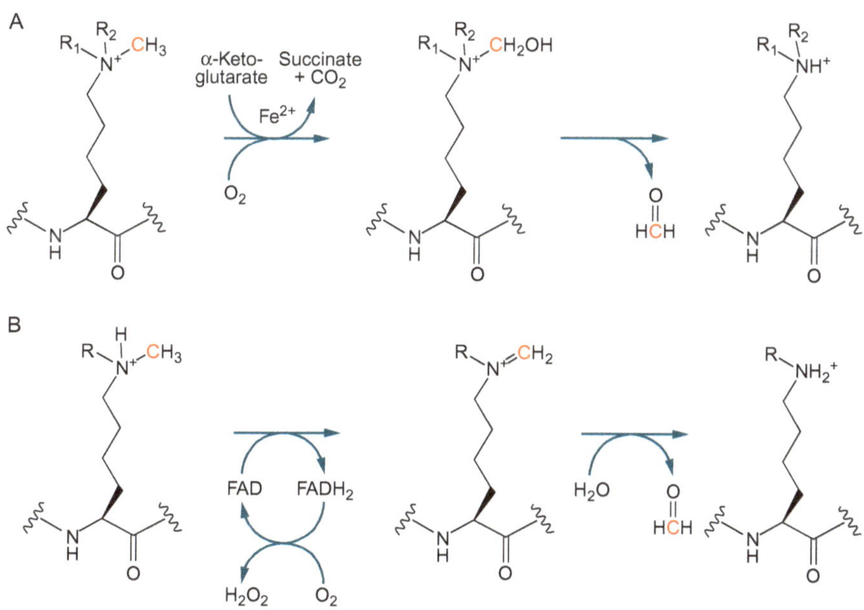

B. Laurent and Y. Shi, Fig. 1 (A) Chemical mechanism of lysine demethylation catalyzed by Jumonji C (JmjC) domain-containing proteins. (B) Chemical mechanism of LSD1/KDM1A-catalyzed lysine demethylation. The carbons in *red* are subjected to demethylation in each reaction. *Adapted from Mosammaparast, N., & Shi, Y. (2010). Reversal of histone methylation: Biochemical and molecular mechanisms of histone demethylases.* Annual Review of Biochemistry, *79, 155–179. doi: 10.1146/annurev.biochem.78.070907.103946.*

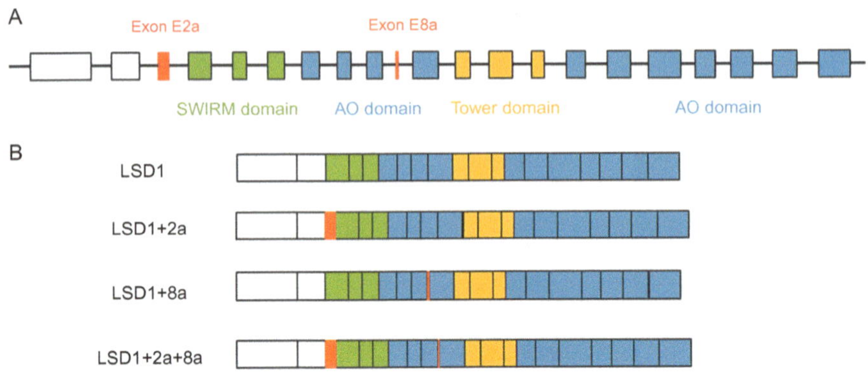

B. Laurent and Y. Shi, Fig. 2 (A) Schematic representation of the human LSD1 gene with its 19 constitutive exons; alternative exons E2a and E8a are represented in *red*. Different colors indicate LSD1 functional domains: the SWIRM domain and the SWIRM-oxidase connector (*green*) coded by exons 3–5, the amine oxidase (AO) domain (*blue*) coded by exons 6–9 and exons 13–19, the tower domain (*orange*) coded by exons 10–12. (B) Exon structure of the four LSD1 transcript variants.

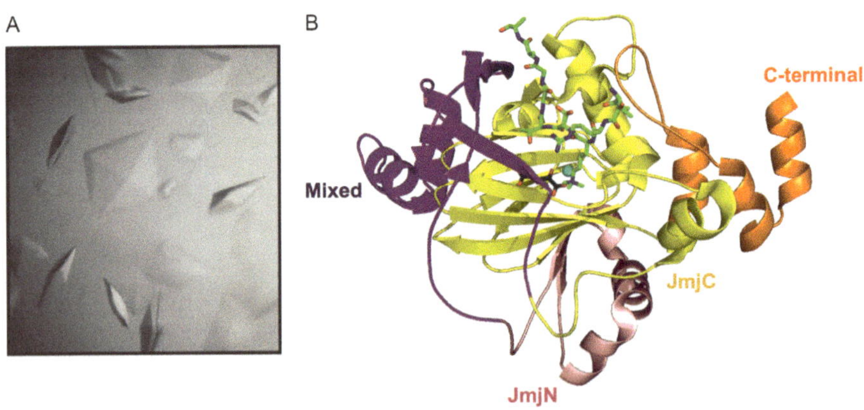

S. Krishnan and R.C. Trievel, Fig. 2 Structure determination of JMJD2D bound to an H3K9me3 peptide. (A) Crystals of the JMJD2D · 2-OG · H3K9me3 peptide complex grown in 7% PEG 3350, 100 m*M* NaSCN, and 350 m*M* KNO_3. (B) Ribbon diagram rendering of the JMJD2D · 2-OG · H3K9me3 peptide ternary complex. The JmjN, mixed, JmjC, and C-terminal domains of JMJD2D are colored and labeled. The H3K9me3 peptide (*green* carbon atoms) and 2-OG (*black* carbon atoms) are shown in stick representation. The active site Ni(II) ion is depicted as a *cyan* sphere. The Zn(II) ion bound in the Zn-binding motif is omitted for clarity.

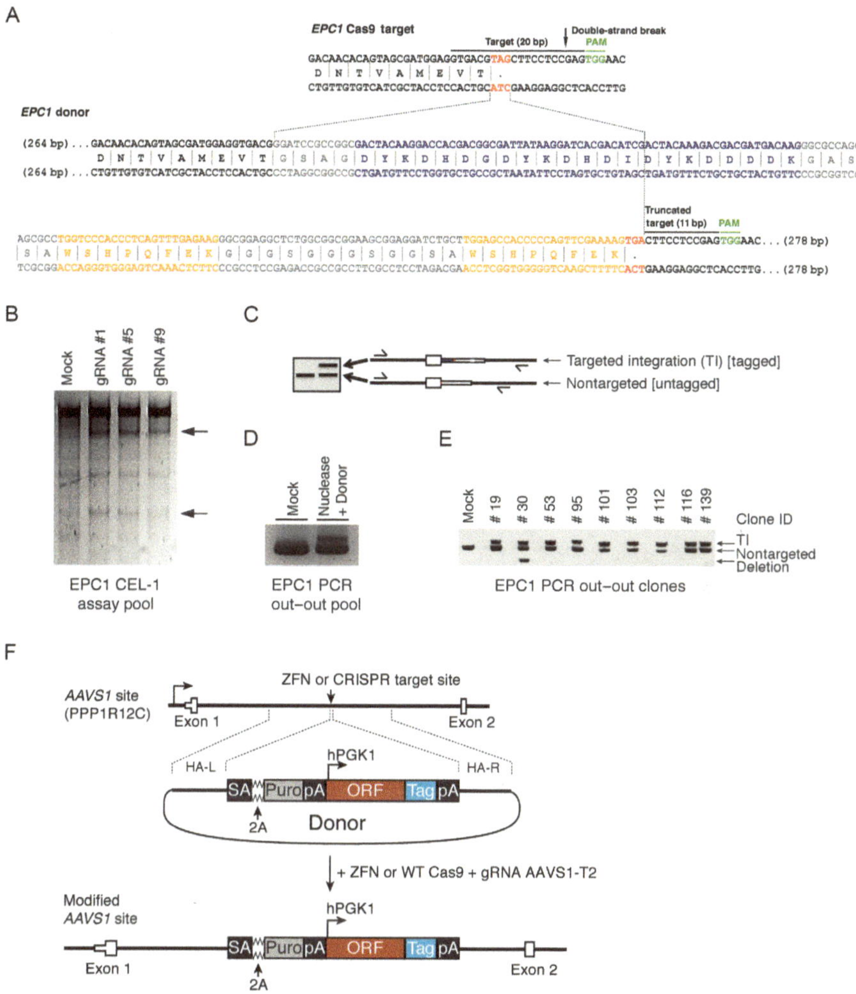

Y. Doyon and J. Côté, Fig. 1 Strategies to establish cell lines expressing tagged proteins from endogenous genes or a genomic safe harbor locus. (A) Schematic of the EPC1 locus, wild-type Cas9 target site, and donor construct. Annotated are the positions of the stop codon (TAG), the target site, and the protospacer adjacent motif (PAM) that specifies the cleavage site. The 3′ end of the left homology arm (HA-L) starts just before the STOP and the 5′ end of the right homology arm (HA-R) starts after the STOP. Note that in this case, the STOP codon in the donor was changed from TAG to TGA to reduce the possibility of cleavage by the CRISPR/Cas9 enzyme. This way, the donor sequence contains a truncated target sequence of just 11 bp, which should be sufficient to prevent DSB formation in the episomal donor and in the chromosome after targeted integration (TI) of the tag. We avoid repetitive sequences in HA-L and HA-R. The optimal guide RNA is selected based on several criteria: (i) high activity based on CEL1 assay, (ii) induction of DSB in proximity to the STOP codon, (iii) option to prevent donor cleavage (see above), and, if possible, (iv) cleavage in the 3′ UTR of the gene to prevent NHEJ-mediated mutagenesis of the coding sequence of the untargeted allele.

(Continued)

Y. Doyon and J. Côté, Fig. 1—Cont'd (B) Result of a Cel-1 assay to detect the activity of different sgRNA targeting EPC1. (C) Schematic of a PCR-based assay (out–out PCR) to detect TI of the tag sequence. Primers are located outside of the homology arms and are designed to yield a longer PCR product if the tag is inserted. (D) Results of a out–out PCR assay conducted on genomic DNA from K562 cells transfected with (i) wild-type Cas9 expression vector and EPC1 donor (Mock) and (ii) wild-type Cas9 expression vector, gRNA #1 and EPC1 donor (nuclease+ donor). Cells were collected 3 days post-transfection. (E) Same as in (D), but on single-cell-derived clones. (F) Schematic of the donor construct and of the *AAVS1* locus following ORF addition. The first two exons of the *PPP1R12C* gene are shown as *open boxes*. Also annotated are the locations of the splice acceptor site (SA), 2A self-cleaving peptide sequence (2A), puromycin resistance gene (Puro), polyadenylation sequence (pA), human phosphoglycerate kinase 1 promoter (h*PGK1*), 3 × FLAG–2 × STREP tandem affinity tag (Tag), and homology arms left and right (HA-L and HA-R are, respectively, 800 and 840 bp).

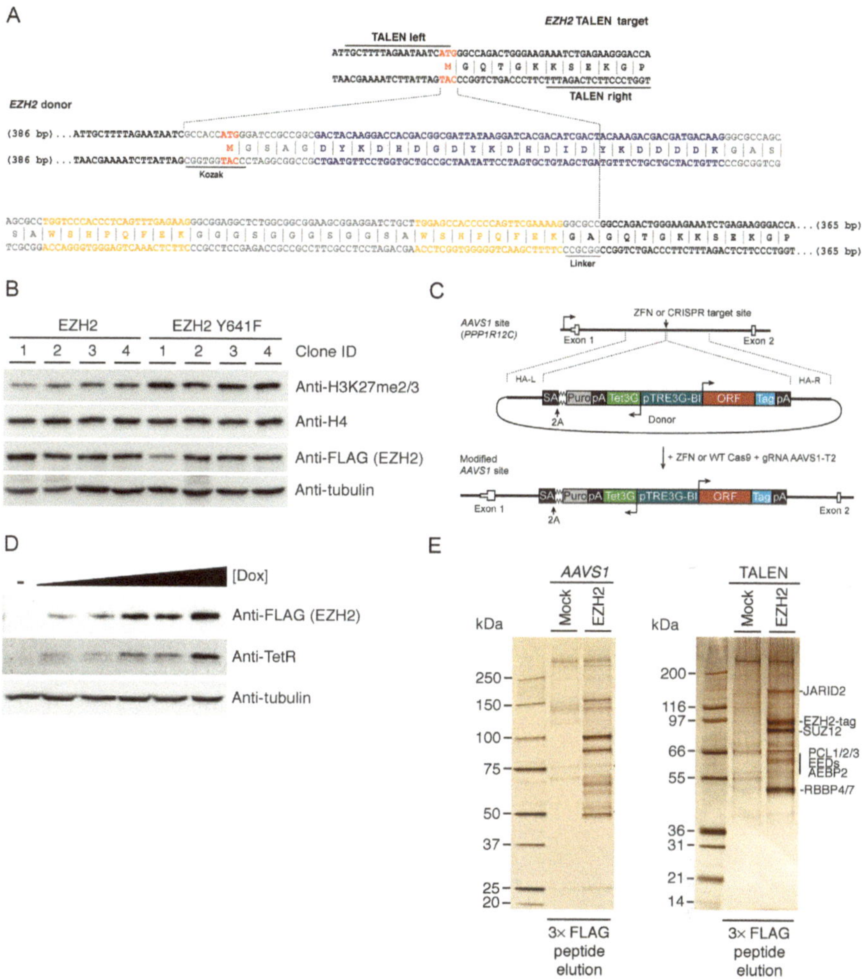

Y. Doyon and J. Côté, Fig. 3 Production of cell lines for tandem affinity purification (TAP) of the EZH2/PRC2 complex. (A) Strategy for TALEN-driven insertion of the TAP tag to the N-terminus of the EZH2 protein within the endogenous gene. Schematic of the EZH2 locus, TALEN target sites, and donor construct. The position of the start codon (ATG) is shown. The 3′ end of the left homology arm (HA-L) starts just before the ATG and the 5′ end of the right homology arm (HA-R) starts after the ATG. A Kozak consensus sequence was placed upstream of the tag sequence for efficient translation initiation at this site. Note that in this case, the TALEN-binding sites are separated by the tag sequence in the donor, which should be sufficient to prevent DSB formation in the episomal donor and in the chromosome after targeted integration of the tag. We avoid repetitive sequences in HA-L and HA-R. (B) Western blots showing the levels of H3K27me3 on bulk chromatin in K562 clones expressing EZH2 ± Y641F mutation from the *AAVS1* locus. (C) Schematic of donor construct and of the AAVS1 locus following ORF addition for the robust induction of target proteins upon doxycycline treatment. The first two exons of the PPP1R12C gene are

(Continued)

Y. Doyon and J. Côté, Fig. 3—Cont'd shown as *open boxes*. Also annotated are the locations of the splice acceptor site (SA), 2A self-cleaving peptide sequence (2A), puromycin resistance gene (Puro), polyadenylation sequence (pA), bidirectional Tet-responsive promoter (pTRE3G-BI), Tet3G transactivator (Tet3G), 3 × FLAG–2 × STREP tandem affinity tag (Tag), and homology arms left and right (HA-L, HA-R). (D) Western blots showing tag-EZH2 and Tet-On 3G transactivator protein expression in a K562 clone treated for 24 h with increasing doses of doxycycline. (E) Silver-stained SDS–PAGE showing the comparison of the EZH2 complex obtained when expressed from the endogenous (TALEN) and the *AAVS1* loci after only the first purification step (FLAG elution, before Streptavidin resin).

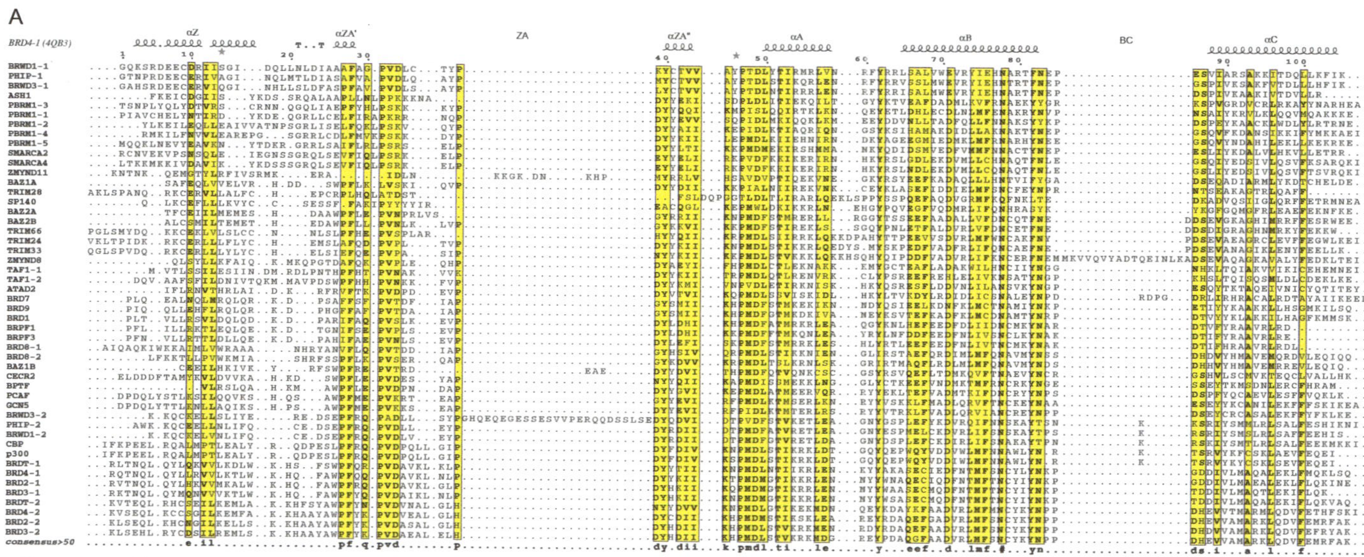

C. Ren *et al.*, Fig. 1 Protein sequence alignment of human bromodomains. (A) Human BrD sequences were aligned using EMBL Clustal Omega (http://www.ebi.ac.uk/Tools/msa/clustalo/). Secondary structure depiction was visualized based on the BRD4-BD1 structure (PDB: 4QB3) in ESPript 3.0. (B) Phylogenetic tree was subsequently generated using EMBL ClustalW2 Phylogeny (http://www.ebi.ac.uk/Tools/phylogeny/clustalw2_phylogeny/; Li et al., 2015). Sequence alignment and phylogenetic tree (unrooted format) were visualized using ESPript 3.0 (http://espript.ibcp.fr/ESPript/cgi-bin/ESPript.cgi; Robert & Gouet, 2014) and iTOL (http://itol.embl.de/index.shtml; Letunic & Bork, 2011), respectively.

(Continued)

B

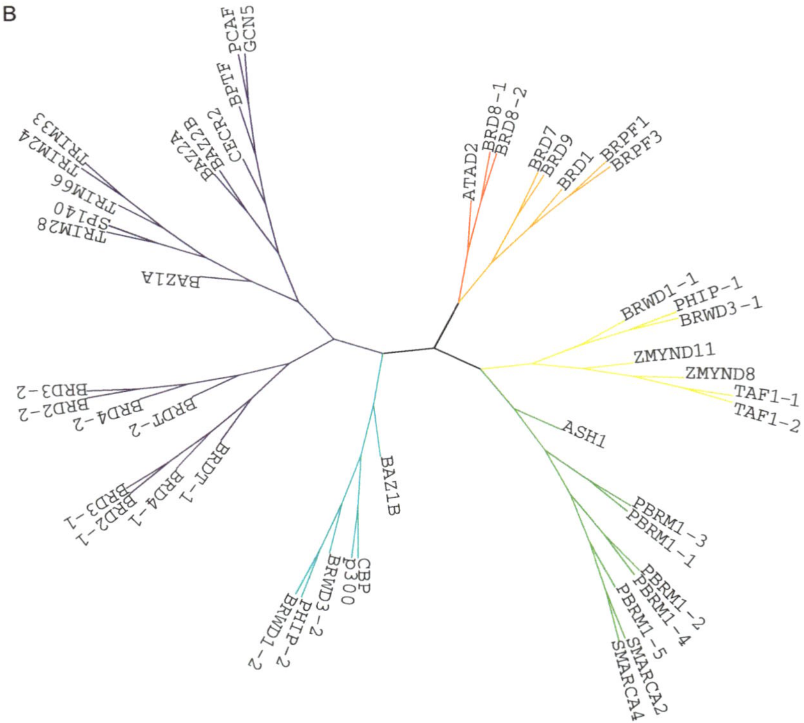

C. Ren *et al.*, Fig. 1—Cont'd

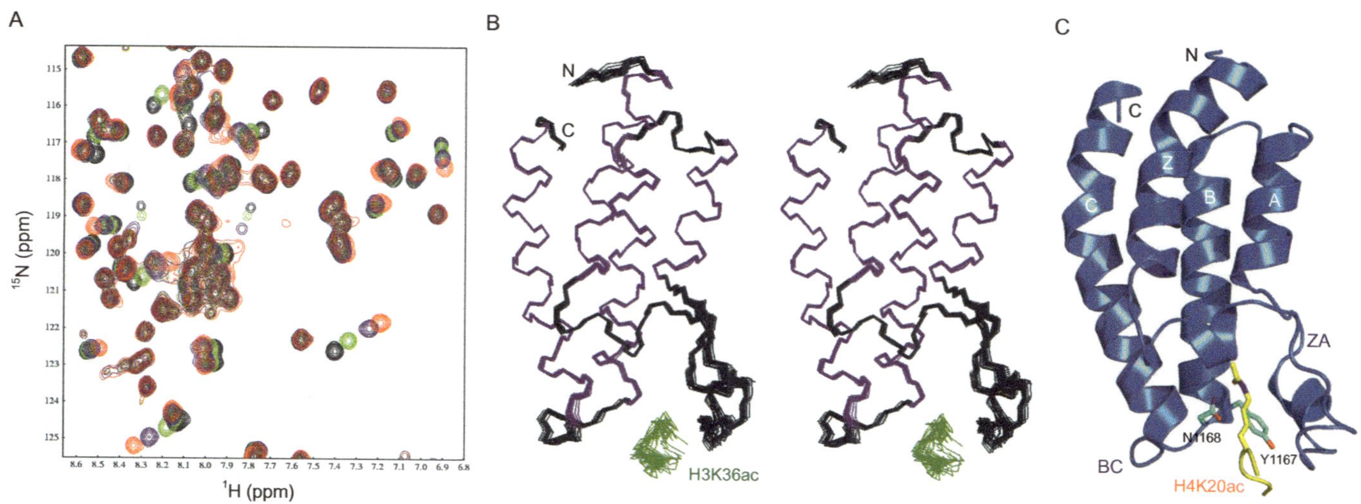

C. Ren *et al.*, Fig. 2 Biochemical analysis of PCAF bromodomain binding to lysine-acetylated histone H3 peptides using NMR spectroscopy. (A) ^{1}H–^{15}N-HSQC spectra of the PCAF bromodomain showing backbone amide resonances in the free state (*black*) or in the presence of a histone peptide containing H3K9ac (*green*), H3K14ac (*red*), and H3K36ac (*blue*). The extent of chemical shift perturbation and line broadening indicates the strength of protein/peptide binding. (B) Stereoview of the backbone (N, Ca, and C′) atoms of 25 superimposed NMR structures of the PCAF bromodomain in complex with an H3K36ac peptide. (C) Crystal structure of CBP bromodomain bound to an H4K20ac peptide. The image was prepared with the structure coordinates of CBP bromodomain–PHD finger tandem module (PDB: 4N3W) in complex with H4K20ac peptide. The PHD finger is not shown for clarity (Plotnikov et al., 2014). *Panel (B) images were modified from Zeng, L., Zhang, Q., Gerona-Navarro, G., Moshkina, N., & Zhou, M. M. (2008). Structural basis of site-specific histone recognition by the bromodomains of human coactivators PCAF and CBP/p300.* Structure, 16, *643–652.*

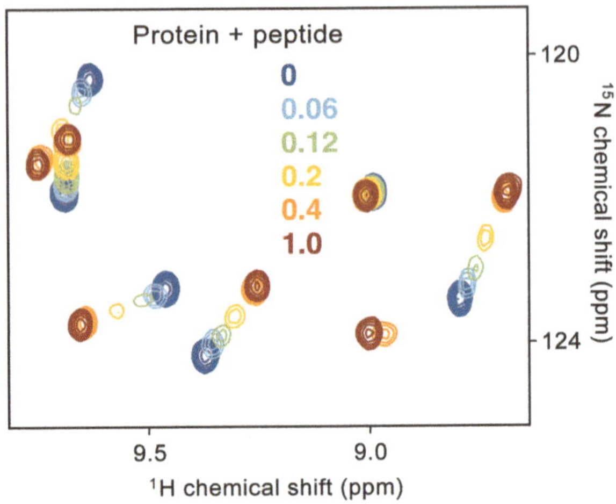

C.A. Musselman and T.G. Kutateladze, Fig. 5 Overlay of successive ^{1}H,^{15}N HSQC spectra of a labeled methyllysine reader domain upon titration of methylated peptide. Arbitrary peptide concentration is shown in legend.

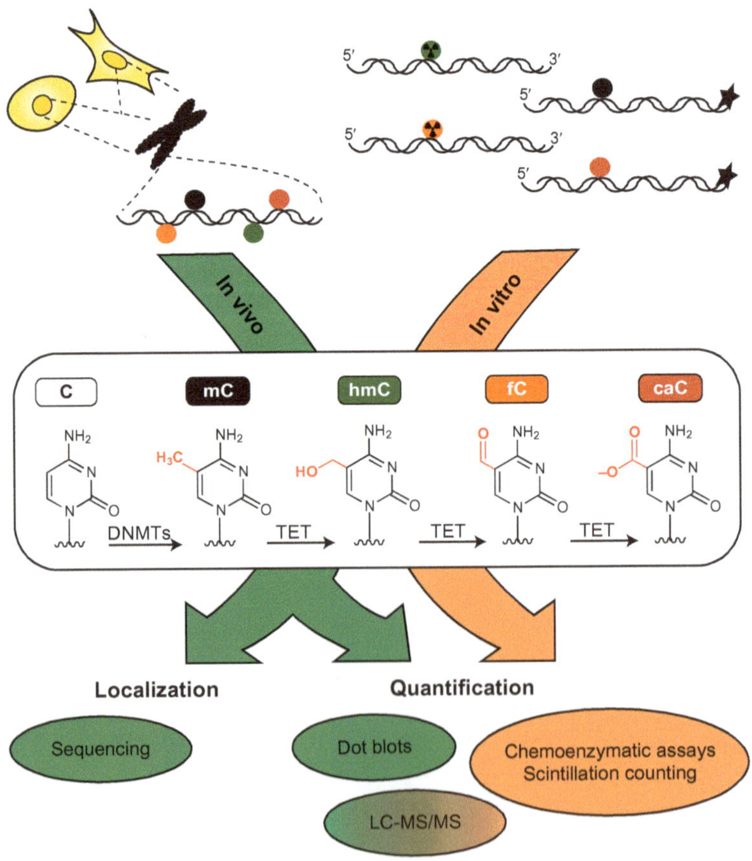

M.Y. Liu *et al.*, Fig. 1 Current methods for static analysis of TET activity via modified cytosines in vivo and in vitro.

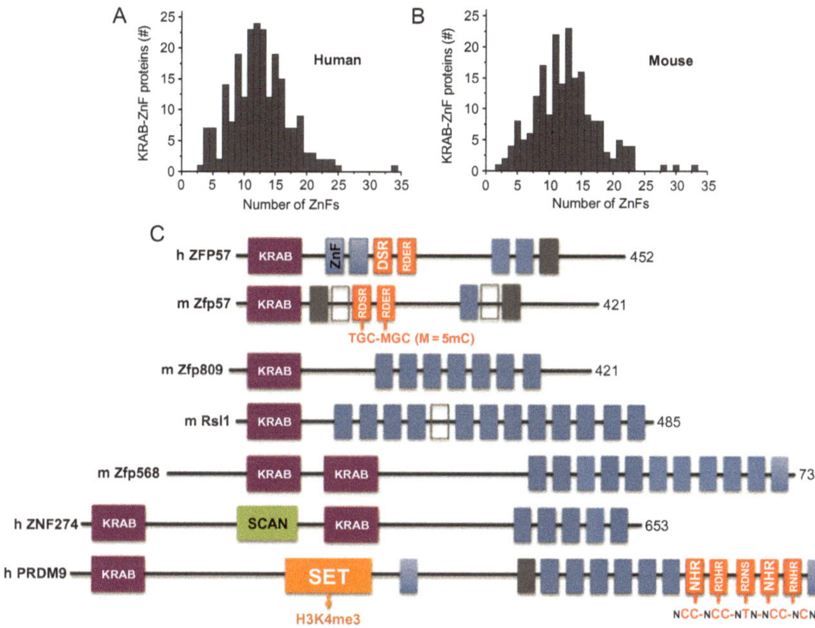

A. Patel *et al.*, Fig. 1 KRAB-ZnF proteins. The SysZNF database was examined for Krüppel-associated box (KRAB)-ZnF proteins in human (A) or mouse (B). In each case, the distribution of KRAB-ZnF proteins containing a given number of ZnF repeats is shown. (C) Examples of mammalian KRAB-ZnF proteins with known biologic roles. The size of each protein (in amino acids) is shown on the right. The classic C2H2 ZnF motifs are shown in *blue boxes* and the *gray* or *open boxes* indicate degenerate ZnFs that contain mutations affecting zinc coordination. The *red boxes* of mouse Zfp57 and human PRDM9 indicate the structurally characterized ZnFs in complex with their recognition sequences shown below. The amino acids within the *red box* indicate the three or four residues of each ZnF involved in base specific interactions. The SCAN box, a leucine-rich region, was named after *S*RE-ZBP, *C*Tfin51, *A*W-1 (ZNF174), *N*umber 18 cDNA (ZnF20) (Williams, Khachigian, Shows, & Collins, 1995). The SET domain was named after *S*u(var)3-9, *E*nhancer of zeste, *T*rithorax (Jenuwein, Laible, Dorn, & Reuter, 1998). *Modified from Liu, Y., Zhang, X., Blumenthal, R. M., & Cheng, X. (2013). A common mode of recognition for methylated CpG.* Trends in Biochemical Sciences, 38, *177–183.*

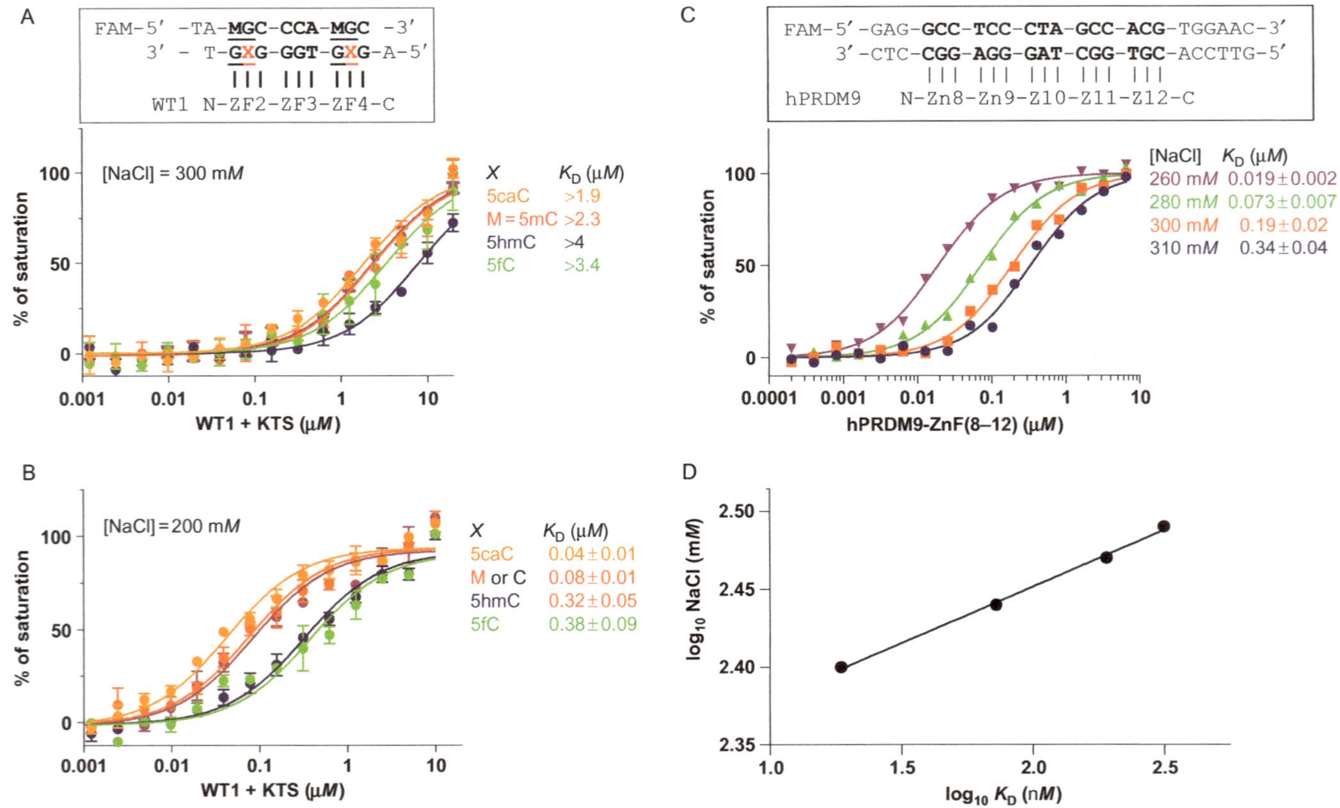

A. Patel *et al.*, Fig. 3 The effect of NaCl on binding affinity. The DNA-binding affinity of WT1 + KTS isoform is uniformly low in 300 m*M* NaCl (A) but increases markedly in 200 m*M* NaCl (B). (C) The DNA-binding affinity (K_D) of human PRDM9 allele-A under NaCl concentrations from 260 to 310 m*M* with 10–20 m*M* increments. (D) The linear correlation of double-logarithmic plot of K_D values and NaCl concentrations. *Panel (B): Adopted from Hashimoto, H., Olanrewaju, Y. O., Zheng, Y., Wilson, G. G., Zhang, X., & Cheng, X. (2014). Wilms tumor protein recognizes 5-carboxylcytosine within a specific DNA sequence.* Genes & Development, 28, *2304–2313.*

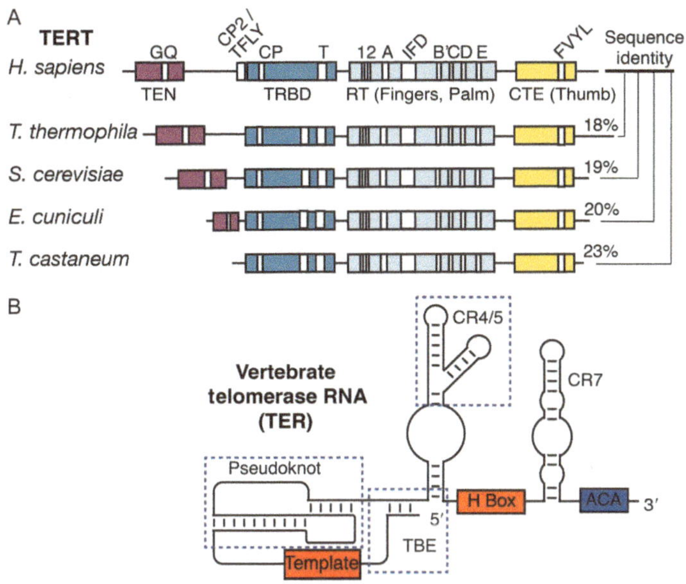

H. Hoffman and E. Skordalakes, Fig. 1 (A) Primary structure and sequence identity of TERT proteins from a diverse group of organisms (Human, *T. thermophila*, *S. cerevisiae*, *E. cuniculi*, and *T. castaneum*). (B) Schematic of the vertebrate telomerase RNA; conserved motifs are shown in color. The activation domain (CR4/5) and the pseudoknot are highlighted with a blue, dashed line.

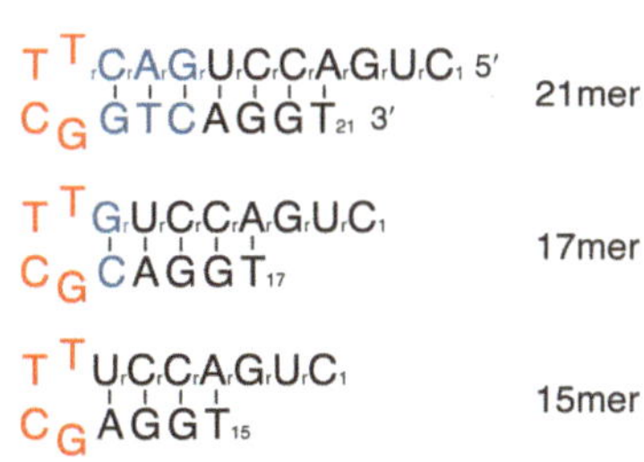

H. Hoffman and E. Skordalakes, Fig. 2 Primary structure and sequence of the RNA–DNA hairpins used in cocrystallization trials with tcTERT. The RNA template and complementary telomeric DNA are shown in *black letters*; the linker in *blue* and the stable tetraloop in *red*.

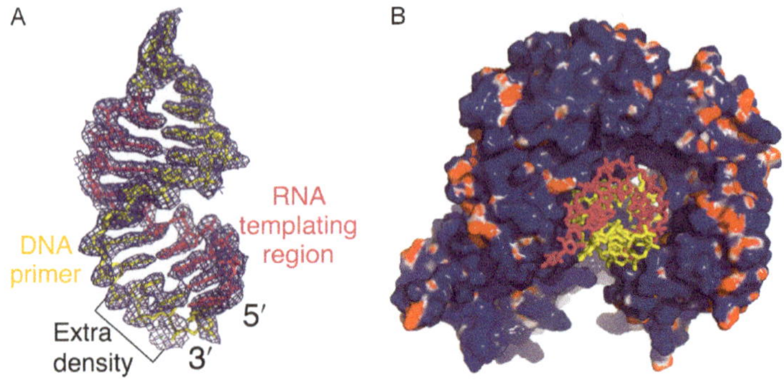

H. Hoffman and E. Skordalakes, Fig. 3 (A) Simulated annealed omit map of the RNA (*magenta stick*)–DNA (*yellow stick*) hairpin cocrystallized with tcTERT at 1.0σ contour level. (B) tcTERT surface charge representation, showing the RNA–DNA hybrid (*stick*) docked in the interior cavity of the TERT ring.

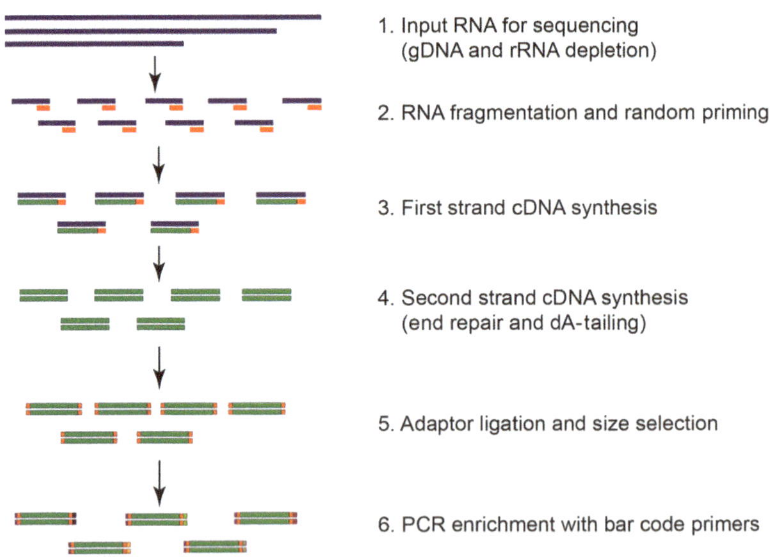

F. Lai *et al.*, Fig. 1 Principle of RNA sequencing library preparation. Following removal of genomic DNA and ribosomal RNA, the remaining RNA is fragmented, and primers are added for cDNA synthesis. At this stage, polyadenylated transcripts can be enriched using oligo $d(T)_{25}$ magnetic beads and oligo $d(T)_6$ primers in the cDNA synthesis. To select for all RNA species, random hexamer primers are used. The PCR product is then ligated with sequencing adaptors, size-selected, and purified for sequencing.

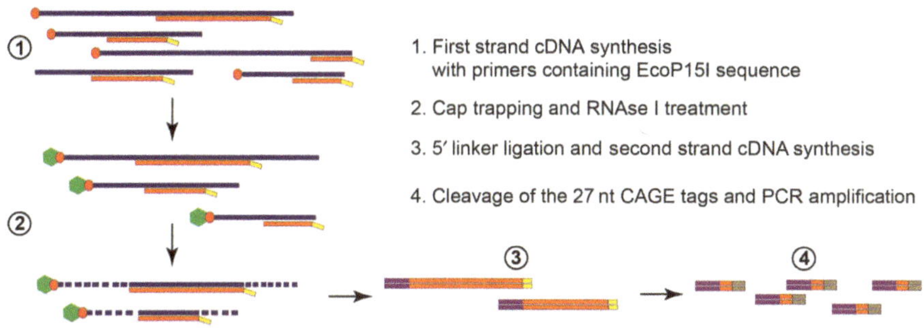

F. Lai *et al.*, Fig. 2 Overview of CAGE RNA library preparation. First strand cDNA synthesis is performed with random hexamer or oligo d(T) primers containing EcoP15I digestion site. Reverse transcriptase without RNase H activity must be used. 5′ Caps are then biotinylated and captured using strepatividin beads. The resultant RNA is subjected to RNase I digestion and purification. Complete second strand cDNA synthesis with modified 5′ linkers and cleave double-stranded cDNA by EcoP15I. After ligation with 3′ linker, the cDNA library is purified with streptavidin beads.

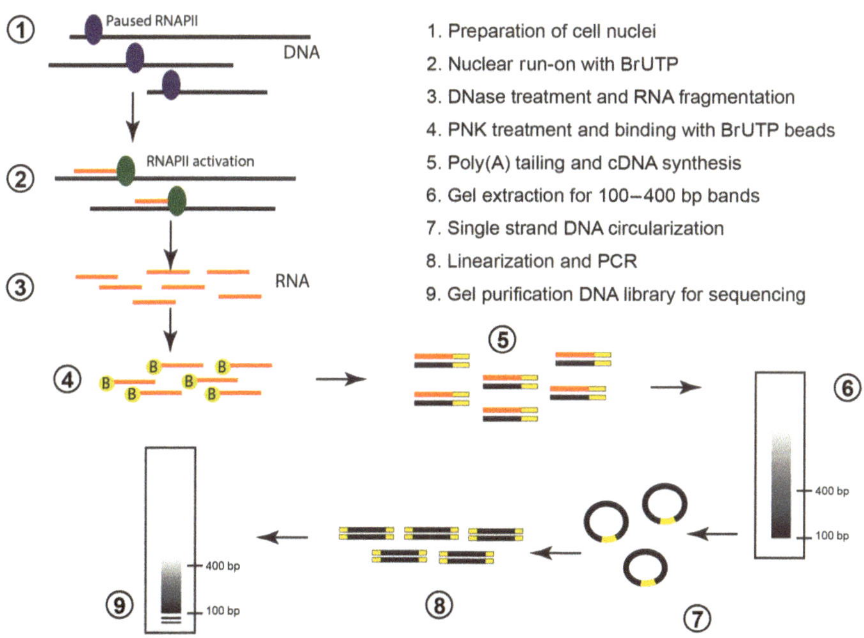

F. Lai *et al.*, Fig. 3 Steps of GRO-seq RNA library preparation. Nuclear run-on is performed by incubating prechilled nuclei at 37°C with NTPs and BrUTP. Following DNAse treatment, RNA fragmentation, and purification, nascent transcripts are treated with PNK and isolated with BrUTP beads. Addition of poly(A) tails provides a sequence for reverse primers used during cDNA synthesis. 100–400 base pair transcripts are then extracted, circularized, linearized, and amplified prior to sequencing.

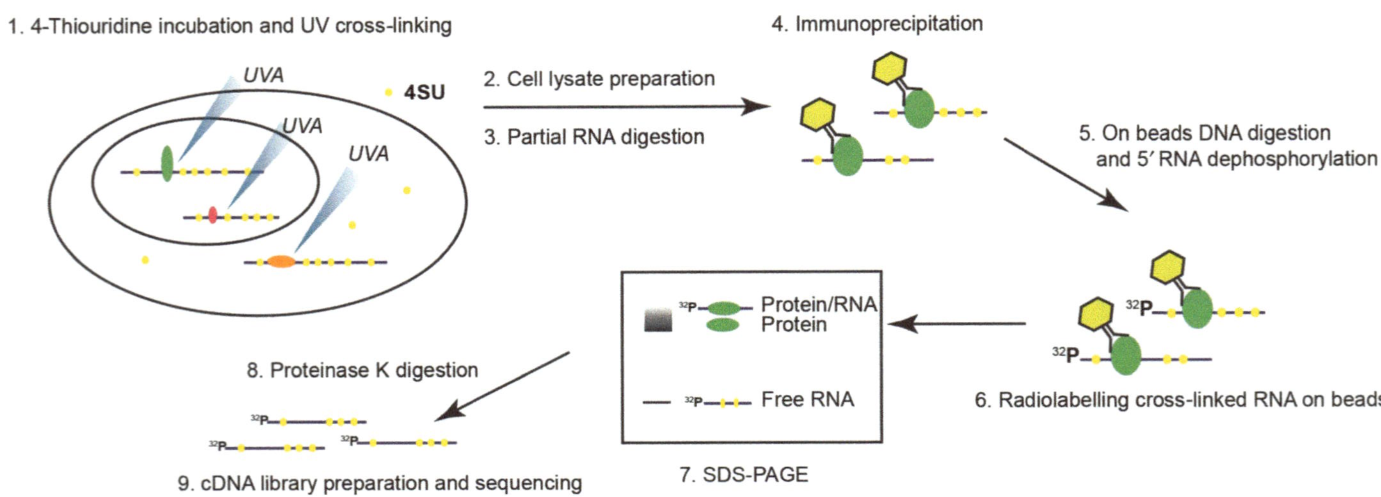

F. Lai *et al.*, Fig. 4 PAR-CLIP workflow. Incubating cells with 4-thiouridine and UV radiation efficiently cross-links RNA to proteins. Cells are lysed, and the extracted RNA is partially digested with RNase. The protein of interest is immunoprecipitated and pulled down using magnetic beads. DNA digestion, 5′ RNA dephosphorylation, and radiolabeling occur on the magnetic beads. Separation of protein-bound RNA from free RNA is achieved with SDS-PAGE electrophoresis. Proteinase K digestion isolates radiolabeled RNA from its immunoprecipitated target for analysis by sequencing.

Kolibri 360 GmbH
Ferdinand-Jühlke-Straße 7
99095 Erfurt, Deutschland
produktsicherheit@kolibri360.de